TERRESTRIAL ECOSYSTEMS THROUGH TIME

TERRESTRIAL ECOSYSTEMS THROUGH TIME

Evolutionary Paleoecology of Terrestrial Plants and Animals

Edited by

Anna K. Behrensmeyer
John D. Damuth
William A. DiMichele
Richard Potts
Hans-Dieter Sues
Scott L. Wing

ETE

The Evolution of
Terrestrial Ecosystems
Consortium

The University of Chicago Press
Chicago and London

ANNA K. BEHRENSMEYER, WILLIAM A. DiMICHELE, and SCOTT L. WING are research curators in the Department of Paleobiology at the National Museum of Natural History, Smithsonian Institution. JOHN D. DAMUTH is research biologist in the Department of Biological Sciences at the University of California at Santa Barbara and research associate in the Department of Paleobiology at the National Museum of Natural History, Smithsonian Institution. RICHARD POTTS is research curator in the Department of Anthropology at the National Museum of Natural History, Smithsonian Institution. HANS-DIETER SUES is associate curator in the Department of Vertebrate Paleontology at the Royal Ontario Museum, Toronto.

The University of Chicago Press, Chicago 60637
The University of Chicago Press, Ltd., London
© 1992 by The University of Chicago
All rights reserved. Published 1992
Printed in the United States of America

00 99 98 97 96 95 94 93 92   5 4 3 2 1

ISBN (cloth): 0-226-04154-9
ISBN (paper): 0-226-04155-7

Library of Congress Cataloging-in-Publication Data
Terrestrial ecosystems through time : evolutionary paleoecology of
   terrestrial plants and animals / edited by Anna K. Behrensmeyer . . .
   [et al.].
      p.   cm.
      Includes bibliographical references and index.
      1. Paleoecology. 2. Paleobiology. I. Behrensmeyer, Anna K.
QE720.T47 1992
   560'.45—dc20   91-44166

⊗ The paper used in this publication meets the minimum requirements of the American National Standard for Information Sciences—Permanence of Paper for Printed Library Materials, ANSI Z39.48-1984.

Contents

Contents

Contents ix

Tables and Figures

TABLES

FIGURES

Preface

The Evolution of Terrestrial Ecosystems program was initiated in 1987 at the National Museum of Natural History as a coordinated effort to study the paleoecology of land-based faunas and floras throughout the Phanerozoic. Its primary goals involve the observation and interpretation of ecological patterns in the fossil record, especially those related to major periods of global change, based on a comparative analysis of species associations or communities through long periods of time. The program was initiated by the ETE Consortium members responsible for this book, and the book itself was conceived as a necessary first step toward the organization of an immense amount of dispersed data and embryonic ideas that are the foundation for the program. These data and ideas are being assembled in a computerized database that will serve the specific purposes of ETE and its network of researchers. A manual for the database is being published as a companion to this book.

The first Evolution of Terrestrial Ecosystems conference to be sponsored by the ETE Program was held near Washington, D.C., in mid-1987. This book is derived from the conference proceedings and from subsequent synthesis of the assembled paleoecological ideas and data. The conferees discussed and summarized the state of Phanerozoic land-based paleoecology, its data, methods, theoretical perspectives, and potential contributions to ecology and evolutionary biology. For this purpose we brought together a wide variety of paleontologists with different specializations who could offer specific information on terrestrial biotas over the past 400 million years and who were willing to try to extract generalities from this information. The contents of this book reflect the dual goals of coordinating a large body of scattered data and demonstrating the potential use of these data in testing hypotheses of general interest to paleontologists, anthropologists, evolutionists, and ecologists.

During the three-day meeting six small working groups were established, three centered on time periods and three on methods. Each group was charged with generating a summary of data and ideas pertaining to their topic, and these were aired before the entire group on the final day of the three-day conference. The chapters that have grown out of these reports inevitably reflect the interests and experience of those involved, and we do not claim an even treatment of all aspects of terrestrial paleoecology. For example, although freshwater biotas fall within the purview of this volume, emphasis has been on terrestrial organisms in the stricter sense. This simply reflects the knowledge and interests of the organizers and the majority of the participants. Given the wide range of time periods, taxonomic groups, and philosophical viewpoints that are represented, however, we are confident that the material in this book provides a valid overview of the state of the field.

One of the important discoveries of the three days of intensive discussion was the extent to which each participant's thoughts were shaped by the particular time framework of his or her research specialty. Paleozoic workers deal in tens of millions of years, while Cenozoic and particularly Quaternary workers are concerned with millions to thousands of years. We found that we were speaking of similar patterns of community change over these widely varying time scales, and realized that we must carefully specify our time framework before discussing processes that could be responsible for these patterns. This, along with the fruitful interchange between researchers in plants and animals, invertebrates and vertebrates, Permian specialists and Quaternary anthropologists, made it clear that we share common goals in paleoecology that transcend the normal boundaries of our individual disciplines.

Chapter rapporteurs were responsible for assembling information from the conference and from written contributions by the participants into coherent, substantive, and provocative pieces that fairly represent our combined knowledge and perspectives. Although the chapter coordinators must take a large portion of the responsibility for these perspectives, the other conference participants had significant input at various stages in the development of this volume. We hope that readers will acknowledge all contributors to each chapter in bibliographic citations to this work. Furthermore, to underline the primary responsibility of rapporteurs, when there is more than one, we would like for formal text citations to name both, e.g., "Wing and Sues, et al."

The range of information presented in the chapters to follow is broad, but it is focused on problems concerning assemblages of fossil organisms rather than individual lineages. As objects of study, these assemblages begin as groups of species that were preserved together in a particular time and place, and taphonomic analysis then makes it possible to judge how they represent the original, living species associations or communities. The roles of species in these communities are assessed through functional anatomy as well as paleoenvironmental context. The sequence of chapters reflects this idealized progression toward a synthesis of ecological history throughout the Phanerozoic. Chapters 5–7 summarize present understanding of this history, although many of the newer ideas in the earlier chapters have yet to contribute significantly to the historical overviews.

ACKNOWLEDGMENTS

This book is the result of the collaborative effort of members of the Evolution of Terrestrial Ecosystems Program (ETE) at the National Museum of Natural History (NMNH), Smithsonian Institution, and the Biology Department of the University of California, Santa Barbara. The Smithsonian Institution provided funding for the 1987 conference and continuing support for the ETE Program. We thank Dr. David Challinor, then the Smithsonian's assistant secretary for science, Dr. Robert S. Hoffmann, then director of the National Museum of Natural History, and Ms. Roberta Rubinoff, director of the SI Office of Fellowships and Grants, for their encouragement and help. We thank Elizabeth Bailey, then with ETE, for her assistance in organizing and running the conference, Richard Bateman, David Greenwood, Robert Hoffmann, and Richard Kay for helpful comments on the manuscript, three anonymous paleobiologists for thorough and thoughtful reviews, and Susan Abrams of the University of Chicago Press for her enthusiasm and her patience. We are grateful to Chris R. Scotese and the PaleoMap Project of the University of Texas, Arlington, for generously providing paleogeographic reconstructions for the time chapters, to Ann Chipperfield for help with the index, and to Mary Parrish of the NMNH Paleobiology Department and T. Britt Griswold III for their artistic contributions. Most of all we thank the conference participants for their free exchange of ideas and contributions in preparing this volume.

This is the Evolution of Terrestrial Ecosystems Program Contribution #2.

1987 Evolution of Terrestrial Ecosystems Conference Participants
Top row: Bob Hook, Steven Scheckler, Scott Wing, Bill Shear, Hans Sues, Paul Olsen, David Weishampel, Mark Wilson, Ralph Chapman, Tom Phillips, Warren Kovach, Nicholas Hotton III, Russ Graham, Richard Beerbower, Bob Spicer, Larry Martin, Peter Dodson.
Center row: Bob Gastaldo, Geoff Spaulding, Bill DiMichele, Judith Harris, Rob Foley, Dan Fisher.
Lower row: Bruce Tiffney, Ralph Taggart, Bruce Winterhalder, Kay Behrensmeyer, Liz Bailey (ETE database administrator), Catherine Badgley, Rick Potts, Susan Mazer, John Damuth, David Jablonski, Blaire Van Valkenbergh, Richard Stucky. Not in photograph: Jürgen Boy.

CONFERENCE PARTICIPANTS

Catherine E. Badgley
Museum of Paleontology
University of Michigan
Ann Arbor, MI 48109

Richard Beerbower
Department of Geological Sciences
and Environmental Studies
State University of New York at
Binghamton
Binghamton, NY 13901

Anna K. Behrensmeyer
Department of Paleobiology
MRC NHB 121
Smithsonian Institution
Washington, DC 20560

Jürgen A. Boy
Institut für Geowissenschaften
Johannes Gutenberg-Universität
Postfach 3980
D-6500 Mainz
Germany

Ralph E. Chapman
Morphometrics Laboratory
MRC NHB 107
Smithsonian Institution
Washington, DC 20560

John D. Damuth
Department of Biological Sciences
University of California
Santa Barbara, CA 93106

William A. DiMichele
Department of Paleobiology
MRC NHB 121
Smithsonian Institution
Washington, DC 20560

Peter Dodson
Department of Animal Biology
School of Veterinary Medicine
University of Pennsylvania
3800 Spruce Street
Philadelphia, PA 19104

Daniel C. Fisher
Museum of Paleontology
University of Michigan
Ann Arbor, MI 48109

Robert Foley
Department of Physical
Anthropology
University of Cambridge
Downing Street
Cambridge CB2 3DZ
England

Robert A. Gastaldo
Department of Geology
Petrie Hall
Auburn University
Auburn, AL 36849-3501

Russell W. Graham
Quaternary Studies Center
Illinois State Museum
Springfield, IL 62706

Judith A. Harris
University of Colorado Museum
Boulder, CO 80302

Robert W. Hook
11215 Applewood Drive
Austin, TX 78758-4205

Nicholas Hotton III
Department of Paleobiology
MRC NHB 121
Smithsonian Institution
Washington, DC 20560

David Jablonski
Department of Geological Sciences
University of Chicago
5734 South Ellis Avenue
Chicago, IL 60637

Warren L. Kovach
Institute of Earth Studies
University College of Wales
Aberystwyth, Wales SY23 3DB
United Kingdom

Larry D. Martin
Department of Systematics and
    Ecology
University of Kansas
Lawrence, KS 66045

Susan J. Mazer
Department of Biological Sciences
University of California
Santa Barbara, CA 93106

Paul E. Olsen
Lamont-Doherty Geological
    Observatory
Columbia University
Palisades, NY 10964

Tom L. Phillips
Department of Plant Biology
289 Morrill/505 S. Goodwin
University of Illinois
Urbana, IL 61801

Richard Potts
Department of Anthropology
MRC NHB 112
Smithsonian Institution
Washington, DC 20560

Stephen E. Scheckler
Department of Biology
Virginia Polytechnic Institute and
    State University
Blacksburg, VA 24601

William A. Shear
Biology Department
Hampden-Sydney College
Hampden-Sydney, VA 23943

W. Geoffrey Spaulding
Dames and Moore
4220 South Maryland Parkway
Suite 108
Las Vegas, NV 89119

Robert A. Spicer
Department of Earth Sciences
Oxford University
Parks Road
Oxford OX1 3PR
England

Richard K. Stucky
Department of Earth Sciences
Denver Museum of Natural History
City Park
Denver, CO 80205

Hans-Dieter Sues
Department of Vertebrate
    Palaeontology
Royal Ontario Museum
Toronto, Ontario M55 2C6 Canada

Ralph E. Taggart
Department of Botany
Michigan State University
East Lansing, MI 48824

Bruce H. Tiffney
Department of Geological Sciences
University of California
Santa Barbara, CA 93106

Blaire Van Valkenburgh
Biology Department
University of California
405 Hilgard Ave.
Los Angeles, CA 90024

David B. Weishampel
Department of Cell Biology and
    Anatomy
The Johns Hopkins School of
    Medicine
725 North Wolfe Street
Baltimore, MD 21205

Mark V. H. Wilson
Department of Zoology
University of Alberta
Edmonton, Alberta T6G 2E9
Canada

Scott L. Wing
Department of Paleobiology
MRC NHB 121
Smithsonian Institution
Washington, DC 20560

Bruce Winterhalder
Department of Anthropology
Alumni Bldg. 004 A
University of North Carolina
Chapel Hill, NC 27514

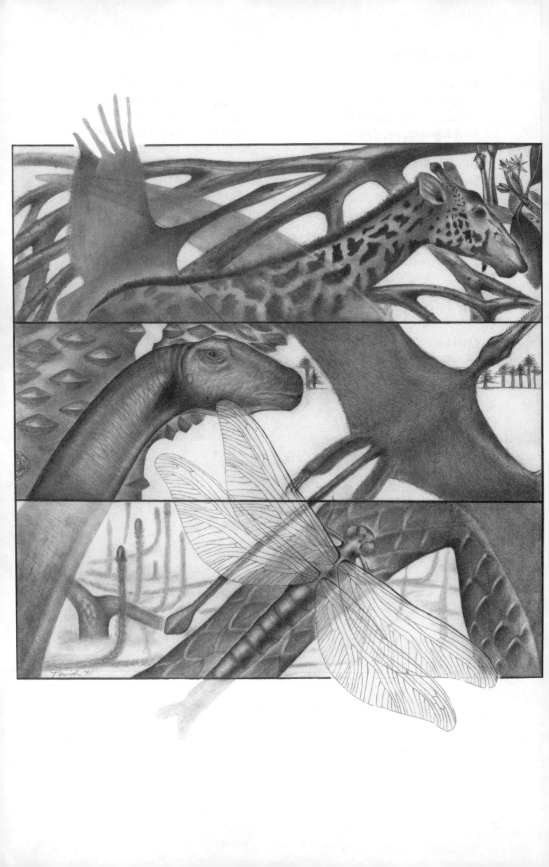

O N E   Evolutionary Paleoecology

Scott L. Wing, Hans-Dieter Sues, Richard Potts,
   William A. DiMichele, and Anna K. Behrensmeyer

1 INTRODUCTION

The objective of paleontology is to discover and analyze biological patterns in the history of lineages and biotas over long periods of time. A primary role for paleontologists in the neo-Darwinian synthesis was to show that such patterns, the "major features" in the history of life, are consistent with and amplify our understanding of population-level processes that can be observed over short time periods in laboratory and natural populations (Simpson 1944, 1953). More recently some paleontologists have explained patterns of diversity and cladogenesis in the fossil record in terms of unique macroevolutionary processes operating on time scales and units of selection that are quite different from those observable for living organisms (Stanley 1975; Vrba 1983; Gould 1985). While the existence and independence of these "higher-order" processes is hotly debated, the attempt to develop theories based on paleontological patterns of biotic change and stasis has stimulated new thinking in evolutionary biology.

In a parallel fashion, fresh examination of ecosystems and species associations in the fossil record, with an emphasis on patterns exhibited over long time spans, will provide new theoretical insights in both ecology and evolutionary biology. Species-level adaptations and the histories of individual clades have been the traditional interests of evolutionary biology. Even studies that have examined patterns of diversification in different environments (Bambach 1977; Jablonski et al. 1983) or the long-term effects of specific predator-prey interactions (Vermeij 1987) have been oriented toward specific clades rather than whole assemblages. Relatively little attention has been paid to sets of unrelated species assembled into ecological associations (communities) that may persist for millions of years.

The foundation of paleoecology comprises inferences about these communities (paleosynecology) or the habits and habitats of their individual members (paleoautecology). Although much has been learned about particular, time-specific faunas and floras, relatively few workers in either the marine or terrestrial realm have gone beyond this to develop a comparative paleoecology, the study of the associations among species and their ecological attributes through time. Since paleoecology includes the study of patterns of ecological organization and change over the long periods of time during which evolution can happen, it is a logical bridge between ecology and evolutionary biology. A primary goal of this book is to outline the material and conceptual basis for a systematic study of this cross-disciplinary realm, which we call "evolutionary paleoecology."

The concern of evolutionary paleoecology is with the biotic and abiotic context of extinct organisms and the effects of this context on the evolution of lineages and morphotypes. Aims of the comparative approach are: (1) to examine the extent to which neo-ecological concepts can account for structural properties and changes within ancient communities, (2) to determine what underlying ecological interactions are independent of species composition and time period, (3) to investigate the long-term responses of ecosystems to changing abiotic and biotic conditions, and (4) to ascertain how ecological relationships have influenced the evolutionary histories of groups of organisms.

The ecological and evolutionary sides of this hybrid mission have somewhat different questions and sources of data. The ecological portion of paleoecology infers original species associations, or communities, and makes comparisons with other fossil or recent associations. At present little is known about how the processes inferred and observed in modern ecosystems, and thought to contribute to the structure of these systems, translate into longer intervals of time. What is the signature of competition through millions of years? Is it compatible with patterns of stasis in morphology and with long-term stability in plant and animal assemblages? Are succession, gap formation, patchiness of food resources, short-term population distributions, etc., detectable in the fossil record? Do these processes operate on such a small spatiotemporal scale that their effects can be averaged and treated as constant over geologically long periods? How does the magnitude of these effects compare with that of disturbances with much longer return times and larger spatial scales that can only be studied in the geological record? If ecological processes unique to long time intervals do indeed act upon biotas, comparison of species associations inferred from the fossil record is the only route available to identify and investigate such processes.

Although there is enough information on fossil faunas and floras to establish that some degree of constancy in species associations can exist through long periods of time, at present we know relatively little else about these associations. What processes are involved in this long-term persistence? Are there regularities in community organization, in energy flow, guild structure, diversity, and biomass distributions, that apply across time, space, and genealogy? Regularities that can be perceived in extant communities are left open to a wide range of conflicting interpretations, in part because of the relatively short periods of time available for observation, and because these communities occur within the Quaternary, a period of earth history subject to unusual climatic perturbations and unique anthropogenic effects. There is discussion about what is pattern and what is "noise" in modern communities, as well as what controls niche partitioning, species diversity, and the integrity of communities themselves (Connor and Simberloff 1979; Connell 1980; Pimm 1984; Cale et al. 1989). To the extent that such debates truly concern the long-term (diachronous) features of communities, we expect that they will benefit greatly from the evidence and perspective provided by the fossil record.

The basic premise of the evolutionary side of paleoecology is that lineages evolve in the context of other organisms and their physical environment, even though endogenous processes and constraints (e.g., development, mechanical properties of organic materials) also play important roles. There are many interesting questions at the interface between evolution and ecology that could draw upon paleoecological data. For instance, at this point we do not know: (1) whether there are consistent relationships between the ecological role of a lineage and its rate or pattern of evolution; (2) at a larger scale, whether there are consistent relationships between the structure of species associations and the rates or patterns of evolution displayed by all or most of their component species; (3) if rates of evolution vary as the structure of a species association changes; and (4) whether geographical scale and return time of environmental disturbances affect evolutionary and ecological patterns in concert or asynchronously.

Ecologists often see their systems as independent of history or genealogy, and evolutionists have tended to see their clades as independent of ecological context. A complete investigation of the history of life requires a synthesis of ecological history and genealogy. Although recent ecosystems provide essential information on ecological patterns and processes, evolutionary paleoecology considers the manner in which ecological processes, including those acting over longer intervals than can be observed today, play out over spans of time relevant to the history of life on earth. In the chapters to follow, these themes are developed using currently available information on terrestrial ecosystems, within which we include both nonaquatic and aquatic continental biotas.

2   THE METHODS OF EVOLUTIONARY PALEOECOLOGY

Although the fossil record has been portrayed as a series of "natural experiments" that provide evidence allowing hypothesis testing in a manner similar to that in the experimental sciences, this analogy is not true in a strict sense. We cannot manipulate the past or design experiments with appropriate controls, and we are limited to observations of historical phenomena. There is, however, considerable power in an approach that uses these observations to test predictions based on ecological or evolutionary theory. We can, for example, propose and test a hypothesis about what happens to ecological structures in floras or faunas on either side of an environmental crisis. Although we cannot generally observe the short-term processes, we can record the longer-term patterns which, if repeated at different points in time and in different biotas, may attest to underlying ecological processes. The effects of time averaging and other taphonomic biases may mask, modify, or mimic the patterns created by these underlying processes, and too often paleoecology has suffered from the temptation to ignore these problems, using paleontological data as if they were collected in a modern ecosystem.

In order to make informative comparisons among extinct communities and between extinct and living communities, there are two requirements: (1) reliable inferences and reconstructions of species associations from fossil samples, and (2) indices of ecologically meaningful structural or morphological features that can be compared even when species associations are taxonomically dissimilar. The first requirement means that we must understand the relationship of fossil assemblages to original species associations through taphonomic analysis. The second requirement means that consistent and informative measures of community-level characteristics need to be developed.

2.1 Taphonomy

There is no denying that taphonomic processes have created discrepancies between original species associations and fossil assemblages, and otherwise have made the paleoecological record difficult to interpret. We see taphonomy, however, as playing a necessary and positive role as well in the goals of comparative paleoecology. The conditions and processes leading to preservation of terrestrial organisms can be delimited and categorized (chap. 2). This permits the identification of similar (isotaphonomic) assemblages among which ecological comparisons can be justified. Recent research has made dramatic progress in defining taphonomic variability within environmentally based categories and establishing standardized procedures for taphonomic inference (Spicer 1988; Gastaldo 1988; Behrensmeyer 1991). Sedimentology and taphonomy must be used to infer not only the local environmental, climatic, and hydraulic equivalence of fossil samples but also the temporal duration represented by each sample. Appropriate process-based explanations for paleoecological patterns are heavily dependent on estimates of the amount of time that passed during the accumulation of individual samples (hours or millenia?) as well as the temporal interval between samples (chap. 2).

Fossil assemblages that are autochthonous, or that show few signs of transport or winnowing, enable examination of the structure and dynamics of ancient species associations with some confidence. Furthermore, certain aspects of species associations, such as rank-order abundance of dominant forms, appear to be fairly robust in spite of various kinds of taphonomic modification. The information summarized in chapter 2 demonstrates that although the evolution of life has changed the biological aspects of taphonomy (e.g., Robinson 1990), some taphonomic contexts occur throughout most of the terrestrial fossil record. This allows samples with relatively similar taphonomic histories to be compared across very long time spans. Understanding change in the biological processes of degradation and recycling of organic remains through geological time is another positive contribution that taphonomic studies can make to evolutionary paleoecology.

Taphonomic evidence makes it clear that the fossil record cannot be accepted literally, and that samples of fossils cannot be treated as equivalent to samples taken from modern ecosystems. Taphonomy also reveals that a remarkable amount of original ecological information is preserved in some assemblages and that the record, though biased, is frequently interpretable. Although there is much to be done in defining recurring isotaphonomic modes, taphonomy provides the essential criteria for judging comparability among samples and will continue to refine and place limits on paleoecological inferences and hypotheses about the relationships between ecology and evolution.

2.2   Ecological Characterization of Extinct Species

Unraveling the ecological preferences or behavioral attributes of extinct species is perhaps the oldest and most familiar part of paleoecology, but too often paleoecology has begun and ended with such autecological reconstructions. We see autecological reconstructions as the building blocks for descriptions of extinct communities that go beyond species lists or relative abundance distributions. Here we briefly describe the general methods by which the ecological characteristics of extinct plants and animals are inferred and outline how autecological reconstruction contributes to evolutionary paleoecology (fig. 1.1). Chapters 3 and 4 contain more detailed discussion of the techniques and data involved in both autecological and synecological interpretation of the fossil record.

The autecology of an extinct species can be inferred in three basic ways: from functional interpretation of morphology or anatomy or chemistry, by analogy with living relatives, or from the sedimentary context and distribution of the fossils of that species (fig. 1.1). Functional morphology is probably the most widely used of these methods, and functional interpretations of form are generally the basis for ecological categorization and morphometric approaches at the community level. Basically functional morphology relies on the analysis of plants and animals as simple machines whose functional characteristics can be inferred from the physical and mechanical properties of their bodies and from their size and shape. In recent years chemical analyses of preserved organic compounds or of stable carbon isotopes have been used to

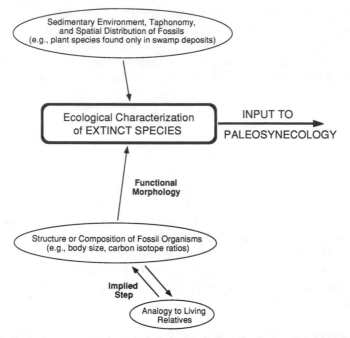

Fig. 1.1 The basic components in ecological characterization of extinct species, which provides the foundation for evolutionary paleoecology. Ecomorphic characterization is based on functional morphology and analogy to living species; other elements in ecological characterization include chemical analysis, taphonomy, and sedimentology.

infer ecologically significant traits like the diets of extinct animals (Ericson et al. 1981) and the photosynthetic pathways of extinct plants (Nambudiri et al. 1978).

Where extinct species have close living relatives, they may be assumed to resemble them in ecological traits. Inference based on living relatives has been applied most extensively to late Tertiary and Quaternary biotas. Logically, analogy with living relatives is a derivative of functional morphology; species have been placed in the same higher taxon because they have similar morphological traits, and species with similar morphologies are assumed to have similar ecological characteristics. (Note that this method implicitly relies on higher taxa that have been defined on similarity, and that may be paraphyletic, rather than on the cladistic criterion of strict monophyly.)

Aspects of the autecology of extinct species also can be derived from their sedimentary context. This approach has been used most widely in plants, because plant fossil assemblages are commonly hypautochthonous and thus preserve the original context and spacing of individuals. Accumulations of vertebrate fossils in paleosols also have the potential to reveal habitat preferences by the correlation of the presence of a species with paleosol characteristics (Bown and Beard 1990). Dietary preferences of animals can be inferred in rare cases where gut contents are preserved (Collinson and Hooker 1987).

2.3 Ecological Characterization of Past Communities: Synchronic Features

Once the taphonomic comparability of fossil assemblages has been established or accounted for and the ecological characteristics of the species have been inferred, we are faced with the problem of deducing and comparing features at the community level (fig. 1.2). The parameters compared must be both accessible and biologically meaningful. Typical characteristics compared in ecological studies (i.e., those of a single time plane) include species composition, richness, evenness, relative abundances of higher taxa and growth habits, guild structure, spatial heterogeneity, and disturbance regime. All of these synchronic features can be known more precisely for living communities than for fossil ones, but we see them largely as a prerequisite for describing diachronic features that only manifest themselves over time, e.g., openness to invasion by immigrants (invasibility), stability, persistence, and rate of change in community composition. Such diachronic characteristics may have greater importance for evolutionary processes than synchronic ones, yet they are only accessible in the fossil record.

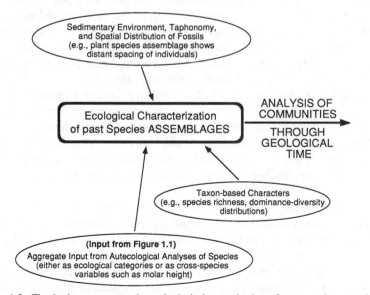

Fig. 1.2. The basic components in ecological characterization of past species assemblages, which may be deemed communities. Autecological analysis from figure 1.1 is one source of input, along with assemblage-oriented sedimentary and taphonomic analysis and ecological variables such as diversity.

Taxon-based Approaches

The simplest way to characterize a species association is to list its members. Lists can be used to compare associations that are (or were) similar taxonomically. There is, however, little basis for comparison when the associations share few species. Measures of species number and the distribution of their relative abundances do not require taxonomic similarity and have been used widely by ecologists to infer processes such as competition (Hairston et al. 1960; Ehrlich and Birch 1967). All three of these approaches have been used to compare fossil assemblages. Although dominance and diversity measures provide an objective way of comparing taxonomically dissimilar associations and recording change through time, studies in recent communities indicate that it is not possible to infer much about structuring processes from these measures alone (May 1975). Furthermore, dominance-diversity measures convey nothing about the kinds of ecological roles occupied by species or the interactions between them.

Ecological Categories and Continuous Variables

Although determining the patterns of change in taxonomic composition and diversity of communities is an important part of evolutionary paleoecology, it is equally important to ascertain the changes in ecological dynamics that are not reflected in such parameters. One approach to studying changes in ecological interactions through time involves placing species in ecological categories based on their inferred autecologies and then observing how the diversity of species within categories changes over time. This approach, sometimes called the ecomorphic approach, is an extension of the concept of guild developed by ecologists (Root 1967). Ecological guilds were defined as groups of species belonging to the same higher taxon that share similar trophic resources; but in paleoecology the term has been used more broadly to refer to species or even higher level taxa that had similar modes of life, including habitat preference, feeding, and locomotion. Theoretically any kind of information could be used to place extinct species into ecological categories or guilds, although functional morphology has been used most widely (chaps. 3 and 4; Andrews et al. 1979; Collinson and Hooker 1987).

Ecological categorization of the species in a fossil assemblage provides another dimension of information that is not captured in simple community-wide diversity measures, but in practice it also has presented several problems: (1) species can be incorrectly assigned to functional categories because of misinterpreted or imperfect form-function correlation; (2) guilds have been based on just a few features and thus may miss significant details; (3) use of species as the smallest recognized unit disguises any intraspecific variation in ecological role; and (4) it is unclear how to interpret change in species diversity within guilds (e.g., what is the significance of one very abundant species as opposed to many rare species?).

An alternative to placing species in ecological categories is to characterize a fossil assemblage along one or more functionally significant morphological gradients. Quantifiable features (e.g., body size, length of tooth cusps) are measured, and then communities are characterized by the shape or position of their distribution along an axis or in a multidimensional space defined by the morphological measurements (e.g., cenograms of Legendre 1987). This kind of morphometric approach can be based either on measurements of individual specimens or on mean values of species. If specimens are measured and plotted independently of their membership in species, then both intraspecific variation and relative abundances of different morphological types will influence the community's profile. However, the morphometric approach has its own limitations. The most important is that it can be used only on assemblages of organisms that are homogeneous enough to allow the same functional characters to be measured, and it thus cannot be applied to dissimilar taxa across long periods of time. Moreover, if measurements of individual specimens rather than species averages are used, then the amount of variability in ecologically significant traits is influenced by the number of specimens.

Synecology from Context and Distribution of Fossils

Spatial distribution of individuals and populations is another characteristic of the structure of species associations. This has seldom been investigated among fossil terrestrial animals, although some paleosol accumulations of vertebrate fossils do preserve patterns of fairly small-scale spatial variation in species abundances and composition (Bown and Beard 1990). Regional-scale ecological variation also has been demonstrated for isochronous faunas in East Africa (Behrensmeyer 1985; Harris et al. 1988) and Eurasia (Bernor 1984). This approach can yield data on the existence and nature of paleoecotones or larger-scale biogeographic boundaries.

The local distributions of individuals and species are major aspects of vegetational structure. Studies of autochthonous assemblages of fossil tree trunks have yielded data on spacing of individuals and also on ecotones between different vegetational types (Jefferson 1982; DiMichele and DeMaris 1987). Similar phenomena have been observed in the distribution of compression fossils (Hickey 1980; Wing 1984). Study of the distribution of plant fossils in sedimentary layers may permit direct measurement of species/area relationships, vegetational heterogeneity in species composition, and perhaps even gap sizes and aspects of canopy structure (chap. 3).

2.4 Ecological Characterization of Past Communities: Diachronic Features

Many diachronic characteristics are related to the stability and persistence of species associations through time: their openness to immigrants, responsiveness to environmental perturbation, and stasis in species or guild composition. One area of controversy in which paleoecological data should have direct bearing is the debate over whether communities act as well-integrated "superorganisms," or whether they are chance juxtapositions of species that operate more or less independently of one another. The data assembled in this volume suggest that it is relatively meaningless to stake out a claim on the side of "individualistic" or "superorganismic" theories; species, or rather local populations of species (avatars of Damuth 1985), can and do act as independent units. It is also true that some species associations are the result of chance events and local history, persist for a very short time, and have weak ecological interactions. On the other hand, studies of living organisms have produced abundant evidence for diffuse and specific coevolution (Herrera 1984; Futuyma and Slatkin 1983), and the fossil record demonstrates that some species associations persist, though perhaps not continuously, for periods of millions of years. The important question is not individualism vs. superorganism but rather: do particular species associations and interactions persist long enough to produce important long-term evolutionary effects on lineages and ecological structure? This is a question that can be pursued only by studying fossil assemblages.

3   CONCLUSION

Ecology, or the "economic" side of the evolutionary history of Earth's biotas, is beginning to receive more attention as a missing piece in the puzzle of pattern and process in evolution (Eldredge 1989). Ecology deserves a stronger role in evolutionary thinking, and paleoecology can greatly expand the universe of patterns that must be explained by ecological theories. To the extent that theories based on living communities do not explain the dynamics, structure, or patterns of change in ancient ones, we will need to modify, expand, or reject these theories and move toward a more comprehensive understanding of species associations in time and space. The fossil record permits observations of communities over a wide range of time intervals ($10^2 - 10^7$ yrs), and patterns of community change in the long term may differ from those observed over shorter intervals in much the same way that some patterns of genealogical change are observable over millions of years but not over a few generations. Cross-disciplinary bridges need to be strengthened, and paleoecological methodologies need to be developed in order to expand the contributions that evolutionary paleoecology is capable of making to the evolutionary synthesis.

The chapters that follow represent the taking-off point for this long-term goal and also serve as a source of information on the current state of terrestrial paleoecology. Many data have been assembled from studies with systematic or biostratigraphic objectives and thus are less than ideal for paleoecological purposes. New phases of data collection could greatly improve the factual basis for the field, but a significant amount of existing data in field notes, museum collections, and sedimentological studies can be utilized as well, as long as the biases inherent in these data are understood.

There has for some time been a pervasive metaphor within evolutionary theory, that species evolve in an "ecological theater," implying that the scenes are the slowly changing, passive elements, and the species are the dynamic, creative actors (Hutchinson 1965). Whether or not this was Hutchinson's original intent, it appears to characterize a great deal of evolutionary thinking that has emphasized the species or the clade with little attention to ecological context. The evidence from the fossil record, as summarized in this book, demonstrates that there is a fundamentally interactive relationship of the theater and the actors; the scenes can change rapidly, the actors can eliminate each other or pieces of the set, the lights can go off and the whole theater close down for unspecified periods of time, only to reopen with different scripts and scenery. The actors are not only actors but authors, directors, and stage managers, yet they still must respond to the changes they help to create by adapting, moving, or becoming extinct. In the history of life on earth, the pace and unpredictability of this drama may never have changed so radically as it has since the advent of our own species.

REFERENCES

Andrews, P., J. M. Lord, and E. M. Nesbit-Evans. 1979. Patterns of ecological diversity in fossil and modern mammalian faunas. *Biological Journal of the Linnean Society* 11:177–205.

Bambach, R. K. 1977. Species richness in marine benthic habitats through the Phanerozoic. *Paleobiology* 3:152–67.

Behrensmeyer, A. K. 1985. Taphonomy and the paleoecological reconstruction of hominid habitats in the Koobi Fora Formation. 309–24 in Y. Coppens, ed., *L'environnement des Hominides au Plio-Pleistocene*. Paris: Foundation Singer-Polignac/Masson.

———. 1991. Terrestrial vertebrate accumulations. In P. A. Allison and D. E. G. Briggs, eds., *Taphonomy: Releasing the Data Locked in the Fossil Record*. New York: Plenum Press.

Bernor, R. 1984. A zoogeographic theatre and biochronologic play: The time/biofacies phenomena of Eurasian and African Miocene mammal provinces. *Paléobiologie Continentale* 14:121–42.

Bown, T. M., and K. C. Beard. 1990. Systematic lateral variation in the distribution of fossil mammals in alluvial paleosols, Lower Eocene Willwood Formation, Wyoming. 135–51 in T. M. Bown and K. D. Rose, eds., *Dawn of the Age of Mammals in the northern part of the Rocky Mountain Interior, North America*. Geological Society of America Special Paper 243.

Cale, W. G., G. M. Henebry, and J. A. Yeakley. 1989. Inferring process from pattern in natural communities. *Bioscience* 39:600–605.

Collinson, M. E., and J. J. Hooker. 1987. Vegetational and mammalian faunal changes in the Early Tertiary of southern England. 259–304 in E. M. Friis, W. G.

Chaloner, and P. R. Crane, eds., *The Origins of Angiosperms and Their Biological Consequences*. Cambridge: Cambridge University Press.

Connell, J. 1980. Diversity and the coevolution of competitors, or the ghost of competition past. *Oikos* 35:131–38.

Connor, E. F., and D. Simberloff. 1979. The assembly of species communities: Chance or competition. *Ecology* 60:1132–40.

Damuth, J. 1985. Selection among "species": a formulation in terms of natural functional units. *Evolution* 39:1132–46.

DiMichele, W. A., and P. J. DeMaris. 1987. Structure and dynamics of a Pennsylvanian-age *Lepidodendron* forest: Colonizers of a disturbed swamp habitat in the Herrin (No. 6) Coal of Illinois. *Palaios* 2:146–57.

Ehrlich, P. R., and L. C. Birch. 1967. The balance of nature and population control. *American Naturalist* 101:97–107.

Eldredge, N. 1989. *Evolutionary Macrodynamics*. New York: McGraw-Hill.

Ericson, J. E., C. H. Sullivan, and N. T. Boaz. 1981. Diets of Pliocene mammals of Omo, Ethiopia, deduced from carbon isotope ratios in tooth apatite. *Palaeogeography, Palaeoclimatology, Palaeoecology* 36:69–73.

Futuyma, D. J., and M. Slatkin. 1983. *Coevolution*. Sunderland, Mass.: Sinauer Press.

Gastaldo, R. A. 1988. Conspectus of phytotaphonomy. 14–28 in W. A. DiMichele and S. L. Wing, eds., *Methods and Applications of Plant Paleoecology*. Paleontological Society Special Publication 3.

Gould, S. J. 1985. The paradox of the first tier: An agenda for paleobiology. *Paleobiology* 11:2–12.

Hairston, N. G., F. E. Smith, and L. B. Slobodkin. 1960. Community structure, population control, and competition. *American Naturalist* 94:421–25.

Harris, J. M., F. H. Brown, M. G. Leakey, A. C. Walker, and R. E. Leakey. 1988. Pliocene and Pleistocene hominid-bearing sites from west of Lake Turkana, Kenya. *Science* 239:27–33.

Herrera, C. M. 1984. Determinants of plant-animal coevolution: The case of mutualistic dispersal of seeds by vertebrates. *Oikos* 44:132–41.

Hickey, L. J. 1980. Paleocene stratigraphy and flora of the Clark's Fork Basin. 33–49 in P. D. Gingerich, ed., *Early Cenozoic Paleontology and Stratigraphy of the Bighorn Basin, Wyoming*. University of Michigan Papers on Paleontology 24.

Hutchinson, G. E. 1965. The Ecological Theater and the Evolutionary Play. New Haven: Yale University Press.

Jablonski, D., J. J. Sepkoski, D. J. Bottjer, and P. M. Sheehan. 1983. Onshore-offshore patterns in the evolution of Phanerozoic shelf communities. *Science* 222:1123–25.

Jefferson, T. H. 1982. The Early Cretaceous fossil forests of Alexander Island, Antarctica. *Palaeontology* 25:681–708.

Legendre, S. 1987. Analysis of mammalian communities from the late Eocene and Oligocene of southern France. *Palaeovertebrata* 16:191–212.

May, R. M. 1975. Patterns of species abundance and diversity. 81–120 in M. L. Cody and J. M. Diamond, eds., *Ecology and Evolution of Communities*. Cambridge, Mass.: Belknap Press.

Nambudiri, E. M. V., W. D. Tidwell, B. N. Smith, and N. P. Hebert. 1978. A C_4 plant from the Pliocene. *Nature* 276:816–17.

Pimm, S. L. 1984. The complexity and stability of ecosystems. *Nature* 307:321–26.

Robinson, J. M. 1990. The burial of organic carbon as affected by the evolution of land plants. *Historical Biology* 3:189–202.

Root, R. B. 1967. The niche exploitation pattern of the Blue-Gray Gnatcatcher. *Ecological Monographs* 37:317–50.

Simpson, G. G. 1944. Tempo and Mode in Evolution. New York: Columbia University Press.

————. 1953. The Meaning of Evolution. New York: Columbia University Press.

Spicer, R. A. 1988. Quantitative sampling of plant megafossil assemblages. 29–51 in W. A. DiMichele and S. L. Wing, eds., *Methods and Applications of Plant Paleoecology*. Paleontological Society Special Publication 3.

Stanley, S. M. 1975. A theory of evolution above the species level. *Proceedings of the National Academy of Science* 72:646–50.

Vermeij, G. J. 1987. *Evolution and Escalation: An Ecological History of Life*. Princeton: Princeton University Press.

Vrba, E. S. 1983. Macroevolutionary trends: New perspectives on the roles of adaptation and incidental effect. *Science* 221:387–89.

Wing, S. L. 1984. Relation of paleovegetation to geometry and cyclicity of some fluvial carbonaceous deposits. *Journal of Sedimentary Petrology* 54:52–66.

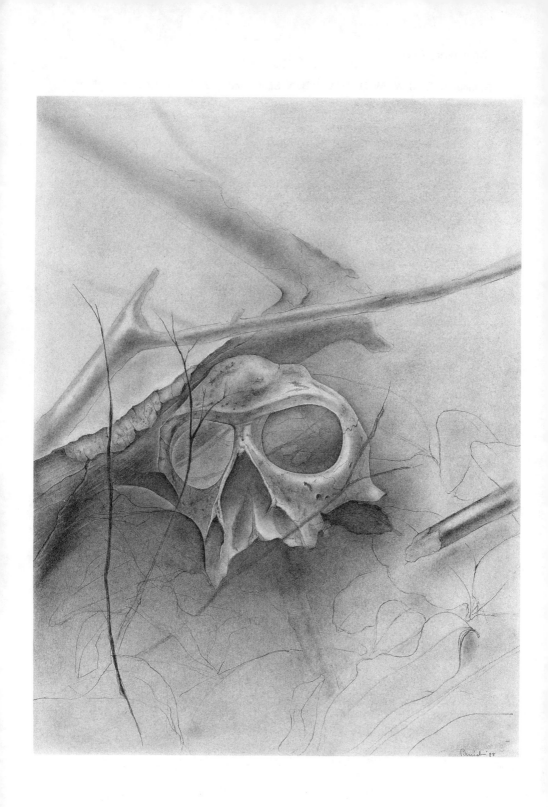

T W O   Paleoenvironmental Contexts
──────   and Taphonomic Modes

Anna K. Behrensmeyer and Robert W. Hook, RAPPORTEURS

IN COLLABORATION WITH CATHERINE E. BADGLEY, JÜRGEN A. BOY, RALPH E.
CHAPMAN, PETER DODSON, ROBERT A. GASTALDO, RUSSELL W. GRAHAM,
LARRY D. MARTIN, PAUL E. OLSEN, ROBERT A. SPICER, RALPH E. TAGGART,
AND MARK V. H. WILSON.

1 INTRODUCTION

Sedimentological analysis of paleoenvironments and taphonomic study of organic remains preserved in these environments are essential components of paleoecology. Sedimentology provides information on the physical settings where organisms lived and were buried and on interactions between biotic and abiotic components of ancient ecosystems. Taphonomy illuminates biases that affect paleoecological reconstructions and provides additional information on biological processes represented in a fossil assemblage. Both taphonomy and sedimentology are necessary for the reconstruction of the original ecological relationships among fossil organisms (fig. 2.1).

Throughout the Phanerozoic (see Appendix), there are particular sedimentary contexts that preserve land organisms, and within such contexts there are recurring patterns of preservation for different organisms. The purpose of this chapter is to identify and document these recurring taphonomic modes. We define a taphonomic mode as a set of fossil occurrences that result from similar physical, chemical, and biological processes. The processes may be only broadly similar, to the extent that they occur within the same physical setting, or they may be more specifically defined according to detailed sedimentological and taphonomic evidence.

The taphonomic modes discussed below are organized in a series of tables that reflect environments or circumstances defined by sedimentological or biological criteria. These tables show the types of organic remains that are found in each subenvironment and include most known preservational circumstances for terrestrial organisms. Thirty-four preservational contexts for five different categories of organic remains are discussed in this chapter. Table 2.1 provides a summary of information recorded in more detail in other tables.

Our classification of taphonomic modes is based on a broad assessment of environmental or biogenic contexts. Each cell within the tables could be subdivided into many other modes with more specific taphonomic features; the

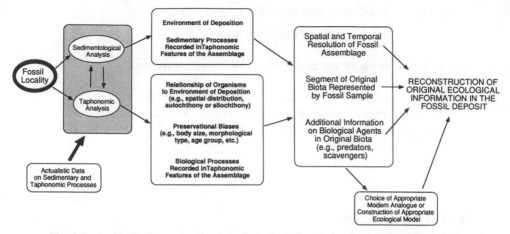

Fig. 2.1. An idealized structure for the paleoecological analysis of paleontological sites, in which sedimentologic and taphonomic analysis provides the foundation for successive levels of inference. Reference to modern processes and analogues can occur throughout this progression but is most common during the initial and final stages of analysis.

Table 2.1   Summary of Preservational Contexts for Terrestrial Organisms

Environmental Context	Occurrences of Terrestrial Organisms				
	Macroplant	Microplant	Invertebrates	Vertebrates	Ichnofossils
Coastal:					
Offshore	+	+ +	((+))	(+)	−
Beach	((+))	−	−	(+)	+
Lagoon	+	+ +	((+))	+ +	+ −
Estuary	+ +	+ −	−	+	+ −
Fluvial/deltaic:					
Channel lags and bars	+ +	+	+	+ +	+
Abandoned channel fill	+ + +	+ + +	+ +	+ +	+ +
Levee	(+)	(+)	−	(+)	+ + +
Floodplain: poorly drained	+ + +	+ + +	+	+	+ +
Floodplain: well-drained	+	(+)	(+)	+	+ +
Crevasse splay	+	+ +	+	+ −	(+)
Interdistributary bay	+ +	+ +	(+)	+ −	+ +

Table 2.1 *(Continued)*

Environmental Context	Occurrences of Terrestrial Organisms				
	Macroplant	Microplant	Invertebrates	Vertebrates	Ichnofossils
Lacustrine:					
Low oxygen:					
Large lake, deep	((+))	+ +	(+)	+ +	−
Small lake, deep	+ +	+ +	(+)	+ +	−
Small lake, shallow	+ +	+ +	(+)	+ +	(+)
Oxygenated:					
Large lake	(+)	+ +	+ +	+ +	+ + +
Small lake	(+)	+ +	+ + +	+ +	+ + +
Volcanigenic:					
Explosive events	+	((+))	−	((+))	((+))
Primary airfall	+	(+)	−	(+)	+ −
Lacustrine	+ +	+ + +	+ +	+ + +	(+)
Eolian:					
Dune	−	−	−	((+))	((+))
Interdune	+	+	(+)	(+)	+
Loess	((+))	+	+	+	+ +
Other contexts:					
Tar seeps	+	+	+	+ + +	−
Amber	+ +	+ +	+ +	((+))	−
Karst (caves, fissures, sinkholes)	(+)	(+)	((+))	+ + +	(+)
Tree trunks	+	+	(+)	+	(+)
Burrows	+	+	+	+ +	((+))
Feces, regurgitates	(+)	+	+	+ +	−
Middens	+ +	+ + +	(+)	(+)	−
Archeological sites	+	+	+	+ +	+
Springs	+ +	+ + +	+ +	+ +	+ +
Peat	+ + +	+ + +	+ −	(+)	−
Charcoal	+ +	+	(+)	(+)	−
Mud slides, flash floods	+	−	−	+ −	−

Note: + = present (no assessment of relative abundance); + + = common; + + + = very common; (+) = uncommon; ((+)) = rare; − = very rare or absent; + − = highly variable; ? = no data. The assessments of "present," "common," "rare," etc., relate to the expected abundance of the specified organic remains or traces (columns) within each context (rows). Qualitative assessments of relative abundance are based on published information and judgments of the 1987 ETE conference participants. They do not refer to how rare or common a particular context is in the geological record as a whole, and they also assume optimal diagenetic circumstances for preservation. "Present" carries no assessment of relative abundance and is used for occurrences where there is insufficient data to judge abundance, even qualitatively. Microplant remains may include pollen, spores, cuticles, phytoliths, and diatom frustules.

scale of a mode is determined by what is useful for comparing particular fossil assemblages. Although the different types of taphonomic evidence pertaining to each mode are discussed in the text, we do not include a detailed treatment of taphonomic methodology in this chapter (see Shipman 1981; Gifford 1981; Gastaldo 1988; Wilson 1988b; Spicer 1989; Behrensmeyer 1991).

Within each of the broadly defined modes (table 2.1), available evidence indicates that at least some taphonomic biases are similar through time. In contrast, different taphonomic modes characteristically have different sets of biases. The organic samples used for the study of ancient ecosystems often have been unsorted with respect to taphonomic history and are subject to unknown taphonomic biases. Comparing fossil assemblages from the same taphonomic mode should increase the probability that observed similarities and differences reflect original ecological or evolutionary factors rather than disparate taphonomic histories (fig. 2.2). Deliberate use of taphonomic modes thus can help to control "taphonomic noise" in paleoecological signals.

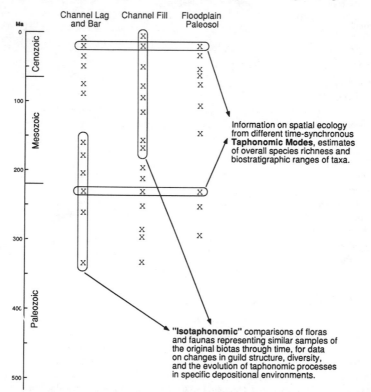

Fig. 2.2. Hypothetical examples of the uses of taphonomic modes. The *X*'s indicate paleontological sites that can be assigned to equivalent taphonomic modes, in this case defined by three subenvironments of fluvial systems. Sequences of sites from the same mode can be compared through time ("isotaphonomy"), and synchronous sites from different modes can provide information on spatial variation in contemporaneous faunas and floras.

Entries in the tables represent hypotheses about recurring preservational patterns for groups of organisms or their traces (plant, invertebrate, and vertebrate remains and ichnofossils). These hypotheses are supported by documentation from present-day environments, examples from the fossil record, or both. Terms such as "common" and "rare" pertain to the relative abundance of a specific type of organic sample (e.g., "macroplant") through the spectrum of subenvironments or taphonomic circumstances. Estimates of relative abundance were reached through consensus within our working group, supplemented by literature reviews. In addition to providing a basis for faunal and floral comparisons through time, the tables can be used to predict the potential for different types of fossil occurrences and co-occurrences in particular environmental settings.

In the tables and the following discussion, the terms *autochthonous* and *allochthonous* are used to describe the inferred transport status of organic remains. Autochthonous remains are preserved at the death site of an organism or the site where parts were discarded. For sessile organisms, this death or discard site coincides with the life habitat; for mobile forms, the death or discard site usually is within the original habitat. Allochthonous refers to remains that have been moved from the site of death and out of the original habitat. The term *parautochthonous* may be used for remains that are transported from the death or discard site but stay within the original habitat.

Distinct patterns emerge from an overview of taphonomic modes for continental organisms. In some environments there is abundant documentation of fossil preservation (e.g., abandoned channel fills, anoxic lake deposits), whereas in others there is comparatively little evidence for fossils (e.g., levees, dunes, burrows). This scarcity of evidence for some depositional environments may be genuine, or it may reflect a variety of factors including inadequate sedimentological analysis or failure to document fossil context. We have included some poorly known modes that appear to be potentially significant, as well as those that have been well documented.

We have attempted to separate taphonomic and sedimentological information derived from recent analogues and that derived from the fossil record. Actualistic data (i.e., observations and experimentation in present-day environments) are available only for some of the organisms and environments to be discussed. Many of the identifications of paleoenvironments and taphonomic processes are based on interpretations from the geological record. Although some such interpretations are well substantiated, others are more speculative and require further investigation. Correctly assigning a particular fossil occurrence to a specific environmental category may require extensive sedimentological studies, including reconstruction of the geometry and geochemistry of the fossil-bearing deposits. Such detailed work is available for some of the examples given below, and others are categorized according to opinions expressed in the referenced articles or on the basis of our own assessment of the published evidence. Readers should note that assignments of the examples to follow represent the most plausible hypotheses based on presently available evidence.

References give specific examples for individual taphonomic modes but do not constitute a comprehensive review of the literature. For more detailed information on sedimentological attributes of the environments discussed here, refer to general works such as Reineck and Singh 1980; Scholle and Spearing 1982; Galloway and Hobday 1983; Scholle et al. 1983; Walker 1984; Davis 1985; Reading 1986; and Selley 1986.

2   COASTAL ENVIRONMENTS

Coastal settings that preserve fossil remains of continental organisms include offshore environments, beaches, lagoons, and estuaries (table 2.2). They are often biologically highly productive, and both sediments and organic remains may have a high probability of permanent burial, compared with inland continental areas, because of the proximity of base level. The physical and chemical processes of coastal areas reflect the complex interaction between marine and continental environments. In the geological record these environments are identified on the basis of lithofacies characteristics, but the distinction between marine and nonmarine deposits is often based on fossil biotas. Because of tidal and fluvial influences, plant and animal accumulations may come from a wide range of subenvironments and be averaged over time periods that are long from a modern ecological perspective (e.g., $>10^2$ yrs) (see §8.4). In contrast, storms and other nearly instantaneous events can produce fossil assemblages characterized by a high degree of temporal and areal resolution (Seilacher 1982).

The coastal environments listed in table 2.2 are characteristic of the ocean-land boundary where coastal processes predominate, although beaches and lagoons also occur in lacustrine settings. Deltas usually are associated with coastal areas, but because many of the important fossil-preserving deltas were constructive (i.e., dominated by fluvial processes), we discuss them later, along with fluvial environments. Similarly, although raised bogs often occur in coastal settings, they are known in other situations and are covered in §7.

2.1   Offshore

Offshore situations that preserve continental organisms include intertidal, shelf, subaquatic fan, and deep benthic environments. These depositional contexts are recognized by using sediment texture and bedding characteristics, as well as marine fossil content. Fossils of continental organisms often occur in black, organic-rich shales, implying anoxic bottom conditions that inhibited destructive biological activity and favored preservation of relatively undisturbed skeletons and, in some instances, soft parts.

There exists only a limited amount of taphonomic and sedimentological in-

Table 2.2 Coastal Environments

Environmental context	Occurrences of Terrestrial Organisms				
	Macroplant	Microplant	Invertebrates	Vertebrates	Ichnofossils
Offshore	Present Locally common logs, seeds, fruits, charcoal	Common	Rare Plankton, nekton	Uncommon Flying forms, tetrapods, aquatics, bone lags	Absent
Beach	Rare	Absent	Very Rare (Terrestrial mollusks)	Uncommon Abraded bone lags, rare articulated skeletons	Present Burrows, roots, tracks
Lagoon	Present Fragmented leaves, woody debris	Common (ALLO + AUTO)	Rare Arthropods	Common Fish, flying forms (AUTO + ALLO)	Variable Bioturbation can be intense
Estuary	Common Logs, highly fragmented debris (ALLO + AUTO)	Variable ALLO: size sorted; AUTO: salt marsh, mangrove	Absent	Present Fish (disarticulated bones, scales); others ALLO	Variable

Note: Remains of marine worms, crustacea, and mollusks may be mixed with nonmarine organisms in these environments. Refer to table 2.1 for explanation of terms. The abbreviations "ALLO" and "AUTO" refer to "allochthonous" (transported) and "autochthonous" (in situ).

formation on present-day occurrences of continental organic remains in off-shore settings. Floating debris from floods of major river systems has been sighted far from land, and it is clear that large quantities of continental organic matter are washed into marine environments (Moseley 1879:368). Studies of palynomorphs in offshore sediments of the Gulf of California indicate that such assemblages are subject to biases due to sedimentation patterns, as well as to distance from shore (Cross et al. 1966).

Occurrences of remains of continental plants and animals are well known in ancient near-shore marine deposits. Cretaceous angiosperm wood has been found in offshore sediments (Stopes 1912). Fragmentary plant material representing at least two onshore communities occurs in Middle Jurassic marine strata in Scotland (R. Bateman pers. comm. 1990). Debris rafted 25 km offshore was deposited with fully marine biotas; dissolution of invertebrate tests apparently helped to generate rapid calcitic permineralization of the plant material. In the Eocene London Clay, plants are preserved in intertidal to shelf deposits, where pyrite permineralization is the dominant mode of preservation for fruits, leaves, seeds, leaves, and wood (King 1981; Collinson 1983; Collinson and Hooker 1987; Allison 1988). There is little evidence of reworking of these remains after initial deposition. Lignitic, largely unmineralized wood and cones occur with Miocene vertebrate fossils in the marine Calvert Formation of eastern North America in depositional contexts where water depths are estimated at 50 m or more and distance to land at 30 km (Myrick 1979).

Examples of well-preserved nonmarine vertebrate remains in marine contexts include the Upper Permian Kupferschiefer deposits in Germany, the Copper Marl (Haubold et al. 1985), Triassic black shales (Martill and Dawn 1986), and Jurassic Oxford Clay (Martill 1985) in England, and possibly the Sakamena Formation in Madagascar (Carroll 1981). In the Upper Permian of Germany, black marine shales preserve remains of vascular plants, rare insects, and associated skeletal parts of terrestrial reptiles (Boy 1977, Haubold et al. 1985). The Lower Jurassic Posidonia Shale of southern Germany records alternating anoxic and oxic conditions of a stable marine shelf and includes rare pterosaurs and terrestrial reptiles as well as plants (especially conifers), in addition to marine organisms (Brenner 1976). Middle Triassic deposits in Switzerland consist of organic-rich marly shales and limestones deposited in a marine embayment; marine fossils predominate, but remains of vascular plants and articulated terrestrial reptiles also occur (Kuhn-Schnyder 1974). Nonmarine dinosaurs are known from Late Cretaceous marine strata in North America (e.g., Horner 1979; Fiorillo 1987, 1989a, 1989b, 1991). In the Miocene of eastern North America, bone and tooth concentrations are associated with fine-grained sediment and condensed sequences of marine invertebrate fossils (Myrick 1979), indicating concentration during periods of sediment starvation associated with transgressions on the continental shelves (Kidwell 1989).

Plant debris is often associated with fine-grained turbidite sequences in submarine fans. Large phytoclasts (e.g., tree limbs, fragments of trunk and bark) are concentrated in turbidite sequences of the Upper Cretaceous of northern California (Page 1981) and have also been noted in Middle Cretaceous deposits in the same area (W. Miller pers. comm. to Gastaldo 1987). In some cases transport down the slopes of subaquatic fans results in minimal abrasion of plant and other organic remains (Allison 1986; Spicer and Thomas 1987). Fern pinnae have been reported in turbidite sequences from DSDP (Deep Sea Drilling Program) cores (Spicer 1989).

In addition to occurrences of articulated or associated land vertebrate remains, a second pattern of preservation is found in shallow, offshore, marine and lacustrine environments. These deposits include mixtures of abraded, fragmentary, nonmarine, and marine vertebrate remains, as well as wood (Gallagher et al. 1989). Offshore bone lags are common in the European Eocene, Oligocene, and Miocene (Tobien 1968). Some of these lag deposits may result from sediment winnowing and stratigraphic condensation during marine regressions or transgressions, but in general little is known about the processes responsible for them. Remains of terrestrial vertebrates also occur in late Cenozoic shelf deposits off the coast of eastern North America and apparently are derived from animals that occupied these areas during periods of lowered sea level (Whitmore et al. 1967; Oldale et al. 1987). Such occurrences demonstrate that remains of continental organisms can be incorporated into marine deposits by shoreline movements as well as transport of organic debris from land to sea.

2.2 Beaches

Sandy beach deposits can be recognized on the basis of mature, well-sorted sands combined with proximity to other marine facies, internal structures, and lateral facies geometry. Beaches are obvious sites for the temporary concentration and burial of organic debris in modern environments (Schäfer 1972), but they are not considered to be favorable contexts for preservation of ancient nonmarine faunas and floras (Scheihing and Pfefferkorn 1984). In addition to the high-energy sediment reworking characteristic of this setting, beach deposits are preserved only under circumstances of rapid transgression (Dubois 1990) or shoreline progradation. Some of the bone lags in offshore environments may be reworked, hydraulic concentrations of beach debris, modified by storms or transgressive and regressive events. Beaches in lower-energy lacustrine settings are probably more favorable contexts for preservation of organic remains than marine beaches, though there has been only limited study of present-day examples (Weigelt 1989).

There is ample documentation of marine invertebrate fossils (e.g., Kidwell 1989) and marine vertebrates (e.g., Spillmann 1959; Whitmore and Gard 1977;

Harington 1977) in beach or other coarse-grained, nearshore deposits but little information for continental organisms. This may be in part because beach facies often are not recognized or differentiated within coastal, fluvio-deltaic, and lacustrine sedimentary sequences. Muizon (1981) has documented Miocene marine and terrestrial vertebrates in beach and other shoreline sediments in Peru. Some vertebrate remains in the Upper Jurassic of Tanzania may occur in beach deposits, as suggested by an association of dinosaur remains within a sandstone that contains marine bivalves (Russell et al. 1980).

2.3 Lagoons and Other Back-Barrier Settings

Present-day lagoons are bodies of water that are partially or completely cut off from open marine water by beach or barrier bars, reefs, or other types of physical barriers. Sediments are characteristically fine-grained, reflecting a low-energy, backwater setting. Lagoons may be similar to shallow lakes in sedimentary facies and taphonomic features; they can be fresh to hypersaline, stable to seasonally ephemeral. The unique physical and biological features of lagoons relate to their characteristically quiet, shallow water and proximity to open marine or lacustrine settings. These features include possible tidal influence, potential variation in water chemistry and temperature, the presence of marine organisms, and susceptibility to strong storm effects. Storms can produce mixed assemblages of marine and nonmarine organic materials in lagoonal settings, most commonly in coarse-grained washover deposits that contain reworked shell and bone fragments.

The diagnostic features of lagoons may not be obvious in the rock record, and distinguishing them from other quiet-water coastal environments can be problematic. Lagoons have been confused with back-barrier estuarine settings, but recent work in modern-day systems has helped to delimit features of estuary fills (e.g., Frey and Howard 1986). The taphonomy of recent terrestrial organisms in lagoons has received little attention, although studies of quiet-water accumulations of plant debris by Zangerl and Richardson (1963) may be applicable to lagoonal settings.

Many organic-rich, fine-grained deposits that contain mixed assemblages of marine and terrestrial organisms have been regarded traditionally as lagoons, although relatively few interpretations are supported by independent sedimentological analysis. The most famous example is the Jurassic Solnhofen Limestone of Germany (Barthel 1970, 1978; Gall 1983), which contains plants, flying vertebrates and invertebrates, and fishes. Allochthonous marine benthic and planktonic organisms are mixed with terrestrial flora and fauna (Viohl 1985; DeBuisonje 1985). On the basis of paleogeographic reconstructions and sedimentology, the deposit is interpreted as a deep lagoon between sponge reefs. Certain trace fossils and other evidence once thought to indicate a shallow lagoonal setting have been discounted by recent sedimentological studies (Seilacher et al. 1985).

In the Lower Carboniferous of Alabama, most of the identifiable plant material in lagoonal sediments belongs to highly resistant, woody aerial parts (Haas and Gastaldo 1986; Haas 1988). Dinosaur trackways are documented in carbonate muds of the Glen Rose Formation of the Lower Cretaceous of Texas (Bird 1944; Farlow 1984, 1987). These tracks are thought to occur in intertidal, supratidal, and even subtidal facies of a shallow lagoon. A broad swath of Lower Cretaceous rocks extending from west Texas to western Arkansas represents a range of coastal environments, including lagoons, and preserves a wide variety of amphibians, reptiles and mammals (Langston 1974). Lagoonal deposits with plant remains occur in Late Jurassic deposits in France (Barale 1981) and also appear to be represented in the Jurassic of Tanzania, where coral colonies and oysters are associated with plant debris and rare abraded dinosaur bones (Russell et al. 1980). Coal deposits in back-barrier lagoonal settings have been documented in the Upper Cretaceous of Wyoming (Roehler 1988).

2.4 Estuaries

The interaction of marine and fluvial processes is perhaps most strongly expressed in estuarine environments, which range from tidally influenced, freshwater rivers upstream from coastal areas, to brackish rivers and salt marshes near the open ocean. Estuaries may be very deep, with anoxic bottom layers, as in fjord coastlines, or shallow, as in constructional deltas. The extremes of water chemistry and the interplay of fluvial and tidal activity generate a stressful setting for living organisms and affect the preservation of both allochthonous and autochthonous organic remains. Estuaries can be depositional sites for "tidalites," which record tidal cycles in continuous sequences of fine laminae (Dalrymple and Makino 1989).

There is a strong sedimentological and biological dichotomy between riverine and salt-marsh estuaries that corresponds roughly to allochthonous versus autochthonous plant and animal assemblages (Frey and Howard 1986). Riverine estuaries typically act as focal areas for the accumulation of transported plant and animal remains derived from large drainage areas. In spite of the large input of detritus, observations in recent estuaries indicate that relatively low rates of sedimentation and high rates of biogeochemical and physical degradation tend to prevent accumulations of identifiable plant remains (Moseley 1879; Berry 1906; Gastaldo et al. 1987; Gastaldo 1989a). Soft-bodied detritivores are active bioturbators in some estuarine subenvironments and thus are important agents in the recycling of plant material. Tracks, trails, and feces of amphibious to terrestrial animals are abundant in modern inter- and supratidal margins of estuaries but are generally obliterated by sediment reworking in buried deposits (Frey and Pemberton 1986).

Plant materials in modern estuarine deposits are typically highly fragmented. This results from physical and biological processes within an estuary, but it may also be attributable to pretransport fragmentation (e.g., as litter on a forest floor) and transport as bedload. Physical rounding and biological degradation homogenize the detritus into assemblages of relatively equidimensional, resistant fragments that are often difficult to identify. Rapid burial of rafted remains, however, may preserve intact foliage (Berry 1906). The distribution and composition of palynomorphs in estuarine deposits appear to be influenced by physical sorting and sediment reworking, leading to regional rather than local representation of floras (Cross et al. 1966; Grindrod and Rhodes 1984).

Present-day freshwater peat swamps associated with salt-marsh estuaries provide a model for the local (i.e., autochthonous) accumulation of identifiable macroscopic plant remains in back-barrier settings (Staub and Cohen 1979). Palynomorph assemblages from restricted back-barrier swamps also appear to be locally derived (Grindrod 1985) and, in conjunction with megafloral and petrographic studies, have been used to reconstruct Recent vegetational histories (Rich and Spackman 1979). The preservation of vertebrate skeletal remains in these typically acidic mires (Staub and Cohen 1978; Benner et al. 1985) would not be expected.

There are a number of well-documented examples of fossiliferous estuarine deposits. The Upper Cretaceous of North America is rich in coastal deposits that include estuaries, with possible examples in South Dakota (Waage 1968), Alberta (Koster and Currie 1987), and Montana (Horner 1979). Eocene deposits of the London Basin (Great Britain) interpreted as estuarine to lagoonal environments contain abundant plant and vertebrate remains (Collinson and Hooker 1987), and terrestrial organisms are preserved in estuarine deposits in the Cretaceous of Germany (Pelzer 1984), the Oligocene of Egypt (Bown et al. 1982; Bown and Kraus 1988), and the Miocene of Libya (Boaz 1987; De-Heinzelin and El-Arnauti 1987) and Arabia (Whybrow and McClure 1981). In the examples from Egypt and Arabia, mangroves have been identified from fossil root structures. Lag deposits of reworked bones from both marine and terrestrial vertebrates occur in Miocene through Pleistocene estuarine sequences of eastern and western North America (Frey et al. 1975; Webb et al. 1981; Riggs 1984) and in South Africa (Hendey 1981). Neogene bone-bearing deposits from estuaries or nearshore situations are often associated with phosphorites (Riggs 1984).

3   FLUVIAL AND DELTAIC ENVIRONMENTS

Rivers and deltas have played a pervasive role in the preservation of land organisms throughout the Phanerozoic. Deposits of ancient alluvial systems are the most common and perhaps most extensively studied of all nonmarine sediments. They encompass everything from fully continental valley infillings to deltaic sequences that interfinger with marine deposits. Fluvial channels have been a traditional focus for paleontological and taphonomic investigations, and finer-grained overbank deposits are proving to be an equally rich source of paleoecological data. Because of their distinctive sedimentological attributes and abundance of fossil remains, fluvial and deltaic deposits are especially suited for examining ecological change through time.

Rivers vary greatly in channel size, discharge, and sediment load, factors that are moderated by biotic processes (e.g., plant cover in the drainage basin [Schumm 1977]) and ultimately controlled by tectonics and climate. Nearly all rivers exhibit some degree of seasonal or longer-term fluctuation in discharge, and many are subject to unusual flooding events that have marked effects on the preservation of organic remains. In deltaic systems or other aggrading situations with a high, relatively stable water table, overbank environments are predominantly wet and include interdistributary lakes, floodplain swamps, and interdistributary bays that are open to marine waters. In settings where the water table is seasonally low, overbank areas may alternate between broad expanses of floodwaters and dry, open, or forested plains. For terrestrial organisms, many floodplains fluctuate from being highly productive to hostile, corresponding to the emergent and submerged parts of flood cycles. Mammals that inhabit such floodplains tend to be amphibious, arboreal, or migratory. During the flood season, a largely aquatic habitat supports high fish diversity and productivity. Reproductive cycles of plants, fishes, and small mammals closely track the cycle of inundation and desiccation of floodplains (Sheppe 1972; Lowe-McConnell 1975, 1977; Goulding 1980). Water-table fluctuation is a critical factor in organic preservation, particularly for plant remains that are rapidly oxidized or biodegraded if exposed to dry conditions (Retallack 1984a).

River systems traditionally are divided into channel and overbank environments, with the latter including a complex of subenvironments (Bridge 1984; Farrell 1987). Modern alluvial environments are characterized by a remarkably high degree of substrate variation that corresponds with floral and faunal differences within a mosaic of alluvial habitats (e.g. Bell 1980; Salo et al. 1986). A comparable degree of lithologic and paleontologic variation in ancient fluvial and deltaic deposits complicates biostratigraphic correlations and paleoecological reconstructions. Recent advances in fluvial sedimentology and taphonomy have helped to establish criteria for identifying fluvial subenvironments in ancient deposits. We recognize seven of these as distinctive contexts for organic preservation in table 2.3: active channels, abandoned channels, levees, well-drained and poorly drained floodplains, crevasse splays, and interdistributary bays. Fluvial lacustrine environments other than abandoned channels are included in §4.

3.1 Channels: Basal Lags and Bars

Channels by definition are where the most continuous and energetic confined flow occurs. They have a wide range of flow patterns and width-to-depth ratios and may converge or diverge downslope. Channels are where the most

Table 2.3 Fluvial and Deltaic Environments

Environmental Context	Occurrences of Terrestrial Organisms					
	Macroplant	Microplant	Invertebrates	Vertebrates	Ichnofossils
Channel: basal lags and bars	Common Logs, fruits, seeds, laminated leaf beds, size sorting	Present Pollen with variable size sorting	Present Bivalves	Common Tetrapods, aquatics, articulated and disarticulated; bone lags	Present Roots, bivalve burrows
Abandoned channel fill	Very common All types, peat, stromatolites, plant succession	Very common	Common Mollusks, terrestrial arthropods	Common Tetrapods, aquatics, usually AUTO, mass death, traps	Common Burrows (decrease upward), roots (increase upward)
Levee	Uncommon Wood fragments, leaf impressions	Uncommon (Modern: ALLO + AUTO)	Absent	Uncommon Disarticulated tetrapods, aquatics	Very common Roots, burrows

Environment	Plant fossils		Invertebrates		Vertebrates		Trace fossils	
Floodplain: poorly drained	Very common AUTO + ALLO	Logs, stumps, erect trees, leaf litter, peat, seeds	Present	Mollusks, crustacea, arthropods	Present	Fish (disarticulated bones, scales), tetrapods articulated, disarticulated	Common	Burrows, roots, trackways
Floodplain: well drained	Uncommon	Logs, stumps, erect trees, seeds	Uncommon	Mollusks, crustacea	Present	Tetrapods, articulated, disarticulated; fish uncommon, coprolites	Common	Roots, burrows, insect traces, trackways
Crevasse splay	Common ALLO + AUTO	Erect trees, logs, leaves, AUTO + ALLO debris	Present	Mollusks, insects, crustacea (usually ALLO)	Variable	Tetrapods, aquatics, fish scrap, coprolites	Uncommon	Escape burrows, rooted upper horizons, trackways
Interdistributary bay	Common ALLO + AUTO, moderate size sorting	AUTO stems, leaves, ALLO debris, plant succession	Uncommon	Arthropods, crustacea, mollusks (marine or nonmarine)	Variable	Disarticulated fish, tetrapods	Common	Roots, burrows

Note: Refer to tables 2.1 and 2.2 for explanation of terms.

active erosion, transport, and deposition of coarse clastic materials occur within an alluvial system (Bridge 1985). The occurrence of organic remains in this context is related to many highly variable hydraulic factors, including channel pattern, size, discharge, and sediment load. In active channels the less transportable clasts accumulate in winnowed lags or other high-energy deposits, usually at the bottom of the thalweg. Other bedload sediments form topographically higher deposits known as bars, the distribution and composition of which reflect channel sinuosity and braiding (Bridge 1985; Miall 1985). Upon abandonment, channels may fill with fine-grained sediments that preserve rich assemblages of organic remains. These fills are a distinct taphonomic mode that is discussed in §3.2.

Distributary channels bifurcate downstream, conducting water flow and sediment over a broader area than does a single channel. They are characteristic of deltaic environments and typically occur where there is a decrease in a stream's capacity to transport sediment. Because of lower current velocities and proximity to base level, distributary channels may be favorable sites for the accumulation and permanent burial of organic debris.

Woody detritus and other fragmentary plant debris have been noted in cores of present-day deltaic channel bottoms (Kames 1970; Allen 1970; Gastaldo 1986a). Plant detritus also has been documented in studies of Recent channel bar deposits (Frazier and Osanik 1961; Boaz 1982; Scheihing and Pfefferkorn 1984; Burnham 1989). Plant debris smaller than bedform scale may accumulate between bedforms such as ripples and dunes, while larger debris is swept through the system (Spicer and Greer 1986). This introduces severe sorting of different plant parts in the resulting assemblages (see Spicer 1989 for discussion of fluvial transport distances and biases). Studies in channels of the Mississippi Delta indicate that most palynomorphs are from arborescent plants and are allochthonous with respect to the depositional site (Darrell and Hart 1970; Darrell 1973). The effects of deltaic distributaries (lacustrine or marine) on transported organic material, such as leaf rafting (Berry 1906) and fruit and seed transport (Moseley 1879), are likely to be similar to estuarine situations (see §2.4).

Bone transport and burial in different types of modern river channels have been investigated using controlled experiments and observations of naturally occurring vertebrate remains (Hanson 1980; Behrensmeyer 1982, 1983; Aslan and Behrensmeyer 1987). When fresh, bones have low densities and are moved easily along with sand and gravel bedload, although they may travel at different rates from barforms within the channel. Bones often are deposited on point bars and other bar surfaces during waning flood stages. Prior to the onset of chemical decomposition, both bones and plant material can travel many kilometers without significant breakage or abrasion (Spicer 1980, 1981; Behrensmeyer 1982; Ferguson 1985; Spicer and Greer 1986). Study of modern carcass concentrations in channel bars of an African river demonstrates characteristics of sorting and abrasion for mass-drowning events (Boaz 1982). Tributaries of the Mississippi River south of Saint Louis, Missouri, provide examples of late Pleistocene bone accumulations in channel-lag deposits (Graham and Kay 1988). There is evidence for winnowing in the selective preservation of remains of the larger animals, including mastodont teeth that are similar in size and shape to cobble-sized limestone clasts; such bone fragments are extensively abraded and polished, with scratches and striations on the bone surfaces. These may have been caused in part by trampling, combined with long-term interaction with sediment either during transport or when bones were in more stationary positions. Similar patterns of taphonomic modification are documented for bone-bearing deposits in Late Pleistocene – Holocene channels in Alaska (Guthrie 1967, 1968).

Taphonomic features such as sorting of skeletal parts and abrasion of fossil vertebrate remains in channel lags and bars generally indicate some degree of transport and hydraulic sorting. High densities of unabraded, articulated and associated skeletal parts also occur in this context, however, and imply rapid burial with minimal reworking. One mechanism for this is quicksand, which can trap and bury large animals in channel-bar and also beach deposits (Weigelt 1989). Some vertebrate accumulations with unusually complete skeletons have been attributed to this cause (Abel 1919). Logjams also provide a mechanism for trapping floating carcasses in active channels and may result in an unsorted, unabraded bone accumulation associated with channel-bar deposits (Fiorillo 1991).

Fossil plant material often is abundant in channel lags, and channel deposits may have characteristic leaf assemblages that reflect riparian species rather than floodplain or interfluve species (Lebedev 1976; Hickey and Doyle 1977; Hickey 1980; Wing 1984; Spicer and Parrish 1986; Spicer 1987). Fragmentary plant debris occurs commonly in drill cores of ancient channel lags and bars (Ferm and Smith 1981; Ferm et al. 1985), and there are exceptional occurrences of whole fallen trees (Bown et al. 1982). Invertebrate remains preserved in fluvial channel deposits are primarily molluscan, including unionid bivalves and freshwater oysters (Kolb et al. 1975; DeHeinzelin et al. 1976; Vondra and Bowen 1976; Dodson et al. 1980a; DeHeinzelin 1983; Bridge et al. 1986).

In the Upper Cretaceous of Alberta, Canada, vertebrate remains are preferentially preserved in channel-lag and bar deposits and are rare in floodbasin facies (Dodson 1971; Wood et al. 1988). Dinosaurs of medium-to-large size occur abundantly as both articulated and disarticulated specimens, and plant remains include seed casts as well as wood. Fiorillo (1987, 1989a, 1989b) has documented a bonebed associated with logs in basal channel deposits; the bone concentration is attributed to flow diversion caused by the logs. Other concentrations of sorted and abraded bones and teeth of smaller terrestrial animals occur in channel lags (Dodson 1971, 1973; Banks et al. 1978; Béland

and Russell 1978) or upper-flow-regime planar beds (Eberth 1990). These microfaunas are associated with clay-pebble conglomerates at or near channel
bottoms and typify an important preservational mode for vertebrates that occurs throughout much of the Phanerozoic. The fossil record of Late Cretaceous and Paleocene mammals from western North America is derived
primarily from such occurrences (Clemens 1964, 1973; Lillegraven et al. 1979;
Rose 1981), which include lower vertebrates (Estes 1964; Estes and Berberian
1970; Sahni 1972) as well as dinosaurs (Dodson 1983, 1987). Microfaunas in
channels of Cretaceous-Paleocene sites in North America (e.g., Johnston and
Fox 1984; Sloan et al. 1986; Rigby et al. 1987; Fastovsky 1987) provide important evidence concerning the terminal Cretaceous extinctions. Hydrody-
namically sorted concentrations of small mammal remains also are documented
in Pleistocene distributary channel deposits of California (Wolff 1973).

  Among the more spectacular concentrations of large terrestrial vertebrates
are those preserved in channel-bar deposits of the Upper Jurassic and Upper
Cretaceous of North America (Lawton 1977; Dodson et al. 1980a, 1980b),
the Lower Jurassic of India (Jain 1980), and the Upper Cretaceous of
Mongolia (Gradzinski 1970). In the latter example, point-bar and channel-bar
sediments are the predominant lithofacies, and articulated dinosaur remains
occur exclusively in this context along with fish, aquatic reptiles, mollusks,
ostracods, eggshell fragments, charophytes, and wood. Large vertebrates also
are commonly recorded in Cenozoic channel-bar or lag deposits (Peterson
1906, 1923; Voorhies 1969; Turnbull and Martill 1988). In the upper Miocene
of Pakistan, vertebrate remains are common in channel-lag conglomerates and
typically are fragmentary, dissociated, and abraded (Badgley 1986a). Variable
characteristics of sorting, abrasion, and orientation are recorded in carefully
excavated channel assemblages in the Pleistocene of southern Ethiopia
(Johanson et al. 1976; Boaz 1982).

3.2 Abandoned Channel Fills

Abandoned channels are sites where sustained, energetic flow has ceased,
leaving a linear-to-curvilinear depression to be filled with fine-grained sediment and organic debris. Such fills may represent cutoff segments of a nearby
active channel or entire distributary systems abandoned upon river avulsion.
They form a preservational context distinct from channel lags and bars because a different set of sedimentary and taphonomic processes leads to the
accumulation of locally derived, rather than appreciably transported and reworked, organic remains. Channel fills commonly consist of clays and muds,
often with abundant plant debris, or with increased clastic input, couplets of
sand and silt and lenses of gravel and sand.

  Many channel fills occur in meander or neck cutoffs that form in high-
sinuosity channels of low-gradient alluvial plains (Fisk 1944, 1947; Frazier and

Osanik 1961; Farrell 1987). Abandonment usually is abrupt, and the resulting oxbow lakes are infilled primarily by very fine-grained, often organic-rich materials. Oxygen in the bottom sediments of the deeper portions of such lakes becomes depleted, resulting in reduced benthic fauna and bioturbation, which favors preservation of organic remains. Such lakes can persist for an estimated $10^3 - 10^4$ yrs (Gagliano and Howard 1984) and furnish an essentially autochthonous record of the lacustrine and adjacent floodplain biota. The development of peat profiles in abandoned channel segments has been studied in the Mississippi Delta (Kosters and Bailey 1983), and the biostratinomy of plant remains in Holocene to recent oxbows is under investigation (Farrell 1987; Gastaldo et al. 1989a).

Most of what is known about channel fills with significant coarse-grained components is derived from studies of ancient fluvial deposits (e.g., Hunt 1978; Behrensmeyer 1987b, 1988). Such fills appear to be more commonly associated with low-sinuosity systems, where linear abandoned channel segments result from chute cutoffs or larger-scale avulsion events. These cutoffs are likely to be reactivated during high-water stages, and the resulting fill may consist of multiple fining-upward units. Intermittent channels in strongly seasonal drainage systems may also accumulate fills with alternating coarse and fine-grained sediments.

Plant assemblages in channel fills are almost exclusively derived from vegetation growing along the channel margin (Scheihing and Pfefferkorn 1984; Gastaldo 1987; Burnham 1989), although nonriparian species may be incorporated where cutbanks erode into distal floodplains. There is little actualistic information on vertebrate preservation in channel fills. A recent study of an abandoned channel segment of the Mississippi River reveals a diverse assemblage of Holocene vertebrate remains (many classes and body sizes) associated with peat deposits, which also preserve abundant micro- and macroplant remains (Graham and Graham 1990). The bone accumulation includes articulated parts of skeletons and evidence of carnivore and scavenger activity (chewing, scattering of associated skeletal elements). A survey of fish in modern floodplains and floodplain lakes indicates that high species diversity would be expected in channel-fill situations if taphonomic processes such as scavenging did not destroy most of the fish remains (Smith et al. 1988).

Abandoned channel fills have a long Phanerozoic history and are an important context for organic preservation. The economic importance of determining the extent of both coal and clay deposits has led to extensive coring and mapping, which allows three-dimensional reconstruction of channel fills (e.g., Elliot 1965; Hook and Ferm 1985, 1988). Oxbow fills have yielded foliage compression floras, fructifications, coal balls, and many types of disarticulated and articulated aquatic-to-terrestrial arthropods and vertebrates (Scott 1978, 1980; Hook and Ferm 1988; Hook and Baird 1988). Well-preserved assemblages of leaves (Potter and Dilcher 1980) and flowers (Crepet et al. 1974;

Crepet and Daghlian 1982) occur in clay-rich oxbow deposits. There may be a strong bias in such plant remains due to higher rates of degradation for certain species (Spicer 1981; Ferguson 1985).

Although coarser, less-organic-rich channel fills are not recognized as frequently as oxbow deposits, they are fairly common in many ancient fluvial and deltaic deposits. Channel fills with plants and ichnofossils are recorded in the Upper Devonian of New York (Bridge and Diemer 1983; Gordon and Bridge 1987). Dinosaur preservation in channel fills is an important taphonomic mode in the Upper Cretaceous of Alberta (Dodson 1971), and several quarry sites for massed dinosaur remains in the Upper Jurassic of North America also represent abandoned channels (Madsen 1976; Dodson et al. 1980a). In the Lower Cretaceous of England (Alvin et al. 1981) and Germany (Riegel et al. 1986), plant material occurs in channel fills within braidplain sequences. Large- and small-scale abandoned channel deposits have rich vertebrate fossil occurrences in the Permo-Carboniferous of New Mexico and Texas (Olson 1958; Olson and Mead 1982; Berman et al. 1985; Sander 1989), the Miocene Siwalik sequence of Pakistan (Badgley 1986a, 1986b; Behrensmeyer 1987b, 1988b), and the Miocene of Nebraska (Hunt 1978). This context is also known as a site for fossil lungfish burrows (Dubiel et al. 1987, 1988).

3.3 Levees

Levees are built up along active channels by successive pulses of overbank flow that deposit greater volumes of sediment near the channel. Their growth is curtailed when crevasse channels breach the levee and carry sediment onto the floodplain or into backswamps. Diagnostic features of modern-day levees include their linear topographic expression parallel to channels, good drainage compared to other floodplain subenvironments, dense plant cover, evidence of pedogenesis, and distinctive stratification. Levee deposits are generally fining-upward sequences consisting of coarser-grained, finely laminated bedsets overlain by extensively bioturbated mudstones.

Studies of plant taphonomy in levee environments of modern deltas have shown that leaves, fragmented wood, and in situ roots may be present in the subsurface (Gastaldo 1985). Roots are often horizontal in situations where the water table is high. Actualistic evidence suggests that except for roots, identifiable plant remains found in levee deposits reflect some degree of transport, though only for short distances (usually less than 0.5 km) (Gastaldo 1985). On the Mississippi Delta, Darrell (1973) found that pollen from levees consisted of reworked, primarily arborescent species with few herbaceous forms. Muller (1959) sampled levees in the Orinoco Delta and found that both autochthonous and allochthonous assemblages could be identified in close proximity to one another. Although levees are sites for considerable initial accumulation of organic debris, actualistic studies indicate that chemical and biological processes rapidly degrade plant remains below the soil surface (Scheihing and Pfeffer-korn 1984; Burnham 1989), leaving little except root traces for the fossil record. Many modern fluvial systems are erosional or are aggrading slowly within oxidizing circumstances, however, and more rapidly accumulating levees in poorly drained situations may have greater plant-preserving potential.

Levees are a recognizable subenvironment in ancient fluvial and deltaic deposits, but there are relatively few examples of fossil plants and animals recorded in this context. Fragmented plant debris has been documented in levee facies of the Upper Devonian of New York State (Bridge and Gordon 1985) and Ireland (Bridge et al. 1980) and in the Lower Cretaceous of Germany (Riegel et al. 1986). Where identified, Carboniferous levee deposits in the southern Appalachians appear to be barren of organic remains. Fine-grained levee facies preserve abundant vertebrate fossils and rare plant remains in the Upper Permian strata of South Africa (R. M. H. Smith 1980, 1990). Dinosaurs occur but are uncommon in levee facies in Jurassic and Cretaceous deposits of North America (Dodson 1971; Dodson et al. 1980a). Palynomorph and megafloral assemblages of levee and crevasse splay facies have been distinguished from those of swamp and channel fill in Eocene overbank deposits of the Bighorn Basin, Wyoming (Wing 1984; Farley 1990).

3.4 Floodplains

Present-day floodplains comprise a complex of environments, including open to forested plains, swamps and marshes, and ephemeral to permanent lakes. The extent of floodplain subenvironments and their plant and animal communities can be evaluated in ancient floodplain deposits on the basis of variations in grain size, mineralogy, sedimentary and biogenic features, and in some cases, modes of fossil preservation. Floodplain sequences generally are made up of the finer-grained deposits from overbank flooding, and also variable amounts of coarser-grained tabular units. Crevasse splays generate such units when coarse sediment is carried out onto the floodplain via crevasse channels that are fed by a main channel. These coarser floodplain deposits, as well as those relating to floodplain lakes, are discussed in §§3.5, 4.3–4.5.

Fine-grained floodplain deposits range from unmodified, well-bedded flood sediments to massive units that have lost all primary bedding because of pedogenic modification. Soils are a prominent feature of floodplain environments, and we include them in this section because most of the available information on ancient soils pertains to alluvial examples, aside from those in Quaternary-Recent time. In modern environments, a soil may occur on any land surface modified by biological, physical, and chemical processes. By this definition most bedding planes in subaerial deposits could be identified as paleosol horizons. More conservative definitions require significant pedogenic modification of the parent sediment, such as translocation of clays and mobile ions, bioturbation structures, and formation of clear zonation (Wright 1986; Gerrard 1987). Identification of soils in many ancient settings may be complicated by diagenetic overprinting of original pedogenic features and questions concerning the applicability of modern analogues (Retallack 1983, 1984a, 1988; Wright 1983, 1986; Kraus and Bown 1986; Bown and Kraus 1987).

Organic materials are incorporated into floodplains through bioturbation and sedimentation and are preserved differentially according to subsequent development of the surrounding soil. The survival of plant remains is strongly affected by oxidation, which is a function of whether the floodplain is poorly drained (i.e., primarily wet with low oxidation rates) or well drained (with or without marked fluctuations in the water table). Preservation of organic remains also is controlled by the frequency and magnitude of flooding events, which may retard chemical and biological destruction through burial (Wing 1988). Entire alluvial systems can be characterized by either well-drained or poorly drained conditions, but such conditions also can occur in close proximity within the same floodplain (Retallack 1977; Besly and Turner 1983; Besly and Fielding 1989). In table 2.3 these two categories of floodplains are separated, primarily because of their strong differential effects on plant preservation.

Poorly Drained Floodplains

Surfaces of poorly drained floodplains are saturated or covered by water during significant portions of the annual cycle and often support marsh or swamp vegetation, including forests. The waterlogged soils typically have chemically reducing conditions favorable to the preservation of plant tissues, palynomorphs, and insect parts. Hydromorphic soils are gray to blue in color, carbonaceous, and if acidic, rich in kaolinite (Gerrard 1987). They are usually situated in topographically low areas of the floodplain that receive periodic influxes of sediment and rarely show a high degree of pedogenic development. Plant remains accumulate and decay in the vicinity of their parent vegetation, but may be transported during floods. Burnham (1989) notes that cores recovered from overbank (backswamp) areas in a paratropical forest in southern Mexico have root traces and that plant parts are highly degraded. Study of a present-day backswamp along the Alabama River indicates that organic-rich muds preserve rare entire leaves in the subsurface as well as woody crown detritus (Gastaldo et al. 1989a). Plant-part diversity in the subsurface parallels the low plant diversity in the forest floor litter (Gastaldo et al. 1989a). Biostratinomic study of hollow, prostrate logs in a modern swamp shows how they may be preserved as mud casts with finely laminated and even bioturbated fills (Gastaldo et al. 1989b). Little is known about the modern-day taphonomy of animals in wet floodplain environments, but ancient examples include a variety of arboreal, ground-dwelling, aquatic and semiaquatic as well as gliding forms (see below).

The interpretation of "wet floodplain" environments in the stratigraphic record often is based on the drab, gray-to-green sediment color, fine-grained lithofacies, pedogenic features indicative of poorly drained soil conditions (including gleying), absence of pedogenic carbonates, lack of oxidized minerals, and the presence of organic carbon and plant remains (e.g., Fastovsky and McSweeney 1987; Besly and Fielding 1989). Color and carbonate content sometimes are open to question because they may reflect later diagenetic modification rather than the original characteristics of the floodplain. In spite of controversy over the significance of sediment color, the recurring association of drab hues with other evidence for wet conditions indicates that color can be a valid indicator of original environmental conditions.

Coal-bearing sequences provide abundant examples of plant roots and compressions preserved in wet floodplain and deltaic environments (Scott 1980; Wnuk and Pfefferkorn 1984, 1987; Wnuk 1985; Gastaldo et al. 1987; Besly and Fielding 1989; and many others). Identification of wetland swamp environments in the Carboniferous is based on the sedimentary context and the presence of occasional in situ stigmarian axes and erect lycopsids (Gastaldo 1986b, 1987). Similar criteria have been used in the interpretation of Devonian wetland swamps (Scheckler 1986). Upright tree stumps associated with tetrapod trackways occur in peat-accumulating (coal-forming) floodplain environments in the Upper Cretaceous of Utah (Parker 1975). Whereas plant remains are predictably present in this context, vertebrates are more variable in their patterns of occurrence. Drab overbank sediments in the Lower Cretaceous of Wyoming and Montana preserve solitary complete skeletons of ornithopod dinosaurs, as well as plant material (Ostrom 1970; Dodson et al. 1980b, 1983). A Late Cretaceous inland floodplain facies in New Mexico includes drab mudstones with fragmentary dinosaur remains (Lucas 1981; Lucas and Mateer 1983). Other Late Cretaceous to Paleocene examples that illustrate the range of preservational circumstances for vertebrates, macroplants, and palynofloras include many coastal or deltaic-to-alluvial plain sequences in New Mexico (Lucas and Mateer 1983), Alaska (Phillips 1987; Brouwers et al. 1987; Spicer and Parrish 1987; Parrish et al. 1987), Montana (Carpenter 1982; Fastovsky and McSweeney 1987; Fastovsky 1987), Wyoming (Roehler 1988), Alberta (Sweet 1987) and Romania (Grigorescu 1983). Poorly drained Eocene-age floodplain deposits in Wyoming preserve abundant plant remains but few vertebrates or invertebrates, in contrast to contemporaneous drier floodplain facies (Wing 1984; Wing and Bown 1985; Maas 1985; Bown and Kraus 1987). This appears in part to be the result of post-depositional decalcification, because molluscan steinkerns and compressed bones occasionally are present (S. Wing pers. comm. 1990).

Well-Drained Floodplains

Well-drained floodplains or parts of floodplains typically are characterized by marked water-table fluctuations, with the result that substantial areas are drained and become relatively dry for extended periods of time. Such areas may be subject to flooding, but during the intervening time organic debris is oxidized and the upper layers of sediment are aerated. Bioturbation and other pedogenic processes may be intense and contribute to oxidation of the sediments and recycling of organic materials. Because well-drained floodplains generally have slower sedimentation rates than low-lying, poorly drained areas, mature soil profiles can develop.

Actualistic investigation of taphonomic processes on well-drained floodplains include the classsic work of Weigelt (1989), study of bone destruction and burial on land surfaces (Hill 1975; Gifford 1977; Behrensmeyer 1978, 1983; Behrensmeyer and Dechant Boaz 1980; Fiorillo 1984; Bickart 1984), and research on plant preservation and destruction in the Orinoco Delta of Venezuela (Scheihing and Pfefferkorn 1984) and in a Mexican paratropical forest (Burnham 1989). Molluscan taphonomy has been investigated on braided river floodplains to provide comparative data for Quaternary cold-stage molluscan assemblages (Briggs et al. 1990).

Ancient well-drained floodplain deposits are often distinguished from poorly drained examples on the basis of their red coloration. Reddening from iron oxidation does not necessarily reflect the original environmental conditions (Walker 1967; Glennie 1970), but paleomagnetic studies of redbeds indicate that the pigment can form very early in diagenesis (i.e., within the first 10^3 -10^4 years) and remain stable thereafter (Tauxe and Badgley 1988; Tauxe et al. 1990). This suggests that primary oxidation in the floodplain soils can be faithfully recorded by red coloration. Not all well-drained floodplain deposits are red, and paleoenvironmental interpretations are based on other criteria, including ichnofossils (Bown and Kraus 1983), pedogenic features, including carbonates (Gile et al. 1966; Leeder 1975; Retallack 1984a, 1986; Wright 1987), lack of plant remains (Besly and Turner 1983) except for root casts, and the presence of fusinized (burned) tissues and phytoliths (Retallack 1983).

Organic remains in floodplain sediments often are associated with paleosols, although they also occur in stratified deposits that have not been subjected to pedogenesis. Well-developed paleosols form during periods of lowered sedimentation and thus reflect local-to-regional depositional hiatuses within alluvial floodplains (Bown and Kraus 1981a, 1981b; Behrensmeyer 1982; Kraus and Bown 1988). Organic remains are incorporated into soils through bioturbation as well as slow burial under accumulating sediment; the resulting assemblage may be mixed vertically, enriched in resistant hard parts such as teeth and seeds, and time averaged (Fiorillo 1988a, 1988b; McBrearty 1990). The survival of plant, invertebrate, and vertebrate remains is directly related to soil chemistry. Acidic soils tend to dissolve shell and bone (Krumbein and Garrells 1952; Bass-Becking et al. 1960; Gordon and Buikstra 1981), and oxidizing conditions destroy most plant remains (Retallack

1984a). In the presence of sufficient mobilized carbonate or other minerals, however, vertebrate remains and plant roots may be permineralized rapidly and preserved.

Burrows and coprolites attributed to a wide variety of invertebrates and vertebrates are fairly common in paleosols and other facies of well-drained floodplain sequences (Clark et al. 1967; Bown and Kraus 1981a; Wright 1983, 1987; Retallack 1983, 1984b; R. M. H. Smith 1987; Dubiel et al. 1987). Vertebrate remains range from isolated bones to articulated skeletons, often with evidence of surface weathering, carnivore damage, and other forms of modification prior to burial (Clark et al. 1967; Behrensmeyer 1978, 1987a; Dodson et al. 1980a; R. M. H. Smith 1980; Bown and Kraus 1981b; Winkler 1983; Winkler et al. 1987; Winkler and Murry 1989). All body sizes are represented, but smaller, more easily buried forms appear to be most common in paleosols.

Throughout the Phanerozoic, there are many examples of fossil vertebrates and some examples of plant remains preserved in well-drained floodplain deposits. Footprints also are a significant biogenic feature in floodplain deposits that retain primary bedding (Sarjeant and Mossman 1978; Olsen and Baird 1986; Gillette and Lockley 1989). Terrestrial vertebrates are abundant in floodplain deposits in the Lower Permian of Europe (Boy 1977) and the Upper Permian of South Africa (Hotton 1967; R. M. H. Smith 1980, 1990) and the Soviet Union (Efremov 1953; Olson 1962). Plant remains consist of wood and occasional in situ casts of stumps, except in local lenses of unoxidized sediment where leaves are preserved. Vertebrate and plant remains have been documented in the red floodplain sediments of the Triassic deposits of western North America (Ash 1972a, 1972b; Long and Padian 1986; Parrish 1989), China (Spicer et al. 1988), and eastern North America (Olsen 1980, 1988). Red mudstones in the Upper Jurassic of Colorado bear isolated, articulated skeletons of large sauropods (Dodson et al. 1980a).

Lower Cretaceous floodplain deposits in central Texas include red mudstones with pedogenic carbonates and root casts and hypsilophodontid dinosaur skeletons in various stages of articulation (Winkler et al. 1987, 1988). Deposits of similar age in Wyoming and Montana have poorly preserved ankylosaur and sauropod dinosaur remains in oxidized floodplain facies (Dodson et al. 1980b, 1983). Lower Cretaceous sequences in southeastern through central China include a characteristic flora of conifers and *Classopollis* pollen (from an extinct group of dry-adapted conifers) (Spicer et al. 1988). Upper Cretaceous deposits in Montana are interpreted as an upper coastal plain where nests of hadrosaurid and hypsilophodontid dinosaurs were preserved in oxidized overbank deposits along with mammals and lizards (Horner and Makela 1979; Horner 1982, 1984a, 1984b, 1987; Hanley and Flores 1987). Low-diversity palynofloras occur in paleosols with mature caliches in the Late Cretaceous of Alberta and contrast with rich pollen and spore assemblages from gray mudstones and coal-bearing facies of the overlying latest Cretaceous and early Paleocene (Sweet 1987).

The Eocene deposits of Wyoming provide one of the best-documented examples of fossil vertebrate preservation in floodplain paleosols, which may develop to a high stage of maturity with distinct zonation and composite profiles (Bown and Kraus 1981a, 1981b, 1987; Winkler 1983; Badgley and Gingerich 1988). The fossil assemblages consist of fragmentary but well-preserved remains of many different species, and paleosols of different stages of maturity appear to have characteristic faunas (Maas 1985; Bown and Kraus 1987; Bown and Beard 1990). A diverse assemblage of ichnofossils is also present in these deposits (Bown and Kraus 1983; Kraus 1988). Occurrences of fossils in paleosols also have been documented in the Oligocene of South Dakota (Clark et al. 1967; Retallack 1986), where the fragmentary and often weathered aspect of the vertebrate remains is consistent with attritional accumulation. Plant remains are restricted to silicified parts, chiefly seeds. In the Miocene of Pakistan and western India, vertebrate remains in floodplain paleosols or other fine-grained facies consist of infrequent occurrences of fragmentary material and occasional large burrow fills (Badgley and Behrensmeyer 1980; Badgley 1986a; Behrensmeyer 1987b). In contrast, vertebrates are relatively common in Miocene through Pleistocene paleosols of East Africa (Bishop 1968; Shipman et al. 1981; Behrensmeyer 1985; Pickford 1986; Hay 1986).

Although terrestrial vertebrate remains are often abundant in ancient floodplain deposits, fish are relatively uncommon. A wide variety of fish occupy floodplains that are inundated for several months per year (Goulding 1980; Smith et al. 1988), and their scarcity as fossils has been attributed to ecological processes such as predation combined with taphonomic factors, particularly the differential preservation of heavy-bodied predaceous forms compared with small-bodied prey species (Smith et al. 1988).

3.5 Crevasse Splays

Crevasse splays are virtually instantaneous events that produce widespread, predominantly coarse-grained sedimentary units within floodbasin environments. Such depositional events are important taphonomically because they can rapidly cover existing subaerial and subaqueous surfaces and bury soils, living vegetation, and other organic remains. Crevasse events also introduce allochthonous organic material from channels to the floodplains and may hydraulically concentrate bones derived from the surface or previous deposits (Badgley 1986a, 1986b). In situations where the crevasse channel remains open for a substantial period of time, fans consisting of multiple flood-generated units can build up over a large area of the floodplain (Farrell 1987; Kraus 1987; Smith et al. 1989).

Taphonomic processes affecting plants associated with modern crevasse splays have been documented in studies of the lower delta plain of the Mobile

Delta, Alabama, where interdistributary bays are infilled with crevasse-splay deposits (Gastaldo et al. 1987; Gastaldo 1989a, 1989b). The overwhelming majority of plant parts recovered from splay deposits are allochthonous and originate in upstream environments. Assemblages consist of a mixture of aerial plant parts that have been transported both in suspension and as bed-load. Crevasse splays in levee-bounded areas of the delta can preserve primarily autochthonous species, however, including erect vegetation buried by a crevasse event. If the buried erect vegetation undergoes differential decay so that evidence for stumps and logs is preserved, then forest structure may be reconstructed from splay deposits (Gastaldo 1986a, 1986b, 1987).

Crevasse splays often are lumped together with floodplain deposits in sedimentological reports on fossil-bearing strata, and the specific features of this taphonomic context have yet to receive broad-scale documentation. There are many examples of crevasse-splay deposits with fossil plant remains (e.g., upright trunks) in the Upper Carboniferous of North America (Cavaroc and Saxena 1979; Rust et al. 1984; Gastaldo 1987) and Europe (Brzyski et al. 1976; Fielding 1984) and in the Cretaceous of Antarctica (Jefferson 1982a) and New Mexico (Cross et al. 1988). Vertebrates occur in this context in the Upper Permian of South Africa (R. M. H. Smith 1987, 1990). Abandoned crevasse channels in the Permo-Carboniferous of New Mexico have concentrations of coprolites and articulated aquatic and terrestrial vertebrate remains in drab sediments that indicate locally reducing conditions in a channel environment (Berman et al. 1985). In the Miocene of Pakistan, small-scale crevasse splay deposits with reworked carbonate nodules are one of the richest contexts for fossil bones (Badgley and Behrensmeyer 1980; Badgley 1986a, 1986b). Trackways may also be preserved by crevasse-splay (Platt 1989a) or similar sheet-flooding events (Lockley and Conrad 1989).

3.6 Interdistributary Bays

Interdistributary bays are large, generally submerged areas between the distributaries of lower delta plains. During delta progradation, these marine-to-brackish water bodies typically are infilled by crevasse-splay deposits. The bays become restricted with the abandonment of delta lobes, and with continued subsidence and barrier-bar reworking, they may be transformed into intradeltaic lagoons (Penland et al. 1988). Mud-dominated deltas may have extensive tidal flats in interdistributary areas (Allen et al. 1979). Observations along the Gulf Coast of North America provide actualistic evidence for taphonomic processes affecting plant remains in this setting (Baganz et al. 1975; McManus and Alizai 1983; Gastaldo et al. 1987). High rates of sedimentation provide physical and geochemical conditions favorable for preservation of detrital plant remains. Aquatic vegetation along bay margins, or traces of it, may be preserved at the growth sites. In the more subaerial parts of the bays, marsh or mangrove vegetation may become established and leave recognizable evidence (root traces, characteristic sedimentary fabric) in the sediment (Tye and Kosters 1986). As bay filling progresses, forest communities may be established in marginal areas. Evidence for this vegetational succession may or may not be preserved over a short stratigraphic interval. Partially infilled interdistributary bays can be sites for peat accumulation, as discussed in §7.7.

Aerial parts of the vegetation often are not preserved in Holocene inter-distributary areas because of periodic flushing by tides or floods of the lower delta plain (Scheihing and Pfefferkorn 1984). Identifiable aerial plant remains from crevasse-splay deposits in interdistributary bays are allochthonous, representing upper delta plains or extrabasinal communities. Brackish-to-marine interdistributary bays are characterized by abundant invertebrates. Vertebrate remains observed in recent bay deposits consist primarily of disarticulated fish; other animals, such as alligators, live in these settings, but the fate of their remains is unknown.

The Upper Carboniferous Mazon Creek localities of Illinois are interpreted as marine interdistributary bay to prodelta environments. Siderite concretions preserve a remarkable array of nonmarine organisms, including vascular plants, crustaceans, bivalves, fish, tetrapods, and insects (Baird et al. 1985, 1986). In the Lower Pennsylvanian of Indiana, tidally laminated mudstones associated with peat deposits and upright trees appear to represent inter-distributary deposition (Kvale and Archer 1990). In Illinois, a *Lepidodendron* forest floor on interdistributary peat was buried by bay-fill and crevasse-splay deposits (DiMichele and DeMaris 1987). A possible interdistributary context for fossil vertebrates has been documented by Lehman (1982, 1990) for the Upper Cretaceous of Texas, where a dark gray, carbonaceous mudstone interpreted as an interdistributary swamp bears a series of monospecific bone-beds of ceratopsian dinosaurs. The bone assemblages are disarticulated but unwinnowed and include poorly preserved plant remains. Brackish-water coal "swamps" that may have been interdistributary bays are sites for dinosaur remains and pollen in the Cretaceous of Alberta (Russell and Chamney 1967).

4 LACUSTRINE ENVIRONMENTS

Lakes occur over a wide range of sizes and in many different tectonic and climatic settings. They vary from seasonally ephemeral to "permanent" (for periods of 10^2-10^6 years) and have widely different water chemistries (Symoens et al. 1981; Beadle 1981; Burgis and Morris 1987). Whatever their overall context and scale is, however, lacustrine deposits have characteristics that distinguish them from sediments that represent other continental environments. These features include a predominance of fine grain sizes and chemical rock types, a high degree of lateral continuity of individual beds, and a generally tabular geometry (Picard and High 1972). Sedimentary and biogenic features are related to water depth and chemistry, and fine-scale laminations may be developed extensively. Lake deposits can be richly fossiliferous, with detailed preservation of soft and hard parts of both plants and animals.

Recent lakes traditionally are divided into eutrophic (abundant nutrients, high productivity) and oligotrophic (low nutrient content, low productivity), but these distinctions are of limited applicability in interpreting ancient lakes. The oligotrophic-eutrophic dichotomy is based on examples from temperate climates with low climatic equability, and these often are not good analogues for paleolakes from other climatic regimes. High-productivity lakes have diverse biota and can generate a rich fossil record, but organic remains are also subject to intensive recycling. Low-productivity lakes have low organic diversity, but their depressed levels of biotic activity can result in relatively undisturbed preservation of indigenous organisms or any that are carried into such systems. Therefore the most productive lakes or parts of lakes do not necessarily produce rich fossil records.

Modern-day lakes also are classified according to whether they are chemically and thermally stratified (Burgis and Morris 1987), a useful distinction from the standpoint of the fossil record. Stratified lakes have a relatively stable thermocline that forms a boundary between an upper, warm, oxygenated zone (epilimnion) and a deeper, cooler, often anoxic zone (hypolimnion). Unstratified lakes lack this distinct thermal and chemical separation and typically are oxygenated throughout their depth. Lacustrine biotas are influenced strongly by the proportion of warmer, oxygenated water volume and the benthic habitats that lie within this realm. Organic preservation also is a function of water temperature and chemistry; undisturbed organic remains, including soft parts, are more likely to survive and become mineralized under anoxic conditions (Wuttke 1983a; Allison 1986). Seasonal turnover of lake waters occurs in stratified lakes and may cause massive die-offs of fish and other aquatic fauna. Stratified lakes typically are deep (>10 m) and large (>10 km^2), whereas unstratified lakes tend to be shallow and restricted in area.

The appearance and chemistry of lacustrine deposits often indicate whether they accumulated under oxygenated or anoxic conditions. The two major preservational modes for lakes in table 2.4 correspond to these depositional contexts, as determined from sedimentological evidence. A significant proportion of the bottom sediments of stratified lakes with an anoxic hypolimnion are carbon and sulfide rich, dark gray to black in color, and well laminated with little evidence of bioturbation. Bottom sediments deposited within the oxygenated zone are often highly bioturbated and lack carbon and sulfides. Subdivisions according to lake size and depth also are given in table 2.4, although size determinations for many ancient lakes are very problematic. Limnologists often refer to lakes as large or small and shallow or deep, but relatively few quantify these terms (Symoens et al. 1981; Burgis and Morris

Table 2.4 Lacustrine Environments

Environmental Context	Occurrences of Terrestrial Organisms					
	Macroplant	Microplant	Invertebrates	Vertebrates	Ichnofossils
Low oxygen: Large lake, deep	Rare Logs, seeds	Common Pollen ALLO, sorted; phytoplankton	Uncommon Mollusks, zooplankton, benthos	Common Fish, articulated, disarticulated; coprolites, rare flying forms, tetrapods	Very rare
Small lake, deep	Common Logs, seeds, leaves plant succession	Common Local pollen, phytoplankton	Uncommon Benthos, insects, zooplankton	Common Articulated fish, rare aquatic tetrapods, flying forms	Very rare
Small lake, shallow	Common Leaves, seeds, logs, stromatolites, plant succession	Common Local pollen	Uncommon Arthropods, rare benthos	Common Aquatics, tetrapods	Uncommon Roots
Oxygenated: Large lake	Uncommon Wood, debris, stromatolites	Common AUTO + ALLO pollen, benthic phytoplankton, charophytes	Common benthos, plankton, rare insects	Common Fish (disarticulated bones, scales), tetrapods in marginal areas	Very common Burrows, roots, tracks
Small lake	Uncommon Logs, leaves, debris, stromatolites	Common Benthic phytoplankton, charophytes	Very common Mollusks, crustaceans, rare insects, sponge spicules	Common Disarticulated fish, tetrapods; may be articulated in carbonates	Very common Roots, burrows, trackways

Note: The arbitrary division between large and small lakes is set at 10 km², and the division between deep and shallow at 10 m. In a lacustrine context, microplants include phytoplankton such as diatoms. Refer to tables 2.1 and 2.2 for explanation of terms.

1987; Tilzer and Serruya 1990). We use an arbitrary area limit of 10 km² to distinguish between large and small lakes, and a water depth of 10 m to distinguish between deep and shallow. Based on both modern and ancient examples, lakes on either side of these boundaries appear to have different characteristics of preservation that are at least partly independent of water chemistry, including the representation of terrestrial organisms and the degree of sorting and mixing of organic remains from different sources.

The taphonomy of plant and pollen preservation in Holocene to Recent lakes has received considerable attention. The allochthonous plant (including pollen) record has been intensively investigated from the standpoint of paleofloral reconstruction (e.g. Birks 1973; Jacobson and Bradshaw 1981; Spicer 1981; M. B. Davis et al. 1984; Spicer and Wolfe 1987; Wainman and Mathewes 1990). The relationship between the preserved pollen record and the original vegetation has been studied in North America (Davis and Webb 1975). Data on late Pleistocene (Wisconsin) plant macrofossils in the North American Great Lakes indicate that long-distance transport, selective sorting, and reworking can significantly affect the climatic signal represented by these remains (Warner and Barnett 1986; Warner et al. 1987). Studies of invertebrate taphonomy in modern East African rift lakes (Cohen 1981, 1986, 1989) and various ancient lacustrine rift settings (Crisman 1978; Hanley and Flores 1987; Gore 1988) are also under way. Wuttke (1983a, 1983b) has investigated the taphonomy of frogs and soft-tissue preservation in lakes. Other experimental work has shown that the fate of fish carcasses is partly a function of water temperature: above 16°C bacterial decay will cause whole fish to float and disperse, but below this temperature they tend to remain on the bottom (Elder and Smith 1988). Studies by Behrensmeyer (1975a), Hill (1975) and Gifford (1977) on the vertebrate taphonomy of modern lake-margin settings underscore the importance of trampling, vegetation, and rapid burial for preservation in this setting.

The sedimentology of Holocene lakes has also been investigated on many continents, with particular emphasis on climatic signals recorded in varves and stable isotopes (e.g., Birks and Birks 1980; Halfman and Johnson 1988; Talbot and Livingstone 1989). Such work provides sedimentological analogues for interpreting time resolution in many lacustrine deposits. Recent increases in the understanding of ancient lakes as biological systems are reflected in volumes edited by Gray (1988), Talbot and Kelts (1989), and in a paleolacustrine theme issue in *Palaios* (2(5):1987).

4.1 Oxygen-Depleted Large Lakes

The deeper, offshore parts of large lakes are usually stratified, and the waters below the hypolimnion are cooler, anoxic, and also aphotic, thus forming a unique environment with little biological activity but conducive to fossil preservation. The chief preservational difference between the anoxic portions of large and small lakes is the sorting effect of distance from shore on the numbers and kinds of terrestrial organisms introduced into the lake sediments. Also, large lakes are more likely to have a diverse benthic fauna because they generally have a deeper epilimnion.

Although many deposits of large, deep lakes lack fossils, significant assemblages of aquatic organisms can be preserved in this context, especially under conditions of high productivity above an anoxic hypolimnion. Diverse assemblages from more biologically active near-shore zones also can be transported into the deposits of the deeper portions of such lakes. Organic-rich black shales of the German Lower Permian represent eutrophic lakes with an anoxic hypolimnion (Boy 1977, 1987); catastrophic sedimentation events in the epilimnion transported and buried remains of fish, some vascular plants, poorly preserved palynomorphs, a few near-shore tetrapods, and abundant coprolites. Upper Carboniferous lake deposits in France that were proximal to major deltas preserve abundant plants, some benthic invertebrates, fish, insects, and rare tetrapods in sideritic concretions (Rolfe et al. 1982; Langiaux 1984; Heyler 1987). In the Lower Permian Odernheim locality and the Niederhaesslich locality (both in Germany), organic-rich, laminated shales and limestones preserve abundant aquatic tetrapods and palynomorphs (but few plant macrofossils), coprolites, and locally, assemblages of aquatic invertebrates (Boy 1977, 1987).

Throughout the Phanerozoic, rift valleys provide many examples of lacustrine deposits from large, stratified lakes. The oldest well-known fish massmortality assemblages are preserved in deep lacustrine deposits of the Middle Devonian of northern Great Britain (Donovan 1975, 1980; Trewin 1986). The deposits of the Newark Supergroup of eastern North America (Mid-Triassic to Lower Jurassic) represent rift-valley lakes of varying scales and water chemistries (Olsen 1988). In the larger lakes, black muds deposited in relatively deep, anoxic conditions preserved abundant crustaceans, fish, and some insect and plant remains (Olsen et al. 1978; Hentz 1985; Olsen 1980, 1986, 1988; McCune 1987; Gore 1988; Olsen et al. 1989). The sedimentary sequence as a whole records some 40 million years and includes depositional sequences interpreted as Milankovitch Cycles (Olsen 1986; Smoot and Olsen 1988).

Many finely laminated, kerogen-rich, carbonate deposits appear to have accumulated in the anoxic portions of permanently stratified large lakes. These may contain remains of fish as well as insects, terrestrial vertebrates, plants, and skin impressions of lower vertebrates (Grande 1984). Although the best-known examples are Eocene in age, such as the "Fossil Lake" portions of the Green River Formation of Wyoming (Bradley 1948; Grande 1984), similar deposits with abundant remains of well-preserved fish are known from the Lower Carboniferous of eastern Canada (Lambe 1910; Greiner 1962).

Pliocene rift-lake deposits in southwestern Idaho also preserve rich fish assemblages (Smith 1975; G. A. Smith 1987).

4.2 Oxygen-Depleted Small, Deep Lakes

Small lakes that are depeleted in oxygen may be shallow or deep, and their chemistry is usually controlled by decaying organic matter and patterns of water circulation. Small, deep lakes with anoxic zones are typical of rift valleys and also occur in floodplains of major rivers (considered above with abandoned channels), volcanic craters, sinkholes, and glacial settings. Deposition in deeper lakes is often steady but slow, although proximity to land and steep slopes can generate gravity flows that increase local rates of sedimentation. Such deposits are reworked and may contain rip-up clasts, and organic remains can be part of the reworked clastic material. Because biologically productive shoreline environments are relatively close to deeper waters with lower biotic activity, remains transported by water or sediment flow are relatively common in the deeper-water deposits.

Upper Devonian deposits in Quebec represent lakes with at least periodically anoxic bottom waters, as indicated by dark, pyritic mudstones with plant and fish remains (Carroll et al. 1972). Permo-Triassic lakes of the Karoo, South Africa, are documented by finely laminated, plant- and animal-bearing carbonaceous sediments indicative of deposition in anoxic conditions (depths estimated at 150 m maximum) (Van Dijk et al. 1978). The Lower Cretaceous Koonwarra Beds in Australia provide an additional example of insect, fish, and plant preservation in a small, stratified lake (Waldman 1971; Elder 1985).

The famous Messel deposit (middle Eocene) of Germany represents a small, relatively deep lake where bottom-water chemistry strongly inhibited organic activity (Franzen et al. 1982; Wuttke 1983a; Franzen 1985; Heil et al. 1987; Franzen and Michaelis 1988; Schaal and Ziegler 1988). The strata consist primarily of laminated bituminous claystones with some thin sideritic layers that are thought to have accumulated in a warm, subtropical lake. According to this interpretation, the lake itself was unstratified, but the sediments indicate bottom conditions unfavorable to benthic life. The deposits preserve insects and other invertebrates, vertebrates, and plants with remarkable fidelity. Benthic organisms are rare (sponge spicules are the main record of aquatic invertebrates), and most of the fossils represent allochthonous input from nearby terrestrial settings. Among the nonaquatic fossils, air-transported elements (leaves, insects, bats, birds) predominate. Most of the amphibious animals are concentrated near the mouth of a shifting drainage that fed the lake (Franzen et al. 1982).

Other Cenozoic examples of relatively small, deep lakes include the Oligocene Florissant Beds in Colorado (McLeroy and Anderson 1966), Eocene deposits in British Columbia (Wilson 1977, 1978a), and a Miocene lake in Idaho (Gray 1985; Smith and Elder 1985). These deposits include diatomites and preserve abundant plant macrofossils, pollen, insects and fish. The Clarkia deposits of northern Idaho show extraordinary preservation of leaf tissue, with intact microstructure and membrane-bound organelles such as chloroplasts (Niklas and Brown 1981). The cytologic preservation is somewhat better in offshore as compared with onshore deposits. The oldest-known DNA sequences have been recovered from chloroplasts of a species of *Magnolia* in these deposits, which are dated at 17–20 Ma (Golenberg et al. 1990).

Some of the examples given above may have exceeded the dimensions of small lakes as defined in this chapter, because their exact dimensions cannot be reconstructed from the available evidence.

4.3   Oxygen-Depleted Small, Shallow Lakes

Shallow lakes with reducing chemical environments are common in alluvial settings (see §3.2). The chemical conditions of such lakes are controlled by abundant decaying organic material, combined with low rates of clastic influx. Despite shallow water depth, decay rates can rapidly deplete the oxygen content of the water in this setting. Plant assemblages and palynomorphs in such lakes can record an autochthonous sequence that occurs during progressive infilling as the aquatic environment is transformed to dry land (Swain 1973, 1978; Spicer 1981; Wing 1984; Green and Lowe 1985). In sediment-starved environments, sulfide-rich sapropels may accumulate, depending on the water chemistry and the nature of organic input (Boy 1977; Hook and Hower 1988). Plant detritus and remains of fish and aquatic-to-amphibious tetrapods are common, and small numbers of terrestrial arthropods and vertebrates may occur as well. Mollusks are present as shelly remains or steinkerns (S. Wing, pers. comm. 1990), although benthic forms are rare.

4.4   Oxygenated Large Lakes

The warmer, more oxygenated waters of shallow lakes and the shallow margins of deep lakes promote greater productivity and also more sediment reworking through the bioturbating effects of the benthic fauna. Deep or shallow lakes with closed drainage basins are also subject to fluctuating water levels and water chemistries, which can lead to biotic stress and restrict organic activity. Large, shallow lakes can have relatively stable water levels in interdistributary bays and cratonic depressions but may dry out completely in other settings, such as playas, often in response to climatic conditions (Eugster and Hardie 1975). The margins of large, fluctuating lakes are generally poor in organic remains, although wood and small vertebrate remains may occur; but they provide an optimal context for ichnofossils, including invertebrate and vertebrate tracks, trails, fish nests, and root traces (Krynine 1935; Sarjeant and Mossman 1978; Olsen 1980; Olsen and Baird 1986; Feibel 1987; Gillette and Lockley 1989).

In examples from the Permo-Carboniferous of Europe, deposits of perennial shallow lakes include marl and limestone (bioturbated and irregularly bedded, deposited in shallow, oxygenated water above wave base or subject to bottom currents), stromatolites and oncolites, aquatic invertebrates, disarticulated fish remains, local plant debris, palynomorphs, and very rare nonaquatic animal remains (Boy 1977, 1987). In addition to these organic remains, allochthonous terrestrial mollusks occur in similar deposits in the Tertiary (Lotze 1968).

The Permo-Triassic rift lakes of the eastern Karoo include examples of oxygenated deep-to-shallow-water deposits with preserved plant material, tetrapods, and rare insects (Van Dijk et al. 1978). In contrast, deposits of large, shallow Permian lakes of Malawi are highly bioturbated and lack in situ faunal or floral remains (Yemane et al. 1989). In shallow portions of the large rift lakes of the Newark Supergroup, diverse plants, insects, fishes, and rare aquatic reptiles are incorporated within asymmetrical sedimentary sequences that include both anoxic and oxygenated facies (Olsen et al. 1978; Olsen 1980). Marginal areas of intermittent lakes apparently provided the context for the preservation of Upper Cretaceous dinosaur nesting grounds in Mongolia (Brown and Schlaikjer 1940; Osmolska 1980) and Montana (Horner 1987).

Lower Cretaceous rift lake deposits of Brazil and West Africa are known for their abundant ostracods and mollusks (Cunha and Moura 1979) and well-preserved fish (Martill 1988). Within another Cretaceous rift setting, the Cameros Basin of northern Spain, ostracods, gastropods, fragmentary vertebrate remains, and footprints occur in extensively bioturbated, limestone-marl-mudstone sequences that are assigned to shallow, low-energy lacustrine systems (Platt 1989b). Portions of the Eocene Green River Formation that represent shallow playa conditions preserve a diverse fossil record of aquatic and terrestrial vertebrates, invertebrates, and plants in marls and oil shales (Bradley 1964; McGrew 1971; Surdam and Wolfbauer 1975; Eugster and Hardie 1975; Buchheim and Surdam 1977; Grande 1984). Miocene-through-Pleistocene lakes in the East African Rift System and the Benue Trough of Cameroon and Zaire have plant, invertebrate, and vertebrate remains preserved in lacustrine-to-marginal lacustrine contexts (Behrensmeyer 1975a, 1975b, 1985; Hay 1976; Pickford 1983; Brunet et al. 1986; Cohen 1989). A mixed vertebrate assemblage of terrestrial and aquatic forms is documented in marginal lake facies in Plio-Pleistocene deposits of East Africa (Behrensmeyer 1975a; Schwartz 1983).

4.5 Oxygenated Small Lakes

Most oxygenated small lakes are shallow and usually are subject to rapid clastic sedimentation or periodic desiccation. Deposits from such lakes are common throughout the Phanerozoic record and often occur in fluvial-deltaic sequences. These lakes are relatively short-lived and are populated by cosmopolitan species that typically are capable of adjusting to moderately high levels of environmental stress.

Several types of deposits of small and shallow lakes can be recognized in the fossil record. These may be discrete temporally and spatially, or single sequences may grade laterally and vertically from one type of deposit to another. In fluvial and deltaic settings, lakes are often dominated by influxes of clastic material (Olson 1958; Scott 1978; Wilson 1980, 1988a; Kerp 1982; Fielding 1984; Fox 1984; Haszeldine 1984; Yuretich et al. 1984; Crane and Stockey 1985; Wighton and Wilson 1986; Boy 1987; Warwick and Flores 1987), and the deposits may resemble the coarsening-upward sequences of bayfills. Lakes with high clastic input tend to preserve plant remains, autochthonous mollusks and crustaceans, disarticulated vertebrates, and rare allochthonous insects. At the opposite end of the spectrum are lakes with low rates of clastic influx, where organic accumulation and decomposition may result in anoxic conditions (see §§3.2, 4.2, 4.3). Low rates of clastic sedimentation also promote increased representation of freshwater carbonates, including algal deposits (Masson and Rust 1984; Berman et al. 1985; DeMicco et al. 1987), as well as mollusks (Hanley and Flores 1987), ostracods (Cohen 1987), and concentrations of fish and other aquatic vertebrates (Tobien 1986).

The Lower Jurassic of India (Jain 1980) includes fluvial sandstones and associated lacustrine limestones and clays. The limestones contain plant remains, ostracods, conchostracans, insects, and diverse fishes. Studies of shallow lacustrine deposits of the Upper Jurassic of North America document abundant charophytes and mollusks. Remains of nonaquatic dinosaurs also are found in this context (Dodson et al. 1980a). Well-preserved skeletons of small vertebrates (birds, bats, and small mammals) occur in Eocene freshwater limestones of Nevada (Emry 1990) and Wyoming (Stucky et al. 1990). Ponds and small lakes in Paleocene fluvial systems of Alberta include insects (Wilson 1982; Wighton and Wilson 1986), mollusks (Hanley and Flores 1987), fish (Wilson 1980), and mammals (Fox 1984).

Ephemeral lakes are recognized on the basis of sedimentological evidence, such as desiccation features and evaporite deposits. Small ephemeral ponds with fossil vertebrates have been recorded in well-drained Lower Permian floodplains of Europe (Boy 1987), and similar deposits in the Permian of the southwestern United States are characterized by articulated lungfish and lysorophid amphibian remains in aestivation assemblages (Olson 1939; Romer and Olson 1954; Olson and Bolles 1975; Berman et al. 1988). Although intermittent lakes generally contain few animal remains and virtually no plant materials, such lakes can preserve assemblages of short-lived aquatic organisms (conchostracans, hydromedusae) and diverse footprint assemblages (Walter 1978; Van Dijk et al. 1978; Boy and Hartkopf 1983). The absence of plant remains other than root traces may be attributed primarily to oxidation during dry periods.

5 VOLCANIGENIC ENVIRONMENTS

Volcanic events have unique effects on ecology, physical sedimentary processes, and organic preservation. Volcaniclastics include fine ash to house-sized blocks—anything generated by volcanoes that has moved by air or water as discrete particles. Ash and pumice are low-density materials that are easily transported over long distances. Volcanoes tend to produce large quantities of loose sediment that overwhelm local depositional systems, destroy biotic habitats, change river courses, and transform topography (Lipman and Mullineaux 1981; G. A. Smith 1987; Lockley and Rice 1990). Often this leads to the accumulation of considerable thicknesses of relatively pure volcanic sediments near the eruptive center. Farther away, ash can be reworked by water, wind, and soil processes but still retain distinctive sedimentary and chemical features over large areas. Both hot and cold volcanic mudflows are more localized but can travel tens of kilometers downslope, instantaneously killing, transporting, and burying plants or animals in their path (Lipman and Mullineaux 1981; Lockley and Rice 1990).

Volcanic deposition can quickly and permanently remove organic remains from the biologically active zone near the soil surface and result in relatively complete preservation and fine levels of time resolution, particularly for sessile populations such as the fossil forests of Yellowstone Park in North America (Yuretich 1984; Fritz 1984). However, the sheer volume of sediment also can dilute organic remains, such as pollen, so that samples are difficult to retrieve. Explosive events can cause severe fragmentation of macroplants (Spicer 1989) and vertebrates (Lyman 1984, 1989). Vertebrate remains are seldom concentrated in primary volcanic sediments unless the densities of the living animals were increased immediately before death by herding behavior or by locally restricted resources, as apparently occurred in a mass death of the Pliocene rhinoceros *Teleoceras* in Nebraska (Voorhies and Thomasson 1979).

Volcanic deposits provide samples of habitats that are poorly represented or otherwise absent in the sedimentary record. Carbonititic volcanics are especially effective in creating alkaline soils and preserving vertebrate remains that otherwise might be destroyed in tropical settings (Bishop 1968; Pickford 1986; Hay 1986). Moreover, volcanic deposits are relatively independent of latitude and climatic zone and may sample organisms in desert-to-tundra settings. Throughout the Phanerozoic, volcaniclastic preservation has occurred more commonly in tectonically active continental areas, and the biotas of such areas may have been affected by frequent and major environmental perturbation. Weathering of volcanic rocks also releases mobile elements (e.g., Fe, P, Ca, F) that could have influenced local ecology. Thus, there may be a bias within the volcanigenic fossil record toward faunas and floras adapted to these active tectonic settings.

Table 2.5 includes two categories that are uniquely volcanic in nature (ex-

Table 2.5 Volcanigenic Environments

Environmental Context	Occurrences of Terrestrial Organisms					
	Macroplant	Microplant	Invertebrates	Vertebrates	Ichnofossils
Explosive events	Present Pulverized-to-complete leaves, wood, fusain	Rare (Extreme dilution)	Absent	Rare	Rare
Primary airfall	Common Compressions, fusain, replaced woody parts	Rare	Absent	Uncommon Articulated, disarticulated tetrapods	Variable Trackways, roots, burrows
Lacustrine	Common Leaves, flowers, seeds (especially at lake margins), fusain	Very common Especially phytoplankton	Common Arthropods	Very common Especially fish, others less common	Rare (Bioturbators may be suffocated)

Note: Explosive events are those that involve direct, often forceful deposition. "Primary airfall" relates to ash that covers or otherwise affects land surfaces, and "lacustrine" is a special case of lakes that are subject to significant volcaniclastic input. Refer to tables 2.1 and 2.2 for explanation of other terms.

plosive event and air fall) and one (lacustrine) in which volcaniclastic material significantly changes the taphonomic features discussed elsewhere in this chapter. The preservational features of alluvial channels are also altered radically by influxes of volcaniclastic materials, as demonstrated by actualistic studies of plant remains in the vicinity of Mount Saint Helens in Oregon (Fritz and Harrison 1985; Fritz 1986; Karowe and Jefferson 1987; Spicer 1989).

5.1 Explosive Events

The fossil products of explosive events are distinguished by taphonomic features such as fragmentation and orientation (e.g., for trees) and sedimentologic evidence indicating high-energy, short-duration deposition. Such evidence is important because it can demonstrate that the preserved organisms were killed instantaneously and buried or, in some cases, transported or oriented by the explosive event. Few of the recurring volcanic eruptions of recent times have been studied with an eye toward the fossil record, but the accessibility of the Mount Saint Helen's explosion in 1983 has generated new information on explosive taphonomic processes (Waitt 1981; Janda et al. 1981; Lyman 1984, 1989; Spicer and Greer 1986; Karowe and Jefferson 1987; Spicer 1989).

Examples of macroplant preservation in explosive deposits are known in New Zealand (Froggatt et al. 1981) and in the Eocene of Wyoming (Fritz 1980). There appear to be very few examples of whole or partial vertebrates preserved in explosive deposits. Remains of vertebrates within the perimeter of a volcanic explosion may be too dispersed or too poorly preserved to attract paleontological attention.

5.2 Primary Airfall

Primary airfall represents the initial fallout of volcanic material, which can kill vegetation as well as aquatic and nonaquatic organisms. The resulting deposits lack evidence for explosive forces (e.g., volcanic bombs, poor sorting, extreme lateral variability) and are often lithologically homogeneous and laterally extensive. Extreme effects on biotas usually occur only in sites close to a vent, which are often susceptible to erosion rather than long-term preservation. Changes in the physical and biological character of land surfaces are more widespread, and these changes can have profound effects on plant and animal communities and the preservation of organic remains. In contrast to explosive death and burial, airfall events generally allow differential survival of organisms and cause less extreme changes in local ecology.

Study of the effects of modern ash falls on plant communities in Mexico shows plants preserved in growth positions in their original ecological associations (Burnham and Spicer 1986). The earliest part of an airfall event buries the litter and ground-cover vegetation, and succeeding layers cover leaves predisposed to abscission or dropped because of the ash fall. Leaf falls during successive eruptions produce a series of plant beds that reflect synchronous species diversity and vertical stratification in the forest. Only in extreme conditions, such as exposure to hot blasts of gas, are the tree species actually killed. Actualistic study of the effects of ash fall on colonies of sea gulls indicates that eggshells and other organic debris may be well preserved in these circumstances (Hayward et al. 1989).

The preservation of an autochthonous gymnosperm community within a coal seam in the Late Permian of Australia has been attributed to burial by an airfall tuff (Diessel 1985). In the Miocene of Kenya, whole leaves, fronds, and fruits with intact microstructure are preserved three-dimensionally in a lapilli tuff, with some of the smaller herbaceous plants in growth positions (Jacobs and Kabuye 1987). Examples of compression plant fossils and fusain (interpreted as primary charcoal) occur commonly in the Tertiary deposits of the Pacific Northwest, where Taggart and Cross (1980) have documented ecological (possibly fire) succession in the floras of the Succor Creek Formation. A large airfall of ash apparently led to the destruction and burial of a group of Pliocene rhinoceroses, who died with grass anthoecia between their teeth and in their throat cavities (Voorhies and Thomasson 1979).

Primary carbonatitic ash provided an optimal setting for the autochthonous preservation of plants and animals, as well as coprolites and trackways, in Pliocene Laetoli deposits in Tanzania (Leakey and Harris 1987). The ash falls were relatively thin and occurred at intervals that allowed plant growth and animal activity to continue with minimal disruption. The composition of the ash led to rapid cementation, thereby preserving spectacular sets of vertebrate tracks, including those of the earliest known bipedal hominid (Leakey 1987; Hay 1987). Laetoli represents a type of dry, upland habitat that is seldom sampled in other depositional settings. Other carbonatitic ash centers in the East African Rift System are known for their excellent preservation of invertebrate, vertebrate, and plant remains (Bishop 1968; Pickford 1983, 1986; Hay 1986).

Primary airfalls often grade into wind-and-water-reworked ash, and the different modes of deposition can be difficult to distinguish. Interpretations of the sedimentary event affect not only the paleoecological resolution of the organic remains, especially the degree of time averaging, but also the precision and accuracy of radiometric dating (see §8).

5.3 *Lacustrine Volcaniclastics*

The differences between the taphonomic features enumerated under the major heading "Lacustrine" (table 2.4) and those influenced by volcaniclastics reflect increased depositional rates and chemical changes in the lake environment. An influx of volcanic material can cause significant increases in sedimentation in both deep and shallow areas of a lake, disrupt the aquatic ecosystem, and result in mass death and burial of benthic and nektonic organisms. Because an increase in dissolved silica and other nutrients promotes diatom blooms, ash and diatomite are frequently associated in Cenozoic lake deposits. Volcaniclastics can also dam existing drainages and create posteruption lateral-lake systems (Spicer and Greer 1986). Fine-grained, often diatomaceous, sediments accumulate during posteruption vegetational recovery and can provide a detailed record of ancient succession, vegetation dynamics (Spicer and Greer 1986), and the reestablishment of aquatic faunas.

Chemical aspects of preservation of compression floras in lacustrine volcaniclastic sediments of the Lower Cretaceous of Antarctica are discussed by Jefferson (1982b). The Tertiary deposits of western North America include examples of lacustrine volcaniclastic preservation of plant and animal remains (McLeroy and Anderson 1966; Taggart and Cross 1980; Lockley and Rice 1990). Lacustrine tuffs of the Miocene through Pleistocene Baringo sequence in Kenya preserve abundant plant and fish remains, as well as insects (Pickford 1986; Hill et al. 1986; Jacobs and Kabuye 1987). The extraordinary preservation of plant tissue and cell organelles in the Miocene Clarkia deposits of Idaho and the Succor Creek deposits of Oregon is attributed in part to the rapid aggradation of fine-grained volcaniclastic sediment in lacustrine and floodplain environments (Niklas and Brown 1981; Smiley 1985; Golenberg et al. 1990).

6 EOLIAN ENVIRONMENTS

Eolian deposits known to preserve organic remains include dunes, interdunes, and loess (wind-transported silts and sandy silts) (table 2.6). In general these deposits are laterally extensive and indicate dry climates or glacial conditions where large expanses of sediment are unprotected by vegetation. Because eolian settings can occur at any altitude, they provide one of the few preservational contexts for drier-habitat "upland" organisms in the terrestrial fossil record.

6.1 Dunes and Interdunes

Subaerial dunes composed of medium-to-coarse-grained sand are uncommon sites for organic preservation, but interdune or dune-slack deposits are recognized as important sources of fossils. Dunes represent a taphonomic setting in which rapid burial of organic remains can occur but where reworking and sand abrasion also are expected. Because active dunes are unfavorable habitats for most organisms, it is likely that fossils associated with dune deposits indicate biologically productive interdune areas that alternated with the dunes (Salisbury 1952). Interdunes tend to be relatively moist and accumulate finer sedi-

Table 2.6 Eolian Environments

Environmental Context	Occurrences of Terrestrial Organisms				
	Macroplant	Microplant	Invertebrates	Vertebrates	Ichnofossils
Dune	Very rare	Absent	Absent	Rare	Rare
				Articulated tetrapods	Trackways, roots, burrows
Interdune	Present	Present	Uncommon	Uncommon	Present
	Stems, roots, woody debris	Pollen	Ostracodes, insects	Aquatic tetrapods, eggshell	Burrows, roots, rare trackways
Loess	Rare	Present	Present	Present	Common
	Roots, charcoal stumps	Primarily phytoliths	Terrestrial mollusks	Tetrapods, articulated and disarticulated	Roots, burrows, rare trackways

Note: Loess refers to all fine-grained eolian material, including reworked volcanic ash. Refer to tables 2.1 and 2.2 for explanation of other terms.

ment grain sizes, and they may also be rich in carbonate (Salisbury 1952; Driese 1985).

Upper Triassic barchan dunes in Scotland preserve a variety of small and medium-sized reptiles (Benton and Walker 1985). The degree of articulation is generally high, indicating that desiccated, solitary specimens were buried by wind-blown sand. Similar taphonomic features occur in eolian dune deposits of the Upper Cretaceous of Mongolia (Gradzinski and Jerzykiewicz 1974). Tetrapod trackways are known from eolian dunes in the Permian of the Grand Canyon (Breed et al. 1974). Fossiliferous interdune deposits are recognized in the Permian of Wyoming (McKee and Bigarella 1979), the Lower Jurassic Navajo Sandstone of Arizona (McKee and Bigarella 1979; Winkler et al. 1987), the Upper Cretaceous of Mongolia, and numerous other eolian deposits (McKee and Bigarella 1979). A wide range of organisms occurs in this context, including plant remains (roots, stems), ostracods, insects, eggs, and vertebrates (including aquatic forms, though fishes are rare). Both sediments and organisms indicate wet areas, springs, or oases, and facies include mud-cracked clays, fine burrowed sand, and freshwater carbonates. Buried soils and peats from interdunal areas in Scotland have produced palynological evidence for Pleistocene and Holocene paleoenvironments (Wilson and Bateman 1986, 1987).

6.2 Loess

Loess deposits are composed of fine-grained sediments derived from eolian processes. The distinguishing characteristics of loess are primarily sedimentological, including particle size that is predominantly silt, angular grain shape, and good sorting. Taphonomic features often reflect more persistantly dry conditions than those of other soil-surface deposits.

Much of what is known about loess deposition and taphonomy is based on Pleistocene deposits of North America, Europe, and China. Windblown silts typically cover large areas and create a new surface that is buried in the next phase of sedimentation. Whereas depositional phases reflect seasonal aridity, an open source of fine-grained sediment (e.g., a desert, glacial outwash), and wind activity, paleosols indicate a return to land-surface stability. Loess deposits usually thicken toward source areas, such as river valleys in late Pleistocene North America (Ruhe 1983) or deserts (Wu and Gao 1985). Because loess deposits typically grow outward from their source, basal contacts and included organic remains can be time transgressive within lithologically correlated units (Ruhe 1983). Correlations can be made using radiometric dating and paleosols over large areas (10^3 km^2). Laterally extensive paleosol horizons may represent distinct periods of regionally moist climatic conditions during which eolian deposition slowed temporarily or ceased altogether.

Loess and windblown ash are associated with the development of open grasslands in the Oligocene-Miocene of central North America (Retallack

1982; Winkler 1987; Emry et al. 1987; Lockley and Rice 1990) and Pleistocene steppes and desert margins in China and eastern Europe (Wu and Gao 1985; Kukla 1975, 1989). Older fossil-bearing loessic sediments are poorly known, perhaps because there are few established sedimentological criteria for distinguishing ancient loess deposits from other settings in which fine-grained sediment may accumulate (Johnson 1989). It seems likely that more of these deposits exist than have been recognized.

Terrestrial mollusks are common in Pleistocene loess; plant macrofossils, namely, root structures, charcoal, and phytoliths, also occur, but pollen is generally poorly preserved (Kukla 1975, 1989). Original ecological associations of organisms may be well preserved in loess because wind cannot effectively transport and rearrange large skeletal parts; however, bioturbation can disturb and mix remains from different stratigraphic levels (Graham 1981). In laterally extensive loess deposits, plant and animal remains would be expected to represent different ecological associations along environmental gradients. Grasslands and steppes are considered to be the typical loess environments, but in central North America, woodlands also are represented (Gruger 1972; Leonard and Frye 1960).

Vertebrates preserved in loess occur as isolated, articulated, or associated skeletons with evidence of land-surface exposure, including scavenging, weathering, mummification, and fragmentation (Graham 1981; Winkler 1987). Skeletal remains are often associated with an attenuated *A* horizon of a buried soil, and fossorial forms may occur within the soil. Such taphonomic features are similar to those of other land-surface assemblages. Eolian deflation can create local bone lags that may be time averaged over long periods of time (Graham 1981) (see §8.4). Invertebrate remains (primarily terrestrial gastropods) represent the more mesic areas on such land surfaces (Evans 1972; Thomas 1985).

7 OTHER CONTEXTS FOR ORGANIC PRESERVATION

Many fossil occurrences represent preservational contexts that do not fall comfortably into the environmental categories of tables 2.2–2.6. These may occur across different physical settings because particular processes are largely independent of environment. Such taphonomic processes involve particular circumstances that concentrate organic remains, enhance burial, or both. Included in this broad category are natural traps (tar, natural cavities, amber, tree trunks), burrows, springs, feces and regurgitates, middens, mud slides, fusain deposits, and peat bogs (table 2.7). Most of these taphonomic modes are more narrowly defined than the other, environmentally based modes because they represent a specific set of biological, physical, and/or chemical processes. A number of these biologically generated modes, such as burrows, feces, regurgitates, and middens, may have been subject to evolutionary change during the Phanerozoic.

7.1   Traps

A trap is a place that can accumulate organic remains by detaining animals and causing their death, or by enclosing and protecting the organic remains themselves regardless of the mode of death. Many traps are natural cavities or holes in the ground, such as caves, fissures, erosional pipes, or sinkholes, with sides too steep for most terrestrial animals to climb. Other traps are formed by soft or adhesive substrate (mud, quicksand, tar, resin) that can impound or immobilize animals, or by plant, snow, ice, or sediment cover that collapses into water or an underground cavity. Because traps can simultaneously kill and bury animals, and because this may continue to occur for an extended period of time, large numbers of relatively complete remains may accumulate. Trap assemblages are frequently diverse and represent relatively unbiased samples of species that were part of a local community. Some traps are taxon specific and accumulate only locomotor types or body sizes unable to escape. Traps also may be self-enhancing because dead or dying animals attract predators and scavengers, some of which, in turn, are trapped. The accumulation of large numbers of bones helps to balance the depredations of local scavengers, and in some cases high concentrations may help to chemically buffer the deposit and prevent diagenetic dissolution (Bass-Becking et al. 1960; Gordon and Buikstra 1981). Plant remains and pollen can also accumulate in traps, depending on whether the chemical composition of the deposits is conducive to preservation.

Table 2.7 includes four general categories of traps (tar seeps, amber, karst, and tree trunks) that are recurring modes for fossil preservation in specific environmental circumstances. Others, such as quicksand and desiccated ponds, are special cases within the context of other environmental categories, primarily fluvial-deltaic examples.

Tar Seeps

Tar seeps are unusual geological phenomena that occur when low-density petroleum derivatives find access to the land surface. The well-known late Pleistocene tar pits of Rancho La Brea in southern California typify this mode of preservation (Merriam 1911; Marcus 1960; Howard 1962; Stock 1972; Woodard and Marcus 1973, 1976; Swift 1979). Remains of vertebrates of all sizes occur, along with insects and plant parts (Miller 1983; Marcus and Berger 1984). The assemblage as a whole includes unusual numbers of predators and scavengers, indicating that entrapped animals attracted carnivores to the same fate (Stock 1972). Preservation of bones and other hard organic material is excellent, and although the slow movement of the viscous tar tends to separate and mix parts of different skeletons, some stratigraphic relationships are preserved. Movement of bones against one another in the tar causes unusual patterns of abrasion known as "pit wear" (Stock 1929; Reynolds 1985).

Table 2.7 Taphonomic Features of Other Preservational Contexts

Environmental Context	Occurrences of Terrestrial Organisms				
	Macroplant	Microplant	Invertebrates	Vertebrates	Ichnofossils
Tar seeps[a]	Present Wood, seeds, fruits	Present Pollen	Present Insects, mollusks	Very common Tetrapods, flying forms, fish less common	Absent
Amber[a]	Common Flowers, other delicate parts	Common	Common Arthropods	Rare Articulated lizards, salamanders, frogs,	Absent
Karst (caves, fissures, sinkholes)	Uncommon (In feces or flood debris)	Uncommon (In feces)	Rare Arthropods, terrestrial mollusks	Very common Tetrapods, articulated and disarticulated; flying forms	Uncommon Tracks, burrows
Tree trunks[a]	Present Wood, debris	Present	Uncommon Terrestrial mollusks, rare arthropods	Present Tetrapods, flying forms, articulated and disarticulated	Uncommon Burrows in log casts
Burrows[b]	Present	Present	Present	Common Tetrapods, fish, snakes, articulated and disarticulated	Rare Scratch marks
Feces, regurgitates[b]	Uncommon (Pleistocene middens, dung)	Present	Present Intestinal parasites	Common All kinds, fragmented disarticulated	Absent

Middens[b]	Common Highly fragmented debris	Very common May be selected by herbivore	Uncommon Arthropods	Uncommon Disarticulated bone fragments	Absent
Archeological sites[b]	Present Wood, fiber	Present	Present (Abundant in shell middens)	Very common Whole and broken parts; ALLO	Present
Springs[b]	Common logs, leaf detritus	Very common	Common Arthropods, mollusks	Common Tetrapods, aquatics	Common Root, stem casts, tracks
Peat[b]	Very common AUTO wood, roots, stems	Very common Mostly AUTO, some sorting	Variable Insects, scorpions, Coleoptera (in Pleistocene)	Uncommon Articulated tetrapods especially in late Cenozoic	Very rare Roots
Charcoal[b]	Very common Wood, seeds, flowers	Common	Uncommon	Uncommon	Absent
Mud slides, flash floods[b]	Present Wood, logs, stumps	Absent (Too dilute)	Very rare	Variable Occasional mass death, articulated skeletons	Absent

Note: Refer to tables 2.1 and 2.2 for explanation of terms.

[a] Situations in which living animals or plants or organic remains are trapped and thereby protected from the usual processes of recycling on land surfaces.

[b] Other contexts involving biogenic or localized situations and events that preserve organic remains.

In addition to La Brea, Cenozoic tar seeps include those of Talara, Peru (Churcher 1959, 1965; Campbell 1979) and La Carolina, Ecuador (Campbell 1976). Such sites provide large and taxonomically important samples of late Pleistocene biotas. The low density and fluid nature of tar does not predispose it to long-term survival in the geologic record.

Amber

Amber forms a rare but recurring context for the extraordinary preservation of intact invertebrates, small vertebrates, and plant material. Amber may be autochthonous within lignites or reworked as clasts in other sedimentary facies. Because amber is relatively impermeable, it protects organic parts from oxidation and bacterial decay and preserves unusually complete anatomical and biochemical information. Fresh resin traps or covers small organisms, thus providing attritional, short-term samples of pollen, flowers, and insects in localized areas within forest habitats (Brues 1933). The organisms usually become trapped on living, upright trees, which may be identifiable from the chemical characteristics of the resin. Thus, this is virtually the only taphonomic mode that preferentially preserves arboreal flora and fauna of relatively small body size. Taphonomic work on amber has been summarized by Dietrich (1975, 1976).

The Baltic ambers of late Eocene–early Oligocene age are the classic examples of this preservational mode (Conwentz 1886; Laarson 1978). In the Dominican Republic, many unusual remains have been recovered from amber, including an Eocene frog (Polnar and Cannatella 1987), early Miocene terrestrial invertebrates, and a nearly complete lizard (Rieppel 1980; Baroni-Urbani and Saunders 1980). Amber with occasional insects also has been reported from several Cretaceous deposits in North America and from the Eocene of the Pacific Coast (Wilson et al. 1967; Mustoe 1985). The earliest fossil bee occurs in Late Cretaceous amber in New Jersey (Michener and Grimaldi 1988). Triassic amber from western North America preserves macroplant remains and palynomorphs (Litwin and Ash 1991).

Natural Cavities: Fissures, Caves, Sinkholes and Talus

Any natural cavity or hole is a potential trap for live animals or a possible burial site for organic remains, which are often protected thereafter from normal processes of disintegration. Because fissures, caves, and sinkholes often occur inland, they may provide evidence of faunas and floras from habitats that are not sampled in lowland depositional settings. Karst topography, however, can develop on carbonate bedrock near sea level (e.g., on exposed coral reefs) as well as at higher elevations. Lowland fissures and cave fills are frequent sites of preservation, because permanent burial can occur with only slight rises of the water table. Most cavities in carbonates or volcanic rocks at higher elevations are destroyed by continuing erosion, but a few have been permanently preserved in depositional settings subject to renewed subsidence and aggradation (e.g., lava tubes in rift-valley settings). Natural cavities are better known for animal than plant remains, but the latter also occur, especially in fecal middens (see §7.4).

Caves, sinkholes, and fissures act as fatal traps for some animals, but they also serve as denning, nesting, perching, and hiding places for vertebrates and invertebrates, as well as sites where carcasses are consumed and skeletal remains modified. Under these circumstances, faunal remains reflect the behavior of animals that utilized caves or other natural cavities. This behavior can result in unusually dense concentrations of bones from a wide variety of animal species. In modern ecosystems, animals of widely varying size (including elephants) enter caves (Sutcliffe 1986), and leopards, hyenas, porcupines, packrats, owls, and humans all are known to transport bones into caves (Zapfe 1954; Alexander 1956; Simons 1966; Sutcliffe 1970, 1986; Skinner et al. 1980; Brain 1980, 1981; Hill 1983). Roosting owls may create especially large concentrations of bone-rich pellets (Andrews 1990). Such bones may bear distinctive patterns of breakage and surface modification that reflect the different accumulating agents and surrounding chemical conditions (Rensberger and Krentz 1988; Andrews 1990; Fernandez-Jalvo in press).

Identification of karst deposits in the stratigraphic record is based on characters such as the irregular shape of the sediment body and the presence of breccias, talus cones, and carbonate deposits such as flowstones and travertines. Underground lacustrine and fluvial processes may also influence deposition and taphonomic features of karst deposits. Reworking and redeposition of bones by fluvial activity is relatively common in this context (Sutcliffe 1970, 1986; Brain 1981). Karst deposits typically include both small and large vertebrates and are an important source of information for small species that are less easily preserved in other depositional contexts.

Sinkholes often form steep-sided depressions with lakes, and these can become traps for some species. At the Pleistocene Hot Springs Mammoth site in South Dakota, the periodic trapping of mammoths in a sinkhole lake resulted in a large concentration of relatively complete skeletons (Agenbroad 1984). Deep footprints and bioturbation are associated with the bones in the lacustrine muds of the sinkhole, indicating considerable churning of the bottom by the mammoths. Other species occur in the deposit but are uncommon and fragmentary.

Collapse of karst terrain was traditionally considered to be part of the taphonomic history of *Iguanodon* remains in the Lower Cretaceous of Bernissart, Belgium, although it has been proposed recently that the animals were concentrated by other processes and were not trapped alive in fissures or sinkholes (Norman 1986, 1987) (see §7.9). A similar explanation is advanced for an *Iguanodon* accumulation in Nehden, Germany (Norman 1987).

Fissure and sinkhole fills preserve a substantial amount of vertebrate material throughout the record of terrestrial ecosystems. One of the oldest-known sites for terrestrial tetrapods occurs in a Lower Carboniferous collapse structure in Iowa (Bolt et al. 1988). Lower Permian fissure fills in marine limestones in southern Oklahoma include remains of numerous small-to-intermediate-sized terrestrial tetrapods (Peabody 1961; Olson 1967; Bolt 1977a, 1980). The Upper Triassic and Lower Jurassic fissures of the Bristol Channel area are famous for their mammalian assemblages and also include a great variety of mostly small nondinosaurian reptiles (Robinson 1957; Halstead and Nicoll 1971; Fraser 1987).

Tertiary fissure fills in Europe include the Paleocene Walbeck fauna (Germany) (Russell 1964) and the late Eocene and Oligocene phosphorites from Quercy, France (Filhol 1876, 1877; Rose 1984). The large ratio of carnivores to herbivores in these deposits is typical of many other karst sites. Sinkhole deposits and fissure fills in Florida range from Miocene (e.g., Thomas Farm: Webb 1981; Hulbert 1984; Pratt 1986, 1990) through the Quaternary (Webb 1974). The South African australopithecine caves date back to the Pliocene and contain abundant faunal remains, as well as stone artifacts (Brain 1958, 1981; Vrba 1980; Maguire et al. 1980). Quaternary cave sites with abundant vertebrate remains are also common in China (e.g., Choukoutien: Young 1932), Australia (Wells 1978), North America (Guilday et al. 1964), Europe (Kurtén 1971), and the Mediterranean region (Boekschoten and Sondaar 1972; Fernandez-Jalvo in press) and occur predictably where there are active karst regions. Late-Pleistocene-to-Holocene "lava blisters" (White et al. 1984) and sinkholes (Wang 1988) provide additional examples of faunas preserved in passive trap situations.

Plant remains are also known in fissure fills (Harris 1957; Kampmann 1983; Hölder and Norman 1986) and sinkholes (Leary and Pfefferkorn 1977; Leary 1981) but are less common than animals. Plant fossils from these depositional circumstances may represent unusual "upland" floras (Leary 1981). In late Quaternary caves of southwestern North America, plant debris and pollen occur in dung from packrats (i.e., middens) and from extinct species such as the giant ground sloth and mountain sheep (Martin et al. 1961; Thompson et al. 1980).

Cavities in lava flows and volcanic talus of Upper Triassic deposits in Nova Scotia are an unusual but rich source of fossil vertebrate remains (Sues et al. 1987; Olsen et al. 1989). The cavities and fissures are filled with silt, and the vertebrate material consists of mostly disassociated skeletal parts. Sedimentary context and the preservation of delicate and articulated skeletal material indicate that the animals were buried in or near the environment where they lived, presumably on the rocky, irregular surfaces of the lava (Sues, pers. comm., 1989).

Tree Trunks

Organic remains including coprolites may be concentrated within hollow stumps and tree trunks. Hollowed-out trunks or stumps occasionally remain open as sediment accumulates around them to form traps for some organisms and hiding or denning areas for others. Remains are somewhat protected from scavenging and weathering and eventually can be permanently buried as sediment fills in the hollows.

Upright tree stumps in overbank sandstone bodies preserve terrestrial invertebrates and vertebrates in the Upper Carboniferous of Nova Scotia (Carroll 1967; Carroll et al. 1972; Rust et al. 1984). Fragmentary plant remains occur within hollow trunks of arborescent lycopsids in the Carboniferous of Scotland (Walton 1935). Rich pockets of mammal and bird remains have been found in fills of hollow stumps of large trees in Miocene deposits in Kenya (Walker and Teaford 1989). In the early Eocene of North America, occurrences of exceptionally well-preserved mammals, birds, and gastropods in carbonate "pods" have been attributed to replaced hollow tree trunks (Gingerich 1987; Kraus 1988).

This taphonomic mode might be more common than is currently recognized, but it is difficult to identify in the absence of obvious, well-preserved tree stumps. Dense pockets of medium-to-small vertebrate bones that occur sporadically throughout the continental record could have resulted from this mechanism.

7.2 Burrows

As places where animals live, hibernate, or aestivate, burrows can provide optimal circumstances for burial of individuals that die underground as well as for bones that are brought in by carnivores or scavengers. Preservation of complete skeletons or associated partial skeletons is common in this context (Voorhies 1975). Recent hyena dens in Africa may be used by different animals (e.g., hyenas, mongooses, porcupines) over periods of tens of years or longer, and the resulting bone accumulations represent time averaged but fairly localized samples of a living community (Brain 1980; Skinner et al. 1980; Bunn 1982; Hill 1983) (see §8.4).

Lungfish aestivation burrows are recognized in Carboniferous and younger rocks and are particularly well known in the Lower Permian (Romer and Olson 1954; Carroll 1965) and Triassic (Dubiel et al. 1987). In the absence of skeletal material, the organism responsible for the burrow structures is subject to considerable debate (Dubiel et al. 1987, 1988). Aquatic amphibians occur in burrows in the Lower Permian of Texas, Oklahoma, Kansas, and New Mexico (Olson and Bolles 1975; Schultze 1985; Berman et al. 1988), where they form closely spaced clusters in floodplain facies. Terrestrial reptiles preserved in burrows are known from Upper Permian paleosols of South Africa (R. M. H. Smith 1987).

In the Cenozoic there are a number of well-documented examples of burrows containing animal remains. A concentration of articulated snake skeletons in the Oligocene White River Group appears to be an example of a preserved hibernaculum (Breithaupt and Duvall 1986). Excavation and sedimentological analysis of a Miocene site in Nebraska (Hunt et al. 1983) indicates that a dense cluster of bones was accumulated by a hyenalike canid in a complex burrow used as a breeding den. Helical burrows of a beaver (*Palaeocastor*) are characteristic of Miocene deposits of midcontinental North America (Martin and Bennett 1977). Pocket-mouse burrows with preserved skeletal material (Voorhies 1974) and arthropod burrows with grass anthoecia (Thomasson 1982) also are known from these deposits. Other documented "bone pockets" in Cenozoic deposits of North America are likely candidates for dens used by bone-collecting animals (e.g., Flagstaff Rim, Wyoming: R. Emry, pers. comm., 1988). In middle Pleistocene deposits in East Africa, clusters of bones with preferred orientations and abundant tooth marks are interpreted as resulting from hyena accumulations in burrows (Potts et al. 1988).

7.3   Feces and Regurgitates

Feeding and digestive processes concentrate organic remains and produce materials that may become fossils. Recent small vertebrate and plant remains can be highly concentrated in feces and regurgitated matter, and there have been several studies of damage patterns and species representation in recent and subrecent bone assemblages accumulated by such processes (Mellett 1974; Mayhew 1977; Dodson and Wexlar 1979; Korth 1979; Fisher 1981; Andrews and Nesbit Evans 1983; Andrews 1990).

Peculiar masses of fragmented bones ("vomite") in Permian deposits of South Africa (R. M. H. Smith, pers. comm., 1987) and North America (N. Hotton pers. comm. 1987) may be fecal or regurgitated remains generated by carnivores. These are of particular interest because there are no known animals with recognized bone-crushing dentitions from this time period (N. Hotton, pers. comm., 1987). Many small amphibian skulls from an Early Permian locality in Oklahoma are distorted uniformly into subrounded shapes that suggest that they passed through the digestive tracts of predators and accumulated as coprolites (Bolt 1977b). In the Cenozoic (especially the Pleistocene-Holocene), concentrations of small vertebrate remains are found in circumstances indicating that they are derived from owl pellets or the scat of carnivores (Mellett 1974; Brain 1981; Gawne 1975; Chaney 1988). Coprolites with fish remains have been found in a wide range of Eocene lake deposits and are attributed to crocodiles or predatory birds (Wilson 1987a).

Fossilized feces from larger predators and scavengers are also known (Jepsen 1963; Clark et al. 1967; Hunt 1978).

Dung of late Pleistocene mammals is preserved in dry deposits in caves in both North and South America. The dung of an extinct species of sloth (*Nothrotheriops shastensis*) have been found in cave deposits throughout the southwestern United States (Martin et al. 1961; Long et al. 1974; Thompson et al. 1980), and dung deposits of mylodon sloths also occur in South America (Margraf 1985). Blankets of dung derived from mammoths, ground sloths, an extinct mountain goat species (*Oreamnos harringtoni*) (O. K. Davis et al. 1984) and other species including humans (Watson 1969; Williams-Dean 1978) have been found in caves in the southwestern and eastern United States. These deposits provide important information about diets of extinct taxa (Hansen 1978), but their survival in the geological record is limited because the dung accumulations degrade quickly in moist environments and require dry conditions or rapid permanent burial for preservation.

Plant-animal interactions reflected in coprolites have been reported in a preliminary way in the Carboniferous of North America and Britain (Baxendale 1979; Scott 1977; Taylor and Scott 1983; Labandiera 1990) and in the Eocene of North Dakota (Jepsen 1963). Analysis of intestinal contents, coprolites, or regurgitates provides evidence for the paleodiets of fish (McAllister 1987), birds (Wilson 1987a), and humans (Jones 1986). As a potential source of relatively direct evidence for animal-plant or animal-animal interactions, coprolites clearly deserve more attention and systematic study.

7.4 Middens

Middens are known only from the later part of the Pleistocene, but they are an important source of information about ancient floras and environments for this time interval. Packrats (*Neotoma* sp.) of the American Southwest create large concentrations (up to several cubic meters) of fecal material and bones which they collect in caves, rock shelters, talus piles, and other places where they live and breed (Mead et al. 1983). The feces are cemented into hardened masses of organic material through diagenetic interaction with packrat urine (Van Devender 1987) and can survive for periods of 10^4 years in semiarid environments. They contain abundant pollen and comminuted macroplant remains, as well as bones and arthropods, and can be readily dated by the radiocarbon method (Van Devender 1986, 1987; Hall et al. 1988). The reconstruction of late Pleistocene through Holocene vegetational changes in the North American southwest is based primarily on midden assemblages (Spaulding 1985; Cole 1986; Van Devender 1986, 1987; Spaulding and Graumlich 1986; Van Devender et al. 1987). Individual middens are regarded as representative of the local vegetation where the packrats lived, with minimal time averaging within each sample.

7.5 Archeological Accumulations

Since approximately 2.5 Ma, humans and their ancestors have been responsible for accumulations of bones, artifacts, and other biotic remains and traces. Archeological accumulations occur in a variety of sedimentary contexts, including caves and rock shelters, and densities of organic remains can be very high. Many such accumulations are similar to middens and other taphonomic modes in which a biological agent collects and concentrates organic remains in a small area. The term "midden" is used for human-generated concentrations of invertebrate shells, but we do not include archeological remains under "middens" because the word is not commonly used for concentrations of bones or plant remains formed by human activity. Archeological accumulations have distinct taphonomic features, which become more diverse in the latest Pleistocene and Holocene with the increase in material culture.

A great deal of actualistic and paleontological taphonomic research has sought to delimit features of bone assemblages that are distinctly human (Brain 1981; Binford 1981; Gifford 1981; Shipman and Rose 1983; Fisher 1984, 1987; Blumenschine 1989; O'Connell et al. 1988; see also review in Behrensmeyer 1987a). Diagnostic features include cutmarks and other damage uniquely associated with human manipulation and tool use. Other attributes of archeological bone accumulations, such as a high degree of fragmentation, differential representation of body parts, and high-density concentrations associated with cultural artifacts, can involve a mixture of human and nonhuman taphonomic processes (Binford 1981; Klein and Cruz-Uribe 1984; Behrensmeyer 1987a; Potts 1988). Ethnoarcheological studies of material remains indicate that certain types of fragmentation (especially of marrow-bearing bones), spatial arrangements indicating refuse heaps and dwelling areas, and selectivity with regard to body part and species are characteristic of modern humans (Brain 1967; Yellen 1977). These may or may not be applicable to human behavior patterns in the distant past, however. Studies of human hunter-gatherers, hyenas, leopards, and other present-day bone collectors indicate considerable variation in the characteristics of bone modification, body-part representation, and species selection (Skinner et al. 1980; Bunn 1983; Blumenschine 1986; Potts 1988; O'Connell et al. 1988).

Because of the interest in human evolution, archeological accumulations have probably received more intensive study than any other taphonomic mode. Recognition of the importance of "site formation processes" has grown rapidly among archeologists over the past two decades. Excavation techniques and methods of analysis that take into account taphonomic information have been used extensively in East and South Africa (Brain 1981; Isaac and Crader 1981; Leakey 1971; Bunn et al. 1980; Klein 1976, 1980; Bunn 1986; Isaac and Isaac in press), Europe (During 1986; Stiner 1991), North America (Stanford and Graham 1978; Morlan 1980; Todd 1987; Saunders et al. 1990; Fisher 1992; Jodry and Stanford 1992), South America (Nunez

1984; Borrero 1986; Caviglia et al. 1986; Politis et al. 1987) and Australia (Horton 1978; Horton and Connah 1981; Horton and Wright 1981; Solomon et al. 1991). This has resulted in new information not only about taphonomic patterns generated by humans but also about nonhuman processes that can be confused with archeological processes.

In the late Pleistocene and Holocene, remains of dwellings, cultural articles made from wood and fiber, and plant foods may be preserved as components of archeological assemblages. Studies of these materials lead to detailed reconstructions of human behavior and environmental parameters, often at high levels of temporal resolution (Hillam 1985; Ward 1987; Smart and Hoffman 1988; Pearsall 1989; Robinson 1990).

7.6 Springs

Springs occur wherever water emerges from underground sources onto the land surface. They may be hot or cold, fresh or mineral, high or low pressure, perennial, ephemeral, or cyclic, and of varying size. Some types of springs concentrate organic remains, and this can be attributed to three factors: (1) they form a source of water that may be attractive to both plants and animals, (2) both physical and chemical burial conditions may facilitate preservation, and (3) animals can become mired in the soft substrates proximal to spring vents. Some springs are poisonous, adding a fourth factor to this list.

Spring deposits generally are lenticular sedimentary bodies with various microfacies (vent, pond, etc.) encapsulated within other sedimentary units (C. Haynes 1985). Vertical, well-sorted clastic sediment bodies representing vent deposits formed by the conduits of water movement are one of the diagnostic characters of spring systems. Because of the vertical water movement, organic remains in vent deposits may be abraded, redeposited, and significantly time averaged. Spring systems generally contain autochthonous accumulations of bones rather than allochthonous assemblages, however. The pond facies of spring systems are similar in terms of taphonomic processes and characteristics to other types of pond deposits. Aquatic invertebrates (mollusks, ostracods) can be abundant in this context (Quade 1986). Bone accumulations in spring deposits can result from the miring of individuals, but concentrations that result from either attritional or catastrophic deaths around water holes are probably more common (Saunders 1977, 1988; G. Haynes 1985a, 1985b) and may form extensive bone beds (Saunders 1977). In contrast to cave faunas, carnivores are characteristically rare or absent from Quaternary spring accumulations.

At East Kirkton, Scotland, well-preserved Lower Carboniferous plants, terrestrial invertebrates, and vertebrates occur in laminated carbonates and cherts thought to be hydrothermal spring deposits (Wood et al. 1985; Rolfe et al. 1990). In the Miocene of Bottingen, Germany, a thermal spring within a fissure created a "kalktuff" limestone that preserves impressions of leaves, insects, myriapods, arachnids, feathers, frogs, lizards, snakes, and mammals (Westphal 1959; Tobien 1965, 1968). Jurassic dinosaur deposits such as those at Howe Quarry in Wyoming (Colbert 1968) may represent ancient spring systems in an abandoned channel context. Fossiliferous Quaternary spring deposits are relatively common throughout the midlatitudes of North America (Holman et al. 1988; Saunders 1988). Mammal faunas such as Keefe Canyon (Hibbard and Riggs 1949), Rexroad, and Deer Park (Hibbard 1970) occur in late Tertiary (Blancan) spring deposits in southwestern Kansas.

7.7   Peat-Forming Environments

Peat is an accumulation of plant detritus. It generally is mixed with varying amounts of mineral matter and grades from nearly pure plant materials to organic mucks (Clymo 1983). Peat may accumulate in a wide variety of depositional settings, including fluvial floodplains, abandoned channels, interdistributary swamps, coastal lowlands, shallow lakes, interdunal areas, and arctic or periglacial settings (Cohen 1974; Gore 1983; Clymo 1983, 1987). The common factor uniting these environments is that water, available either as rainfall or groundwater, permits organic materials to accumulate faster than they can be removed through decay.

Peat-forming environments can be divided broadly into rheotrophic and ombrotrophic (Clymo 1987). In rheotrophic peatlands the vegetation is sustained by nutrients from groundwater, and such peats often accumulate in swamps and marshes that are flooded either permanently or seasonally. Peat lands sustained solely by nutrients carried in by rainfall are described as ombrotrophic and are referred to as blanket peats, bogs, or fens. The term "mire" can be applied to any peat-forming environment. Rheotrophic peats, because they are supported by nutrients from surface waters, have elevations at or below the local water table and generally are "planar." Rain-fed ombrotrophic peats, which are most common in modern environments, have centers elevated above local base level and often are called "domed." This morphology prevents the addition of nutrients by floodwaters. The two geometries may grade into one another within a single peat body (Esterle and Ferm 1986; Cohen et al. 1990; Calder in press), and both may be precursors to coal (Cecil 1990).

Actualistic studies recently have shown that peat can form under a variety of environmental conditions. Domed peat deposits of tropical Southeast Asia have received considerable attention as analogues for thick, low-ash, low-sulfur coals. Studies of these peat deposits and environments reveal that much of the nonwoody detritus is decayed at the surface but that woody material survives at depth (Anderson 1964a, 1964b; Andriesse 1974; Dittus 1985; Cohen et al. 1987; Cohen et al. 1990; Esterle et al. 1989). Distinct vegetational zonation has been recognized across the surface of the peat dome (Anderson

1964b), which may be manifested as spatial-temporal succession during an interval of peat accumulation (Anderson and Muller 1975). In the extreme center of peat domes, nutrients are very low, resulting in plant stunting and reduced input of organic detritus to the peat and significant levels of decay. The biotic and physical characteristics of these modern domed peats are used as guides to the identification of similar habitats in the geological past (Smith 1963; McCabe 1984, 1987; Cecil et al. 1985; Esterle and Ferm 1986; Littke 1987; Warwick and Flores 1987; Pocknall and Flores 1987; Winston 1990; Eble and Grady 1990). Prior to the focus on ombrotrophic peat-forming environments, rheotrophic or planar swamps such as the Okefenokee swamp (Georgia, USA) and the Everglades (Florida) (see summary in Cohen et al. 1987) served as the preferred analogues for ancient peats, resulting in extensive knowledge of such habitats. Major coal deposits, particularly in the Middle-Late Pennsylvanian, appear to have formed as planar peats (Horne et al. 1978; Phillips and DiMichele 1981; Grady and Eble 1990; Willard in press).

Ancient peat deposits, particularly those of pre-Pleistocene age, are represented as lignites and coals (Stach et al. 1982; Teichmuller 1987). The economic importance of these deposits has generated great interest in their physical characteristics and sedimentary context (see papers in Murchison and Westoll 1968; Scott 1987). Exposures created by mining have provided sedimentologists and paleontologists with opportunities to examine coal-bearing strata and build a detailed understanding of their depositional histories.

Studies of pollen and spores have been most commonly used to reconstruct the vegetation of ancient mires because these fossils are abundant throughout the deposit. Lignites, mostly of Tertiary age, and certain bituminous coals retain the microfossil remains of the parent vegetation unless coalification has reached the low-volatile bituminous or anthracite rank. The biostratigraphy of coal-bearing sequences has relied heavily on palynological analyses (e.g., Saulnier 1950; Peppers 1970, 1979, 1985; Nichols and Traverse 1971), and much of our understanding of ancient peat-forming forests and their depositional settings is derived from pollen and spore data (Winslow 1959; Smith 1962, 1968; Smith and Butterworth 1967; Scott and King 1981; Mahaffy 1985; Bartram 1987; Fulton 1987; Grady and Eble 1990; Eble and Grady 1990; Willard in press). Palynology has been important in the study of Holocene climates and ecology (Moore and Bellamy 1974; Birks and Birks 1980; Webb 1985; Middledorp 1986; Delcourt and Delcourt 1987).

Macrofossils also occur in ancient peats and coals but are usually restricted areally and stratigraphically. In Tertiary lignites, such as the Brandon Lignite of Vermont (Spackman 1949; Tiffney 1981) or the brown coals of Germany (Tobien 1968), leaves, fruits, and seeds have been preserved with anatomical detail sufficient for comparisons with modern plants. Microscopic examination of blocks of polished coal has led to detailed knowledge of plant tissues and microstructure (Stach et al. 1982; Winston 1988, 1989, 1990; Teich-müller 1990). This method has permitted macrofossil analysis for Pennsyl-vanian coals on an areal scale comparable to that achieved by palynological studies. Macrodetritus also is preserved as permineralized nodules of calcium carbonate or silica within the coal body. The entombed plant debris is chemi-cally coalified to the same degree as the surrounding coal (Lyons et al. 1985) but retains the structural fabric of the peat. The best-known forms of per-mineralized peat are calcium carbonate "coal balls," which are typical of Carboniferous coals of North America and Europe (Phillips and Peppers 1984; Scott and Rex 1985). Coal balls have provided a detailed understanding of the biology and ecology of Carboniferous plants and serve as a baseline for comparison with compression-impression fossils from clastic rocks of similar age.

It is generally agreed that carbonate coal balls form after peat deposition and following the introduction of marine or freshwater carbonate to the drowned peat-forming habitat (Stopes and Watson 1909; Anderson et al. 1980; DeMaris et al. 1983; Scott and Rex 1985). Silica permineralization of peats occurs throughout the geological record (Ting 1972; Knoll 1985) but is less common than carbonate in the Upper Carboniferous. A range of taphonomic information can be derived or inferred from permineralized peats, including levels of decay prior to burial, estimates of EH/pH characteristics of the mire, degree of peat subaerial exposure, and proximity to marine conditions (Phillips et al. 1977; Raymond 1987; Anderson et al. 1980; Phillips and Di-Michele 1990).

Animal remains are preserved in peat, lignite, and clastic-rich coal depos-its. Arthropods occur in coals of several ages (Bartram et al. 1987), but the paleoenvironmental context of these records is poorly known. Mesozoic lig-nites (brown coals) are generally uncommon, but an Upper Jurassic lignite in Portugal contains abundant remains of fish, amphibians, reptiles, and mam-mals, including articulated skeletons (Henkel and Krebs 1977; Krebs 1980, 1987). Vertebrate remains are known from lignites in Miocene deposits of Italy (Hürzeler 1975; Azzaroli et al. 1986; Harrison and Harrison 1989) and China (Badgley et al. 1988) and from sapropels associated with Carboniferous coals (Hook and Ferm 1988), but not from any of the coal deposits currently interpreted as originating from domed peats. There are numerous examples of Pleistocene-through-Holocene peat deposits with abundant, well-preserved plant and insect remains (Coope 1979; Wilson and Bateman 1987), and there are some instances of soft-tissue preservation of vertebrates, including hu-mans (Glob 1971).

7.8   Charcoal Deposits

When plant material is burned, it often is transformed into chemically inert charcoal or pyrofusain, which then may be transported and concentrated in sedimentary deposits. There has been some question concerning whether some of the fusain found in the fossil record could have formed through postburial diagenetic processes rather than burning (Beck et al. 1982), but recent studies indicate that most is pyrolitic in origin (Cope and Chaloner 1985; Scott 1989; Sander and Gee 1990). Charcoal-rich sedimentary units can form in fluvial floodplains, deltaic plains, and coastal settings, and such occurrences are recognized as a recurring taphonomic mode in the Phanerozoic record (Sander and Gee 1990; Gastaldo 1989b). Charcoal preserves the structure of wood, often in fine detail, as well as the morphology of seeds, cones, and flowers (Cope and Chaloner 1985; Sander and Gee 1990). It records potentially important paleoecological information concerning the influence of fire on the structure of ancient ecosystems (Cope and Chaloner 1985; DiMichele and Phillips 1988). The increase in runoff and sediment transport following wildfire also may affect depositional rates and burial of organic remains.

Charcoal is a low-density material that is readily transported and widely distributed by water. It is commonly preserved in coal seams and organic-rich mudstones that represent deposition in reducing aquatic environments, but it also occurs in other sedimentary contexts, such as fissure fills (Harris 1958; Kampmann 1983). It is subject to crushing and fracturing after burial and may be compressed to one-fifth of its original thickness (Sander and Gee 1990). When not severely crushed, it is often cubic in shape, with distinctive fracture and luster. Charcoal may be permineralized with pyrite or calcite.

Charcoal and other charred plant remains have received considerable attention in archeology, where they provide evidence for human food resources and building materials (Bar-Yosef and Kislev 1986; Schoch 1986). Fossil charcoal is known throughout the Phanerozoic, with the earliest occurrence documented in the Lower Devonian (Edwards et al. 1986). There is evidence for cyclicity in the occurrence of fusain in the brown coals of Europe (Teichmüller 1975) and bituminous coals of the Paleozoic (Schopf 1952). This may reflect cyclic sedimentary processes or ecological controls on wildfire frequency (Cope and Chaloner 1985). Study of fusain in Jurassic-Cretaceous deep-sea sediments also indicates periodic fluctuations suggesting long-term changes in the production of charcoal on land (Summerhayes 1981).

7.9 Mud Slides and Flash Floods

Catastrophic events such as mud slides or flash floods are physical processes, but they lead to fossil accumulations that also reflect biological factors, such as herding in vertebrates or tree spacing in a forest. Mud slides provide an obvious mechanism for burying organic remains quickly and permanently and are therefore somewhat analogous to explosive volcanic events or ash falls. Mud slides often are associated with volcanic and seismic activity, but both mud slides and mudflows occur in other circumstances as well.

Flash floods are generally associated with ephemeral or strongly seasonal rivers. Catastrophic floods often are invoked as a cause of unusual fossil occurrences, perhaps because they offer a simple explanation that is easy to relate to recent processes. However, there are relatively few documented fossil examples with convincing sedimentological evidence for this mode of formation.

Preservation of erect trees in Holocene valley fills in Oklahoma has been attributed to mud slides (Hall and Lintz 1983). Distinguishing trunks preserved in situ versus transported (but still erect) trunks in a mud-slide context is critical to paleoecological interpretations (Karowe and Jefferson 1987). A possible example of vertebrate preservation in a mud slide occurs in the Upper Triassic of Trossingen, Germany. Remains of *Plateosaurus* occur over a 10-m stratigraphic interval and are largely disarticulated except for a 1-m interval that contains complete and partially articulated skeletons. The peculiar sediment ("Marmomergel") is interpreted as a mudflow in a semiarid environment (Seemann 1933; Weishampel 1984; Norman 1987). Norman (1987) also favors a series of catastrophic flash floods as the explanation for massed *Iguanodon* remains in lacustrine deposits at Bernissart, Belgium. Concentrations of mostly fragmentary vertebrate remains occur in a Jurassic debris flow deposit in northern Mexico (Fastovsky et al. 1987). In this case it appears that bones on alluvial land surfaces were incorporated into a series of sheet flood events.

In modern ecosystems, herding behavior makes some species of terrestrial animals more vulnerable to mass deaths due to floods (Boaz 1982), and it seems likely that similar behavior in extinct species had similar effects. There are examples of massed remains dominated by single species including *Coelophysis* (Schwartz and Gillette in press), *Iguanodon* (Norman 1987), *Centrosaurus* (Currie 1984) and titanothere (Turnbull and Martill 1988) that may represent flash flood events, making them potentially important as instantaneous samples of single populations. However, other causes for monotypic or low-diversity massed remains, such as traps, burrows, and drought, can produce similar assemblages that are time averaged rather than instantaneous (Rogers 1990). Sedimentological evidence and the age structure of the preserved animals are critical in distinguishing among the possible causes of massed vertebrate remains (Voorhies 1969; Sander 1987; Turnbull and Martill 1988; Haynes 1988; Rogers 1990; Winkler and Murry 1989; Behrensmeyer 1991).

8   TIME AND SPACE RESOLUTION IN THE FOSSIL RECORD

All fossil samples represent some amount of time and space, in the sense that the preserved animals and plants lived within a particular interval of time and occupied a finite area of land surface. To resolve time or space in the fossil record is to determine the meaning of these variables in terms of years or area for a particular fossil assemblage. In biology as well as in paleobiology, the scale of temporal and spatial resolution in samples of faunas or floras is crucial to the interpretation of ecosystem structure, evolutionary patterns and rates, and responses to environmental perturbations. Resolution plays a critical role in determining how paleobiological evidence relates to evolutionary processes that can be observed and measured in the present, and it also is important in comparisons of patterns of diversification and extinction within the fossil record itself. Although the need for specifying levels of resolution is appreciated generally in paleobiology, it is difficult to obtain precise measures of these variables. We begin this section with a review of resolution and follow it with some estimates of levels of resolution that can be expected in the different taphonomic modes.

8.1 Time Resolution and Completeness

Paleoecologists often seek to determine the time resolution (or time interval) represented by a fossil assemblage. For most single assemblages of fossils, the amount of time represented is a function of the sedimentary and taphonomic processes that formed the fossil-bearing stratum or concentration. This interval of time is the result of "taphonomic time averaging" (fig. 2.3); it may amount to only a few seconds in the case of a fossil footprint, or $10^3 - 10^5$ or more years for reworked assemblages of durable organic remains. Time averaging can mix

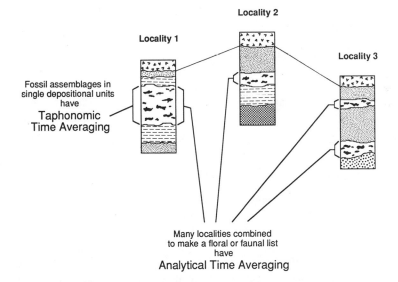

Fig. 2.3. Diagrammatic explanation for the difference between taphonomic time averaging and analytical time averaging. Symbols in stratigraphic columns indicate differing lithologies, as is typical of separate fossil localities. The volcanic unit at the top of each column is used in this example to correlate the localities.

organic remains from different environments or evolutionary stages. Because of taphonomic time averaging, species in a fossil assemblage can be regarded as contemporaneous only within the inferred time interval, which may be very long compared with a biologist's concept of ecological contemporaneity. This poses problems in comparing characteristics of communities, such as standing diversity, in ancient and present-day floras and faunas (Badgley 1982).

Paleoecologists also are concerned with identifying approximately time-synchronous assemblages of plant or animal communities and calibrating how much time these combined assemblages represent. In addition to taphonomic time averaging, there are additional factors influencing temporal resolution for faunas or floras from multiple localities or depositional environments. The accuracy of temporal correlations among localities is limited by analytical errors in radiometric dating, magnetostratigraphy, and other methods used in chrono-, litho-, and biostratigraphy. Typically the net result is a collection of assemblages that can be bracketed temporally, usually within intervals on the order of $10^3 - 10^6$ yrs. This "analytical time averaging" (fig. 2.3) limits the level of time resolution that can be achieved for regional- to continental-scale floras and faunas. Analytical time averaging generally is more pronounced for time periods further from the present, because dating methods are less reliable and because fossil localities generally are more dispersed in time and space.

Another important aspect of the paleontological record is completeness, which depends on the length of time that intervenes between fossil samples (Sadler 1981; Tipper 1983) (fig. 2.4). In older parts of the record the average interval between samples increases, thus decreasing relative completeness. Although this incompleteness is responsible for lower resolution of evolutionary changes for more ancient time periods, particular assemblages that show only a limited amount of taphonomic time averaging can be compared to much younger examples with similar time averaging. The selective use of such data allows comparisons of relatively short-term samples of ecological communities, time averaged over periods of $10^2 - 10^5$ yrs, throughout the Phanerozoic.

8.2 Methods for Assessing Time Resolution

Time resolution in continental sediments has received considerable attention from the standpoint of expected completeness for broad environmental categories, such as fluvial and lacustrine (Schindel 1980; Sadler 1981; Behrensmeyer 1982; Sadler and Dingus 1982; Badgley 1986a; Anders et al. 1987; Tauxe and Badgley 1988). For a specified rock interval, completeness is the proportion of time represented by sediments versus the proportion of time not represented because of nondeposition of erosion. Average sedimentation rates from specific Recent environments are available for estimating the amount of time in a stratigraphic sequence. These rates also can be used to estimate time in a particular rock unit, if compaction is taken into account. Difficulties

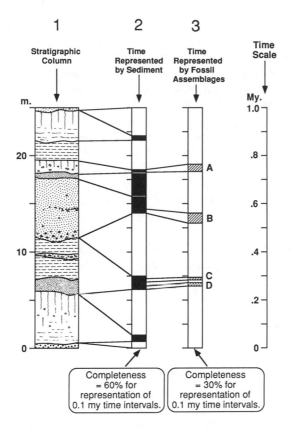

Fig. 2.4. Diagrammatic explanation of stratigraphic completeness. Columns 2 and 3 show how time actually is represented in hypothetical stratigraphic column 1. The black portions of column 2 show intervals when sedimentation occurred and account for about 23% of the total time represented by the stratigraphic column. Completeness must be calculated with respect to a given time interval, however, which is 0.1 my in this example. Because of the vertical distribution of times of active sedimentation, 6/10 of the 0.1 my time intervals in column 2 are represented, for a completeness of 60%. Fossiliferous deposits in column 3 represent less time (8% of the 1.0 my), and also somewhat different intervals of time. Fossil deposit *A* is in a paleosol, part of which formed after sedimentation ceased. Fossil deposit *B* is in a channel lag and includes organic remains reworked from earlier sediments. The four fossil deposits represent 3/10 of the 0.1 my time intervals, for a completeness of 30%. Time resolution *within* each deposit is controlled by taphonomic time averaging; for *A* and *B* it is on the order of 10^4 yrs, and for *C* and *D* it is 10^3–10^4 yrs. In an actual study, time resolution with respect to an absolute scale, i.e., how each deposit is placed in time, depends on how accurately radiometric or other chronostratigraphic methods can resolve real time relationships.

arise, however, because of the wide range of potential error in such estimates. Another approach to estimating time in specific depositional units is based on sedimentary characters (texture, bedding structures, geometry) and modern analogues. Retallack (1984a) compared these two methods for Oligocene paleosols and found an order-of-magnitude difference in time estimates for the same rock units. In some cases the internal features of a fossil-bearing unit may be a more reliable guide to time resolution than estimates based on average rates of sedimentation.

Although the degree of temporal and spatial averaging is partly a function of sedimentological processes, organic remains may represent different amounts of time and space from their enclosing sedimentary unit. For instance, a crevasse splay is often a nearly instantaneous depositional event, but it can bury organic debris that represents tens to hundreds of years as it sweeps onto a floodplain surface. On the other hand, a paleosol formed over a period of 10^4 years might include an organic sample representing a much shorter time period, if older remains were destroyed by soil processes. Taphonomic processes other than those involved in the events that form a sedimentary unit may exert the most important controls on time averaging in a fossil assemblage.

Taphonomic processes affecting temporal resolution have been studied to varying degrees in the different contexts shown in table 2.1. Much of the work is actualistic and indicates the kinds of taphonomic evidence for time averaging that can be sought in the fossil record. Because of both sedimentological and taphonomic processes, however, it appears that only order-of-magnitude estimates are possible in all but exceptional circumstances.

8.3 Spatial Resolution

Most organic remains are potentially mobile and can be transported away from the site of death or the place of discard, in the case of shed parts such as leaves, exoskeletons, and teeth. This affects paleoecological information concerning an organism's original habitat and community associations with other taxa. The problem of transport and mixing of different components of fossil assemblages is well known, and there are numerous studies of taphonomic variables such as abrasion and sorting that can be used to infer the degree of transport and reworking in plant and animal remains. Such information places general limits on the area of land surface that is represented by a fossil assemblage. The taphonomic and sedimentological processes that affect spatial resolution are often the same as those that cause taphonomic time averaging.

The terms autochthonous and allochthonous are used to express the most general interpretation of spatial resolution, i.e., whether or not remains are transported with respect to the site of death or discard and the habitat of the living organism (see §1). "Parautochthonous" is an intermediate term for transport of remains within the boundaries of an organism's habitat. In order to conclude that a particular fossil is autochthonous, parautochthonous, or allochthonous, it is necessary to gather taphonomic and sedimentological evidence concerning transport distance and to make general inferences about the size of the original habitat, based on lateral facies variation and modern ecological analogues.

Paleoecological consideration of spatial resolution seldom generates quantitative estimates of the square meters of an ecosystem that are represented by a fossil assemblage. The problems most often involve relative degrees of allochthony or autochthony in different assemblages, or questions such as whether co-occurring fossils are derived from one environment or a mixture of environments. There is great variability in spatial sampling both within and between the different taphonomic modes, and this can have important consequences for paleoecological reconstructions.

8.4   Resolution in the Taphonomic Modes

Each of the preservational modes listed in table 2.8 encompasses processes and circumstances that affect the degree of temporal resolution represented by assemblages of organic remains. Likewise, each mode represents different areas of the land surface where the organisms lived, and samples may be autochthonous or allochthonous to varying degrees. Emphasis in the following sections is on time and space resolution *within* individual assemblages from different contexts, resolution that reflects taphonomic time averaging and taphonomic space averaging. Estimates for the approximate amount of temporal and spatial resolution that can be expected in the different taphonomic modes are presented as *preliminary hypotheses* to be discussed further and tested, if possible, in present-day or other situations that are appropriate for high-resolution dating.

Coastal Environments

Coastal settings tend to sample plant and animal remains from large geographic areas, with allochthonous components often dominating those from local environments. Fossils from these contexts are more likely to represent the habitats and organisms from large parts of drainage basins, rather than individual communities, and the total preserved diversity may exceed any local life association of plants or animals. An exception to the predominantly allochthonous nature of coastal fossil assemblages occurs in salt-marsh estuaries, where actualistic work indicates autochthonous preservation of macroplants and palynomorphs (Staub and Cohen 1979; Rich and Spackman 1979; Frey and Howard 1986).

Storm-generated deposits in all four coastal contexts represent very brief sedimentation events, but such events often mix hard parts from recently dead animals with much older, reworked remains, and time averaging may be on the order of $10^3 - 10^6$ yrs (Kidwell 1982; Staff and Powell 1988). Concentrations of vertebrate materials formed during periods of maximum transgression and sediment starvation on continental shelves may represent $10^5 - 10^6$ yrs (Kidwell 1989). Remains of mid-Tertiary through Recent terrestrial vertebrates occur together in Recent estuaries, indicating that a large degree of time averaging can occur in this environment (Frey et al. 1975). Similar degrees of

Table 2.8   Hypothesized Levels of Temporal and Spatial Resolution in the Taphonomic Modes

Environments	Hypothesized Time Resolution (yrs)	Hypothesized Spatial Resolution
Coastal	$10^3 - 10^6$ (Lagoons: $10^0 - 10^2$)	Regional; local in lagoons, estuaries
Fluvial/deltaic		
Channel lags	$10^3 - 10^5$	Regional to drainage basin
Channel fill	$10^0 - 10^4$	Local
Floodplain	$10^0 - 10^4$	Local
Lacustrine	Varved deposits: $<10^0 - 10^2$ Other: $10^1 - 10^3$	Large lakes: regional and local Small lakes: local
Volcanic	$<10^{-2} - 10^0$	Local Mudflows, lakes: local to regional
Eolian		
Dune	$10^4 - 10^6$	Local
Interdune	$10^3 - 10^4$	Local
Loess	$10^1 - 10^6$	Local
Other contexts		
Tar seeps	$10^2 - 10^4$	Local
Amber	$10^{-3} - 10^4$	Local
Karst (caves, fissures, sinkholes)	$10^0 - 10^4$	Local; predator/scavenger collections may be regional
Tree trunks	$10^0 - 10^2$	Local; larger areas for predator collections
Burrows	$10^{-1} - 10^1$	Local; larger areas for predator collections
Feces, regurgitates	$10^{-2} - 10^2$	Local, territorial areas
Middens	$10^0 - 10^3$	Local, territorial areas
Springs	$10^1 - 10^4$	Local
Archeological sites	$10^{-3} - 10^3$	Local to regional
Peat	$10^0 - 10^4$	Local
Charcoal	$10^{-3} - 10^0$	Local; regional if transported
Mud slides, flash floods	$10^{-3} - 10^1$	Local to regional

Note: Hypotheses concerning the *finest* levels of temporal and spatial resolution represented in individual fossil assemblages from the specified taphonomic modes. "Local" means that the assemblages represent areas and life habitats proximal to the depositional site. "Regional" means that organisms in the assemblages are derived from larger areas, up to entire drainage basins.

time averaging for macroplant remains such as wood would be expected only if they were mineralized prior to reworking.

Where input of transported, attritional organic remains and bioturbation are relatively continuous at a depositional site, the time represented by individual fossil assemblages probably can be resolved no more finely than 10^4 yrs (Gastaldo 1989a). In lagoonal settings and other organic-rich coastal environments, however, relatively rapid sedimentation of fine-grained material and low rates of bioturbation may lead to high levels of time resolution (10^0-10^2 yrs), comparable to those of some lacustrine settings (Zangerl and Richardson 1963).

Fluvial and Deltaic Environments

In most cases, channels bars and lags provide allochthonous samples of terrestrial plant and animal communities. Although deposition of individual sedimentary units usually is quite rapid, the organic materials typically represent longer periods of time in transit or in temporary storage. Given the distance traveled by plant debris, carcasses, and bones in present-day rivers, ancient channel accumulations are likely to represent regional, rather than local, samples of the original biotas (Frazier and Osanik 1961; Darrell 1973; Behrensmeyer 1982, 1988; Scheihing and Pfefferkorn 1984). Moreover, because palynomorphs and bones can be reworked from previous channel and overbank deposits, samples in channel lags or bars may be time averaged over periods of 10^3-10^5 yrs (Behrensmeyer 1982).

In contrast to channel-bar and lag deposits, most channel fills offer a sample of organic remains from the immediate vicinity of an abandoned channel (Behrensmeyer 1988). Although individual units within channel fills may sample shorter time intervals than depositional units in channel lags and bars, the overall period of accumulation for sediment and organic remains within channels exceeding several meters in depth may be similar, on the order of 10^2-10^4 yrs (Behrensmeyer 1982; Gagliano and Howard 1984; Farrell 1987; Fiorillo 1988b). The filling sequence of abandoned channels may also record a true ecological succession from aquatic to fully terrestrial environments (Wing 1984). If such successions occurred at rates comparable to those in modern environments, this would indicate a time interval of 10^2-10^4 years.

Although fossil assemblages from floodplain deposits are affected by a large number of biological, chemical, and physical processes, many of these biases can be assessed using sedimentological and taphonomic data (Behrensmeyer and Dechant Boaz 1980; Bown and Kraus 1981a, 1981b; Retallack 1984a; Badgley and Gingerich 1988). Floodplain fossil assemblages appear to be predominantly autochthonous, with attritional accumulations of plant and animal debris incorporated gradually in paleosols, wet areas, and overbank deposits. Estimates of time intervals represented by organic assemblages of single sedimentary units range from 10^0-10^4 yrs, and time intervals between samples from 10^4-10^5 my (Behrensmeyer 1982; Gingerich 1982; Winkler 1983; Badgley and Gingerich 1988; Wing and Farley 1990). Because plant remains in many climatic settings are rapidly degraded unless covered by sediment, an assemblage of undegraded plant parts may represent periods of accumulation of 1 year or even less prior to burial (Wing 1984, 1988).

Taphonomic samples from successive paleosols that have similar time-space resolution have been used as a basis for examining sampling effects in reconstructing faunal turnover (Badgley and Gingerich 1988; Badgley 1990). Actualistic estimates of rates of soil development, pedogenic carbonate formation, and floodplain aggradation (Leeder 1975; Retallack 1984a, 1986, 1988; Behrensmeyer 1987b) can be integrated with paleontological data to estimate the amount of time represented in paleosols and to address questions concerning rates of environmental and faunal or floral change.

Intraclast conglomerates are commonly associated with assemblages of disarticulated vertebrate fossils in fluvial deposits including channel lags and bars, crevasse splays, and channel fills (Voorhies 1969; Badgley and Behrensmeyer 1980; Rose 1981; Badgley 1986a; Rigby et al. 1987; Wood et al. 1988; Eberth 1990). The presence of large or angular mudclasts, unabraded bones, and poorly sorted sediment fabrics implies erosion of channel banks or floodplain surfaces followed by rapid deposition with little reworking and winnowing of the sediment. In such cases, the sources of bones and mudclasts are likely to have been close to the site of deposition. In cases where bone assemblages show considerable reworking, sorting, and abrasion, a longer period of interaction with fluvial processes is implied (Dodson 1987; Eberth 1990). Although many vertebrate remains and mudclasts are roughly hydraulically equivalent in intraclast conglomerates, most such conglomerates are unfossiliferous, indicating localized sources for the organic debris (Eberth 1990). Modern analogues for the mudclast-bone association currently are lacking, but there is little doubt that this kind of deposit represents a wide range of temporal and spatial scales. Mudclasts and bones may be derived from the same local sources, but erosion and concentration of bones from pre-existing deposits could lead to time averaging on the order of $10^2 - 10^4$ yrs (Behrensmeyer 1982).

Lacustrine Environments

Because large lakes generally receive allochthonous organic remains (e.g., leaves, pollen, insects), the resulting land-derived fossil record can be regional in scale. Smaller lakes, or discrete clastic fillings within larger ones, preserve more localized samples that may provide detailed insights into the source communities (Spicer 1981).

The varved sediments of deep lake deposits provide some of the highest levels of time resolution available in the fossil record, on the scale of single years or even seasons (Bell and Hagland 1982; Bell et al. 1987; Wilson 1987b; Anderson and Dean 1988). For example, in one 700-year varved sequence, it was possible to determine exactly when, to the nearest varve, an individual fish died (Wilson 1987b). Because there is no guarantee that putative varves really are annual, however, difficulties arise in establishing the amount of time represented by each one (Wilson 1977, 1984, 1988a). Radiocarbon dating of subrecent deposits in the Dead Sea and in Lake Turkana indicates that varves often do not record a strictly annual cycle (Halfman and Johnson 1988).

Although the amount of time represented by a single fossiliferous unit can be estimated from average rates of sediment accumulation calculated for long stratigraphic sequences, this method incorporates hiatuses and underestimates actual sedimentation rates (Sadler 1981). The mean rate of sediment accumulation for a sample of 291 North American Holocene lakes has been calculated at 91 cm/10^3 yrs, with a range of <1 to >3500 cm/10^3 yrs, implying that time resolution within units of centimeter-to-decimeter thicknesses can be on the order of 10^1–10^3 yrs (Webb and Webb 1988). This would apply to typical, nonvarved, fossiliferous lacustrine beds, especially bioturbated units. Other sedimentation events, such as turbidity flows or storm deposits, as well as chemical precipitates, may preserve instantaneous mass-death samples of aquatic and nearshore biotas, but remains from such catastrophic assemblages may be mixed with older, reworked biological materials (Warner and Barnett 1986; Cohen 1987).

Volcanigenic Deposits

Because volcanic depositional events are unusually rapid and extensive, they provide some of the best-preserved autochthonous assemblages of contemporaneous organisms, especially plants. Explosive and airfall events represent minutes to weeks and provide a fine level of time resolution, < 10^{-2}–10^0 years. Death and burial may be simultaneous, resulting in a preserved snapshot of an autochthonous assemblage that may be compared directly with present-day ecosystems (Voorhies and Thomasson 1979; Burnham and Spicer 1986). Reworked volcaniclastic material generally provides lower temporal and spatial resolution, but this still may exceed average levels of resolution in nonvolcanic lacustrine and fluvial deposits (Taggart and Cross 1980; Vicars and Breyer 1981; Lockley and Rice 1990). Some volcanic deposits also yield radiometric dates that allow rates of sediment accumulation to be calculated over short stratigraphic intervals.

Eolian Environments

Wind has limited power to move larger organic remains, and assemblages of fossils in eolian deposits generally appear to be untransported, with the exception of pollen and other microplant remains. Occurrences of vertebrate remains in dunes are often isolated, and assemblages may consist of collections of individual fossils from throughout a particular rock interval or formation, rather than from a single fossiliferous sedimentary unit. In such cases, an assemblage could represent 10^4–10^6 years, resulting from analytical time averaging caused by the collecting strategy of the paleobiologist. Interdune deposits represent moist environments that are attractive to a wide variety of organisms, and fossil assemblages formed by the gradual accumulation of organic remains could represent ecological ranges that overlapped only at this focal resource. Thus, the total area covered by these ranges would be represented by the organisms in the resulting fossil assemblage. Time averaging in interdune assemblages might represent periods on the order of $10^3 - 10^4$ yrs or shorter intervals, depending on rates of dune movement and water table fluctuations (Wendorf and Schild 1980; Driese 1985).

Loessic sequences can be subdivided by biostratigraphic methods and the physical tracing of paleosols into units representing periods of $10^3 - 10^5$ yrs. This level of time resolution reflects analytical time averaging, in which fossils from the same loess unit (e.g., between the same paleosols) are combined by the paleontologist to make a faunal or floral list. Under ideal circumstances, such as the burial of organic remains on a soil surface by renewed eolian deposition, fossil assemblages from loess deposits could represent periods of as little as $10^1 - 10^2$ yrs. Eolian deflation in loess deposits creates local bone lags that may be time averaged over $10^4 - 10^6$ yrs, such as in Pleistocene accumulations from deflation basins in eastern Colorado (Graham 1981). Bioturbation also can destroy bedding structures and mix biological and archeological materials from different depositional units, resulting in taphonomic time averaging, probably on the order of $10^2 - 10^4$ yrs. Most macrofossil assemblages in loess are assumed to be autochthonous or parautochthonous because wind and slope wash would not be strong enough to cause significant transport.

Other Contexts

Many of the taphonomic modes in other contexts involve short-term, biologically mediated concentration and burial of organic remains. Tar pits, traps formed by fissures, caves, or other cavities, as well as springs, tree trunks, peat swamps, fusain, and amber, all represent situations where remains of organisms may accumulate because of various ecological, behavioral, chemical or accidental factors. When tar pools, caves, sinkholes, and springs offer water, shelter, or other resources, animals from a wide range of habitats may be drawn into the area and produce accumulations of remains from both local and distant sources. If, on the other hand, these circumstances act simply as passive traps without an attraction factor, assemblages of animal and plant remains represent a sample of a local ecosystem.

Fecal concentrations, regurgitates, and middens directly reflect the behavior and range of animals responsible for them. Packrat middens are known to represent relatively small areas (e.g., a 50-m radius: Van Devender 1987) where the packrat collects plant materials, bones, and other debris. Therefore changes in the vegetation represented in the middens are interpreted as local climate shifts or other ecological disturbances (Van Devender and Spaulding 1979). Owl pellets represent foraging ranges on the order of several km^2 (Andrews and Cooke 1985; Andrews 1990), as do other digested or partially digested remains from aerial or ground-dwelling predators. These samples generally represent relatively small areas but still may incorporate remains from a variety of habitats. Time intervals in such deposits probably range from less than a year to 10^2 yrs, but in cave, fissure, or other protected situations pellets can accumulate for longer intervals of 10^3-10^4 yrs. In karst situations in general, bone assemblages usually represent habitats proximal to the site and may be averaged over time intervals of 10^2-10^4 yrs (Vrba 1980; Brain 1981; Behrensmeyer 1988, 1991; Andrews 1990).

Burrows and tree trunks preserve autochthonous samples of the animals that occupied them or were trapped there and also may include plant remains from local sources. In these contexts, remains of nonfossorial animals indicate the activities of bone collectors that may have transported skeletal remains from distances of hundreds of meters to several kilometers (Skinner et al. 1980; Maguire et al. 1980; Vrba 1980; Brain 1981; Hill 1983). Clusters of burrows with preserved articulated skeletons appear to represent in situ burial of failed aestivation assemblages (Olson 1939; Olson and Bolles 1975; Carlson 1968). Time averaging in these situations is likely to be brief, amounting to a year or season when conditions were unusually dry.

In tar deposits, organic remains are constantly mixed by viscous flow, and the amount of time represented is roughly equivalent to the time span represented by individual stratigraphic units within the tar pits. This is approximately 10^4 yrs for the La Brea Tar Pits (Stock 1972; Woodard and Marcus 1973, 1976). Individual fossil samples in amber represent instants in time, but many assemblages recovered from this context occur in reworked deposits that could combine organic remains from time periods of up to 10^6 years (Mustoe 1985). Bone accumulations in karst sites (fissures, caves, sinkholes) may represent from less than a year to 10^4-10^5 yrs, depending in part on the biological agencies involved. The sedimentary deposits as well as the bone sample often reflect highly variable rates of input, as documented in studies of Pleistocene-through-Holocene examples (Brain 1981; Genoways and Dawson 1984; Pratt 1990).

Spring systems typically have periods of activity and inactivity, and deposition and accumulation of paleobiological remains around springs also may be sporadic. Quaternary spring deposits in western Missouri encapsulate 10^4-30^4 yrs in single sedimentary units, whereas the combination of all such units may span hundreds of thousands of years (C. V. Haynes 1985; Saunders 1988). Although of short geological duration, these samples incorporate long periods of ecological time.

Pre-Holocene archeological deposits generally are assumed to represent occupations ranging from a few days to years, based on ethnographic analogy to modern hunter-gatherers. The duration of site occupation for Plio-Pleistocene hominids has been estimated at 5–10 yrs using bone-weathering stages (Potts 1986). Such occupation is thought to have been sporadic rather than continuous. Archeological materials in caves or shell middens may represent longer time intervals of $10^1–10^4$ yrs if these sites were reoccupied. In many cases, successive occupations over time intervals of $10^3–10^4$ years can be distinguished, because they occur in different rock units. The faunal remains in archeological sites reflect the foraging range of the hominids, which is on the order of a 10–20-km radius for modern hunters (Lee and DeVore 1968; Gould 1980). Lithic materials of the African Pleistocene can represent greater distances of 40–100 km, implying longer expeditions or trade with other populations (Potts 1989).

Peat accumulates for time intervals of up to $10^3–10^4$ yrs (Anderson 1964b; Anderson and Muller 1975; Wilson and Bateman 1987; Payette 1988). Determinations of peat accumulation rates within present-day systems are problematic (Clymo 1984) and are difficult to apply to rates in ancient peat deposits. If the peat or coal is interbedded with clastics, it may be possible to designate individual units representing intervals of $10^2–10^3$ yrs (Phillips et al. 1977). Layers of coal balls are believed to preserve plant accumulations of 10^2 yrs or less, and fusain layers in coal balls may be virtually instantaneous samples with characteristic floral compositions. Polished blocks of coal and microstratigraphic pollen samples can reveal similar short time intervals (Mahaffy 1985; Grady and Eble 1990). Fusain layers preserved in coals or other sedimentary contexts represent virtually instantaneous samples of the standing vegetation and litter (Cope and Chaloner 1985; Sander and Gee 1990), but transported accumulations of charcoal could mix the records of a number of fires.

Mud slides and flash floods can preserve instantaneous samples of organisms, particularly trees and herding animals, (Fritz and Harrison 1985; Norman 1987; Turnbull and Martill 1988). Along with volcanic airfall and explosive deposits, sedimentary units formed during these unusual depositional events contain some of the most anatomically complete and most time-specific samples of populations of fossil organisms. Some mudflow deposits, however, appear to incorporate remains that represent longer periods of time (Fastovsky et al. 1987).

The estimates given above for the different terrestrial taphonomic modes offer an initial framework for understanding levels of time and space resolution within individual samples of fossil faunas and floras. Aside from a few exceptional deposits, this wide range of estimates represents the relatively low levels of resolution that are inherent to nearly all of the terrestrial fossil record.

9 CHANGE THROUGH TIME IN THE TAPHONOMIC MODES

As emphasized in the previous sections, information on ancient ecosystems is controlled by the taphonomic features of individual fossil assemblages and ultimately by any broad-scale processes that affect these attributes. It is theoretically possible that change through time in the taphonomic modes has generated important directional effects on the fossil record and that these changes result in taphonomic megabiases (Behrensmeyer and Kidwell 1985). Such changes might reflect significant biological shifts within the food web, such as a decreased role for decomposers, which affected the kinds of organic remains that were available for fossilization. Large-scale tectonic and climatic forces may have altered the frequency of different taphonomic modes and perhaps changed the taphonomic or depositional processes that control each mode. Evolution of taphonomic modes has been proposed for marine invertebrate faunas (Kidwell and Jablonski 1983; Kidwell and Aigner 1985) but has not yet received serious attention in the terrestrial realm.

The mosaic nature of the fossil record can be likened to the periodic illumination of parts of the evolutionary theater by spotlights that play over the stage as a whole (extending the metaphor of chap. 1). It is not yet possible to say whether the uneven highlighting is random or directional for parts or all of the past 450 Ma. Inevitably, however, more information comes from the parts of the stage that are illuminated longer or more frequently, information pertaining not only to the ecosystems as a whole but to the evolutionary patterns of the constituent organisms. If patterns of diversification, extinction, and adaptation vary across the different environments, then our broad perceptions of cladogenesis and community change also are affected by this spotlight. Moreover, if most evolutionary innovation takes place offstage, in the shadows, or when the lights are down, then the illuminated parts of the play record primarily the fine-tuning episodes of evolutionary history, when groups of organisms interacted with relatively stable environments over extended periods of time. Taphonomic biases that are inherent in these most frequently sampled environmental settings also could affect our perceptions of broad-scale evolutionary and ecological patterns.

9.1 Effects of Area

Change through time in the frequency of different taphonomic modes could result simply from the decreased or increased area covered by particular environments, such as coastal swamps or deserts, and also from differential preservation of these environments in the geological record. The greater the area of present-day outcrop, the larger the representation of taphonomic modes from these environments, resulting in accentuated effects of their biases with respect to the original biotas. This may seem obvious, but it is important to note that the uneven representation of different environments (and different geographic areas) through time inevitably affects interpretations of evolutionary paleoecology.

One recognized time-dependent bias results from the effect of having more of a fossil record for intervals closer to our own time (Raup 1976; Sepkoski 1976), a function of the general increase in the volume of available sedimentary rocks nearer to the present. Lacustrine deposits provide a good example, with approximately 85% of the record being Cenozoic (65 my), 11% Mesozoic (180 my), and 4% Paleozoic (325 my) (Picard and High 1972). The greater availability of late Cenozoic deposits of all kinds leads to more information on the biotas of drier and cooler environments than is possible for wetter, warmer environments of the preceding epochs. It is likely that reconstructions of tempo and mode in plant and animal evolution might differ if the most recent time interval sampled another climatic regime.

9.2   Consequences of Environmental Change

The distribution of various taphonomic modes through the Phanerozoic reflects environmental variations at the regional-to-global scale. Fundamental to the comparative use of fossil assemblages from different taphonomic modes in paleoecology is the recognition of these larger-scale controls on their occurrence. Though most of these controls are presently anecdotal and in need of rigorous documentation, they may be tied to environmental conditions that ultimately were governed by tectonic factors and continental configurations. For example, it has been suggested that tectonic and climatic conditions that promoted avulsion-dominated fluvial systems affected the vertebrate record in abandoned channels (Behrensmeyer 1988), and that large-scale sea-level changes exerted important controls on dinosaur ecology and preservation in coastal settings (Horner 1984a). Many paleoecologic studies have emphasized the role of global climate as a major control on biotic change (see §4 of chap. 7), but climate alone neither explains nor predicts the composition of subsequent biotas.

Floral and faunal turnovers often coincide with lithofacies changes that indicate environmental shifts, and this leads to problems in interpreting the scale and even the reality of the biotic event. For example, in the Permian faunas of central North America, the transition from one major terrestrial faunal assemblage (chronofauna) to another could reflect a local shift in environments and modes of fossil preservation or a true, larger-scale evolutionary event (Olson 1952). Fossil samples from a variety of sedimentary contexts, all of which show similar biotic change, and correlations to similar transitions in other regions were necessary to demonstrate the scale of the Permian faunal events and their temporal resolution (Olson 1984a, 1984b). Thus, the availability of information from different taphonomic modes and from time-equivalent deposits in other regional settings limits the inferences that can be drawn from apparent biotic change through time in any given rock sequence.

The record of peat and coal formation provides another example of large-scale bias. Coals are not uniformly distributed through the Phanerozoic but show distinct variations in the volumes of tropical peat that accumulated under different climatic and tectonic conditions (e.g. Parrish 1982; Ziegler et al. 1987; Ziegler 1989). During the Carboniferous, the northward movement of Euramerica gradually caused the landmass to move into, through, and out of the wet tropics, shifting lowland areas from perennially wet to increasingly seasonal climates (Ziegler 1989; Cecil 1990). The degree of seasonality may have had a strong impact on the relative proportions of domed-versus-planar peat. On a finer level of time resolution, broadly contemporaneous variations in Carboniferous coal abundance and depositional characteristics across the Euramerican paleocontinent (Phillips and Peppers 1984; Cecil et al. 1985) may reflect the overprint of climatic fluctuations on a scale of hundreds to thousands of years.

Taphonomic megabiases resulting from long-term climatic and tectonic change are poorly understood at present, but there are clear indications that they exist and that they may have a major impact on our perception of terrestrial ecosystems. More intensive study is needed to test the relationships between changes in taphonomic modes through time and regional-to-global-scale environmental processes.

9.3   Changes in Taphonomic Processes

Change through time in taphonomic processes could affect the preservation of individual taxa, specific body plans, or organisms inhabiting particular environments. These changes might be primarily biological, such as the evolution of mechanisms for cellulose digestion in arthropods or bone-crushing dentitions in mammalian predators. They could also involve secondary chemical or physical effects of organisms, such as soil acidity and tannin release caused by plant decomposition, or changes in rates of physical-versus-chemical weathering and sedimentation due to landscape stabilization by vegetation (Schumm 1977).

Nonbiological chemical effects also may control preservation in particular environments by influencing the abundance of basic elements that comprise organic parts, especially carbon, nitrogen, calcium, and phosphorus. When one or more of these elements are abundant, as in the Neogene carbonatitic volcanic centers of East Africa (Bishop 1968; Pickford 1986) extraordinary preservation of both hard and soft parts can result. When any of these elements is in short supply, biological recycling or chemical disequilibriums (both before and after burial) tend to destroy organic remains. Local ecology, and the organisms that are available for preservation, are also controlled by nutrient cycling. There is no reason at present to suspect a directional change in the chemistry of land ecosystems through the Phanerozoic, but shorter-term shifts in the critical elements could be correlated with episodes of volcanicity or climate change, especially when these affected atmospheric CO_2.

The fossil record is in large part a function of physical conditions that promote rapid and permanent burial of organic remains on local–to–regional scales. It is possible that such conditions have fluctuated throughout the Phanerozoic along with episodes of seafloor spreading or global climate change. Periods of continental collision or accretion, with or without volcanicity, generate regionally higher rates of sediment accumulation in fluvial, lacustrine, and deltaic systems. Glacial conditions also promote eolian silt deposition (loess), alluvial outwash, and lakes.

Almost any change in biological processes could have some effect on the fossil record, because dead organisms or shed parts are dynamic components of the ecological web that supports living plants and animals. Fossils are possible only when some of the organic remains are not recycled. In the broad overview of the terrestrial fossil record, which consists of relatively small samples that survived recycling, it will be important to determine how evolutionary changes in biological taphonomic processes and circumstances have affected information bearing on phylogeny and paleoecology.

There can be little doubt that evolution in body size and shape, and changes in the biological agents and processes that affect organic remains, have altered the characteristics of the fossil record through time. There is, however, little empirical evidence to document the taphonomic consequences of evolution in body form, hard parts, food-processing strategies, or behavior. A number of possible large-scale biological effects on fossil preservation can be proposed:

1. Evolution in macro- and microbiodegraders that permitted more complete destruction of plant remains, possibly combined with adaptations in plants to encourage or discourage such destruction;
2. Changing capabilities among vertebrate predators and scavengers for dismembering carcasses, collecting, masticating, and digesting bones;
3. Increased size, number, or durability of hard parts or other supportive tissues in plants and animals, resulting in slower rates of postmortem destruction and (in some environments) increased probabilities of burial;
4. Evolution of burrowing behavior that predisposed animals to subterranean death, or herding behavior that resulted in group drownings or other mass-mortality events;
5. Enhanced preservation of lacustrine organisms in diatomaceous sediments after the invasion of diatoms into freshwater environments during the Eocene (Wilson 1988a);
6. Evolution of deciduous plants, resulting in periodic flushes of leaf debris, and insect pollination, which affected volume of pollen production and the representation of some plant groups in the palynological record.

At present the above suggestions are for the most part hypothetical. There are enough data on land ecosystems to test some of them, but these data have not yet been organized and focused on the problem of preservational mega-biases in the fossil record. One strategy for investigating this problem involves study of fossil assemblages in similar sedimentary contexts through time. This "isotaphonomic" approach (Behrensmeyer 1991) would allow researchers to compare assemblages that have at least some of the same biases. If some taphonomic features, such as those resulting from sedimentary transport and burial, can be held constant, then these would help to identify changes in other features that might reflect evolution in biologically mediated taphonomic processes.

10 CONCLUSION

The examples and discussion in this chapter portray the fundamental nature of the fossil record, namely, that samples of past ecosystems have been subjected to highly variable modification by taphonomic processes, represent single and mixed habitats, often are averaged over time periods that seem very long to modern ecologists, and are highly patchy in time and space throughout the Phanerozoic. In spite of this, there are consistent patterns of preservation within particular environments or related to specific biological circumstances that represent recurring taphonomic modes. Delineation of these modes provides a basis for comparative study of ancient terrestrial ecosystems.

The patterns of ecomorphic reiteration or change, guild structure, response to environmental perturbation, and biological diversity can be examined within particular taphonomic modes and contrasted between different ones. Because fossil organisms represent the results of evolutionary processes acting within particular environmental settings, careful documentation of paleo-environmental contexts can provide a basis for testing the relationship between evolutionary patterns and these environmental settings. Analysis of specific taphonomic modes through time could reveal consistent patterns of evolution and extinction within a particular environmental setting or differences between settings. Lack of environmentally correlated patterns would indicate controls that were independent of local habitats or ecosystems, while the presence of such patterns would indicate strong ecosystem control. This would be a useful approach in examining evolutionary convergence (the emergence of similar ecomorphs from different phylogenetic origins).

As emphasized in §1, the entries in tables 2.1–2.7 should be regarded as testable hypotheses that may change with continued research. In their present form, however, these tables, along with supporting discussions, provide the following perspectives on the continental fossil record:

1. Although there is great variability in preservational modes and biases for different types of organisms, recurring patterns within particular taxonomic groups imply that similar organic samples are available

through time. Most nonbiological taphonomic processes have been relatively constant through the Phanerozoic, but those processes mediated
by living organisms undoubtedly have changed as organisms evolved.
Environmentally based taphonomic modes thus contain components that
represent different sets of biologically driven processes in different parts
of the rock record.

2. The available taphonomic modes provide limited, environmentally specific samples of the terrestrial realm through the Phanerozoic. Most of
these samples represent sites of active deposition, such as coastal areas,
alluvial plains, and rift valleys. Tectonic and other physical controls
limit the occurrence of some taphonomic modes to younger time periods, many others occur only sporadically, and a few are relatively common throughout the rock record. The large-scale biases inherent in this
nonuniform sampling of past biotas must be evaluated in studies of long-
term ecosystem change for different groups of organisms.

3. The systematic examination of taphonomic modes shows where different types of organic records can be expected to occur together (table 2.1).
Plant and vertebrate remains, for example, co-occur fairly frequently in
wet floodplain settings but rarely in drier floodplain deposits. Environments where there is information on different kinds of organisms can be
targeted for more comprehensive ecosystem reconstructions and for
comparing the effects of possible climatic changes on different components of the paleocommunities.

4. Information from broad-scale environmental settings indicates whether
local or regional samples are likely to be represented in a particular
taphonomic mode. The degree of allochthony or autochtony can be
tested further using taphonomic attributes to specify how the scale
of paleoecological reconstruction relates to the scale of the original
ecosystem.

5. Paleobiologists traditionally make the best of the evidence provided by
the fossil record. Often this has involved comparison of samples with
similar preservational histories, as in research on floras preserved in
coal balls. Increased understanding of the types of samples that are
available in particular environmental settings encourages a more rigorous approach that deliberately selects taphonomic modes for specific
types of organic records. For diversity studies, samples could be chosen
from different taphonomic modes to maximize the chances for documenting species from different habitats. For investigations of ecosystem
evolution, samples from similar taphonomic modes can be compared to
minimize the biases inherent in different preservational histories.

6. The evidence presented in this chapter suggests that change in faunas or
floras through a rock sequence may correspond to local shifts in environments or taphonomic modes rather than larger-scale evolutionary
or ecological trends. Establishing the latter, and showing concordance
or discordance in time for ecological trends within different taxonomic
groups, requires sampling of different taphonomic modes over a broad
paleogeographic area, preferably in different depositional basins. Both

the environments themselves and the preserved organic record must be subjected to careful analysis in order to reveal true patterns of change and stasis in past ecosystems.

REFERENCES

Abel, O. 1919. *Die Stämme der Wirbeltiere*. Berlin and Leipzig: Walter de Gruyter.

Agenbroad, L. D. 1984. Hot springs, South Dakota: Entrapment and taphonomy of Columbian mammoth. 113–27 in P. S. Martin and R. G. Klein, eds., *Quaternary Extinctions*. Tucson: University of Arizona Press.

Alexander, J. E. 1956. Bone carrying by a porcupine. *South African Journal of Science* 52:257–58.

Allen, J. R. L. 1970. Sediments of the modern Niger Delta: A summary and review. *Society of Economic Paleontologists and Mineralogists Special Publication* 15: 138–52.

Allen, J. R. L., D. Laurier, and J. Thouvenin. 1979. Etude sédimentologique du Delta de la Mahakam. *Compagnie Française des Pétroles, Notes et Mémoires* 15:1–156.

Allison, P. A. 1986. Soft-bodied animals in the fossil record: The role of decay in fragmentation during transport. *Geology* 14:979–81.

———. 1988. Taphonomy of the Eocene London Clay biota. *Palaeontology* 31:1079–1100.

Alvin, K. L., C. J. Fraser, and R. A. Spicer. 1981. Anatomy and palaeoecology of *Pseudofrenelopsis* and associated conifers in the English Wealden. *Palaeontology* 24:759–78.

Anders, M. H., S. W. Krueger, and P. M. Sadler. 1987. A new look at sedimentation rates and the completeness of the stratigraphic record. *Journal of Geology* 95:1–14.

Anderson, J. A. R. 1964a. The structure and development of the peat swamps of Sarawak and Brunei. *Journal of Tropical Geography* 18:7–16.

———. 1964b. Observations on climatic change in peat swamp forests in Sarawak. *Empire Forestry Review* 43:145–58.

———. 1983. The tropical peat swamps of western Malaysia. 181–99 in A. J. P. Gore, ed., *Ecosystems of the World, 4B, Mires: Swamp, Bog, Fen, and Moor*. Regional Studies. Amsterdam: Elsevier Scientific Publications.

Anderson, J. A. R., and J. Muller. 1975. Palynological study of a Holocene peat and a Miocene coal deposit from NW Borneo. *Review of Palaeobotany and Palynology* 19:291–351.

Anderson, R. Y., and W. E. Dean. 1988. Lacustrine varve formation through time. *Palaeogeography, Palaeoclimatology, Palaeoecology* 62:215–37.

Anderson, T. F., M. E. Brownlee, and T. L. Phillips. 1980. A stable isotope study on the origin of permineralized peat zones in the Herrin Coal. *Journal of Geology* 88:713–22.

Andrews, P. 1990. *Owls, Caves, and Fossils*. London: British Museum (Natural History).

Andrews, P., and J. Cooke. 1985. Natural modifications to bones in a temperate setting. *Man* 20:675–91.

Andrews, P., and E. M. Nesbit Evans. 1979. The environment of *Ramapithecus* in Africa. *Paleobiology* 5:22–30.

————. 1983. Small mammal bone accumulations produced by mammalian carnivores. *Paleobiology* 9:289–307.

Andriesse, J. P. 1974. *Tropical lowland peats in Southeast Asia.* Amsterdam: Koniklijk Instituut voor de Tropen.

Ash, S. R. 1972a. Plant megafossils from the Chinle Formation. *Museum of Northern Arizona Bulletin* 47:23–43.

————. 1972b. Late Triassic plants from the Chinle Formation in northeastern Arizona. *Palaeontology* 15:598–618.

Aslan, A., and A. K. Behrensmeyer. 1987. Vertebrate taphonomy in the East Fork River, Wyoming. *Geological Society of America Abstracts with Program* 19 (7):575.

Azzaroli, A., M. Boccaletti, E. Delson, G. Moratti, and D. Torre. 1986. Chronological and paleogeographical background to the study of *Oreopithecus bambolii. Journal of Human Evolution* 15:533–40.

Badgley, C. E. 1982. How much time is represented in the present? The development of time-averaged modern assemblages as models for the fossil record. *Proceedings of the Third North American Paleontological Convention* 1:23–28.

————. 1986a. Taphonomy of mammalian fossil remains from Siwalik rocks of Pakistan. *Paleobiology* 12:119–42.

————. 1986b. Counting individuals in mammalian fossil assemblages from fluvial environments. *Palaios* 1:328–38.

————. 1990. A statistical assessment of last appearances in the Eocene record of mammals. *Geological Society of America Special Paper* 243:153–68.

Badgley, C. E., and A. K. Behrensmeyer. 1980. Paleoecology of Middle Siwalk sediments and faunas, northern Pakistan. *Palaeogeography, Palaeoclimatology, Palaeoecology* 30:133–55.

Badgley, C. E., and P. D. Gingerich. 1988. Sampling and faunal turnover in early Eocene mammals. *Palaeogeography, Palaeoclimatology, Palaeoecology* 63: 141–58.

Badgley, C. E., Q. Guoqin, C. Wanyon, and H. Defen. 1988. Paleoecology of a Miocene, tropical, upland fauna: Lufeng, China. *National Geographic Research* 4:178–95.

Baganz, B. P., J. C. Horne, and J. C. Ferm. 1975. Carboniferous and Recent Mississippi lower delta plains: Comparison. *Transactions of the Gulf Coast Association of Geological Societies* 25:183–91.

Baird, G. C., C. W. Shabica, J. L. Anderson, and E. S. Richardson. 1985. Biota of a Pennsylvanian muddy coast: Habitats within the Mazonian delta complex, northeast Illinois. *Journal of Paleontology* 59:253–81.

Baird, G. C., S. D. Sroka, C. W. Shabica, and G. J. Kuecher. 1986. Taphonomy of Middle Pennsylvanian Mazon Creek area fossil localities, Northeast Illinois: Significance of exceptional fossil preservation in syngenetic concretions. *Palaios* 1:271–85.

Banks, M. R., J. W. Cosgriff, and N. R. Kemp. 1978. A Tasmanian Triassic stream community. *Australian Natural History* 19:150–57.

Barale, G. 1981. La paléoflore Jurassique du Jura Francais: Etude systématique, aspects stratigraphique et paléoécologiques. *Documents de Laboratoire de Géologie, Lyon* 81:1–467.

Barber, K. E. 1981. *Peat Stratigraphy and Climatic Change.* Rotterdam: A. A. Balkema.

Baroni-Urbani, C. and J. B. Saunders. 1980. The fauna of the Dominican Republic amber: The present status of knowledge. *Transactions of the Ninth Caribbean Geological Conference (1980), Santo Domingo, Dominican Republic,* vol. 1:213–23.

Barthel, K. W. 1970. On the deposition of the Solnhofen lithographic limestone (Lower Tithonian, Bavaria, Germany). *Neues Jahrbuch für Geologie und Paläontologie, Abhandlungen* 135:1–18.

———. 1978. *Solnhofen: Ein Blick in die Erdgeschichte.* Thun, Switzerland: Ott-Verlag.

Bartram, K. M. 1987. Lycopod succession in coals: An example from the Low Barnsley Seam (Westphalian B), Yorkshire, England. In. A. C. Scott, *Coal and Coal-Bearing Strata: Recent Advances.* Geological Society Special Publication 32:187–99.

Bartram, K. M., A. J. Jeram, and P. A. Selden. 1987. Arthropod cuticles in coal. *Journal of the Geological Society of London* 144:513–17.

Bar-Yosef, O., and M. E. Kislev. 1986. Earliest domesticated barley in the Jordan Valley. *National Geographic Research* 2:257.

Bass-Becking, L. G. M., I. R. Kaplan, and D. Moore. 1960. Limits of the natural environment in terms of pH and oxidation-reduction potential. *Journal of Geology* 68:243–84.

Baxendale, L. W. 1979. Plant-bearing coprolites from North American Pennsylvanian coal balls. *Palaeontology* 22:537–48.

Beadle, L. C. 1981. *The Inland Waters of Tropical Africa.* 2nd ed. London: Longman.

Beck, C. B., K. Coy, and R. Schmid. 1982. Observations on the fine structure of *Callixylon* wood. *American Journal of Botany* 69:54–76.

Behrensmeyer, A. K. 1975a. The taphonomy and paleoecology of Plio-Pleistocene vertebrate assemblages east of Lake Rudolf, Kenya. *Bulletin of the Museum of Comparative Zoology* 146:473–578.

———. 1975b. Taphonomy and paleoecology in the hominid fossil record. *Yearbook of Physical Anthropology* 19:36–50.

———. 1978. Taphonomic and ecological information from bone weathering. *Paleobiology* 4:150–62.

———. 1982. Time resolution in fluvial vertebrate assemblages. *Paleobiology* 8:211–27.

———. 1983. Patterns of natural bone distribution on recent land surfaces: Implications for archeological site formation. *British Archeological Series* 163:93–106.

———. 1985. Taphonomy and the paleoecologic reconstruction of hominid habitats in the Koobi Fora Formation. 309–24 in Y. Coppens, ed., *L'environment des Hominidés au Plio-Pléistocène.* Paris: Foundation Singer-Polignac/Masson.

———. 1987a. Taphonomy and Hunting. 423–50 in M. H. Nitecki and D. V. Nitecki, eds., *The Evolution of Human Hunting.* New York: Plenum Press.

———. 1987b. Miocene fluvial facies and vertebrate taphonomy in Northern Pakistan. *Society of Economic Paleontologists and Mineralogists Special Publication* 39:169–76.

———. 1988. Vertebrate preservation in fluvial channels. *Palaeogeography, Palaeoclimatology, Palaeoecology* 63:183–99.

———. 1991. Terrestrial vertebrate accumulations. 291–335 in P. A. Allison and D. E. G. Briggs, eds., *Taphonomy: Releasing the Data Locked in the Fossil Record.* New York: Plenum.

Behrensmeyer, A. K., and D. E. Dechant Boaz. 1980. The Recent bones of Amboseli Park, Kenya, in relation to East African paleoecology. 72–92 in A. K. Behrensmeyer and A. P. Hill, eds., *Fossils in the Making.* Chicago: University of Chicago Press.

Behrensmeyer, A. K., and S. M. Kidwell. 1985. Taphonomy's contributions to paleobiology. *Paleobiology* 11:105–19.

Béland, P., and D. A. Russell. 1978. Paleoecology of Dinosaur Provincial Park, Alberta, interpreted from the distribution of articulated vertebrate remains. *Canadian Journal of Earth Sciences* 15:1012–24.

Bell, D. T. 1980. Gradient trends in the streamside forest of central Illinois. *Bulletin of the Torrey Botanical Club* 107:172–80.

Bell, M.A., and T. R. Haglund. 1982. Fine-scale temporal variation of the Miocene stickleback *Gasterosteus doryssus. Paleobiology* 8:282–92.

Bell, M.A., M. S. Sadagursky, and J. V. Baumgartner. 1987. Utility of lacustrine deposits for the study of variation within fossil samples. *Palaios* 2:455–66.

Benner, R., M. A. Moran, and R. E. Hodson. 1985. Effects of pH and plant source on lignocellulose biodegradation rates in two wetland ecosystems, the Okefenokee Swamp and a Georgia salt marsh. *Limnology and Oceanography* 30:489–99.

Benton, M. J., and A. D. Walker. 1985. Palaeoecology, taphonomy, and dating of Permo-Triassic reptiles from Elgin, north-east Scotland. *Palaeontology* 28:207–34.

Berman, D. S, D. A. Eberth, and D. B. Brinkman. 1988. *Stegotretus agyrus:* A new genus and species of microsaur (amphibian) from the Permo-Pennsylvanian of New Mexico. *Annals of the Carnegie Museum* 57:293–323.

Berman, D. S, R. R. Reisz, and D. A. Eberth. 1985. *Ecolsonia cutlerensis,* an Early Permian dissorophid amphibian from the Cutler Formation of north-central New Mexico. *New Mexico Bureau of Mines and Mineral Resources Circular* 191:1–31.

Berry, C. W. 1906. Leaf rafts of fossil leaves. *Torreya* 6:246–49.

Besly, B. M., and C. R. Fielding. 1989. Palaeosols in Westphalian coal-bearing and red-bed sequences, central and northern England. *Palaeogeography, Palaeoclimatology, Palaeoecology* 70:303–30.

Besly, B. M., and P. Turner. 1983. Origin of red beds in a moist tropical climate (Eturia Formation, Upper Carboniferous, UK). *Geological Society of London Special Publication* 11:131–47.

Bickart, K. J. 1984. A field experiment in avian taphonomy. *Journal of Vertebrate Paleontology* 4:525–35.

Binford, L. R. 1981. *Bones: Ancient Men and Modern Myths.* New York: Academic Press.

Bird, R. 1944. Did *Brontosaurus* ever walk on land? *Natural History* 53:61–67, 151–62. Reprint, 1983.

Birks, H. J. B. 1973. *Past and Present Vegetation of the Isle of Skye: A Paleoecological Study.* Cambridge: Cambridge University Press.

Birks, H. J. B., and H. H. Birks. 1980. *Quaternary Palaeoecology.* Baltimore: University Park Press.

Bishop, W. W. 1968. The evolution of fossil environments in East Africa. *Transactions of the Leicester Literary and Philosophical Society* 62:22–44.

Blumenschine, R. J. 1986. Carcass consumption sequences and the archaeological distinction of scavenging and hunting. *Journal of Human Evolution* 15:639–59.

————. 1989. A landscape taphonomic model of the scale of prehistoric scavenging opportunities. *Journal of Human Evolution* 18:345–71.

Boaz, D. D. 1982. Modern Riverine Taphonomy: Its Relevance to the Interpretation of Plio-Pleistocene Hominid Paleoecology in the Omo Basin. Ph.D. diss., University of California, Berkeley.

————. 1987. Taphonomy and paleoecology at the Pliocene site of Sahabi, Libya. 337–48 in N. T. Boaz, A. El-Arnauti, A. W. Gaziry, J. DeHeinzelin, and D. D. Boaz, eds., *Neogene Paleontology and Geology of Sahabi*. New York: Alan R. Liss.

Boekschoten, G. J., and P. Y. Sondaar. 1972. On the fossil Mammalia of Cyprus. *Nederlandse Akademie van Wetenschappen Proceedings B* 75:306–38.

Bolt, J. R. 1977a. *Cacops* (Amphibia: Labyrinthodontia) from the Fort Sill locality, Lower Permian of Oklahoma. *Fieldiana, Geology* 37:61–73.

————. 1977b. Dissorophoid relationships and ontogeny, and the origin of the Lissamphibia. *Journal of Paleontology* 51:235–49.

————. 1980. New tetrapods with bicuspid teeth from the Fort Sill locality (Lower Permian, Oklahoma). *Neues Jahrbuch für Geologie und Paläontologie Monatshefte* 1980:449–59.

Bolt, J. R., R. M. McKay, B. J. Witzke, and M. P. McAdams. 1988. A new Lower Carboniferous tetrapod locality in Iowa. *Nature* 333:768–70.

Borrero, L. A. 1986. Cazadores de *Mylodon* en la Patagonia Austral. 281–94 in A. L. Bryan, ed., *New evidence for the Pleistocene Peopling of the Americas*. Orono: Center for the Study of Early Man, University of Maine.

Bown, T. M. and K. C. Beard. 1990. Systematic lateral variation in the distribution of fossil mammals in alluvial paleosols, Lower Eocene Willwood Formation, Wyoming. *Geological Society of America Special Paper* 243:135–51.

Bown, T. M., and M. J. Kraus. 1981a. Lower Eocene alluvial paleosols (Willwood Formation, northwest Wyoming, USA) and their significance for paleoecology, paleoclimatology, and basin analysis. *Palaeogeography, Palaeoclimatology, Palaeoecology* 34:1–30.

————. 1981b. Vertebrate fossil-bearing paleosol units (Willwood Formation, Lower Eocene, northwest Wyoming, USA): Implications for taphonomy, biostratigraphy, and assemblage analysis. *Palaeogeography, Palaeoclimatology, Palaeoecology* 34: 31–56.

————. 1983. Ichnofossils of the alluvial Willwood Formation (Lower Eocene), Bighorn Basin, northwest Wyoming. *Palaeogeography, Palaeoclimatology, Palaeoecology* 43:95–128.

————. 1987. Integration of channel and floodplain suites, I: Developmental sequence and lateral relations of alluvial paleosols. *Journal of Sedimentary Petrology* 57:587–601.

————. 1988. Geology and paleoenvironment of the Oligocene Jebel Qatrani Formation and adjacent rocks, Fayum Depression, Egypt. *United States Geological Survey Professional Paper* 1452:1–60.

Bown, T. M., M. J. Kraus, S. L. Wing, J. G. Fleagle, B. H. Tiffney, E. L. Simons, and C. F. Vondra. 1982. The Fayum primate forest revisited. *Journal of Human Evolution* 11:603–32.

Boy, J. A. 1977. Typen und Genese jungpaläozoischer Tetrapoden-Lagerstaetten. *Palaeontographica, A* 156:111–67.

————. 1987. Die Tetrapoden-Lokalitäten des saarpfälzischen Rotliegenden (?Ober-Karbon-Unter-Perm; SW-Deutschland) und die Biostratigraphie der Rotliegend-Tetrapoden. *Mainzer Geowissenschaftliche Mitteilungen* 16:31–65.

Boy, J. A., and C. Hartkopf. 1983. Paläontologie des saarpfälzischen Rotliegenden. *Exkursion C, Paläontologische Gesellschaft* 53:1–85.

Bradley, W. H. 1948. Limnology and the Eocene lakes of the Rocky Mountain Region. *Geological Society America Bulletin* 59:635–48.

————. 1964. Geology of the Green River Formation and associated Eocene rocks in southwestern Wyoming and adjacent parts of Colorado and Utah. *United States Geological Survey Professional Paper* 496-A:1–86.

Brain, C. K. 1958. *The Transvaal Ape-Man-Bearing Cave Deposits.* Transvaal Museum Memoir 11:1–131.

————. 1967. Hottentot food remains and their bearing on the interpretation of fossil bone assemblages. *Scientific Papers of the Namid Desert Research Station* 32:1–11.

————. 1980. Some criteria for the recognition of bone-collecting agencies in African caves. 108–30 in A. K. Behrensmeyer and A. P. Hill, eds., *Fossils in the Making.* Chicago: University of Chicago Press.

————. 1981. *The Hunters or the Hunted?* Chicago: University of Chicago Press.

Breed, W. J., B. T. Foster, and P. Lunge. 1974. Paleozoic fossils of the Grand Canyon. 65–75 in W. J. Breed and E. C. Roat, eds., *Geology of the Grand Canyon.* Flagstaff: Museum of Northern Arizona.

Breithhaupt, B. H., and D. Duvall. 1986. The oldest record of serpent aggregation. *Lethaia* 19:181–85.

Brenner, K. 1976. Biostratinomische Untersuchungen im Posidonienschiefer (Lias epsilon, Unteres Toarcium) von Holzmaden (Württemberg, Süd-Deutschland). *Zentralblatt für Geologie und Paläontologie.* Teil 2, 1976:223–26.

Bridge, J. S. 1984. Large-scale facies sequences in alluvial overbank environments. *Journal of Sedimentary Petrology* 54:583–88.

————. 1985. Paleochannel patterns inferred from alluvial deposits: A critical evaluation. *Journal of Sedimentary Petrology* 55:579–89.

Bridge, J. S., and J. A. Diemer. 1983. Quantitative interpretation of an evolving ancient river system. *Sedimentology* 30:599–623.

Bridge, J. S., and E. A. Gordon. 1985. Quantitative reconstructions of ancient river systems in the Oneonta Formation, Catskill Magnafacies. *Geological Society of America Special Paper* 201:143–61.

Bridge, J. S., E. A. Gordon, and R. C. Titus. 1986. Non-marine bivalves and associated burrows in the Catskill magnafacies (Upper Devonian) of New York State. *Palaeogeography, Palaeoclimatology, Palaeoecology* 55:65–77.

Bridge, J. S., P. M. Van Veen, and L. C. Matten. 1980. Aspects of the sedimentology, palynology, and paleobotany of the Upper Devonian of southern Kerry Head, Co. Kerry, Ireland. *Geological Journal* 15:143–70.

Briggs, D. J., D. D. Gilbertson, and A. L. Harris. 1990. Molluscan taphonomy in a braided river environment and its implications for studies of Quaternary cold-stage river deposits. *Journal of Biogeography* 17:623–37.

Brouwers, E. M., W. A. Clemens, R. A. Spicer, T. Ager, D. L. Carter, and W. V. Sliter. 1987. Dinosaurs on the North Slope, Alaska: Reconstructions of high latitude, latest Cretaceous environments. *Science* 237:1608–10.

Brown, B., and E. M. Schlaikjer. 1940. The structure and relationships of *Protoceratops*. *Annals of the New York Academy of Science* 40:133–266.

Brues, C. T. 1933. Progressive changes in the insect population of forests since the early Tertiary. *The American Naturalist* 67:383–406.

Brunet, M., Y. Coppens, D. Pilbeam, D. Djallo, A. K. Behrensmeyer, A. Brillanceau, W. R. Downs, M. Duperon, G. Ekodeck, L. Flynn, E. Heinz, J. Hell, Y. Jehenne, L. Martin, C. Mosser, M. Salard-Cheboldaeff, S. Wenz, and S. L. Wing. 1986. Les formations sédimentaires continentales du Crétacé et du Cénozoïque Camerounais: Premiers resultats d'une prospection paléontologique. *Comptes Rendus d'Academie des Sciences Paris,* Serie 2 303:425–28.

Brzyski, B., R. Gradzinski, and R. Krzanowska. 1976. Stodace pnie kalamitow w odslonieciu cegielni Brynow i warunki ich pogrzebania (Upright calamite stems from Brynow and conditions of their burial). *Annales de la Société Géologique de Pologne* 46:159–82.

Buchheim, H. P., and R. C. Surdam. 1977. Fossil catfish and the depositional environment of the Green River Formation, Wyoming. *Geology* 5:196–98.

Bunn, H. T. 1982. Meat-eating and Human Evolution: Studies on the Diet and Subsistence Patterns of Plio-Pleistocene Hominids in East Africa. Ph.D. diss., University of California, Berkeley.

———. 1983. Comparative analysis of modern bone assemblages from a !San hunter-gatherer camp in the Kalahari Desert, Botswana, and from a spotted hyena den near Nairobi, Kenya. *British Archaeological Series* 163:143–48.

———. 1986. Patterns of skeletal representation and hominid subsistence activities at Olduvai Gorge, Tanzania, and Koobi Fora, Kenya. *Journal of Human Evolution* 15:673–90.

Bunn, H. T., J. W. K. Harris, G. Ll. Isaac, Z. Kaufulu, E. Kroll, K. Schick, N. Toth, and A. K. Behrensmeyer. 1980. FxJj50: An Early Pleistocene site in northern Kenya. *World Archaeology* 12:109–44.

Burgis, M. J., and P. Morris. 1987. *The Natural History of Lakes.* Cambridge: Cambridge University Press.

Burnham, R. J. 1989. Relationship between standing vegetation and leaf litter in a paratropical forest: Implications for paleobotany. *Reviews of Palaeobotany and Palynology* 58:5–32.

Burnham, R. J., and R. A. Spicer. 1986. Forest litter preserved by volcanic activity at El Chichon, Mexico: Potentially accurate record of the pre-eruption vegetation. *Palaios* 1:158–61.

Calder, J. In press. The evolution of a groundwater-influenced (Westphalian B) peat-forming ecosystem in a piedmont setting: The No. 3 Seam, Springhill Coal Field, Cumberland Basin, Nova Scotia. *Geological Society of America Special Paper.*

Campbell, K. E. 1976. The late Pleistocene avifauna of La Carolina, Ecuador. *Smithsonian Contributions to Paleobiology* 27:155–68.

———. 1979. The non-passerine Pleistocene avifauna of the Talara Tar Seeps, northwestern Peru. *Life Science Contribution Royal Ontario Museum* 118:1–203.

Carlson, K. J. 1968. The skull morphology and estivation burrows of the Permian lungfish, *Gnathorhiza serrata. Journal of Geology* 76:641–63.

Carpenter, K. 1982. Baby dinosaurs from the Late Cretaceous Lance and Hell Creek formations and a description of a new species of theropod. *University of Wyoming Contributions to Geology* 20:123–34.

Carroll, R. L. 1965. Lungfish burrows from the Michigan coal basin. *Science* 148:963–64.

———. 1967. Labyrinthodonts from the Joggins Formation. *Journal of Paleontology* 41:111–42.

———. 1981. Plesiosaur ancestors from the Upper Permian of Madagascar. *Philosophical Transactions of the Royal Society of London, B* 293:315–83.

Carroll, R. L., E. S. Belt, D. L. Dineley, D. Baird, and D. C. McGregor. 1972. Vertebrate paleontology of eastern Canada. *Proceedings Twenty-fourth International Geological Congress, Field Guide for Excursion* A59:1–113.

Cavaroc, V. V., Jr., and R. S. Saxena. 1979. Crevasse splay deposits in the state line area of northwestern Pennsylvania of northeastern Ohio. 160–72 in J. C. Ferm and J. C. Horne, eds., *Carboniferous Depositional Environments in the Appalachian Region*. Columbia, S.C.: Carolina Coal Group.

Caviglia, S. E., H. D. Yacobaccio, and L. A. Borrero. 1986. Las Buitreras: Convivencia del Hombre con Fauna Extincta en Patagonia Meridional. 295–313 in A. L. Bryan, ed., *New evidence for the Pleistocene Peopling of the Americas*. Orono: Center for the Study of Early Man, University of Maine.

Cecil, C. B. 1990. Paleoclimate controls on stratigraphic repetition of chemical and siliciclastic rocks. *Geology* 18:533–36.

Cecil, C. B., R. W. Stanton, S. G. Neuzil, F. T. Dulong, L. F. Ruppert, and B. S. Pierce. 1985. Paleoclimate controls on late Paleozoic sedimentation and peat formation in the central Appalachian basin (USA). *International Journal of Coal Geology* 5:195–230.

Chaloner, W. G. 1985. Problems with permineralization of peat. *Philosophical Transactions of the Royal Society of London, B* 311:139–41.

Chaney, D. S. 1988. The Jacona microfauna (Late Barstovian–Early Clarendonian) Pojoaque Member, Tesuque Formation, North Central New Mexico. Part 1. Geology, taphonomy, paleontology, Insectivora. M.Sc. thesis, University of California, Riverside.

Churcher, C. S. 1959. Fossil *Canis* from the tarpits of La Brea, Peru. *Science* 130:564–65.

———. 1965. Camelid material of the genus *Palaeolama* Gervais from the Talara tar seeps, Peru, with a description of a new subgenus *Astylolama*. *Proceedings of the Zoological Society of London* 145:161–205.

Clark, J., J. R. Beerbower, and K. K. Kietzke. 1967. Oligocene sedimentation, stratigraphy, paleoecology, and paleoclimatology in the Big Badlands of South Dakota. *Fieldiana, Geology Memoirs* 5:1–158.

Clemens, W. A. 1964. Fossil mammals of the type Lance Formation, Wyoming. Part 1: Introduction and Multituberculata. *University of California Publications in Geological Sciences* 48:1–105.

———. 1973. Fossil mammals of the type Lance Formation, Wyoming. Part 3: Eutheria and summary. *University of California Publications in Geological Sciences* 94:1–102.

Clymo, R. S. 1983. Peat. 159–224 in A. P. J. Gore, ed., *Ecosystems of the World, 4A(B). Mires: Swamp, Bog, Fen and Moor, General and Regional Studies*. Amsterdam: Elsevier.

———. 1984. The limits of peat bog growth. *Philosophical Transactions of the Royal Society of London, B* 303:605–54.

————. 1987. Rainwater-fed peat as a precursor of coal. 17–23 in A. C. Scott, ed., *Coal and Coal-Bearing Strata: Recent Advances*. Geological Society Special Publication no. 32.

Cohen, A. D. 1974. Petrography and paleoecology of Holocene peats from the Okefenokee swamp-marsh complex of Georgia. *Journal of Sedimentary Petrology* 44:716–26.

Cohen, A. D., R. R. Raymond, Jr., A. Ramirez, C. Morales, and F. Ponce. 1990. The Chanquinola peat deposit of northwestern Panama: A tropical, domed, back-barrier coal-forming environment. *International Journal of Coal Geology* 16:139–42.

Cohen, A. D., W. Spackman, and R. R. Raymond, Jr. 1987. Interpreting characteristics of coal seams from chemical, physical, and petrographic studies of peat deposits. 107–25 in A. C. Scott, ed., *Coal and Coal-Bearing Strata: Recent Advances*. Geological Society Special Publication no. 32.

Cohen, A. S. 1981. Paleolimnological research at Lake Turkana, Kenya. 61–82 in J. A. Coetzee and E. M. Van Zinderen Bakker, eds., *Palaeoecology of Africa*, vol. 13. Rotterdam: A. A. Balkema.

————. 1986. Distributions and faunal associations of benthic invertebrates at Lake Turkana, Kenya. *Hydrobiologia* 141:179–97.

————. 1987. Fossil ostracodes from Lake Mobutu (Lake Albert): Paleoecological and taphonomic implications. 271–81 in J. A. Coetzee and E. M. Van Zinderen Bakker, eds., *Palaeoecology of Africa*, vol. 13. Rotterdam: A. A. Balkema.

————. 1989. The taphonomy of gastropod shell accumulations in large lakes: An example from Lake Tanganyika, Africa. *Paleobiology* 15:26–45.

Colbert, E. H. 1968. *Men and Dinosaurs: The Search in Field and Laboratory*. New York: E. P. Dutton.

Cole, K. L. 1986. The Lower Colorado Valley: A Pleistocene desert. *Quaternary Research* 25:392–400.

Collinson, M. E. 1983. Palaeofloristic assemblages and palaeoecology of the Lower Oligocene Bembridge Marls, Hamstead Ledge, Isle of Wight. *Botanical Journal of the Linnean Society* 86:177–225.

Collinson, M. E., and J. J. Hooker. 1987. Vegetational and mammalian faunal changes in the early Tertiary of southern England. 259–304 in E. M. Friis, W. G. Challoner, and P. R. Crane, eds., *The Origins of Angiosperms and Their Biological Consequences*. Cambridge: Cambridge University Press.

Conwentz, H. 1886. *Die Flora des Bernsteins*. Vol. 2, *Die Angiospermen des Bernsteins*. Leipzig: Wilhelm Engelmann.

Coope, G. R. 1979. Late Cenozoic fossil Coleoptera: Evolution, biogeography, and ecology. *Annual Review of Ecology and Systematics* 10:247–68.

Cope, M. J., and W. G. Chaloner. 1985. Wildfire: An interaction of biological and physical processes. 257–77 in B. H. Tiffney, ed., *Geological Factors and the Evolution of Plants*. New Haven: Yale University Press.

Crane, P. R., and R. A. Stockey. 1985. Growth and reproductive biology of *Joffrea spearsii* gen. et sp. nov., a *Cercidiphyllum*-like plant from the Late Paleocene of Alberta, Canada. *Canadian Journal of Botany* 63:340–64.

Crepet, W. L., and C. P. Dughlian. 1982. Euphorbioid inflorescences from the Middle Eocene Claiborne Formation. *American Journal of Botany* 69:258–66.

Crepet, W. L., D. L. Dilcher, and F. W. Potter. 1974. Eocene angiosperm flowers. *Science* 185:781–82.

Crisman, T. L. 1978. Reconstruction of past lacustrine environments based on the remains of aquatic invertebrates. 69–102 in D. Walker and J. C. Guppy, eds., *Biology and Quaternary Environments*. Canberra: Australian Academy of Science.

Cross, A. T., R. E. Taggart, A. Jameossanaie, and K. C. Kelley. 1988. Reconstruction of a fossil forest, Menefee Formation, Late Cretaceous, New Mexico. *American Journal of Botany* 75 (6, pt. 2): 106.

Cross, A. T., G. C. Thompson, and J. B. Zaitzeff. 1966. Source and distribution of palynomorphs in bottom sediments, southern part of Gulf of California. *Marine Geology* 4: 467–524.

Cunha, M., and J. Moura. 1979. Especies novas de ostracodes Nao-Marinhos da serie do Reconcavo: Paleontologia e biostratigrafia. *Bollettino Technico Petrobras* 22: 87–100.

Currie, P. J. 1984. Mass death of a herd of ceratopsian dinosaurs. 61–66 in W. E. Reif and F. Westphal, eds., *Third Symposium on Mesozoic Terrestrial Ecosystems, Short Papers*. Tübingen: Attempto Verlag.

Dalrymple, R. W., and Y. Makino. 1989. Description and genesis of tidal bedding in the Cobequid Bay: Salmon River estuary, Bay of Fundy, Canada. 151–77 in A. Taira and F. Masuda, eds., *Sedimentary Facies in the Active Plate Margins*. Tokyo: Terra Scientific Publishing.

Darrell, J. H. 1973. Statistical evaluation of palynomorph distribution in the sedimentary environments of the modern Mississippi River Delta. Ph.D. diss., Louisiana State University, Baton Rouge.

Darrell, J. H., and G. F. Hart. 1970. Environmental determinations using absolute miospore frequency, Mississippi River Delta. *Geological Society of America Bulletin* 81: 2513–18.

Davis, M. B., R. E. Moeller, and J. Ford. 1984. Sediment focusing and pollen influx. 261–93 in E. Y. Haworth and J. W. G. Lund, eds., *Lake Sediments and Environmental History*. Leicester, England: Leicester University Press.

Davis, M. B., and T. Webb III. 1975. The contemporary distribution of pollen from eastern North America: A comparison with vegetation. *Quaternary Research* 5: 395–434.

Davis, O. K., L. D. Agenbroad, P. S. Martin, and J. I. Mead. 1984. The Pleistocene Dung Blanket of Bechan Cave, Utah. *Carnegie Museum of Natural History Special Publication* 8: 267–82.

Davis, R. J., ed. 1985. *Coastal Sedimentary Environments*. 2d ed. New York: Springer-Verlag.

DeBuisonje, P. 1985. Climatological conditions during deposition of the Solnhofen Limestones. 45–65 in M. K. Hecht, J. H. Ostrom, G. Viohl, and P. Wellnhofer, eds., *The Beginning of Birds*. Eichstätt: Jura–Museum.

DeHeinzelin, J., ed. 1983. The Omo Group: Archives of the International Omo Research Expedition. *Annales Musée Royal de l'Afrique Centrale Tervuren, Sciences Géologiques* 85: 1–365.

DeHeinzelin, J., and A. El-Arnauti. 1987. The Sahabi Formation and related deposits. 1–22 in N. T. Boaz, A. El-Arnauti, A. W. Gaziry, J. DeHeinzelin, and D. D. Boaz, eds., *Neogene Paleontology and Geology of Sahabi*. New York: Alan R. Liss.

DeHeinzelin, J., P. Haesaerts, and F. C. Howell. 1976. Plio-Pleistocene formations of the Lower Omo Basin, with particular reference to the Shungura Formation. 24–49 in Y. Coppens, F. C. Howell, G. L. Isaac, and R. E. Leakey, eds., *Earliest Man and Environments in the Lake Rudolf Basin: Stratigraphy, Paleoecology, and Evolution.* Chicago: University of Chicago Press.

Delcourt, P. A., and H. R. Delcourt. 1987. *Long-Term Dynamics of the Temperate Zone.* New York: Springer-Verlag.

DeMaris, P. J., R. A. Bauer, R. A. Cahill, and H. H. Damberger. 1983. Geologic investigation of roof and floor strata: Longwall demonstration, Old Ben mine no. 24—prediction of coal balls in the Herrin Coal. *Illinois State Geological Survey, Contract/Grant Report* 1983-2:1–69.

DeMicco, R. V., J. S. Bridge, and K. C. Cloyd. 1987. A unique freshwater carbonate from the Upper Devonian Catskill magnafacies of New York State. *Journal of Sedimentary Petrology* 57:327–34.

Diessel, C. F. K. 1985. Tuffs and tonsteins in the coal measures of New South Wales, Australia. *Dixième Congrès International de Stratigraphie et de Géologique du Carbonifère* Compte Rendu 4:197–210.

Dietrich, H. G. 1975. Zur Entstehung und Erhaltung von Bernstein-Lagerstätten. *Neues Jahrbuch für Geologie und Paläontologie Abhandlungen* 144:39–72.

———. 1976. Bernstein-Lagerstätten. *Zentralblatt für Geologie und Paläontologie.* Teil 2, 1976:261–67.

DiMichele, W. A., and P. J. DeMaris. 1987. Structure and dynamics of a Pennsylvanian–age *Lepidodendron* forest: Colonizers of a disturbed swamp habitat in the Herrin (no. 6) coal of Illinois. *Palaios* 2:146–57.

DiMichele, W. A., and T. L. Phillips. 1988. Paleoecology of the Middle Pennsylvanian–age Herrin coal swamp (Illinois) near a contemporaneous river system, the Walshville paleochannel. *Review of Palaeobotany and Palynology* 56:151–76.

Dittus, W. P. J. 1985. The influence of cyclones on the dry evergreen forest of Sri Lanka. *Biotropica* 17:1–14.

Dodson, P. 1971. Sedimentology and taphonomy of the Oldman Formation (Campanian), Dinosaur Provincial Park, Alberta (Canada). *Palaeogeography, Palaeoclimatology, Palaeoecology* 10:21–74.

———. 1973. The significance of small bones in paleoecological interpretation. *University of Wyoming Contributions to Geology* 12:15–19.

———. 1983. A faunal review of the Judith River (Oldman) Formation, Dinosaur Provincial Park, Alberta. *Mosasaur* 1:89–118.

———. 1987. Microfaunal studies of dinosaur paleoecology, Judith River Formation of southern Alberta. *Occasional Paper of the Tyrrell Museum of Palaeontology* 3:70–75.

Dodson, P., and D. Wexlar. 1979. Taphonomic investigations of owl pellets. *Paleobiology* 5:275–84.

Dodson, P., A. K. Behrensmeyer, R. T. Bakker, and J. S. McIntosh. 1980a. Taphonomy of the Upper Jurassic Morrison Formation. *Paleobiology* 6:208–32.

Dodson, P., A. K. Behrensmeyer, and R. T. Bakker. 1980b. Taphonomy of the Morrison Formation (Kimmeridgian-Portlandian) and Cloverly Formation (Aptian-Albian) of the western United States. *Mémoires de la Société Géologique de France* n. s. 139:87–93.

Dodson, P., R. T. Bakker, and A. K. Behrensmeyer. 1983. Paleoecology of the dinosaur-bearing Morrison Formation. *National Geographic Society Research Reports* 15:145–56.

Donovan, R. N. 1975. Devonian lacustrine limestone at the margin of the Orcadian

Basin, Scotland. *Quarterly Journal of the Geological Society of London* 131: 489–510.

———. 1980. Lacustrine cycles, fish ecology, and stratigraphic zonation in mid-Devonian of Caithness (Orkney Islands). *Scottish Journal of Geology* 16:35–50.

Driese, S. G. 1985. Interdune pond carbonates, Weber sandstone (Pennsylvanian-Permian), northern Utah and Colorado. *Journal of Sedimentary Petrology* 55:187–95.

Dubiel, R. F., R. H. Blodgett, and T. M. Bown. 1987. Lungfish burrows in the Upper Triassic Chinle and Dolores Formations, Colorado Plateau. *Journal of Sedimentary Petrology* 57:512–21.

———. 1988. Lungfish burrows in the Upper Triassic Chinle and Dolores Formations, Colorado Plateau: Reply. *Journal of Sedimentary Petrology* 58:367–69.

Dubois, R. N. 1990. Barrier-beach erosion and rising sea level. *Geology* 18: 1150–52.

During, E. 1986. The fauna of Alvastra. *Ossa* 12, Supplement 1:1–210.

Eberth, D. A. 1990. Stratigraphy and sedimentology of vertebrate microfossil localities in uppermost Judith River Formation (Campanian) of Dinosaur Provincial Park. *Palaeogeography, Palaeoclimatology, Palaeoecology* 78:1–36.

Eble, C. F., and W. C. Grady. 1990. Paleoecological interpretation of a Middle Pennsylvanian coal bed from the central Appalachian basin, USA. *International Journal of Coal Geology* 16:255–86.

Edwards, D., U. Fanning, and J. B. Richardson. 1986. Stomata and sterome in early land plants. *Nature* 323:438–40.

Efremov, I. A. 1953. Taphonomie et Annales Géologiques. *Annales du Centre d'Etude et de Documentation Paléontologiques* 4:1–164. Translation of Russian by S. Ketchian and J. Roger.

Elder, R. L. 1985. Principles of aquatic taphonomy with examples from the fossil record. Ph.D. diss., University of Michigan, Ann Arbor.

Elder, R. L., and G. R. Smith. 1988. Fish taphonomy and environmental inference in paleolimnology. *Palaeogeography, Palaeoclimatology, Palaeoecology* 62:577–92.

Elliot, R. E. 1965. Swilleys in the Coal Measures of Nottinghamshire interpreted as palaeo-river courses. *Mercian Geologist* 1:133–42.

Emry, R. J. 1990. Mammals of the Bridgerian (Middle Eocene) Elderberry Canyon Local Fauna of eastern Nevada. *Geological Society of America Special Paper* 243: 187–210.

Emry, R. J., L. S. Russell, and P. R. Bjork. 1987. The Chadronian, Orellan, and Whitneyan North American Land Mammal Ages. 118–52 in M. O. Woodburne, ed., *Cenozoic Mammals of North America.* Berkeley: University of California Press.

Esterle, J. S., and J. C. Ferm. 1986. Relationship between petrographic and chemical properties and coal seam geometry, Hance seam, Breathitt Formation, southeastern Kentucky. *International Journal of Coal Geology* 6:199–214.

———. 1987. Spatial distribution of peat types in selected deposits in Indonesia and Malaysia as analogues to ancient coal types. *Geological Society of America Abstracts with Program* 19(7):657.

Esterle, J. S., J. C. Ferm, and T. Yau-Liong. 1989. A test for the analogy of tropical domed peat deposits to "dulling up" sequences in coal beds: Preliminary results. *Journal of Organic Chemistry* 14:333–42.

Estes, R. 1964. Fossil vertebrates from the Late Cretaceous Lance Formation, eastern Wyoming. *University of California Publications in Geological Sciences* 49:1–180.

Estes, R., and P. Berberian. 1970. Paleoecology of a Late Cretaceous vertebrate community. *Breviora* 343:1–35.

Eugster, H. P., and L. A. Hardie. 1975. Sedimentation in an ancient playa-lake complex: The Wilkins Peak Member of the Green River Formation of Wyoming. *Geological Society of America Bulletin* 86:319–34.

Evans, J. G. 1972. *Land Snails in Archaeology.* London: Seminar Press.

Farley, M. B. 1990. Vegetation distribution across the Early Eocene depositional landscape from palynological analysis. *Palaeogeography, Palaeoclimatology, Palaeoecology* 79:11–27.

Farlow, J. O. 1984. Dinosaur footprints from Texas. *American Philosophical Society Year Book:* 1–25.

———. 1987. Lower Cretaceous dinosaur tracks, Paluxy River Valley, Texas. *Geological Society of America Field Guide, South Central Section:* 1–70.

Farrell, K. M. 1987. Sedimentology and facies architecture of overbank deposits of the Mississippi River, False River region, Louisiana. *Society of Economic Paleontologists and Mineralogists Special Publication* 39:111–20.

Fastovsky, D. E. 1987. Paleoenvironments of vertebrate-bearing strata during the Cretaceous-Paleogene transition, eastern Montana and western North Dakota. *Palaios* 2:282–95.

Fastovsky, D. E., J. M. Clark, and J. A. Hopson. 1987. Preliminary report of a vertebrate fauna from an unusual paleoenvironmental setting, Huizachal Group, Early or Mid-Jurassic, Tamaulipas, Mexico. *Occasional Paper of the Tyrrell Museum of Palaeontology* 3:82–87.

Fastovsky, D. E., and K. McSweeney. 1987. Paleosols spanning the Cretaceous-Paleogene transition, eastern Montana and western North Dakota. *Geological Society of America Bulletin* 99:66–77.

Feibel, C. S. 1987. Fossil fish nests from the Koobi Fora Formation (Plio-Pleistocene) of northern Kenya. *Journal of Paleontology* 61:130–34.

Ferguson, D. K. 1985. The origin of leaf assemblages: New light on an old problem. *Review of Palaeobotany and Palynology* 46:117–88.

Ferm, J. C., and G. C. Smith. 1981. *A Guide to Cored Rocks in the Pittsburgh Basin.* Lexington and Columbia: University of Kentucky and University of South Carolina.

Ferm, J. C., G. C. Smith, G. A. Weisenfluh, and S. B. DuBois. 1985. *Cored Rocks in the Rocky Mountain and High Plains Coal Fields.* Lexington: University of Kentucky.

Fernandez-Jalvo, Y. In press. Tafonomia de micromamiferos de Gran Dolina, Atapuerca (Burgos). In E. Aguirre, E. Carbonell, and J. M. Bermudez de Castro, eds., *El Hombre fossil de Ibeas y el Pleistoceno de la Sierra de Atapuerca II: Investigaciones Arqueologicas de Castilla y Leon.* Madrid: Consejeria de Cultura y Bienestar Social.

Fielding, C. R. 1984. Upper delta plain lacustrine and fluviolacustrine facies from the Westphalian of the Durham coalfield, NE England. *Sedimentology* 31:547–67.

Filhol, H. 1876. Recherches sur les phosphorites du Quercy. Etude des fossiles qu'on y recontre et spécialement des mammifères. *Annales des Sciences Géologiques* 7:1–220.

———. 1877. Recherches sur les phosphorites du Quercy: Etude des fossiles qu'on y recontre et specialement des mammifères. *Annales des Sciences Geologiques* 8: 1–340.

Fiorillo, A. R. 1984. An introduction to the identification of trample marks. *Current Research in the Pleistocene* 1:47–48.

———. 1987. Significance of juvenile dinosaurs from Careless Creek Quarry (Judith River Formation, Wheatland County, Montana). *Occasional Paper of the Tyrrell Museum of Palaeontology* 3:88–95.

———. 1988a. Taphonomy of Hazard Homestead Quarry (Ogallala Group), Hitchcock County, Nebraska. *University of Wyoming Contributions to Geology* 26: 57–97.

———. 1988b. Aspects of bone modification applied to time resolution in the fossil record: An example from the Miocene of western Nebraska. *Current Research in the Pleistocene* 5:103–9.

———. 1989a. Taphonomy and paleoecology of the Judith River Formation (Late Cretaceous) of south-central Montana. Ph.D. diss., University of Pennsylvania, Philadelphia.

———. 1989b. The vertebrate fauna from the Judith River Formation (Late Cretaceous) of Wheatland and Golden Valley Counties, Montana. *Mosasaur* 4: 127–42.

———. 1991. Taphonomy and depositional setting of Careless Creek Quarry (Judith River Formation), Wheatland County, Montana, USA. *Palaeogeography, Palaeoclimatology, Palaeoecology* 81:281–311.

Fischer, A. G. 1981. Climatic oscillations in the biosphere. 103–31 in M. H. Nitecki, ed., *Biotic Crises in Ecological and Evolutionary Time*. New York: Academic Press.

Fisher, D. C. 1981. Crocodilian scatology, microvertebrate concentrations, and enamel-less teeth. *Paleobiology* 7:262–75.

———. 1984. Taphonomic analysis of Late Pleistocene mastodon occurrences: Evidence of butchery by North American Paleo-Indians. *Paleobiology* 10:338–57.

———. 1987. Mastodont procurement by Paleoindians of the Great Lakes Region: Hunting or scavenging? 309–422 in M. H. Nitecki and D. V. Nitecki, eds., *The Evolution of Human Hunting*. New York: Plenum.

Fisher, J. 1992. Observations on the Late Pleistocene bone assemblage from the Lamb Spring site. 52–82 in D. J. Stanford and J. Day, eds., *Ice Age Hunters of the Rockies*. Denver: Denver Museum of Natural History.

Fisk, H. N. 1944. *Geological Investigation of the Alluvial Valley of the Lower Mississippi River.* Vicksburg: Mississippi River Commission.

———. 1947. *Fine Grained Alluvial Deposits and their Effects on Mississippi River Activity.* Vicksburg: Mississippi River Commission.

Fox, R. C. 1984. *Melaniella timosa,* n. gen. and sp., an unusual mammal from the Paleocene of Alberta, Canada. *Canadian Journal of Earth Sciences* 21:1335–38.

Franzen, J. L. 1985. Exceptional preservation of Eocene vertebrates in the lake deposit of Grube Messel (West Germany). *Philosophical Transactions of the Royal Society of London, B* 311:181–86.

Franzen, J. L., and W. Michaelis, eds. 1988. Der eozaene Messelsee: Eocene Lake Messel. *Courier Forschungsinstitut Senckenberg* 107:1–452.

Franzen, J. L., J. Weber, and M. Wuttke. 1982. Senckenberg-Grabungen in der Grube

Messel bei Darmstadt 3: Ergebnisse 1979–1981. *Courier Forschungsinstitut Senckenberg* 54:1–118.

Fraser, N. C. 1987. Biostratigraphy of north-west European Late Triassic reptile assemblages. *Occasional Paper of the Tyrrell Museum of Palaeontology* 3:101–5.

Frazier, D. E., and A. Osanik. 1961. Point-bar deposits, Old River locksite, Louisiana. *Transactions of the Gulf Coast Geological Societies* 11:121–37.

Frey, R. W., and J. D. Howard. 1986. Mesotidal estuarine sequences: A perspective from the Georgia Bight. *Journal of Sedimentary Petrology* 56:911–24.

Frey, R. W., and S. G. Pemberton. 1986. Vertebrate *Lebensspuren* in intertidal and supratidal environments, Holocene barrier islands, Georgia. *Senckenbergiana Maritima* 18:45–95.

Frey, R. W., M. R. Voorhies, and J. D. Howard. 1975. Estuaries of the Georgia Coast, USA: Sedimentology and biology, 8. Fossil and Recent skeletal remains in Georgia Estuaries. *Senckenbergiana Maritima* 7:257–95.

Fritz, W. J. 1980. Reinterpretation of the depositional environment of the Yellowstone "fossil forests." *Geology* 8:309–13.

———. 1984. Comment on "Yellowstone fossil forests: New evidence for burial in place." *Geology* 12:638–39.

———. 1986. Plant taphonomy in areas of explosive volcanism. *University of Tennessee Department of Geological Sciences Studies in Geology* 15:1–9.

Fritz, W. J., and S. Harrison. 1985. Transported trees from the 1982 Mount Saint Helens sediment flows: Their use as paleocurrent indicators. *Sedimentary Geology* 42:49–64.

Froggatt, P. C., C. J. N. Wilson, and G. P. I. Walker. 1981. Orientation of logs in the Taupo Ignimbrite as an indicator of flow direction and vent position. *Geology* 9: 109–11.

Fulton, I. M. 1987. Genesis of the Warwickshire Thick Coal: A group of long-residence histosols. 201–18 in A. C. Scott, ed., *Coal and Coal-Bearing Strata: Recent Advances.* Geological Society Special Publication 32.

Gagliano, S. M., and P. C. Howard. 1984. The neck cutoff lake cycle along the lower Mississippi River. *Proceedings Conference on Rivers 1983, Waterway, Port, Coastal, and Ocean Division, ASCE, New Orleans:* 147–58.

Gall, J. C. 1983. *Ancient Sedimentary Environments and the Habitats of Living Organisms: Introduction to Paleoecology.* New York: Springer-Verlag.

Gallagher, W. B., D. C. Parris, B. S. Grandstaff, and C. DeTample. 1989. Quaternary mammals from the continental shelf off New Jersey. *Mosasaur* 4: 101–10.

Galloway, W. E., and D. K. Hobday. 1983. *Terrigenous Clastic Depositional Systems: Applications to Petroleum, Coal, and Uranium Exploration.* New York: Springer-Verlag.

Gastaldo, R. A. 1985. Plant accumulating deltaic depositional environments: Mobile Delta, Alabama. *Alabama Geological Survey Reprint Series* 66:1–35.

———. 1986a. Selected aspects of plant taphonomic processes in coastal deltaic regimes. *University of Tennessee Department of Geological Sciences, Studies in Geology* 15:27–43.

———. 1986b. Implications on the paleoecology of autochthonous Carboniferous lycopods in clastic sedimentary environments. *Palaeogeography, Palaeoclimatology, Palaeoecology* 53:191–212.

———. 1987. Confirmation of Carboniferous clastic swamp communities. *Nature* 326:869–71.

———. 1988. Conspectus of phytotaphonomy. *Paleontological Society Special Publication* 3:14–28.

———. 1989a. Preliminary observations on phyto-taphonomic assemblages in a subtropical/temperate Holocene bayhead delta: Mobile Delta, Gulf Coastal Plain, Alabama. *Review of Palaeobotany and Palynology* 58:61–83.

———. 1989b. Processes of incorporation of plant parts in deltaic clastic sedimentary environments: Implications for paleoecological restorations. *Compte Rendu. Eleventh International Carboniferous Congress of Stratigraphy and Geology* (Beijing, China) 3:109–20.

Gastaldo, R. A., D. P. Douglas, and S. M. McCarroll. 1987. Origin, characteristics, and provenance of plant macrodetritus in a Holocene crevasse splay, Mobile Delta, Alabama. *Palaios* 2:229–40.

Gastaldo, R. A., S. C. Bearce, C. W. Degges, R. J. Hunt, M. W. Peebles, and D. L. Violette. 1989a. Biostratinomy of a Holocene oxbow lake: A backswamp to mid-channel transect. *Reviews of Palaeobotany and Palynology* 58:47–59.

Gastaldo, R. A., T. M. Demko, Y. Liu, W. E. Keefer, and S. L. Abston. 1989b. Biostratinomic processes for the development of mud-cast logs in Carboniferous and Holocene swamps. *Palaios* 4:356–65.

Gawne, C. E. 1975. Rodents from the Zia Sand Miocene of New Mexico. *American Museum Novitates* 2586:1–25.

Genoways, H. H., and M. R. Dawson, eds. 1984. *Contributions in Quaternary vertebrate paleontology: Volume in Memorial to John E. Guilday. Carnegie Museum of Natural History Special Publication* 8:1–538.

Gerrard, J., ed. 1987. *Alluvial Soils.* New York: Van Nostrand Reinhold.

Gifford, D. P. 1977. Observations of contemporary human settlements as an aid to archaeological interpretation. Ph.D. diss., University of California, Berkeley.

———. 1981. Taphonomy and paleoecology: A critical review of archaeology's sister disciplines. *Advances in Archaeological Method and Theory* 4:365–438.

Gile, L. H., F. F. Peterson, and R. B. Grossman. 1966. Morphological and genetic sequences of carbonate accumulation in desert soils. *Soil Science* 101:347–60.

Gillette, D. D. and M. G. Lockley, eds. 1989. *Dinosaur Tracks and Traces.* New York: Cambridge University Press.

Gingerich, P. D. 1982. Time resolution in mammalian evolution: Sampling, lineages, and faunal turnover. *Proceedings Third North American Paleontological Convention* 1:205–10.

———. 1987. Early Eocene bats (Mammalia, Chiroptera) and other vertebrates in freshwater limestones of the Willwood Formation, Clark's Fork Basin, Wyoming. *University of Michigan Contributions from the Museum of Paleontology* 27:275–320.

Glennie, K. W. 1970. *Desert Sedimentary Environments:* Developments in Sedimentology 14. Amsterdam: Elsevier.

Glob, P. V. [1969] 1971. *The Bog People.* New York: Ballantine Books.

Golenberg, E. M., D. E. Giannasi, M. T. Clegg, C. J. Smiley, M. Durbin, D. Henderson, and G. Zurawski. 1990. Chloroplast DNA sequence from a Miocene *Magnolia* species. *Nature* 344:656–58.

Gordon, C. C., and J. E. Buikstra. 1981. Soil pH, bone preservation, and sampling bias at mortuary sites. *American Antiquity* 46:566–71.

Gordon, E. A., and J. S. Bridge. 1987. Evolution of Catskill (Upper Devonian) river systems: Intra- and extrabasinal controls. *Journal of Sedimentary Petrology* 57: 234–49.

Gore, A. P. J., ed. 1983. *Ecosystems of the World*. Vol. 4A(B), *Mires: Swamp, Bog, Fen and Moor, General and Regional Studies*. Amsterdam: Elsevier.

Gore, P. J. W. 1988. Lacustrine sequences in an early Mesozoic rift basin: Culpeper Basin, Virginia, USA. *Geological Society Special Publication* 40:247–78.

Gould, R. A. 1980. *Living Archaeology*. Cambridge: Cambridge University Press.

Goulding, M. 1980. *The Fishes and the Forest*. Berkeley: University of California Press.

Grady, W. C., and C. F. Eble. 1990. Relationships among macerals, miospores, and paleoecology in a column of Redstone coal (Upper Pennsylvanian) from north-central West Virginia (USA). *International Journal of Coal Geology* 15:1–26.

Gradzinski, R. 1970. Sedimentation of dinosaur-bearing Upper Cretaceous deposits of the Nemegt Basin, Gobi Desert. *Palaeontologia Polonica* 21:147–229.

Gradzinski, R., and T. Jerzykiewicz. 1974. Dinosaur- and mammal-bearing aeolian and associated deposits of the Upper Cretaceous in the Gobi Desert (Mongolia). *Sedimentary Geology* 12:249–78.

Graham, M. A., and R. W. Graham. 1990. Holocene records of *Martes pennante* and *Martes americana* in Whiteside County, northwestern Illinois. *American Midland Naturalist* 124:81–92.

Graham, R. W. 1981. Preliminary report on late Pleistocene vertebrates from the Selby and Dutton archeological/paleontological sites, Yuma County, Colorado. *University of Wyoming Contributions to Geology* 20:33–56.

Graham, R. W., and M. Kay. 1988. Taphonomic comparisons of cultural and non-cultural faunal deposits at the Kimmswick and Barnhart sites, Jefferson County, Missouri. *Bulletin of the Buffalo Society of Natural Sciences* 33:227–40.

Grande, L. 1984. Paleontology of the Green River Formation, with a review of the fish fauna. 2d ed. *Geological Survey of Wyoming Bulletin* 63:1–333.

Gray, J. 1985. Interpretation of co-occurring megafossils and pollen: A comparative study with *Clarkia* as an example. 185–244 in C. J. Smiley, ed., *Later Cenozoic History of the Pacific Northwest*. San Francisco: American Association for the Advancement of Science, California Academy of Sciences.

Gray, J., ed. 1988. Paleolimnology: Aspects of freshwater paleoecology and biogeography. *Palaeogeography, Palaeoclimatology, Palaeoecology* 62:1–623.

Green, K. D., and D. J. Lowe. 1985. Stratigraphy and development of c. 17,000-year-old Lake Maratoto, North Island, New Zealand, with some inferences about postglacial climatic change. *New Zealand Journal of Geology and Geophysics* 28:675–99.

Greiner, H. R. 1962. Facies and sedimentary environments of Albert Shale, New Brunswick. *Bulletin of the American Association of Petroleum Geologists* 46: 219–34.

Grigorescu, D. 1983. A stratigraphic, taphonomic, and paleoecologic approach to a "forgotten land": The dinosaur-bearing deposits from the Hateg Basin (Transylvania, Romania). *Acta Palaeontologica Polonica* 28:103–21.

Grindrod, J. 1985. The palynology of mangroves on a prograded shore, Princess Charlotte Bay, North Queensland, Australia. *Journal of Biogeography* 12:323–95.

Grindrod, J., and E. G. Rhodes. 1984. Holocene sea-level history of a tropical estuary: Missionary Bay, north Queensland. 151–78 in B. G. Thom, ed., *Coastal Geomorphology in Australia*. New York: Academic Press.

Gruger, 1972. Pollen and seed studies of late Wisconsinan vegetation in Illinois, USA. *Geological Society of America Bulletin* 83:2715–34.

Guilday, J. E., P. S. Martin, and A. D. McCrady. 1964. New Paris No. 4: Late Pleistocene cave deposit in Bedford County, Pennsylvania. *Bulletin of the National Speleological Society* 26:121–94.

Guthrie, R. D. 1967. Differential preservation and recovery of Pleistocene large mammal remains in Alaska. *Journal of Paleontology* 41:243–46.

———. 1968. Paleoecology of the large-mammal community in interior Alaska during the Late Pleistocene. *American Midland Naturalist* 79:346–63.

Haas, C. A. 1988. Barrier island depositional systems in the Black Warrior Basin, Lower Pennsylvanian "Pottsville" Formation, in northwestern Alabama. M.S. thesis, Auburn University.

Haas, C. A., and R. A. Gastaldo. 1986. Barrier island depositional systems in the Black Warrior Basin, Lower Pennsylvanian (Pottsville) in northwestern Alabama. *American Association of Petroleum Geologists Bulletin* 70:597–98.

Halfman, J. D., and T. C. Johnson. 1988. High-resolution record of cyclic climatic change during the past 4 ka from Lake Turkana, Kenya. *Geology* 16:496–500.

Hall, S. A., and C. Lintz. 1983. Buried trees, water table fluctuations, and 3,000 years of changing climate in west central Oklahoma. *Quaternary Research* 22:129–33.

Hall, W. E., T. R. Van Devender, and C. A. Olson. 1988. Late Quaternary arthropod remains from Sonoran Desert pack rat middens, southwestern Arizona and northwestern Sonora. *Quaternary Research* 29:277–93.

Halstead, L. B., and P. G. Nicoll. 1971. Fossilized caves of Mendip. *Studies in Speleology* 2:93–102.

Hanley, J. H., and R. M. Flores. 1987. Taphonomy and paleoecology of nonmarine Mollusca: Indicators of alluvial plain lacustrine sedimentation, upper part of the Tongue River Member, Ft. Union Fm. (Paleocene), northern Powder River Basin, Wyoming and Montana. *Palaios* 2:479–96.

Hansen, R. M. 1978. Shasta ground sloth food habits, Rampart Cave, Arizona. *Paleobiology* 4:302–19.

Hanson, C. B. 1980. Fluvial taphonomic processes: Models and experiments. 156–81 in A. K. Behrensmeyer and A. P. Hill, eds., *Fossils in the Making*. Chicago: University of Chicago Press.

Harington, C. R. 1977. Marine mammals in the Champlain Sea and the Great Lakes. *Annals of the New York Academy of Sciences* 288:508–37.

Harris, T. M. 1957. A Liasso-Rhaetic flora in South Wales. *Proceedings of the Royal Society of London, B* 147:289–308.

———. 1958. Forest fires in the Mesozoic. *Journal of Ecology* 46:447–53.

Harrison, T. S. and Harrison, T. 1989. Palynology of the late Miocene *Oreopithecus*-bearing lignite from Baccinello, Italy. *Palaeogeography, Palaeoclimatology, Palaeoecology* 76:45–65.

Haszeldine, R. S. 1984. Muddy deltas in freshwater lakes, and tectonism in the Upper Carboniferous coalfield of NE England. *Sedimentology* 31:811–22.

Haubold, H., G. Schaumberg, and G. Katzung. 1985. *Die Fossilien des Kupferschiefers*. Wittenberg-Lutherstadt: Neue Brehm-Bücherei, A. Ziemsen Verlag.

Hay, R. L. 1976. *Geology of the Olduvai Gorge*. Berkeley: University of California Press.

———. 1986. Role of tephra in the preservation of fossils in Cenozoic deposits of East Africa. *Geological Society of London Special Publication* 25:339–44.

———. 1987. Geology of the Laetoli area. 23–47 in M. D. Leakey and J. M. Harris, eds., *Laetoli: A Pliocene Site in Northern Tanzania*. Oxford: Oxford University Press.

Haynes, C. V. 1985. Mastodon-bearing springs and late Quaternary geochronology of the lower Pomme de Terre Valley, Missouri. *Geological Society of America Special Paper* 204:1–35.

Haynes, G. 1985a. Age profiles of elephant and mammoth bone assemblages. *Quaternary Research* 24:333–45.

———. 1985b. On watering holes, mineral licks, death, and predation. 53–72 in J. I. Mead and D. J. Meltzer, eds., *Environments and Extinctions: Man in Late Glacial North America*. Orono: Center for the Study of Early Man, University of Maine.

———. 1988. Mass deaths and serial predation: Comparative taphonomic studies of modern large mammal death sites. *Journal of Archaeological Science* 15:219–35.

Hayward, J. L., C. J. Amlaner, and K. A. Young. 1989. Turning eggs into fossils: A natural experiment in taphonomy. *Journal of Vertebrate Paleontology* 9:196–200.

Heil, R., W. v. Koenigswald, and H. G. Lippmann. 1979. *Fossilien der Messeler Schichten*. Darmstadt: Hessisches Landesmuseum.

Heil, R., W. v. Koenigswald, H. G. Lippmann, D. Graner, and C. Heunisch. 1987. *Fossilien der Messel-Formation*. Darmstadt: Hessisches Landesmuseum.

Hendey, Q. B. 1981. Palaeoecology of the late Tertiary fossil occurrences in "E" quarry, Langebaanweg, South Africa, and a reinterpretation of their geological context. *Annals of the South African Museum* 84:1–104.

Henkel, S., and B. Krebs. 1977. Der erste Fund eines Säugetierskeletts aus der Jura-Zeit. *Umschau in Wissenschaft und Technik* 77:217–18.

Hentz, T. F. 1985. Early Jurassic sedimentation of a rift-valley lake: Culpeper Basin, northern Virginia. *Geological Society of America Bulletin* 96:92–107.

Heyler, D. 1987. Vertébrés des bassins stéphaniens et autuniens du Massif Central Français: Paléobiogéographie et paléoénvironments. *Annales de la Société Géologique du Nord* 106:123–38.

Hibbard, C. W. 1970. Pleistocene mammalian faunas from the Great Plains and Central Lowland provinces of the United States. 395–433 in W. Dort and J. K. Jones, eds., *Pleistocene and Recent Environments of the Central Great Plains*. Lawrence: University Press of Kansas.

Hibbard, C. W., and E. S. Riggs. 1949. Upper Pliocene vertebrates from Keefe Canyon, Meade County, Kansas. *Geological Society of America Bulletin* 60:829–60.

Hickey, L. J. 1980. Paleocene stratigraphy and flora of the Clark's Fork Basin. 33–49 in P. D. Gingerich, ed., *Early Cenozoic Paleontology and Stratigraphy of the Bighorn Basin, Wyoming*. University of Michigan Papers on Paleontology 24.

Hickey, L. J., and J. A. Doyle. 1977. Early Cretaceous fossil evidence for angiosperm evolution. *Botanical Review* 43:3–104.

Hill, A. P. 1975. Taphonomy of Contemporary and Late Cenozoic East African vertebrates. Ph.D. diss., University of London.

———. 1983. Hyenas and early hominids. *British Archaeological Series* 163:87–92.

Hill, A., R. Drake, L. Tauxe, M. Monaghan, J. C. Barry, A. K. Behrensmeyer, G. Curtis, B. F. Jacobs, L. L. Jacobs, N. M. Johnson, and D. Pilbeam. 1986. Neogene paleontology and geochronology of the Baringo Basin, Kenya. *Journal of Human Evolution* 14:759–73.

Hillam, J. 1985. Theoretical and applied dendrochronology: How to make a date with a tree. *Council for British Archaeology Research Report* 58:17–23.

Hölder, H., and D. B. Norman. 1986. Kreide-Dinosaurier im Sauerland. *Naturwissenschaften* 73:109–16.

Holliday, V. T. 1989. The Blackwater Draw Formation (Quaternary): A 1.4-plus-m.y. record of eolian sedimentation and soil formation on the Southern High Plains. *Geological Society of America Bulletin* 101:1598–1607.

Holman, J. A., L. M. Abraczinskas, and D. B. Westjohn. 1988. Pleistocene proboscideans and Michigan salt deposits. *National Geographic Research* 4:4–5.

Hook, R. W., and D. Baird. 1988. An overview of the Upper Carboniferous fossil deposit of Linton, Ohio. *Ohio Journal of Science* 88:55–60.

Hook, R. W., and J. C. Ferm. 1985. A depositional model for the Linton tetrapod assemblage (Westphalian D, Upper Carboniferous) and its palaeoenvironmental significance. *Philosophical Transactions of the Royal Society of London B* 311:101–9.

———. 1988. Paleoenvironmental controls on vertebrate-bearing abandoned channels in the Upper Carboniferous. *Palaeogeography, Palaeoclimatology, Palaeoecology* 63:159–81.

Hook, R. W., and J. C. Hower. 1988. Petrography and taphonomic significance of the vertebrate-bearing cannel coal of Linton, Ohio (Westphalian D, Upper Carboniferous). *Journal of Sedimentary Petrology* 58:72–80.

Horne, J. C., J. C. Ferm, F. T. Caruccio, and B. P. Baganz. 1978. Depositional models in coal exploration and mine planning. *Bulletin of the American Association of Petroleum Geologists* 62:2379–2411.

Horner, J. R. 1979. Upper Cretaceous dinosaurs from the Bearpaw Shale (marine) of south-central Montana with a checklist of Upper Cretaceous dinosaur remains from marine sediments in North America. *Journal of Paleontology* 53:566–77.

———. 1982. Evidence of colonial nesting and "site fidelity" among ornithischian dinosaurs. *Nature* 297:675–76.

———. 1984a, Three ecologically distinct vertebrate faunal communities from the Late Cretaceous Two Medicine Formation of Montana, with discussion of evolutionary pressures induced by Interior Seaway fluctuations. *Montana Geological Society 1984 Field Conference:* 299–303.

———. 1984b. The nesting behavior of dinosaurs. *Scientific American* 250:130–37.

———. 1987. Ecologic and behavioral implications derived from a dinosaur nesting site. 51–63 in S. J. Czerkas and E. C. Olson, eds., *Dinosaurs Past and Present*, vol. 2. Los Angeles: Los Angeles County Museum.

Horner, J. R., and R. Makela. 1979. Nest of juveniles provides evidence of family structure among dinosaurs. *Nature* 282:296–98.

Horton, D. R. 1978. Preliminary notes on the analysis of Australian coastal middens. *Australian Institute of Aboriginal Studies Newsletter* 10:30–33.

Horton, D. R., and G. E. Connah. 1981. Man and megafauna at Reddestone Creek, near Glen Innes, northern New South Wales. *Australian Archaeology* 13:35–52.

Horton, D. R. and R. V. S. Wright. 1981. Cuts on Lancefield bones: Carnivorous *Thylacoleo,* not humans, the cause. *Archaeology in Oceania* 16:73–80.

Hotton, N., III. 1967. Stratigraphy and sedimentation in the Beaufort Series (Permian-Triassic), South Africa. *University of Kansas Department of Geology Special Publication* 2:390–428.

Howard, H. 1962. A comparison of avian assemblages from individual pits at Rancho La Brea. *Contributions to Science, Natural History Museum of Los Angeles County* 58:1–24.

Hulbert, R. C. 1984. Paleoecology and population dynamics of the Early Miocene (Hemingfordian) horse *Parahippus leonensis* from the Thomas Farm site, Florida. *Journal of Vertebrate Paleontology* 4:547–58.

Hunt, R. M., Jr. 1978. Depositional setting of a Miocene mammal assemblage, Sioux County, Nebraska (USA). *Palaeogeography, Palaeoclimatology, Palaeoecology* 24:1–52.

Hunt, R. M., Jr., X. X. Xue, and J. Kaufman. 1983. Miocene burrows of extinct bear-dogs: Indication of early denning behavior of large mammalian carnivores. *Science* 221:364–66.

Hürzeler, J. 1975. L'age géologique et les rapports géographiques de la faune des mammifères du lignite de Grosseto. *Colloque Internationale du Centre National de la Recherche Scientifique* 218:873–76.

Isaac, G. Ll., and D. C. Crader. 1981. To what extent were early hominids carnivorous? An archaeological perspective. 37–103 in R. Harding and G. Telecki, eds., *Omnivorous Primates: Gathering and Hunting in Human Evolution.* New York: Columbia University Press.

Isaac, G. Ll., and B. Isaac. In press. *Koobi Fora Research Project Volume 5: Plio-Pleistocene Archaeology.* London: Clarendon Press.

Jacobs, B. F., and C. H. S. Kabuye. 1987. A middle Miocene (12.2 my old) forest in the East African Rift Valley, Kenya. *Journal of Human Evolution* 16:147–55.

Jacobson, G. L., and R. H. W. Bradshaw. 1981. The selection of sites for paleovegetational studies. *Quaternary Research* 16:80–96.

Jain, S. L. 1980. The continental Lower Jurassic fauna from the Kota Formation, India. 99–124 in L. L. Jacobs, ed., *Aspects of Vertebrate History.* Flagstaff: Museum of Northern Arizona Press.

Janda, R. J., K. M. Scott, K. M. Nolan, and H. A. Martinson. 1981. Lahar movement, effects, and deposits. *United States Geological Survey Professional Paper* 1250:461–78.

Jefferson, T. H. 1982a. The preservation of fossil leaves in Cretaceous volcaniclastic rocks from Alexander Island, Antarctica. *Geological Magazine* 119:291–300.

———. 1982b. Fossil forests from the Lower Cretaceous of Alexander Island, Antarctica. *Palaeontology* 25:1–47.

Jepsen, G. L. 1963. Eocene vertebrates, coprolites, and plants in the Golden Valley Formation of western North Dakota. *Geological Society of America Bulletin* 74:673–84.

Jodry, M. A. and D. J. Stanford. 1992. Stewart's Cattle Guard site: An analysis of bison remains at a Folsom kill-butchery campsite. 101–68 in D. J. Stanford and J. Day, eds., *Ice Age Hunters of the Rockies.* Denver: Denver Museum of Natural History, n. s. 1.

Johanson, D. C., M. Splingaer, and N. T. Boaz. 1976. Paleontological excavations in the Shungura Formation, Lower Omo Basin, 1969–73. 402–20 in Y. Coppens, F. C. Howell, G. L. Isaac, and R. E. Leakey, eds., *Earliest Man and Environments in the Lake Rudolf Basin: Stratigraphy, Paleoecology, and Evolution.* Chicago: University of Chicago Press.

Johnson, S. Y. 1989. Significance of loessites in the Maroon Formation (Middle Pennsylvanian to Lower Permian), Eagle Basin, northwest Colorado. *Journal of Sedimentary Petrology* 59:782–91.

Johnston, P. A., and R. C. Fox. 1984. Paleocene and Late Cretaceous mammals from Saskatchewan, Canada. *Palaeontographica, A* 186:163–222.

Jones, A. K. G. 1986. Fish bone survival in the digestive systems of the pig, dog, and man: Some experiments. *British Archaeological Reports International Series* 294:53–61.

Kames, W. H. 1970. Facies and development of the Colorado River Delta in Texas. *Society of Economic Paleontologists and Mineralogists Special Publication* 15: 78–106.

Kampmann, H. 1983. Mikrofossilien, Hölzer, Zapfen, und Pflanzenreste aus der unterkretazischen Sauriergrube bei Brilon-Nehden. *Geologie und Paläontologie von Westfalen* 1:1–146.

Karowe, A. L., and T. H. Jefferson. 1987. Burial of trees by eruptions of Mount Saint Helens, Washington: Implications for the interpretation of fossil forests. *Geology Magazine* 124:191–204.

Kerp, J. H. F. 1982. New palaeobotanical data on the "Rotliegendes" of the Nahe area (FRG). *Courier Forschungsinstitut Senckenberg* 56:7–14.

Kidwell, S. M. 1982. Time scales of fossil accumulations: Patterns from Miocene benthic assemblages. *Proceedings of the North American Paleontological Convention* 1:295–300.

———. 1989. Stratigraphic condensation of marine transgressive records: Origin of major shell deposits in the Miocene of Maryland. *Journal of Geology* 97:1–24.

Kidwell, S. M., and T. Aigner. 1985. Sedimentary dynamics of complex shell beds: Implications for ecologic and evolutionary patterns. 382–95 in U. Bayer and A. Seilacher, eds., *Sedimentary and Evolutionary Cycles.* Berlin: Springer-Verlag.

Kidwell, S. M., and D. Jablonski. 1983. Taphonomic feedback: Ecological consequences of shell accumulation. 195–248 in M. J. S. Tevesz and P. McCall, eds., *Biotic Interactions in Recent and Fossil Benthic Communities.* New York: Plenum Press.

King, C. 1981. The stratigraphy of the London Clay and associated deposits. *Tertiary Research Special Papers* 6:1–158.

Klein, R. G. 1976. The mammalian fauna of the Klasies River Mouth sites, southern Cape Province, South Africa. *South African Archaeological Bulletin* 31:75–98.

———. 1980. The interpretation of mammalian faunas from Stone Age Archaeological sites, with special reference to sites in the southern Cape Province, South Africa. 223–46 in A. K. Behrensmeyer and A. P. Hill, eds., *Fossils in the Making.* Chicago: University of Chicago Press.

Klein, R. G., and K. Cruz-Uribe. 1984. *The Analysis of Animal Bones from Archeological Sites.* Chicago: University of Chicago Press.

Knoll, A. H. 1985. Exceptional preservation of photosynthetic organisms in silicified carbonates and silicified peats. *Philosophical Transactions of the Royal Society London, B* 311:111–22.

Kolb, K. K., M. E. Nelson, and R. J. Zakrezewski. 1975. The Duck Creek molluscan fauna (Illinoian) from Ellis County, Kansas. *Transactions of the Kansas Academy of Science* 78:63–74.

Korth, W. W. 1979. Taphonomy of microvertebrate fossil assemblages. *Annals of the Carnegie Museum* 48:235–85.

Koster, E. H., and P. J. Currie. 1987. Upper Cretaceous coastal plain sediments at Dinosaur Provincial Park, southeast Alberta. *Geological Society of America Centennial Field Guide of the Rocky Mountain Section* 2:9–14.

Kosters, E. C., and A. Bailey. 1983. Characteristics of peat deposits in the Mississippi River delta plain. *Transactions Gulf Coast Association of Geological Societies* 33:311–25.

Kraus, M. J. 1987. Integration of channel and floodplain suites, II: Vertical relations of alluvial paleosols. *Journal of Sedimentary Petrology* 57:602–12.

———. 1988. Nodular remains of early Tertiary forests, Bighorn Basin, Wyoming. *Journal of Sedimentary Petrology* 58:888–93.

Kraus, M. J., and T. M. Bown. 1986. Paleosols and time resolution in alluvial stratigraphy. 180–207 in V. P. Wright, ed., *Paleosols: Their Recognition and Interpretation*. Princeton: Princeton University Press.

———. 1988. Pedofacies analysis: A new approach to reconstructing ancient fluvial sequences. *Geological Society of America Special Paper* 216:143–52.

Krebs, B. 1980. The search for Mesozoic mammals in Spain and Portugal. 23–25 in G. Olshevsky, ed., *New Mesozoic Faunas*. Mesozoic Vertebrate Life no. 1. Astoria, Ill.: Stevens Publishing.

———. 1987. The skeleton of a Jurassic eupantothere and the arboreal origin of modern mammals. *Occasional Paper of the Tyrrell Museum of Palaeontology* 3:132–37.

Krumbein, W. C., and R. M. Garrels. 1952. Origin and classification of chemical sediments in terms of pH and oxidation-reduction potentials. *Journal of Geology* 60:1–33.

Krynine, P. D. 1935. Formation and preservation of desiccation features in a humid climate. *American Journal of Science* 30:96–97.

Kuhn-Schnyder, E. 1974. Die Triasfauna der Tessiner Kalkalpen. *Neujahrsblatt der Naturforschenden Gesellschaft Zürich* 176:1–119.

Kukla, G. 1975. Loess stratigraphy of central Europe. 99–188 in K. W. Butzer and G. Ll. Isaac, eds., *After the Australopithecines*. The Hague: Mouton.

———. 1989. Loess stratigraphy in central China. *Palaeogeography, Palaeoclimatology, Palaeoecology* 72:203–25.

Kurtén, B. 1971. *The Age of Mammals*. New York: Columbia University Press.

Kvale, E. P., and A. W. Archer. 1990. Tidal deposits associated with low-sulfur coals, Brazil Fm. (Lower Pennsylvanian), Indiana. *Journal of Sedimentary Petrology* 60:563–74.

Laarson, S. G. 1978. Baltic amber: A paleobiological study. *Entomonograph* 1:1–192.

Labandeira, C. C. 1990. Rethinking the diets of Carboniferous terrestrial arthropods:

Evidence for a nexus of arthropod/vascular plant interactions. *Geological Society of America Abstracts with Program* 22(7): A265.

Lambe, L. M. 1910. Palaeoniscid fishes from the Albert Shales of New Brunswick. *Memoirs of the Geological Survey Branch of Canada* 3: 7–35.

Langiaux, J. 1984. Flores et faunes des formations supérieurs du Stéphanien de Blanzy-Montceau (Massif Central Français): Stratigraphie et paléoecologie. *Revue périodique de la Physiophile Supplement* 100: 1–270.

Langston, W., Jr. 1974. Nonmammalian Comanchean tetrapods. *Geoscience and Man* 8: 77–102.

Lawton, R. 1977. Taphonomy of the dinosaur quarry, Dinosaur National Monument. *University of Wyoming Contributions to Geology* 15: 119–26.

Leakey, M. D. 1971. *Olduvai Gorge,* vol. 3. London: Cambridge University Press.

———. 1987. The hominid footprints: Introduction. 490–95 in M. D. Leakey and J. M. Harris, eds., *Laetoli: A Pliocene Site in Northern Tanzania.* Oxford: Oxford University Press.

Leakey, M. D., and J. M. Harris, eds. 1987. *Laetoli: A Pliocene Site in Northern Tanzania.* Oxford: Oxford University Press.

Leary, R. L. 1981. Early Pennsylvanian geology and paleobotany of the Rock Island County, Illinois, area. Part 1: Geology. *Illinois State Museum Reports of Investigations* 37: 1–88.

Leary, R. L., and H. W. Pfefferkorn. 1977. An early Pennsylvanian flora with *Megalopteris* and Noeggerathiales from west-central Illinois. *Illinois State Geological Survey, Circular* 500: 1–77.

Lebedev, Y. L. 1976. Evolution of Albian-Cenomanian floras of northeast USSR and the association between their composition and facies conditions. *International Geology Reviews* 19: 1183–90.

Lee, R. B., and I. DeVore, eds. 1968. *Man the Hunter.* Chicago: Aldine.

Leeder, M. R. 1975. Pedogenic carbonates and flood sediment accretion rates: A quantitative model for alluvial arid-zone lithofacies. *Geological Magazine* 112: 257–70.

Lehman, T. M. 1982. A ceratopsian bone bed from the Aguja Formation (Upper Cretaceous) Big Bend National Park, Texas. Master's thesis, University of Texas, Austin.

———. 1990. The ceratopsian subfamily Chasmosaurinae: Sexual dimorphism and systematics. 211–29 in P. J. Currie and K. Carpenter, eds., *Dinosaur Systematics.* Cambridge: Cambridge University Press.

Leonard, A. B., and J. C. Frye. 1960. Wisconsinan molluscan faunas in the Illinois Valley region. *Illinois Geological Survey Circular* 304: 1–32.

Lillegraven, J., Z. Kielan-Jaworowska, and W. A. Clemens, eds. 1979. *Mesozoic Mammals.* Berkeley: University of California Press.

Lipman, P. W., and D. R. Mullineaux. 1981. The 1980 Eruptions of Mount Saint Helens, Washington. *United States Geological Survey Professional Paper* 1250: 1–844.

Littke, R. 1987. Petrology and genesis of Upper Carboniferous seams from the Ruhr region, West Germany. *International Journal of Coal Geology* 7: 147–85.

Litwin, R. J., and S. Ash. 1991. First early Mesozoic amber in the Western Hemisphere. *Geology* 19: 273–76.

Lockley, M. G., and K. Conrad. 1989. The paleoenvironmental context, preservation, and paleoecological significance of dinosaur tracksites in the western USA. 121–34 in D. D. Gillette and M. G. Lockley, eds., *Dinosaur Tracks and Traces*. New York: Cambridge University Press.

Lockley, M. G., and A. Rice, eds. 1990. Volcanism and Fossil Biotas. *Geological Society of America Special Paper* 244:1–125.

Long, A., R. M. Hansen, and P. S. Martin. 1974. Extinction of the Shasta ground sloth. *Geological Society of America Bulletin* 85:1843–48.

Long, R. A., and K. Padian. 1986. Vertebrate biostratigraphy of the Late Triassic Chinle Formaiton, Petrified Forest National Park, Arizona: Preliminary results. 161–69 in K. Padian, ed., *The Beginning of the Age of Dinosaurs*. Cambridge: Cambridge University Press.

Lotze, F. 1968. *Geologie Mitteleuropas*. 4th ed. Stuttgart: Schweizerbart'sche Verlagsbuchhandlung.

Lowe-McConnell, R. H. 1975. *Fish Communities in Tropical Fresh Waters*. London: Longman.

————. 1977. *Ecology of Fishes in Tropical Waters*. London: Edward Arnold.

Lucas, S. G. 1981. Dinosaur communities in the San Juan Basin: A case for lateral variations in the composition of Late Cretaceous dinosaur communities. 337–93 in S. G. Lucas, K. Rigby, Jr., and B. Kues, eds., *Advances in San Juan Basin Paleontology*. Albuquerque: University of New Mexico Press.

Lucas, S. G., and N. J. Mateer. 1983. Vertebrate paleoecology of the late Campanian (Cretaceous) Fruitland Formation, San Juan Basin, New Mexico (USA). *Acta Palaeontologica Polonica* 28:195–204.

Lyman, R. L. 1984. Broken bones, bone expediency tools, and bone pseudotools: Lessons from the blast zone around Mount Saint Helens, Washington. *American Antiquity* 49:315–33.

————. 1989. Taphonomy of cervids killed by the 18 May 1980 volcanic eruption of Mount Saint Helens, Washington, USA. 149–69 in R. Bonnichsen and M. Sorg, eds., *Bone Modification*. Orono, Maine: Center for the Study of the Earliest Americans.

Lyons, P. C., P. G. Hatcher, F. W. Brown, C. L. Thomson, and M. A. Millay. 1985. Coalification of organic matter in a coal ball from the Calhoun coal bed (Upper Pennsylvanian), Illinois Basin, United States of America. *Tenth International Congress of Carboniferous Stratigraphy and Geology, Compte Rendu* 1:155–59.

Maas, M. C. 1985. Taphonomy of a late Eocene microvertebrate locality, Wind River Basin, Wyoming (USA). *Palaeogeography, Palaeoclimatology, Palaeoecology* 52:123–42.

Madsen, J. H. 1976. *Allosaurus fragilis:* A revised osteology. *Utah Geology and Mineralogy Survey Bulletin* 109:1–163.

Maguire, J. M., D. Pemberton, and M. H. Collett. 1980. The Makapansgat limeworks grey breccia: Hominids, hyaenas, hystricids, or hillwash? *Palaeontologia Africana* 23:75–98.

Mahaffy, J. F. 1985. Profile patterns of peat and coal palynology in the Herrin (No. 6) Coal Member, Carbondale Formation, Middle Pennsylvanian of southern Illinois. *Ninth International Congress of Carboniferous Stratigraphy and Geology, Compte Rendu* 5:25–34.

Marcus, L. F. 1960. A census of the abundant large late Pleistocene mammals from Rancho La Brea. *Contributions in Science, Natural History Museum of Los Angeles County* 38:1–11.

Marcus, L. F., and R. Berger. 1984. The significance of radiocarbon dates from Rancho La Brea. 159–83 in P. S. Martin and R. G. Klein, eds., *Quaternary Extinctions: A Prehistoric Revolution.* Tucson: University of Arizona Press.

Margraf, V. 1985. Late Pleistocene extinctions in southern Patagonia. *Science* 228: 1110–12.

Martill, D. M. 1985. The preservation of marine vertebrates in the Lower Oxford Clay (Jurassic) of central England. *Philosophical Transactions of the Royal Society of London, B* 311:155–65.

———. 1988. Preservation of fish in the Cretaceous Santana Formation of Brazil. *Palaeontology* 31:1–18.

Martill, D. M., and A. Dawn. 1986. Fossil vertebrates from new exposures of the Westbury Formation (Upper Triassic) at Newark, Nottinghamshire. *Mercian Geologist* 10:127–33.

Martin, L. D., and D. K. Bennett. 1977. The burrows of the Miocene beaver *Palaeocastor,* western Nebraska, USA. *Palaeogeography, Palaeoclimatology, Palaeoecology* 22:173–93.

Martin, P. S., B. E. Sabels, and D. Shutler. 1961. Rampart Cave coprolite and ecology of the Shasta ground sloth. *American Journal of Science* 259:102–7.

Masson, A. G., and B. R. Rust. 1984. Freshwater shark teeth as paleoenvironmental indicators in the Upper Pennsylvanian Morien Group of the Sydney Basin, Nova Scotia. *Canadian Journal of Earth Sciences* 21:1151–55.

Mayhew, D. F. 1977. Avian predators as accumulators of fossil mammal material. *Boreas* 6:25–31.

McAllister, J. A. 1987. Phylogenetic distribution and morphological reassessment of the intestines of fossil and modern fishes. *Zoologische Jahrbücher, Abteilung für Anatomie und Ontogenie der Tiere* 115:281–94.

McBrearty, S. 1990. Consider the humble termite: Termites as agents of post-depositional disturbance at African Archaeological sites. *Journal of Archaeological Science* 17:111–43.

McCabe, P. J. 1984. Depositional environments of coal and coal-bearing strata. 13–42 in R. A. Rahmani, and R. M. Flores, eds., *Sedimentology of Coal and Coal-Bearing Sequences.* Special Publication of the International Association of Sedimentologists 7. Oxford: Blackwell Scientific Publications.

———. 1987. Facies studies of coal and coal-bearing strata. 51–66 in A. C. Scott, ed., *Coal and Coal-Bearing Strata: Recent Advances.* Geological Society Special Publication no. 32.

McCune, A. R. 1987. Lakes as laboratories of evolution: Endemic fishes and environmental cyclicity. *Palaios* 2:446–54.

McGrew, P. O. 1971. Early and Middle Eocene faunas of the Green River Basin. *University of Wyoming Contributions to Geology* 10:65–68.

McKee, E. D., and J. J. Bigarella. 1979. Ancient sandstones considered to be eolian. *United States Geological Survey Professional Paper* 1052:187–240.

McLeroy, C. A., and R. Y. Anderson. 1966. Laminations in the Oligocene Florissant Lake deposits, Colorado. *Geological Society of America Bulletin* 77:605–18.

McManus, J., and S. A. K. Alizai. 1983. Sediment-filled reed stems as contributors to marsh sediment fabrics. *Journal of Sedimentary Petrology* 53:407–10.

Mead, J. I., T. R. Van Devender, and K. L. Cole. 1983. Late Quaternary small mammals from Sonoran Desert packrat middens, Arizona and California. *Journal of Mammalogy* 64:173–80.

Mellett, J. S. 1974. Scatological origins of microvertebrate fossils. *Science* 185: 349–50.

Merriam, J. C. 1911. The fauna of Rancho La Brea. Part 1, Occurrence. *Memoirs of the University of California* 1:199–213.

Miall, A. D. 1985. Architectural-element analysis: A new method of facies analysis applied to fluvial deposits. *Earth Science Reviews* 22:261–308.

Michener, C. D., and D. A. Grimaldi. 1988. The oldest fossil bee: Apoid history, evolutionary status, and antiquity of social behavior. *Proceedings of the National Academy of Sciences, USA* 85:6424–26.

Middeldorp, A. A. 1986. Functional palaeoecology of the Hahnenmoor raised bog ecosystem: A study of vegetation history, production, and decomposition by means of pollen density dating. *Review of Palaeobotany and Palynology* 49:1–73.

Miller, S. E. 1983. Late Quaternary insects of Rancho La Brea and McKittrick, California. *Quaternary Research* 20:90–104.

Moore, P. D., and D. J. Bellamy. 1974. *Peatlands*. New York: Springer-Verlag.

Morlan, R. E. 1980. Taphonomy and archaeology in the Upper Pleistocene of the northern Yukon Territory: A glimpse of the peopling of the New World. *Archaeological Survey of Canada Paper* 94:1–398.

Moseley, H. N. 1879. *Notes by a Naturalist on the "Challenger."* London: Macmillan.

Muizon, C. de. 1981. Les vértebres fossiles de la Formation Pisco (Perou). *Institut Français d'Etudes Andines Mémoire* no. 6:5–14.

Muller, J. 1959. Palynology of Recent Orinoco Delta and shelf sediments. *Micropaleontology* 5:1–32.

Murchison, D. G., and T. S. Westoll. 1968. *Coal and Coal-Bearing Strata*. Edinburgh: Oliver and Boyd.

Mustoe, G. E. 1985. Eocene amber from the Pacific Coast of North America. *Geological Society of America Bulletin* 96:1530–36.

Myrick, A. C. 1979. Variation, taphonomy, and adaptation of the Rhabdosteidae (Eurhinodelphidae) (Odontoceti, Mammalia) from the Calvert Formation of Maryland and Virginia. Ph.D. diss., University of California, Los Angeles.

Nichols, D. J., and A. Traverse. 1971. Palynology, petrology, and depositional environments of some Early Tertiary lignites in Texas. *Geoscience and Man* 3:37–48.

Niklas, K. J. and R. M. Brown, Jr. 1981. Ultrastructural and paleobiochemical correlations among fossil leaf tissues from the Saint Maries River (Clarkia) area, northern Idaho, USA. *American Journal of Botany* 68:332–41.

Norman, D. B. 1986. On the anatomy of *Iguanodon atherfieldensis* (Ornithischia: Ornithopoda). *Bulletin de l'Institut Royal des Sciences Naturelle Belgique, Sciences de la Terre* 56:281–372.

———. 1987. A mass-accumulation of vertebrates from the Lower Cretaceous of Nehden (Sauerland), West Germany. *Proceedings of the Royal Society of London, B* 230:215–55.

Nunez, L. 1984. The Paleo-Indian occupation at Quereo: Multidisciplinary reconstruction in the semiarid region of Chile. *National Geographic Research Reports* 18:551–61.

O'Connell, J. F., K. Hawkes, and N. Blurton-Jones. 1988. Hadza scavenging: Implications for Plio-Pleistocene hominid subsistence. *Current Anthropology* 29:356–63.

Oldale, R. N., F. C. Whitmore, Jr., and J. R. Grimes. 1987. Elephant teeth from the western gulf of Maine and their implications. *National Geographic Research* 3: 439–46.

Olsen, P. E. 1980. A comparison of the vertebrate assemblages from the Newark and Hartford Basins (Early Mesozoic, Newark Supergroup) of eastern North America. 35–54 in L. L. Jacobs, ed., *Aspects of Vertebrate History*. Tucson: Museum of Northern Arizona Press.

——. 1986. A forty-million-year lake record of early Mesozoic orbital climatic forcing. *Science* 234:842–48.

——. 1988. Paleontology and paleoecology of the Newark Supergroup (early Mesozoic, eastern North America). 185–230 in W. Manspeizer, ed., *Trassic-Jurassic Rifting, Continental Breakup, and the Origin of the Atlantic Ocean and Passive Margins*. New York: Elsevier.

Olsen, P. E., and D. Baird. 1986. The ichnogenus *Atreipus* and its significance for Triassic biostratigraphy. 61–87 in K. Padian, ed., *The Beginning of the Age of Dinosaurs*. Cambridge: Cambridge University Press.

Olsen, P. E., C. L. Remington, B. Cornet, and K. S. Thomson. 1978. Cyclic change in Late Triassic lacustrine communities. *Science* 201:729–33.

Olsen, P. E., R. W. Schlische, and P. J. W. Gore, eds. 1989. Tectonic, depositional, and paleoecological history of early Mesozoic rift basins, eastern North America. *Twenty-eighth International Geological Congress, Guidebook for Field Trip T351*: 1–174. Washington, DC: American Geophysical Union.

Olson, E. C. 1939. The fauna of the *Lysorophus* pockets in the Clear Fork Permian, Baylor County, Texas. *Journal of Geology* 46:389–97.

——. 1952. The evolution of a Permian vertebrate chronofauna. *Evolution* 6: 181–96.

——. 1958. Fauna of the Vale and Choza: 14. Summary, review, and integration of the geology and the faunas. *Fieldiana Geology* 10:397–448.

——. 1962. Late Permian Terrestrial Vertebrates, USA and USSR. *Transactions of the American Philosophical Society* n. s. 52:1–196.

——. 1967. Early Permian vertebrates. *Oklahoma Geological Survey Circular* 74:1–111.

——. 1984a. Permo-Carboniferous vertebrate communities. 331–45 in J. T. Dutro and H. W. Pfefferkorn, eds., *Neuvième Congrès International de Stratigraphie et de Géologie du Carbonifère, Compte Rendu* 5. Carbondale: Southern Illinois University Press.

——. 1984b. Nonmarine vertebrates and Late Paleozoic climates. 403–14 in J. T. Dutro and H. W. Pfefferkorn, eds., *Neuvième Congrès International de Stratigraphie et de Géologie du Carbonifère, Compte Rendu* 5. Carbondale: Southern Illinois University Press.

Olson, E. C., and K. Bolles. 1975. Permo-Carboniferous fresh water burrows. *Fieldiana Geology* 33:271–90.

Olson, E. C., and J. G. Mead. 1982. The Vale Formation (Lower Permian): Its vertebrates and paleoecology. *Texas Memorial Museum Bulletin* 29:1–46.

Osmolska, H. 1980. The Late Cretaceous vertebrate assemblages of the Gobi Desert, Mongolia. *Mémoires de la Société Géologique de France* n.s. 139:145–50.

Ostrom, J. H. 1970. Stratigraphy and paleontology of the Cloverly Formation (Lower Cretaceous) of the Bighorn Basin Area, Wyoming and Montana. *Yale Peabody Museum of Natural History Bulletin* 35:1–234.

Page, V. M. 1981. Dicotyledonous woods from the Upper Cretaceous of central California. *Journal of the Arnold Arboretum* 62:437–55.

Parker, L. R. 1976. The paleoecology of the fluvial coal-bearing swamps and associated floodplain environments in the Blackhawk Formation (Upper Cretaceous) of central Utah. *Brigham Young University Geological Studies* 22:99–116.

Parrish, J. M. 1989. Vertebrate paleoecology of the Chinle Formation (Late Triassic) of the southwestern United States. *Palaeogeography, Palaeoclimatology, Palaeoecology* 72:227–48.

Parrish, J. M., J. T. Parrish, J. H. Hutchinson, and R. A. Spicer. 1987. Cretaceous vertebrates from Alaska: Implications for dinosaur ecology. *Palaios* 2:377–89.

Parrish, J. T. 1982. Upwelling and petroleum source beds, with reference to the Paleozoic. *American Association of Petroleum Geologists Bulletin* 66:750–74.

Payette, S. 1988. Late Holocene development of subarctic ombrotrophic peatlands: Allogenic and autogenic succession. *Ecology* 69:516–31.

Peabody, F. E. 1961. Annual growth zones in living and fossil vertebrates. *Journal of Morphology* 108:11–62.

Pearsall, D. M. 1989. *Paleoethnobotany: A Handbook of Procedures.* New York: Academic Press.

Pelzer, G. 1984. Transition from fluvial to littoral environments in the Wealden facies (lowermost Cretaceous) of northwest Germany. *Occasional Paper of the Tyrrell Museum of Palaeontology* 3:179–84.

Penland, S., R. Boyd, and J. R. Suter. 1988. Transgressive depositional systems of the Mississippi Delta plain: A model for barrier shoreline and shelf sand development. *Journal of Sedimentary Petrology* 58:932–49.

Peppers, R. A. 1970. Correlation and palynology of coals in the Carbondale and Spoon Formations (Pennsylvanian) of the northeastern part of the Illinois Basin. *Illinois State Geological Survey Bulletin* 93:1–173.

———. 1979. Development of coal-forming floras during the early part of the Pennsylvanian in the Illinois Basin. 8–14 in *Guidebook to Field Trip 9, Ninth International Congress on Carboniferous Stratigraphy and Geology*, part 2, *Invited Papers.* Urbana: Illinois State Geological Survey.

———. 1985. Comparison of miospore assemblages in the Pennsylvanian System of the Illinois Basin with those in the Upper Carboniferous of western Europe. *Ninth International Congress on Carboniferous Stratigraphy and Geology, Compte Rendu* 2:483–502.

Peterson, O. A. 1906. The Agate Spring fossil quarry. *Annals Carnegie Museum* 3:487–94.

———. 1923. A fossil-bearing slab of sandstone from the Agate Spring quarries of western Nebraska exhibited in the Carnegie Museum. *Annals Carnegie Museum* 7:91–93.

Phillips, R. L. 1987. Late Cretaceous to early Tertiary deltaic to marine sedimentation, North Slope, Alaska. *American Association of Petroleum Geologists Bulletin* 71:601–2.

Phillips, T. L., and W. A. DiMichele. 1981. Paleoecology of Middle Pennsylvanian age coal swamps in southern Illinois: Herrin Coal Member at Sahara Mine no. 6. 231–84 in K. J. Niklas, ed., *Paleobotany, Paleoecology, and Evolution*, vol. 1. New York: Praeger.

———. 1990. From plants to coal: Peat taphonomy of Upper Carboniferous coals. *International Journal of Coal Geology* 16:151–56.

Phillips, T. L., A. B. Kunz, and D. J. Mickish. 1977. Paleobotany of permineralized peat (coal balls) from the Herrin (no. 6) Coal Member of the Illinois Basin. 18–49 in P. N. Given and A. D. Cohen, eds., *Interdisciplinary Studies of Peat and Coal Origins*. Geological Society of America, Microform Publication 7:18–49.

Phillips, T. L., and R. A. Peppers. 1984. Changing patterns of Pennsylvanian coal-swamp vegetation and implications of climatic control on coal occurrence. *International Journal of Coal Geology* 3:205–55.

Picard, M. D., and L. R. High. 1972. Criteria for recognizing lacustrine rocks. *Society of Economic Paleontologists and Mineralogists Special Publication* 16:108–45.

Pickford, M. H. L. 1983. Sequence and environments of the Lower and Middle Miocene hominoids of western Kenya. 421–39 in R. L. Ciochon and R. S. Corruccini, eds., *Interpretations of Ape and Human Ancestry*. New York: Plenum.

———. 1986. Sedimentation and fossil preservation in the Nyanza Rift System, Kenya. *Geological Society of London Special Publication* 25:345–62.

Platt, N. H. 1989a. Continental sedimentation in an evolving rift basin: The Lower Cretaceous of the western Cameros Basin (northern Spain). *Sedimentary Geology* 64:91–109.

———. 1989b. Lacustrine carbonates and pedogenesis: Sedimentology and origin of palustrine deposits from the Early Cretaceous Rupelo Formation, western Cameros Basin, northern Spain. *Sedimentology* 36:665–84.

Pocknell, D. T., and R. M. Flores. 1987. Coal palynology and sedimentology in the Tongue River Member, Fort Union Formation, Powder River Basin, Wyoming. *Palaios* 2:133–45.

Politis, G. G., E. P. Tonni, F. Fidalgo, M. C. Salemme, and L. M. M. Guzman. 1987. Man and Pleistocene megamammals in the Argentine Pampa: Site 2 Arroyo Seco. *Current Research in the Pleistocene* 4:159–62.

Polnar, G. O., and D. C. Cannatella. 1987. An Upper Eocene frog from the Dominican Republic and its implications for Caribbean biogeography. *Science* 237:1215–16.

Potter, F. W., and D. L. Dilcher. 1980. Biostratigraphic analysis of Eocene clay deposits in Henry County, Tennessee. 211–25 in D. L. Dilcher and T. N. Taylor, eds., *Biostratigraphy of Fossil Plants*. Stroudsburg, Pa: Dowden, Hutchinson and Ross.

Potts, R. 1986. Temporal span of bone accumulations at Olduvai Gorge and implications for early hominid foraging behavior. *Paleobiology* 12:25–31.

Potts, R. 1988. *Early Hominid Activities at Olduvai*. New York: Aldine de Gruyter.

———. 1989. New excavations at Olorgesailie and Kanjera, Kenya. *Journal of Human Evolution* 18:477–84.

Potts, R., P. Shipman, and E. Ingall. 1988. Taphonomy, paleoecology, and hominids of Lainyamok, Kenya. *Journal of Human Evolution* 17:597–614.

Pratt, A. E. 1986. The taphonomy and paleoecology of the Thomas Farm local fauna

(Miocene, Hemingfordian), Gilchrist County, Florida. Ph.D. diss., University of Florida, Gainesville.

————. 1990. Taphonomy of the large vertebrate fauna from the Thomas Farm Locality (Miocene, Hemingfordian), Gilchrist County, Florida. *Bulletin of the Florida Museum of Natural History* 35:35–130.

Quade, J. 1986. Late Quaternary environmental changes in the Upper Las Vegas Valley, Nevada. *Quaternary Research* 26:340–57.

Raup, D. M. 1976. Species diversity in the Phanerozoic: An interpretation. *Paleobiology* 2:289–97.

Raymond, A. 1987. Interpreting ancient swamp communities: Can we see the forest in the peat? *Review of Palaeobotany and Palynology* 52:217–31.

Reading, H. G., ed. 1986. *Sedimentary Environments and Facies*. 2d ed. Oxford, England: Blackwell Scientific Publications.

Reineck, H. E., and I. B. Singh. 1980. *Depositional Sedimentary Environments*. 2d ed. Berlin: Springer-Verlag.

Rensberger, J. M., and H. B. Krentz. 1988. Microscopic effects of predator digestion on the surfaces of bones and teeth. *Scanning Microscopy* 2:1541–51.

Retallack, G. J. 1977. Triassic palaeosols in the upper Narrabeen Group of New South Wales. Part 2: Classification and reconstruction. *Journal of the Geological Society of Australia* 24:19–36.

————. 1982. Paleopedological perspectives on the development of grasslands during the Tertiary. *Proceedings of the Third North American Paleontological Convention* 2:417–21.

————. 1983. A paleopedological approach to the interpretation of terrestrial sedimentary rocks: The mid-Tertiary fossil soils of Badlands National Park, South Dakota. *Geological Society of America Bulletin* 94:823–40.

————. 1984a. Completeness of the rock and fossil record: Some estimates using fossil soils. *Paleobiology* 10:59–78.

————. 1984b. Trace fossils of burrowing beetles and bees in an Oligocene paleosol, Badlands National Park, South Dakota. *Journal of Paleontology* 58:571–92.

————. 1986. Fossil soils as grounds for interpreting long-term controls on ancient rivers. *Journal of Sedimentary Petrology* 56:1–18.

————. 1988. Field recognition of paleosols made simple. *Geological Society of America Special Paper* 216:1–20.

Reynolds, R. L. 1985. Domestic dog associated with human remains at Rancho La Brea. *Bulletin of the Southern California Academy of Sciences* 84:76–85.

Rich, F. J., and W. Spackman. 1979. Modern and ancient pollen sedimentation around tree islands in the Okefenokee Swamp. *Palynology* 3:219–26.

Riegel, W., V. Wilde, and G. Rebzee. 1986. Erste Ergebnisse einer paläobotanischen Grabung in der fluviatilen Wealden-Facies des Osterwaldes bei Hannover. *Courier Forschungs-Institut Senckenberg* 86:137–70.

Rieppel, O. 1980. Green anole in Dominican amber. *Nature* 286:486–87.

Rigby, J. K., Jr., K. R. Newman, J. Smit, S. Van Der Kaars, R. E. Sloan, and J. K. Rigby. 1987. Dinosaurs from the Paleocene part of the Hell Creek Formation, McCone County, Montana. *Palaios* 2:296–302.

Riggs, S. R. 1984. Paleoceanographic model of Neogene phosphorite deposition, U.S. Atlantic continental margin. *Science* 223:123–31.

Robinson, D. E., ed. 1990. *Experimentation and Reconstruction in Environmental Archaeology.* Oxford, England: Oxbow Books.

Robinson, P. L. 1957. The Mesozoic fissures of the Bristol Channel area and their vertebrate fauna. *Journal of the Linnean Society of London, Zoology* 43:260–82.

Roehler, H. W. 1988. The Pintail coal bed and barrier bar G: A model for coal of barrier bar-lagoon origin, Upper Cretaceous Almond Formation, Rock Springs coal field, Wyoming. *United States Geological Survey Professional Paper* 1398:1–60.

Rogers, R. R. 1990. Taphonomy of three dinosaur bone beds in the Upper Cretaceous Two Medicine Formation of northwestern Montana: Evidence for drought-related mortality. *Palaios* 5:394–413.

Rolfe, W. D., F. R. Sohram, G. Pacaud, D. Sotty, and S. Secretan. 1982. A remarkable Stephanian biota from Montceau-les-Mines, France. *Journal of Paleontology* 56:426–28.

Rolfe, W. D. I., G. P. Durant, A. E. Fallick, A. J. Hall, D. J. Large, A. C. Scott, T. R. Smithson, and G. Walkden. 1990. An early terrestrial biota preserved by Visean vulcanicity in Scotland. *Geological Society of America Special Paper* 244:13–24.

Romer, A. S., and E. C. Olson. 1954. Aestivation in a Permian lungfish. *Breviora* 30:1–8.

Rose, K. D. 1981. Composition and species diversity in Paleocene and Eocene mammal assemblages: An empirical study. *Journal of Vertebrate Paleontology* 1:367–88.

———. 1984. Evolution and radiation of mammals in the Eocene, and the diversification of modern orders. *University of Tennessee Department of Geological Sciences Studies in Geology* 8:110–27.

Ruhe, R. V. 1983. Depositional environment of late Wisconsin loess in the midcontinental United States. 130–37 in S. C. Porter, ed., *Late Quaternary Environments of the United States.* Vol. 1, *The Late Pleistocene.* Minneapolis: University of Minnesota Press.

Russell, D. A., P. Béland, and J. S. McIntosh. 1980. Paleoecology of the dinosaurs of Tendaguru (Tanzania). *Mémoires Société Géologique de France*, n.s. 139:169–75.

Russell, D. A. and T. P. Chamney. 1967. Notes on the biostratigraphy of dinosaurian and microfossil faunas in the Edmonton Formation (Cretaceous), Alberta. *National Museum of Canada Natural History Papers* 35:1–22.

Russell, D. E. 1964. Les mammifères paléocenes d'Europe. *Mémoire Musée Nationale d'Histoire Naturelle Paris, C* 13:1–324.

Rust, B. R., M. R. Gibling, and A. S. Legun. 1984. Coal deposition in an anastomosing fluvial system: The Pennsylvanian Cumberland Group south of Joggins, Nova Scotia, Canada. *Special Publication of the International Association of Sedimentologists* 7:105–20.

Sadler, P. M. 1981. Sediment accumulation rates and the completeness of stratigraphic sections. *Journal of Geology* 89:569–84.

Sadler, P. M., and L. W. Dingus. 1982. Expected completeness of sedimentary sections: Estimating a time-scale dependent, limiting factor in the resolution of the fossil record. *Proceedings of the Third North American Paleontological Convention* 2:461–64.

Sahni, A. 1972. The vertebrate fauna of the Judith River Formation, Montana. *American Museum of Natural History Bulletin* 147:321–412.

Salisbury, E. J. 1952. *Dunes and Downs.* London: G. Bell & Sons.

Salo, J., R. Kalliola, I. Hakkinen, Y. Makinen, P. Niemela, M. Puhakka, and P. Coley. 1986. River dynamics and the diversity of Amazon lowland forest. *Nature* 322:254–58.

Sander, P. M. 1987. Taphonomy of the Lower Permian Geraldine Bonebed in Archer County, Texas. *Palaeogeography, Palaeoclimatology, Palaeoecology* 61:221–36.

———. 1989. Early Permian depositional environments and pond bonebeds in central Archer County, Texas. *Palaeogeography, Palaeoclimatology, Palaeoecology* 69:1–21.

Sander, P. M., and C. R. Gee. 1990. Fossil charcoal: Techniques and applications. *Review of Palaeobotany and Palynology* 63:269–79.

Sarjeant, W. A. S., and D. J. Mossman. 1978. Vertebrate footprints from the Carboniferous sediments of Nova Scotia: A historical review and description of newly discovered forms. *Palaeogeography, Palaeoclimatology, Palaeoecology* 23:279–306.

Saulnier, H. S. 1950. The paleopalynology of the (Paleocene) Fort Union coals of Red Lodge, Montana. Master's thesis, University of Massachusetts.

Saunders, J. J. 1977. Late Pleistocene vertebrates of the western Ozark Highland, Missouri. *Illinois State Museum Reports of Investigations* 33:1–118.

———. 1988. Fossiliferous spring sites in southwestern Missouri. *Bulletin of the Buffalo Society of Natural Sciences* 33:127–49.

Saunders, J. J., C. V. Haynes, D. J. Stanford, and G. Agogino. 1990. A mammoth ivory semifabricate from Blackwater Draw locality 1, New Mexico. *American Antiquity* 55:112–19.

Schaal, S., and W. Ziegler, eds. 1988. *Messel: Ein Schaufenster in die Geschichte der Erde und des Lebens*. Frankfurt am Main: Verlag Waldemar Kramer.

Schäfer, W. 1972. *Ecology and Paleoecology of Marine Environments*. Chicago: University of Chicago Press.

Scheckler, S. 1986. Geology, floristics, and paleoecology of late Devonian coal swamps from Appalachian Laurentia (USA). *Annales de la Société Géologique de Belgique* 109:209–22.

Scheihing, M. H., and H. M. Pfefferkorn. 1984. The taphonomy of land plants in the Orinoco Delta: A model for the incorporation of plant parts in clastic sediment of Late Carboniferous age of Euramerica. *Review of Palaeobotany and Palynology* 41:205–40.

Schindel, D. E. 1980. Microstratigraphic sampling and the limits of paleontologic resolution. *Paleobiology* 6:408–26.

Schoch, W. 1986. Wood and charcoal analysis. 619–26 in B. E. Berglund, ed., *Handbook of Holocene Palaeoecology and Palaeohydrology*. New York: Wiley.

Scholle, P. A., D. G. Bebout, and C. H. Moore, eds. 1983. Carbonate depositional environments. *American Association of Petroleum Geologists Memoir* 33:1–708.

Scholle, P. A., and D. Spearing, eds. 1982. Sandstone depositional environments. *American Association of Petroleum Geologists Memoir* 31:1–410.

Schopf, J. M. 1952. Was decay important in the origin of coal? *Journal of Sedimentary Petrology* 22:61–69.

Schultze, H.-P. 1985. Marine to onshore vertebrates in the Lower Permian of Kansas and their paleoenvironmental implications. *University of Kansas Paleontological Contributions* 113:1–18.

Schumm, S. A. 1977. *The Fluvial System*. New York: John Wiley and Sons.

Schwartz, H. L. 1983. Paleoecology of Late Cenozoic fishes from the Turkana Basin, northern Kenya. Ph.D. diss., University of California, Santa Cruz.

Schwartz, H. L., and D. D. Gillette. In press. Geology and taphonomy of the *Coelophysis* quarry, Upper Triassic Chinle Fm, Ghost Ranch, New Mexico. *Journal of Paleontology.*

Scott, A. C. 1977. Coprolites containing plant material from the Carboniferous of Britain. *Palaeontology* 20:59–68.

————. 1978. Sedimentological and ecological control of Westphalian B plant assemblages from West Yorkshire. *Proceedings of the Yorkshire Geological Society* 41:462–508.

————. 1980. The ecology of some Upper Paleozoic floras. 87–115 in A. Panchen, ed., *The Terrestrial Environment and Origin of Land Vertebrates.* London: Academic Press.

————. 1989. Observations on the nature and origin of fusain. *International Journal of Coal Geology* 12:443–75.

Scott, A. C., ed. 1987. Coal and Coal-bearing Strata: Recent Advances. *Geological Society Special Publication* 32:1–332.

Scott, A. C., and G. R. King. 1981. Megaspores and coal facies: An example from the Westphalian A of Leicestershire, England. *Review of Palaeobotany and Palynology* 34:107–13.

Scott, A. C., and G. Rex. 1985. The formation and significance of Carboniferous coal balls. *Philosophical Transactions of the Royal Society of London, B* 311:123–37.

Seemann, R. 1933. Das Saurischierlager in den Keupermergeln bei Trossingen. *Jahresberichte des Vereins für vaterländische Naturkunde* 89:129–60.

Seilacher, A. 1982. General remarks about event deposits. 161–73 in G. Einsele and A. Seilacher, eds., *Cyclic and Event Stratification.* Berlin: Springer-Verlag.

Seilacher, A., W. E. Reif, and F. Westphal. 1985. Sedimentological, ecological, and temporal patterns of fossil Lagerstätten. *Philosophical Transactions of the Royal Society of London, B* 311:5–23.

Selley, R. C. 1986. *Ancient Sedimentary Environments.* 3d ed. Ithaca: Cornell University Press.

Sepkoski, J. J., Jr. 1976. Species diversity in the Phanerozoic: Species-area effects. *Paleobiology* 2:298–303.

Sheppe, W. 1972. The annual cycle of small mammal populations on a Zambian floodplain. *Journal of Mammalogy* 53:445–60.

Shipman, P. 1981. *Life History of a Fossil.* Cambridge: Harvard University Press.

Shipman, P., and J. Rose. 1983. Early hominid hunting, butchering, and carcass-processing behaviors: Approaches to the fossil record. *Journal of Anthropological Archaeology* 2:57–98.

Shipman, P., A. Walker, J. A. Van Couvering, P. J. Hooker, and J. A. Miller. 1981. The Fort Ternan hominoid site, Kenya: Geology, age, taphonomy and paleoecology. *Journal of Human Evolution* 10:49–72.

Simons, J. W. 1966. The presence of leopard and a study of the food debris in the leopard lairs of the Mount Suswa Caves. *Bulletin of Cave Exploration Group of East Africa* 1:51–69.

Skinner, J. D., S. Davis, and G. Ilani. 1980. Bone collecting by striped hyaenas, *Hyaena hyaena,* in Israel. *Palaeontologia Africana* 23:99–104.

Sloan, R. E., J. K. Rigby, Jr., L. M. Van Valen, and D. Gabriel. 1986. Gradual dinosaur extinction and simultaneous ungulate radiation in the Hell Creek Formation. *Science* 232:629–33.

Smart, T. L., and E. S. Hoffman. 1988. Environmental interpretation of archaeological charcoal. 167–205 in C. A. Hastorf and V. S. Popper, eds., *Current Paleoethnobotany: Analytical Methods and Cultural Interpretations of Archaeological Plant Remains*. Chicago: University of Chicago Press.

Smiley, C. J., ed. 1985. *Late Cenozoic History of the Pacific Northwest*. San Francisco: American Association for the Advancement of Science.

Smith, A. H. V. 1962. The palaeoecology of Carboniferous peats based on the miospores and petrography of bituminous coals. *Proceedings of the Yorkshire Geological Society* 33:423–74.

———. 1963. Palaeoecology of Carboniferous peats. 57–66, 74–75 in A. E. M. Nairn, ed., *Problems in Palaeoclimatology*. London: Interscience.

———. 1968. Seam profiles and seam characters. 31–40 in D. G. Murchison and T. S. Westoll, eds., *Coal and Coal-Bearing Strata*. Edinburgh: Oliver and Boyd.

Smith, A. H. V., and M. A. Butterworth. 1967. Miospores in the coal seams of the Carboniferous of Great Britain. *Special Papers in Palaeontology* 1:1–324.

Smith, G. A. 1987. The influence of explosive volcanism on fluvial sedimentation: The Deschutes Formation (Neogene) in Central Oregon. *Journal of Sedimentary Petrology* 57:613–29.

Smith, G. R. 1975. Fishes of the Pliocene Glenns Ferry Fm, southwest Idaho. *Museum of Paleontology University of Michigan Papers on Paleontology* 14:1–68.

Smith, G. R., and R. L. Elder. 1985. Environmental interpretation of Clarkia fishes. 85–93 in C. J. Smiley, ed., *Later Cenozoic History of the Pacific Northwest*. San Francisco: American Association for the Advancement of Science.

Smith, G. R., R. F. Stearley, and C. E. Badgley. 1988. Taphonomic bias in fish diversity from Cenozoic floodplain environments. *Palaeogeography, Palaeoclimatology, Palaeoecology* 63:263–73.

Smith, N. D., T. A. Cross, J. P. Dufficy, and S. R. Clough. 1989. Anatomy of an avulsion. *Sedimentology* 36:1–23.

Smith, R. M. H. 1980. The lithology, sedimentology, and taphonomy of floodplain deposits of the Lower Beaufort (Adelaide Subgroup) strata near Beaufort West. *Transactions of the Geological Society of South Africa* 83:399–413.

———. 1987. Helical burrow casts of therapsid origin from the Beaufort Group (Permian) of South Africa. *Palaeogeography, Palaeoclimatology, Palaeoecology* 60:155–70.

———. 1990. Alluvial paleosols and pedofacies sequences in the Permian Lower Beaufort of the southwestern Karoo Basin, South Africa. *Journal of Sedimentary Petrology* 60:258–76.

Smoot, J. P., and P. E. Olsen. 1988. Massive mudstones in basin analysis and paleoclimatic interpretation of the Newark Supergroup. 249–74 in W. Manspeizer, ed., *Triassic-Jurassic Rifting: Continental Breakup, and the Origin of the Atlantic Ocean and Passive Margins*. New York: Elsevier.

Solomon, S., I. Davidson, and D. Watson, eds. 1991. Problem Solving in Taphonomy. *Tempus* 2. Saint Lucia, Queensland, Australia: Anthropology Department, University of Queensland.

Spackman, W., Jr. 1949. The flora of the Brandon Lignite: Geological aspects and a comparison of the flora with its modern equivalents. Ph.D. diss., Harvard University, Cambridge, Mass.

Spaulding, W. G. 1985. Vegetation and climates of the last 45,000 years in the vicinity of the Nevada Test Site. *United States Geological Survey Professional Paper* 1329:1–83.

Spaulding, W. G., and L. J. Graumlich. 1986. The last pluvial climatic episodes in the deserts of southwestern North America. *Nature* 320:441–44.

Spicer, R. A. 1980. The importance of depositional sorting to the biostratigraphy of plant megafossils. 171–83 in D. L. Dilcher and T. N. Taylor, eds., *Biostratigraphy of Fossil Plants*. Stroudsburg, Pa.: Dowden, Hutchinson and Ross.

————. 1981. The sorting and deposition of allochthonous plant material in a modern environment at Silwood Lake, Silwood Park, Berkshire, England. *United States Geological Survey Professional Paper* 1143:1–77.

————. 1987. The significance of the Cretaceous flora of northern Alaska for the reconstruction of the Cretaceous climate. *Geologisches Jahrbuch* A 96:265–91.

————. 1989. The formation and interpretation of plant fossil assemblages. 95–191 in J. Callow, ed., *Advances in Botanical Research*. New York: Academic Press.

Spicer, R. A., and A. G. Greer. 1986. Plant taphonomy in fluvial and lacustrine systems. *University of Tennessee Department of Geological Sciences, Studies in Geology* 15:10–26.

Spicer, R. A., and J. T. Parrish. 1986. Paleobotanical evidence for cool North Polar climates in middle Cretaceous (Albian-Cenomanian) time. *Geology* 14:703–6.

————. 1987. Plant megafossils, vertebrate remains, and paleoclimate of the Kogosukruk Tongue (Late Cretaceous), North Slope, Alaska. *United States Geological Survey Circular* 993:47–48.

Spicer, R. A., and B. A. Thomas. 1987. A Mississippian Alaska-Siberian connection: Evidence from plant megafossils. 355–358 in P. Weimer and I. L. Tailleur, eds., *Alaskan North Slope Geology*. Society of Paleontologists and Mineralogists and American Association of Petroleum Geologists.

Spicer, R. A., and J. A. Wolfe. 1987. Plant taphonomy of late Holocene deposits in Trinity (Clair Engle) Lake, northern California. *Paleobiology* 13:227–45.

Spicer, R. A., J. Yao, and M. A. Horrell. 1988. Early Cretaceous phytogeography and climate of China based on numerical analysis of megafossils. *Abstracts: International Organization of Paleobotany Conference*. August 1988 (Melbourne).

Spillmann, F. 1959. Die Sirenen aus dem Oligozaen des Linzer Beckens (Oberoesterreich), mit Ausfuehrungen ueber "Osteosklerose" und "Pachyostose." *Akademie der Wissenschaften, Vienna, Denkschriften* 110:1–65.

Stach, E., M.-Th. Mackowshy, M. Teichmüller, G. H. Taylor, D. Chandra, and R. Teichmüller. 1982. *Stach's Textbook of Coal Petrology*. 3d ed. Stuttgart: Gebrüder Borntraeger.

Staff, G. M., and E. N. Powell. 1988. The paleoecological significance of diversity: The effect of time averaging and differential preservation on macroinvertebrate species richness in death assemblages. *Palaeogeography, Palaeoclimatology, Palaeoecology* 63:73–90.

Stanford, D. J. and R. W. Graham. 1978. Archeological investigation of the Selby and

Dutton Sites, Yuma County, Colorado. *National Geographic Society Research Reports* 19:519–41.

Staub, J. R., and A. D. Cohen. 1978. Kaolinite-enrichment beneath coals: A modern analog, Snuggedy Swamp, South Carolina. *Journal of Sedimentary Petrology* 48: 203–10.

———. 1979. The Snuggedy Swamp of South Carolina: A back-barrier estuarine coal-forming environment. *Journal of Sedimentary Petrology* 49:133–44.

Stiner, M. C. 1991. The cultural significance of Grotta Guattari reconsidered. I. The faunal remains from Grotta Guattari: A taphonomic perspective. *Current Anthropology* 32:103–17.

Stock, C. 1929. Significance of abraded and weathered mammalian remains from Rancho La Brea. *Bulletin of the Southern California Academy of Science* 28:1–5.

———. 1972. Rancho La Brea, a record of Pleistocene life in California. *Los Angeles County Museum Science Series* 20:1–81.

Stopes, M. C. 1912. Petrifactions of the earliest European angiosperms. *Philosophical Transactions of the Royal Society of London, B* 203:75–100.

Stopes, M. C., and D. M. S. Watson. 1909. On the present distribution and origin of the calcareous concretions in coal seams known as "coal balls." *Philosophical Transactions of the Royal Society of London, B* 200:167–218.

Stucky, R. K., L. Krishtalka, and A. D. Redline. 1990. Geology, vertebrate fauna, and paleoecology of the Buck Spring Quarries (early Eocene, Wind River Formation), Wyoming. *Geological Society of America Special Paper* 243:169–86.

Sues, H.-D., N. H. Shubin, and P. E. Olsen. 1987. A diapsid faunule from the Lower Jurassic of Nova Scotia, Canada. *Occasional Paper of the Tyrrell Museum of Paleontology* 3:205–7.

Summerhayes, C. P. 1981. Organic facies of Middle Cretaceous black shales in deep North Atlantic. *American Association of Petroleum Geologists Bulletin* 65: 2364–80.

Surdam, R. C., and C. A. Wolfbauer. 1975. Green River Formation, Wyoming: A playa-lake complex. *Geological Society of America Bulletin* 86:335–45.

Sutcliffe, A. J. 1970. A section of an imaginary bone cave. *Studies in Speleology* 2:79–80.

———. 1986. *On the Track of Ice Age Mammals*. London: British Museum (Natural History).

Swain, A. M. 1973. A history of fire and vegetation in northeastern Minnesota as recorded in lake sediments. *Quaternary Research* 3:383–96.

———. 1978. Environmental changes during the past 2,000 years in north-central Wisconsin: Analysis of pollen, charcoal, and seeds from varved lake sediments. *Quaternary Research* 10:55–68.

Sweet, A. R. 1987. Sedimentary facies and environmentally controlled palynological assemblages: Their relevance to floral changes at the Cretaceous-Tertiary boundary. *Occasional Paper of the Tyrrell Museum of Palaeontology* 3:208–13.

Swift, C. C. 1979. Freshwater fish of the Rancho La Brea deposit. *Abstracts Southern California Academy Sciences Annual Meeting:* 44.

Symoens, J. J., M. Burgis, and J. J. Gaudet, eds. 1981. *The Ecology and Utilization of African Inland Waters*. Nairobi, Kenya: United Nations Environment Program.

Taggart, R. E., and A. T. Cross. 1980. Vegetation change in the Miocene Succor Creek Flora of Oregon and Idaho: Case study in paleosuccession. 185–210 in D. L. Dilcher and T. N. Taylor, eds., *Biostratigraphy of Fossil Plants*. Stroudsburg, Pa.: Dowden, Hutchinson and Ross.

Talbot, M. R., and K. Kelts, eds. 1989. The Phanerozoic record of lacustrine basins and their environmental signals. *Palaeogeography, Palaeoclimatology, Palaeoecology* 70:1–304.

Talbot, M. R., and D. A. Livingstone. 1989. Hydrogen index and carbon isotopes of lacustrine organic matter as lake level indicators. *Palaeogeography, Palaeoclimatology, Palaeoecology* 70:121–38.

Tauxe, L., and C. E. Badgley. 1988. Stratigraphy and remanence acquisition of a paleomagnetic reversal in alluvial Siwalik rocks of Pakistan. *Sedimentology* 35: 697–715.

Tauxe, L., C. G. Constable, L. Stokking, and C. Badgley. 1990. The use of anisotropy to distinguish the origin of characteristic remanence in the Siwalik red beds of Northern Pakistan. *Journal of Geophysical Research* 95:4391–4404.

Taylor, T. N., and A. C. Scott. 1983. Interactions of plants and animals during the Carboniferous. *Bioscience* 33:488–93.

Teichmüller, M. 1975. Origin of the petrographic constituents of coal. 176–238 in E. Stach, ed., *Stach's Textbook of Coal Petrology*. Berlin: Gebrüder Borntraeger.

———. 1987. Recent advances in coalification studies and their application to geology. 127–69 in A. C. Scott, ed., *Coal and Coal-Bearing Strata: Recent Advances*. Geological Society Special Publication 32.

———. 1990. The genesis of coal from the viewpoint of coal geology. *International Journal of Coal Geology* 16:121–24.

Thomas, K. D. 1985. Land snail analysis in archaeology: Theory and practice. *British Archaeological Reports* 266:67–92.

Thomasson, J. R. 1982. Fossil grass anthoecia and other plant fossils from arthropod burrows in the Miocene of western Nebraska. *Journal of Paleontology* 56:1011–17.

Thompson, R. S., T. R. Van Devender, P. S. Martin, T. Foppe, and A. Long. 1980. Shasta ground sloth (*Nothrotheriops shastense* Hoffstetter) at Shelter Cave, New Mexico: Environment, diet, and extinction. *Quaternary Research* 14:360–76.

Tiffney, B. H. 1981. Fruits and seeds of the Brandon Lignite, 6: Microdiptera (Lythraceae). *Journal of the Arnold Arobretum* 62:487–516.

Tilzer, M. M., and C. Serruya. 1990. *Large Lakes: Ecological Structure and Function*. Berlin: Springer-Verlag.

Ting, F. T. C. 1972. Petrified peat from a Paleocene lignite in North Dakota. *Science* 177:165–66.

Tipper, J. C. 1983. Rates of sedimentation and stratigraphical completeness. *Nature* 302:696–98.

Tobien, H. 1965. Insekten-Frasspuren an tertiären und pleistozänen Säugetier-Knochen. *Senckenbergiana lethaea (Weiler-Festschrift)* 46a:441–51.

———. 1968. Typen und Genese tertiärer Säugerlagerstätten. *Ecologae geologicae Helvetiae* 61:549–75.

———. 1986. Die jungtertiäre Fossilgrabungsstätte Höwenegg im Hegau (Südwestdeutschland): Ein Statusbericht. *Carolinea* 44:9–34.

Todd, L. C. 1987. Analysis of kill-butchery bonebeds and interpretation of Paleo-indian hunting. 225–66 in M. H. Nitecki and D. V. Nitecki, eds., *The Evolution of Human Hunting*. New York: Plenum Press.

Trewin, N. H. 1986. Palaeoecology and sedimentology of the Achanarras fish bed of the Middle Old Red Sandstone, Scotland. *Transactions of the Royal Society of Edinburgh: Earth Sciences* 77:21–46.

Turnbull, W. D., and D. M. Martill. 1988. Taphonomy and preservation of a mono-specific titanothere assemblage from the Washakie Formation (Late Eocene), southern Wyoming. *Palaeogeography, Palaeoclimatology, Palaeoecology* 63:91–108.

Tye, R. W., and E. C. Kosters. 1986. Styles of interdistributary basin sedimentation: Mississippi Delta Plain, Louisiana. *Transactions Gulf Coast Association of Geological Societies* 36:575–88.

Van Devender, T. R. 1986. Pleistocene climates and endemism in the Chihuahuan desert flora. 1–19 in J. C. Barlow, A. M. Powell, and B. N. Timmermann, eds., *Second Symposium on Resources of the Chihuahuan Desert Region*. Alpine, Tex.: Chihuahuan Desert Research Institute.

———. 1987. Holocene vegetation and climate in the Puerto Blanco Mountains, southwestern Arizona. *Quaternary Research* 27:51–72.

Van Devender, T. R., and W. G. Spaulding. 1979. Development of vegetation and climate in the southwestern United States. *Science* 204:701–10.

Van Devender, T. R., R. S. Thompson, and J. L. Betancourt. 1987. Vegetation history of the deserts of southwestern North America: The nature and timing of the Late Wisconsin-Holocene transition. *Geological Society of America, The Geology of North America* K-3:323–52.

Van Dijk, D. E., D. K. Hobday, and A. J. Tankard. 1978. Permo-Triassic lacustrine deposits in the Eastern Karoo Basin, Natal, South Africa. *Special Publications, International Association Sedimentologists* 2:225–39.

Vicars, R. G., and J. A. Breyer. 1981. Sedimentary facies in air-fall pyroclastic debris, Arikaree Group (Miocene), northwest Nebraska, USA. *Journal of Sedimentary Petrology* 51:909–21.

Viohl, G. 1985. Geology of the Solnhofen lithographic limestones and the habitat of *Archaeopteryx*. 31–44 in M. K. Hecht, J. H. Ostrom, G. Viohl, and P. Wellnhofer, eds., *The Beginning of Birds*. Eichstätt: Jura-Museum.

Vondra, C. F., and B. E. Bowen. 1976. Plio-Pleistocene deposits and environments, East Rudolf, Kenya. 79–93 in Y. Coppens, F. C. Howell, G. L. Isaac, and R. E. Leakey, eds., *Earliest Man and Environments in the Lake Rudolf Basin: Stratigraphy, Paleoecology, and Evolution*. Chicago: University of Chicago Press.

Voorhies, M. R. 1969. Taphonomy and population dynamics of an early Pliocene vertebrate fauna, Knox County, Nebraska. *University of Wyoming Contributions to Geology Special Paper* 1:1–69.

———. 1974. Fossil pocket mouse burrows in Nebraska. *American Midland Naturalist* 91:492–98.

———. 1975. Vertebrate burrows. 325–50 in R. W. Frey, ed., *The Study of Trace Fossils*. New York: Springer-Verlag.

Voorhies, M. R., and J. R. Thomasson. 1979. Fossil grass anthoecia within Miocene rhinoceros skeletons: Diet in an extinct species. *Science* 206:331–33.

Vrba, E. S. 1980. The significance of bovid remains as indicators of environment and predation patterns. 247–71 in A. K. Behrensmeyer and A. P. Hill, eds., *Fossils in the Making*. Chicago: University of Chicago Press.

Waage, K. M. 1968. The type Fox Hills Formation, Cretaceous (Maastrichtian), South Dakota. Part 1: Stratigraphy and paleoenvironments. *Bulletin Yale Peabody Museum of Natural History* 27:1–175.

Wainman, N., and R. W. Mathewes, 1990. Distribution of plant macroremains in surface sediments of Marion Lake, southwestern British Columbia. *Canadian Journal of Botany* 68:364–73.

Waitt, R. B. 1981. Devastating pyroclastic density flow and attendant air fall of May 18: Stratigraphy and sedimentology of deposits. *United States Geological Survey Professional Paper* 1250:439–58.

Waldman, M. 1971. Fish from the freshwater Lower Cretaceous of Victoria, Australia, with comments on the palaeo-environment. *Palaeontological Association Special Papers in Palaeontology* 9:1–124.

Walker, A. W., and M. Teaford. 1989. The hunt for *Proconsul*. *Scientific American* 260:76–82.

Walker, R. G., ed. 1984. *Facies Models*. 2d ed. Toronto: Geological Association of Canada, Geosciences Canada Reprint Series 1:1–317.

Walker, T. R. 1967. Formation of red beds in modern and ancient deserts. *Geological Society of America Bulletin* 78:353–68.

Walter, H. 1978. Zur Paläontologie der Hornburger Schichten (Rotliegendes) unter besonderer Berücksichtigung der Aufschlüsse von Rothenschirmbach und Sittichenbach (DDR). *Freiberger Forschungshefte* C 334:163–75.

Walton, J. 1935. Scottish Lower Carboniferous plants: The fossil hollow trees of Arron and their branches (*Lepidophloios wunshinnus* Caruthers). *Transactions of the Royal Society of Edinburgh* 58:313–37.

Wang, X. 1988. Systematics and population ecology of Late Pleistocene Bighorn Sheep (*Ovis canadensis*) of Natural Trap Cave, Wyoming. *Transactions of the Nebraska Academy of Sciences* 16:173–83.

Ward, R. G. W., ed. 1987. Applications of tree-ring studies: Current research in dendrochronology and related subjects. *British Archaeological Report* 5333.

Warner, B. G., and P. J. Barnett. 1986. Transport, sorting, and reworking of late Wisconsinan plant macrofossil from Lake Erie, Canada. *Boreas* 15:323–29.

Warner, B. G., P. F. Karrow, A. V. Morgan, and A. Morgan. 1987. Plant and insect fossils from Nipissing sediments along the Goulais River, southeastern Lake Superior. *Canadian Journal of Earth Sciences* 24:1526–36.

Warwick, P. D., and R. M. Flores. 1987. Evolution of fluvial styles in the Eocene Wasatch Formation, Powder River Basin, Wyoming. *Society of Economic Paleontologists and Mineralogists Special Publication* 39:303–10.

Watson, P. J. 1969. The prehistory of Salts Cave, Kentucky. *Illinois State Museum Reports of Investigations* 16:1–86.

Webb, R. S., and T. Webb III. 1988. Rates of sediment accumulation in pollen cores from small lakes and mires of eastern North America. *Quaternary Research* 30:284–97.

Webb, S. D. 1974. *Pleistocene Mammals of Florida*. Gainesville: University Presses of Florida.

————. 1981. The Thomas Farm fossil site. *Plaster Jacket* 37:6–25.

Webb, S. D., B. J. MacFadden, and J. A. Baskin. 1981. Geology and paleontology of the Love Bone Bed from the Late Miocene of Florida. *American Journal of Science* 281:513–44.

Webb, T., III. 1985. Holocene palynology and climate. 163–95 in A. D. Hecht, ed., *Paleoclimate Analysis and Modeling*. New York: John Wiley & Sons.

Weigelt, J. 1989. *Recent Vertebrate Carcasses and Their Paleobiological Implications*. Trans. J. Schaefer. Chicago: University of Chicago Press.

Weishampel, D. B. 1984. Trossingen: E. Fraas, F. von Huene, R. Seemann, and the "Schwabische Lindwurm" *Plateosaurus*. 249–53 in W. E. Reif and F. Westphal, eds., *Third Symposium on Mesozoic Terrestrial Ecosystems, Short Papers*. Tübingen: Attempto Verlag.

Wells, R. T. 1978. Fossil mammals in the reconstruction of Quaternary environments with examples from the Australian fauna. 103–24 in D. Walker and J. C. Guppy, eds., *Biology and Quaternary Environments*. Canberra: Australian Academy of Science.

Wendorf, F., and R. Schield. 1980. *Prehistory of the Eastern Sahara*. New York: Academic Press.

Westphal, F. 1959. Neue Wirbeltiere (Fledermäuse, Frösche, Reptilien) aus dem obermiozänen Travertin von Böttingen (Schwäbische Alb). *Neues Jahrbuch für Geologie und Paläontologie, Abhandlungen* 107:341–66.

White, J. A., H. G. McDonald, E. Anderson, and J. M. Soiset. 1984. Lava blisters as carnivore traps. *Special Publication Carnegie Museum of Natural History* 8:241–56.

Whitmore, F. C., Jr., K. O. Emery, H. B. S. Cooke, and D. J. P. Swift. 1967. Elephant teeth from the Atlantic continental shelf. *Science* 156:1477–81.

Whitmore, F. C., Jr., and L. M. Gard. 1977. Steller's Sea Cow (*Hydrodamalis gigas*) of Late Pleistocene Age from Amchitka, Aleutian Islands, Alaska. *United States Geological Survey Professional Paper* 1036:1–19.

Whybrow, P. J., and H. A. McClure. 1981. Fossil mangrove roots and palaeoenvironments of the Miocene of the eastern Arabian peninsula. *Palaeogeography, Palaeoclimatology, Palaeoecology* 32:213–25.

Wighton, D. C., and M. V. H. Wilson. 1986. The Gomphaeschninae (Odonata: Aeshnidae): New fossil genus, reconstructed phylogeny, and geographical history. *Systematic Entomology* 11:505–22.

Willard, D. A. In press. Vegetational patterns in the Springfield Coal (Middle Pennsylvanian, Illinois Basin): Comparison of miospore and coal-ball records. *Geological Society of America Special Paper*.

Williams-Dean, G. J. 1978. Ethnobotany and cultural ecology of prehistoric man in southwest Texas. Ph.D. diss., Department of Anthropology, Texas A&M University, College Station.

Wilson, E. O., F. M. Carpenter, and W. L. Brown, Jr. 1967. The first Mesozoic ants, with the description of a new subfamily. *Psyche* 74:1–19.

Wilson, M. V. H. 1977. Paleoecology of Eocene lacustrine varves at Horsefly, British Columbia. *Canadian Journal of Earth Sciences* 14:953–62.

————. 1978a. Paleogene insect faunas of western North America. *Quaestiones Entomologicae* 14:13–34.

————. 1978b. Evolutionary significance of North American Paleogene insect faunas. *Quaestiones Entomologicae* 14:35–42.

————. 1980. Oldest known *Esox* (Pisces:Esocidae), part of a new Paleocene teleost fauna from western Canada. *Canadian Journal of Earth Sciences* 17:307–12.

————. 1982. Early Cenozoic insects: Paleoenvironmental biases and evolution of the North American insect fauna. *Proceedings Third North American Paleontological Convention* 2:585–88.

————. 1984. Year classes and sexual dimorphism in the Eocene catostomid fish Amyzon aggregatum. *Journal of Vertebrate Paleontology* 3:137–42.

————. 1987a. Predation as a source of fish fossils in Eocene lake sediments. *Palaios* 2:497–504.

————. 1987b. Preliminary study of microevolution and temporal averaging in an Eocene catostomid from varved lacustrine sediments. *American Society of Icthyologists and Herpetologists, Sixty-seventh Annual Meeting Program and Abstracts:* 88.

————. 1988a. Reconstruction of ancient lake environments using both autochthonous and allochthonous fossils. *Palaeogeography, Palaeoclimatology, Palaeoecology* 62:609–23.

————. 1988b. Taphonomic processes: Information loss and information gain. *Geoscience Canada* 15:131–48.

Wilson, P., and R. M. Bateman. 1986. Native and palaeoenvironmental significance of a buried soil sequence at Magilligan Foreland, Northern Ireland. *Boreas* 15: 137–53.

————. 1987. Pedogenic and geomorphic evolution of a buried dune palaeocatena at Magilligan Foreland, Northern Ireland. *Catena* 14:501–17.

Wing, S. L. 1984. Relation of paleovegetation to geometry and cyclicity of some fluvial carbonaceous deposits. *Journal of Sedimentary Petrology* 54:52–66.

————. 1988. Depositional environments of plant-bearing sediments. *Paleontological Society Special Publication* 3:1–13.

Wing, S. L., and T. M. Bown. 1985. Fine-scale reconstruction of late Paleocene-early Eocene paleogeography in the Bighorn basin of northern Wyoming. *Rocky Mountain Section—Society of Economic Paleontologists and Mineralogists, Cenozoic Paleogeography of West-Central United States:* 93–105.

Wing, S. L., and M. B. Farley. 1990. Stability in the composition of Paleogene forests. *Fourth International Congress of Systematic and Evolutionary Biology, Abstracts with Program.*

Winkler, D. A. 1983. Paleoecology of an early Eocene mammalian fauna from paleosols in the Clark's Fork Basin, northwestern Wyoming (USA). *Palaeogeography, Palaeoclimatology, Palaeoecology* 43:261–98.

————. 1987. Vertebrate-bearing eolian unit from the Ogallala Group (Miocene) in northwestern Texas. *Geology* 15:705–8.

Winkler, D. A., L. L. Jacobs, J. D. Congleton, and W. R. Downs. 1987. Taphonomy and paleontology of the Navajo Sandstone. *Journal of Vertebrate Paleontology* 7 (Abstract Supplement, 3):29A.

Winkler, D. A., and P. A. Murry. 1989. Paleoecology and hypsilophodontid behavior at the Proctor Lake dinosaur locality (Early Cretaceous), Texas. *Geological Society of America Special Paper* 238:55–61.

Winkler, D. A., P. A. Murry, L. L. Jacobs, W. R. Downs, J. R. Branch, and P. Trudel.

1988. The Proctor Lake Dinosaur locality, Lower Cretaceous of Texas. *Hunteria* 2:1–8.

Winslow, M. R. 1959. Upper Mississippian and Pennsylvanian megaspores and other plant microfossils from Illinois. *Illinois State Geological Survey Bulletin* 86:1–135.

Winston, R. B. 1988. Paleoecology of Middle Pennsylvanian–age peat-swamp plants in Herrin coal, Kentucky, USA. *International Journal of Coal Geology* 10:203–38.

———. 1989. Identification of plant megafossils in Pennsylvanian-age coal. *Review of Palaeobotany and Palynology* 57:265–76.

———. 1990. Implications of paleobotany of Pennsylvanian-age coal of the central Appalachian basin for climate and coal-bed development. *Geological Society of America Bulletin* 102:1720–26.

Wnuk, C. 1985. Paleoecology of *Lepidodendron rimonsum* and *Lepidodendron bretonenese* trees from the Middle Pennsylvanian of the Bernice Basin (Sullivan County, Pennsylvania). *Palaeontographica, B* 195:153–81.

Wnuk, C., and H. W. Pfefferkorn. 1984. The life habits and paleoecology of Middle Pennsylvanian medullosan pteridosperms based on an in situ assemblage from the Bernice Basin (Sullivan County, Pennsylvania, USA). *Review of Palaeobotany and Palynology* 41:329–51.

———. 1987. A Pennsylvanian-age terrestrial storm deposit: Using plant fossils to characterize the history and process of sediment accumulation. *Journal of Sedimentary Petrology* 57:212–21.

Wolff, R. G. 1973. Hydrodynamic sorting and ecology of a Pleistocene mammalian assemblage from California (USA). *Palaeogeography, Palaeoclimatology, Palaeoecology* 13:91–101.

Wood, J. M., R. G. Thomas, and J. Visser. 1988. Fluvial processes and vertebrate taphonomy: The Upper Cretaceous Judith River Formation, south-central Dinosaur Provincial Park, Alberta, Canada. *Palaeogeography, Palaeoclimatology, Palaeoecology* 66:127–43.

Wood, S. P., A. L. Panchen, and T. R. Smithson. 1985. A terrestrial fauna from the Scottish Lower Carboniferous. *Nature* 314:355–56.

Woodard, G. D., and L. F. Marcus. 1973. Rancho La Brea fossil deposits: A reevaluation from stratigraphic and geological evidence. *Journal of Paleontology* 47:54–68.

———. 1976. Reliability of late Pleistocene correlation using C-14 dating: Baldwin Hills—Rancho La Brea, Los Angeles, California. *Journal of Paleontology* 50:128–32.

Wright, V. P. 1983. A rendzina from the Lower Carboniferous of South Wales. *Sedimentology* 30:159–79.

Wright, V. P. 1987. The ecology of two early Carboniferous paleosols. 345–58 in J. Miller, A. E. Adams, and V. P. Wright, eds., *European Dinantian Environments*. Chichester: John Wiley & Sons.

Wright, V. P., ed. 1986. *Paleosols: Their Recognition and Interpretation*. Princeton: Princeton University Press.

Wu, Z., and F. Gao. 1985. The formation of loess in China. 137–38 in T. Liu, ed., *Quaternary Geology and Environment of China*. Beijing: China Ocean Press.

Wuttke, M. 1983a. "Weichteil-Erhaltung" durch lithifizierte Mikroorganismen bei mittel-eozänen Vertebraten aus den Ölschiefern der "Grube Messel." *Senckenbergiana lethaea* 64:509–27.

————. 1983b. Aktuopaläontologische Studien über den Zerfall von Wirbeltieren. Teil I: Anura. *Senckenbergiana lethaea* 64:529–60.

Yellen, J. 1977. *Archeological Approaches to the Present.* New York: Academic Press.

Yemane, K., C. Siegenthaler, and K. Kelts. 1989. Lacustrine environments during Lower Beaufort (Upper Permian) Karoo deposition in northern Malawi. *Palaeogeography, Palaeoclimatology, Palaeoecology* 70:165–78.

Young, C. C. 1932. On the fossil vertebrate remains from localities 2, 7, and 8 at Choukoutien. *Palaeontologia Sinica,* C 7:1–21.

Yuretich, R. F. 1984. Yellowstone fossil forests: New evidence for burial in place. *Geology* 12:159–62.

Yuretich, R. F., L. J. Hickey, B. P. Gregson, and Y. L. Hsia. 1984. Lacustrine deposits in the Paleocene Fort Union Formation, northern Bighorn Basin, Montana. *Journal of Sedimentary Petrology* 54:836–52.

Zangerl, R., and E. S. Richardson, Jr. 1963. The paleoecological history of two Pennsylvanian black shales. *Fieldiana, Geology Memoirs* 4:1–352.

Zapfe, H. 1954. Beiträge zur Erklärung der Entstehung von Knochenlagerstätten in Karstspalten und Höhlen. *Geologie, Beihefte* 12:1–60.

Ziegler, A. M. 1989. Phytogeographic patterns and continental configurations during the Permian Period. 363–79 in W. S. McKerrow and C. R. Scotese, eds., *Palaeozoic Palaeogeography and Biogeography. Geological Society of London Memoir* no. 12.

Ziegler, A. M., A. L. Raymond, T. C. Gierlowski, M. A. Horrell, D. B. Rowley, and A. L. Lottes. 1987. Coal, climate, and terrestrial productivity: The present and Early Cretaceous compared. 25–49 in A. C. Scott, ed., *Coal and Coal-Bearing Strata: Recent Advances. Geological Society Special Publication* 32.

T H R E E Ecological Characterization of Fossil Plants

Scott L. Wing and William A. DiMichele, RAPPORTEURS

IN COLLABORATION WITH SUSAN J. MAZER, TOM L. PHILLIPS, W. GEOFFREY
SPAULDING, RALPH E. TAGGART, AND BRUCE H. TIFFNEY

1 INTRODUCTION

As discussed in chapter 1, one of the central aims of evolutionary paleoecology is to compare communities of different times with each other and with
living communities. This is done in order to search for underlying ecological
interactions that are independent of taxonomic composition and time period,
to determine the extent to which neoecological principles can explain the
structural properties of ancient communities, and to assess how ecological relationships have influenced the evolutionary history of groups of organisms.

This chapter is a summary of the informational and methodological bases
for comparing ancient plant communities. Plant paleoecology is undertaken at
a variety of spatial and temporal scales that form a loose hierarchy. At the
finest scale of analysis is the functional interpretation of plant organs, which
usually are not preserved connected to one another. Section 2.1 of this chapter
reviews some of the physiological and biomechanical correlations of form and
function that make these interpretations possible.

At the next level of inclusiveness, the autecology of an extinct species can
be inferred from an integration of the functional-ecological analyses of its
component organs in combination with information on its depositional setting
and the ecological characteristics of any close living relatives. The methods
used at this level of analysis are discussed in §§2.2 and 2.3.

The reconstruction of ancient vegetation is an even more inclusive level of
paleoecological analysis. Paleosynecology relies in part on information derived from the individual autecological analyses of the species present in the
fossil assemblage, but there are also sources of data and methodological problems unique to this level. A primary consideration in paleosynecology is the
amount of time over which the fossil assemblage being studied accumulated.
Recently developed understanding of the depositional environments of fossil
plants and how those environments sample the original vegetation (chap. 2),
shows that some plant fossil assemblages are nearly autochthonous and represent very short periods of time (a few years). This means that the plant fossil

record preserves information about ancient vegetation on very fine temporal and spatial scales, and in some cases patterns can be resolved with the refinement attainable in living vegetation. The ability to reconstruct very short-term and small-scale ecological patterns is significant because it permits direct comparisons of fossil and extant vegetation in terms of composition and spatial structure. Fossil assemblages closely spaced in time or associated with particular environmental indicators may permit the description of short-term dynamic aspects of ancient vegetation as well (e.g., phenomena related to succession and disturbance).

In §5 we discuss the means by which comparisons of vegetation from different time periods can be carried out. These methods involve different ways of characterizing plant communities independently of their species composition; they have been referred to as taxon independent. This is still a small part of paleoecology, but eventually it may yield the greatest insights into general ecological organization and the context for major evolutionary trends.

All these approaches are treated in more detail below, but we note here that paleoecology begins in the field. Commitment to ambitious programs of interdisciplinary research and collecting, careful consideration of field-based sampling strategies, and detailed information on depositional and stratigraphic context are all necessary predecessors to more informative paleoecological analyses.

2   AUTECOLOGY OF EXTINCT PLANTS

Three kinds of information have been used to infer the ecological roles of extinct plant species. The first is functional morphology-anatomy; the function, and to some degree the ecological importance, of characteristics seen in fossils can be inferred by using basic principles of physics and chemistry applicable to all plants, or can be interpreted by analogy with the function of similar traits seen in living plants. The second approach to understanding the autecology of extinct species relies on analogy with the ecological preferences of closely related living species. This approach is most reliable for Pleistocene and Neogene floras. As mentioned in chapter 1, analogy with living relatives implicitly assumes that closely related species are morphologically similar, so analogy with living relatives can be viewed as a subset of functional morphology-anatomy. The third kind of information used to infer ecological characteristics of extinct species is the sedimentary context of autochthonous fossils, which may indicate the physical and chemical aspects of the environment in which the plant grew. Each of these data sources is treated in greater detail below.

2.1   *Inferring Autecology from Functional Morphology*

There are two steps in inferring the autecology of an extinct species from the morphological-anatomical traits observed in fossils. The first step is the inference of function from form. We agree that the linkage of form and function is imperfect (Gould and Lewontin 1979) and that plesiomorphy is often overlooked in studies of adaptation (Coddington 1988), but the ubiquity of convergence and the general aptness of plant and animal morphology indicate a substantial role for natural selection in shaping or channeling functional attributes. The inference of function from form is predicated on the operation of natural selection and on an understanding of the constraints placed on function by the basic materials and designs of organisms.

The second step, the inference of autecology from function, may be more difficult for plants than for animals. Plants lack the great variety of trophic modes and locomotor styles seen in animals. Nearly all vascular plants are sessile autotrophs that share the basic requirements of light, mineral nutrients, water, and carbon. However, land plants have diverged extensively in their ways of obtaining these resources, in aspects of their life cycles associated with reproduction and dispersal, and in their defenses against herbivorous animals and each other.

Inferences about the autecology of extinct species are also complicated by the ability of plants to arrive at solutions to similar environmental problems through changes in different organ systems. For example, seasonal water stress can be mitigated by deep roots, vasicentric tracheids, deciduous, succulent, or small leaves, thickened cuticle or sunken stomata, villose epidermis, C4 or CAM photosynthesis, or annual life history (Rury and Dickison 1984; Carlquist 1985; Spicer 1989). Specific environmental conditions do not necessarily produce unique evolutionary responses in morphology, physiology, or behavior. Conversely, morphological and anatomical features are not unambiguous indicators of specific environmental stresses. For instance, sunken stomata and small leaf size can be evolutionary responses to arid or cold climate or to low nutrient substrates (Spicer 1989). Ecological interpretations of fossil plants are further complicated by the tendency for organs to become disconnected prior to fossilization, with the result that interpretations are based on one or few organs rather than the whole plant body.

In spite of these difficulties, much functional and ecological information is preserved in the plant fossil record. Because plant tissues resist decay in many depositional settings, the anatomy and morphology of features used in energy and nutrient capture, gas exchange, circulation, structural support, reproduction, and dispersal are frequently preserved (chap. 2). Morphological and anatomical characters used in arriving at ecological inference are summarized in table 3.1. The form-function correlations discussed below are the basis for these inferences.

Axes

The primary functions of plant axes are support and transport of water and nutrients. Functional interpretations of fossil plant axes are generally based on one or more of the following criteria: diameter and length, branching pattern (e.g., frequency and angle), and xylem anatomy.

Probably the most salient ecological features of an individual plant are its size and growth architecture, which are among the most difficult traits to determine from fossil evidence. Plant stature is of obvious ecological significance both for the autecology of species and for determining the height of the vegetation. Stature is a major factor controlling canopy position, leaf placement, and photosynthetic strategy; it is also a measure of the investment in the sporophytic generation, thus helping to place species along the r-K continuum. For example, species that typically attain very large size are generally late maturing and thus have longer generation times. Such species also tend to be "competitors" (*sensu* Grime 1977). The stature of extinct species can be inferred best in the rare instances where long segments of tree trunks are preserved. These fossils allow estimation of the minimum size of mature individuals (Thomas and Watson 1976; DiMichele and DeMaris 1987) and sometimes reveal aspects of architecture and reproductive biology. Correlations between trunk diameter and height observed in living trees can be used to estimate the original height of fossil trees even if only the basal portion of the trunk is preserved.

Wood is the support tissue of most seed-plant trees, and characteristics of wood tissue can be used to infer aspects of life history such as adult size, sporophyte expense, growth rate, and environmental preferences of species (Loehle 1988). This is particularly true of fossil angiosperms and conifers (Wolfe and Upchurch 1987b; Spicer and Parrish 1986). The only other groups of plants to be wood supported were the arborescent sphenopsids of the Paleozoic and early Mesozoic and some of the progymnospermous groups of the Devonian and Lower Carboniferous. The sphenopsids had little variation in wood structure, expressed mostly as differences in wood thickness. The progymnosperms had wood very similar to that of their seed-plant descendants (Beck 1970).

Features of xylem tissue are strongly correlated with climatic regime, plant size and habit, and height above the base of the plant (Fritts 1976; Carlquist 1975, 1977). The most commonly used of these features are growth rings, which reflect periodic changes in the size of tracheids or vessels, usually in response to seasonal changes in water, temperature, or light. The thickness and regularity of growth rings, in addition to the proportion of late wood, can be sensitive indicators of the length of growing seasons, their interannual consistency, and the cause of seasonal dormancy (Fritts 1976). Growth rings have been widely used to infer paleoclimate and light regimes (Fritts 1976; Creber and Chaloner 1984, 1985; Spicer and Parrish 1986; Parrish and Spicer 1988; Upchurch and Wolfe 1987).

The capacity and redundancy of the vascular system also are related to the habit and habitat of the plant. The water-transport capacity of a tube scales roughly to the fourth power of its radius, so larger vascular tubes transport water far more effectively than narrow tubes (Carlquist 1975; Calkin et al. 1985; Gibson et al. 1985; Zimmermann 1983). Vulnerability to aneurysms induced by freeze-thaw cycles, water stress, and perhaps high negative pressure may increase with vessel diameter (Lewis and Tyree 1985; Ewers 1985; Carlquist 1975), although the diameter of intervessel pores probably plays a more important role (Sperry et al. 1988). Vulnerable vascular systems are most often found in plants growing in mesic tropical forest understories (Carlquist 1975). Vines typically have high conductivities and vulnerabilities, presumably because they support large canopies with a small stem, and because the stem functions only to conduct water, not to support the plant (Carlquist 1985; Ewers 1985; Gartner et al. 1990). In seasonally dry tropical climates, evergreen species generally have lower rates of conductivity than deciduous species (Gartner et al. 1990).

Functional analyses of vascular systems appear to have great potential for yielding paleoecological inferences about fossil plants, particularly if coordinated with studies of leaves or sedimentary indicators of paleoclimate. Calculations of water-transport capacity have permitted ecological inferences to be made about the habit of entirely extinct groups (Niklas 1984a; Cichan 1986), and vulnerability and conductance have been used to infer paleoclimate (Wolfe and Upchurch 1987b).

There is wide variation in the anatomical characteristics of plant support and vascular tissue, and in many plant groups tree habit does not involve stem wood. Nonwoody, arborescent vascular plants are supported in a number of ways, including adventitious roots (ferns), persistent leaf bases (some monocots), and parenchymatous tissue. Because they are largely air space, adventitious roots and persistent leaf bases require less material for the same amount of support than wood and seem to be more typical of early successional or weedy trees.

The aborescent lycopsids of the late Paleozoic, some of them trees 20–30-m tall (Thomas and Watson 1976; Wnuk 1985), were supported by wood-like parenchymatous bark around the periphery of the stem. The wood cylinder itself was narrow and composed of thin-walled, large-diameter tracheids (Cichan 1986). Water-transport and mechanical-support functions were performed by different tissue types in this group; this was possibly a unique feature in vascular plants. A wide variety of support structures occurs within tree-fern groups. Late Palezoic Marattiales were supported by a dense, complex mantle of adventitious roots (Ehret and Phillips 1977). These roots were mostly air space, probably providing support at low metabolic cost to the plant. Like wood, adventitious roots performed the dual functions of water transport and mechanical support. Other methods of support seen in tree ferns include fibrous leaf bases in extant Dicksoniaceae and intertwined stems and roots in the Mesozoic fern *Tempskya*. Even within the flowering plants, the monocots have developed tree habit independently of the dicots. Complex apical growth and the addition of secondary vascular bundles characterize many monocotyledonous trees, such as palms and Pandanaceae.

Within taxonomic groups, such as conifers and dicots, that generally produce wood as the primary means of support, less-dense wood is positively correlated with species that achieve smaller mature size, have shorter generation time and life span, and have smaller seeds (Loehle 1988). These correlations can be used to infer aspects of the life history of species known from fossilized axes.

Buttress roots are basically triangular wings of woody tissue that flare outward from the basal region of tree trunks. They are generally one to several meters high at their point of attachment to the trunk of the tree and probably serve to decrease the chance of blow down (Smith 1972). High degrees of buttressing are associated with softer, wetter soils, in which trees cannot anchor effectively. Large, or plank, buttresses do not occur in regions that experience below-freezing temperatures, probably because they have extremely high surface area to volume ratios and are thus especially susceptible to rapid freezing and thawing in seasonal climates (Smith 1972). Plank buttress morphology is preserved in some mid-Tertiary trunk casts from North Africa (T. M. Bown, pers. comm., 1989).

Branching pattern is also a major determinant of ecological role, both as an index of the plant's habit and because it commonly relates to timing of reproductive output. In the Paleozoic arborescent lycopsids, crown architecture appears to be tied to reproduction rather than light capture, and it correlates with other indications of monocarpic vs. polycarpic reproduction, predisposing species to particular ecological roles (DiMichele and Phillips 1985; Wnuk 1985, 1989). Clonal growth of some fossil forms has been demonstrated on the basis of the gross morphology of the root system (Tiffney and Niklas 1985).

Roots

Roots are commonly preserved as compressions or impressions in fossil soils. Anatomically preserved roots are considerably rarer, although they are perhaps described less often than they occur. Few functional interpretations of fossil roots have been attempted, in large part because it is very difficult to link fossil roots with the aerial portions of the same plant. The most notable exceptions to this are the roots of Carboniferous plants preserved in coal balls, from which anatomical details have been described and, to some degree, interpreted (Frankenburg and Eggert 1969; Ehret and Phillips 1977; Rothwell and Whiteside 1974; Rothwell 1984). The anatomy of roots, their position within fossil soils, and their relationship to overall plant habit offer major clues to physical aspects of the habitat. Pfefferkorn and Fuchs (in press) have developed a typological system for classifying fossil root-bearing soils that may be a first step toward a more synthetic system.

The diameter distribution of root compressions preserved in paleosols has been used to place limits on the stature of the vegetation that occupied that paleosol, on the assumption that absence of large roots indicates herbaceous vegetation (Kortlandt 1980). This approach may not be justified, given the tendency for larger roots of most trees to be shallow and thus more vulnerable to decay and erosion.

Epidermis

Characteristics of the plant epidermis, cuticle, and epicuticular hairs are closely tied to water retention, since most gas exchange and water loss occur at the surface. Many of these characteristics can be observed on fossil cuticle, which has high preservation potential because it resists decay (Upchurch 1989). Mechanisms for reducing water loss include thick cuticle, sunken stomata—which may further be covered by papillae—and dense trichomes on the leaf surface (Kerp 1989; Upchurch 1989). Trichomes create a thicker boundary layer of air around the leaf surface, which then becomes more saturated with water vapor than the surrounding atmosphere, reducing the rate of water loss during transpiration. Xeromorphic features also are present in some species that grow on low-nutrient substrates where excessive transpiration (water loss) may exacerbate loss of nutrients; and they are present in species that grow in areas that experience diurnal subfreezing temperatures, problems not related directly to low water tension in the substrate (Spicer 1989).

A variety of epidermal features can be important ecological indicators. A vast array of trichomes, emergences, spines, etc., are known as far back as the Early Devonian. In some extant plant groups (e.g., Urticaceae) trichomes serve to deter herbivores. Epidermal emergences on the Paleozoic genus *Sphenophyllum* apparently tangled these thin-stemmed plants into dense, mutually supporting thickets (Batenburg 1982). In Devonian–Lower Carboniferous plants, prior to the origin of significant tetrapod and arthropod herbivory, the function of such hairs and trichomes is difficult to interpret. The glandular hairs in *Lyginopteris* (Oliver and Scott 1904), for example, could have attracted seed dispersers, protected dispersed seeds, or secreted chemicals that discouraged the growth of other plants.

Leaves

Leaves are the primary site of photosynthesis for most vascular plants. Consequently size, shape, organization, and internal anatomy all contribute to functional and ecological interpretations. Leaf size (or effective size, taking into account the depth of any marginal sinuses) is a major factor controlling the temperature of the leaf surface, which in turn influences rates of photosynthesis and evapotranspiration. Leaves with larger surface areas dissipate heat more slowly and thus have higher equilibrium temperatures. Smaller size or deep sinuses permit more effective air circulation around the leaf and so decrease leaf temperature. Both within and between species, leaves exposed to full sun tend to be small or of medium size or to be large with deeply incised margins (Parkhurst and Loucks 1972). Very large, thin leaves generally are found in understory trees or shrubs that are exposed to full sun only in the form of transient light flecks that penetrate the canopy. Leaves of sun-adapted species also tend to be thicker and to have more highly columnar palisade tissue than do those of shade-adapted species (Lee et al. 1990), anatomical differences that should be detectable in well-preserved petrified leaves (Schabilion and Reihman 1985). Small, even scale-shaped, leaves characterize many species that undergo continual or seasonal water stress.

Leaf size and shape are also important in determining the display of photosynthetic surfaces in three dimensions. Small or deeply incised leaves cast small shadows. Thus another layer of leaves on the same tree can be arrayed a short vertical distance below these leaves without suffering from substantially decreased light levels (Horn 1971). High degrees of intraspecific and intraindividual variation in degree of lobing are characteristic of species that have multiple layers of leaves or deep canopies, whereas many understory trees subject to low light levels have a single layer of less variable leaves (Horn 1971).

Certain leaf shapes and patterns of organization are associated with particular plant growth habits. Deeply cordate leaves are frequently found on vines or forest floor herbs, although the correlation is far from perfect (Givnish and Vermeij 1976). Leaves of aquatic (particularly floating) plants are often peltate (i.e., the petiole is attached in the center of the roughly circular leaf blade). Many early successional trees and shrubs bear pinnately compound foliage. It has been argued that these leaves are equivalent to "throw-away" branches, permitting successional plants to reduce their investment in wood that does not support active photosynthetic tissue (Givnish 1979). Shedding the supporting rachis might also reduce transpirational water loss during seasonal droughts.

Fossil evidence for the existence of the specialized C4 and CAM photosynthetic pathways has recently been obtained from anatomically preserved specimens and by stable carbon isotope analysis. Silicified grasses from the Miocene of North America have the distinctive Kranz anatomy seen in living C4 plants, indicating the presence of drought-resistant grasses by this time (Thomasson et al. 1986; Tidwell and Nambudiri 1989). A technique that can be widely applicable in the absence of anatomical preservation uses the $^{12}C/^{13}C$ ratio to differentiate plants with C3 photosynthesis from those with CAM or C4 photosynthesis; the latter produce heavier carbon (O'Leary 1988). This approach has been applied to fossil soil carbon in order to track landscape-scale changes in vegetation (e.g., Cerling et al. 1989) and to individual organically preserved fossils in order to determine the photosynthetic pathway of a particular species (e.g., Nambudiri et al. 1978).

Reproductive Organs

Reproductive biology is one of the most important keys to ecological characteristics such as population structure, environmental preferences, tolerance of disturbance, etc. Many aspects of reproductive biology are inaccessible to paleobotanists because of the fragmentary nature of the fossil record and the chemical or dynamic nature of the phenomena involved. However, as knowledge of whole plants is increased, a substantial number of solid inferences can be drawn. Reproductive organs themselves offer clues to modes of dispersal and pollination. This is true broadly of plants at all times in the past. The earliest plants, which had simple homosporous reproduction, had very complex gametophytes that at least matched extant plants in their range of morphological and functional specialization (Remy and Remy 1980; Remy et al. 1980; Schweitzer 1980). However, these features are rarely preserved in detail. The evolution of heterospory and the seed habit is well documented (Chaloner 1967; Rothwell and Scheckler 1988). The earliest seeds had morphological attributes that appear to be related to wind pollination, on the basis of experimental functional-morphological studies (Niklas 1981).

Fossil angiosperms have close living analogues that provide a basis for interpreting fossil morphology. For example, in flowering plants, the size of microspores and pollen grains is related to their mode of dispersal; pollen grains > 40 microns and < 10 microns in diameter are usually animal dispersed. Animal-pollinated angiosperms also commonly have pollen with elaborate surface sculpture. Some undispersed fossil pollen shows the presence of pollen kitt, a sticky substance that holds small grains together for easier transport by insects (Friis and Crepet 1987). Wind-pollinated species typically have relatively smooth pollen in the 15–35-micron size range (Traverse 1988).

Characteristics of the ovules, female strobilus, or inflorescence are also correlated with mode of fertilization. Ovule structure in gymnosperms is related to mode of pollen delivery. In lyginopterid seed ferns, for example, simple pollen chambers and specialized features for trapping pollen and microspores are indicators of wind pollination (Taylor 1988). Some Paleozoic plant fossils even preserve pollen grains trapped in pollen drops (Rothwell 1977). In living conifers bracts surrounding the ovules may enhance capture of wind-borne conspecific male spores on the receptive surface (Niklas 1984b). Highly specialized lower vascular plants such as lepidodendrid lycopsids had complex megasporangiate reproductive organs that trapped microspores in the air or on the surface of the water (Phillips 1979; Thomas 1981).

Features of animal pollination also are seen in female reproductive organs. Typically animal-pollinated species have specialized appendages surrounding the ovary. Among flowering plants, the size, shape and symmetry of these floral parts are correlated with pollination by different animal groups (pollination syndromes). Some pollination syndromes have been recognized in the fossil record (Crepet 1984; Crepet and Friis 1987; Friis and Crepet 1987). Pollination biology in turn has implications for population structure and other aspects of life history. If population density of wind-pollinated species decreases, reproduction declines because of low rates of pollination or high rates of selfing. In contrast, plants pollinated by faithful pollinators (generally bees or vertebrates capable of traveling long distances to feed on flowers of a particular species) can maintain highly dispersed populations. This kind of pollination biology is also thought to enhance the probability of isolation of peripheral populations and potentially to decrease the chances of extinction by increasing effective population size.

Disseminules

Ecologically significant aspects of dispersed megaspores or seeds include the quantity of stored nutrients they contain and the nature of the membranes that surround the megaspore. The amount of stored nutrient included within the dispersed megaspores is correlated with the life form of the plant, generation time, successional status, predators, climate, and dispersal mode (Salisbury 1942, 1974; Janzen 1969; Harper et al. 1970; Baker 1972; Silvertown 1981; van der Pijl 1982; Tiffney 1984; Foster and Janson 1985; Rockwood 1985; Foster 1986; Mazer 1989). Seed size also reflects a strong phylogenetic component (Mazer 1989, 1990).

The properties of the tissues surrounding the dispersed megaspore are correlated with the mode of dispersal. Presence of a fleshy sarcotesta or fruit, especially when it surrounds a thick sclerotesta, implies endozoochory, ingestion and eventual dispersal by animals. Hooks and barbs frequently are associated with ectozoochory, dispersal by clinging to skin, fur, or feathers and subsequently dropping off. In contrast, wings or corky tissue are more commonly seen in abiotically dispersed disseminules (Ridley 1930; Augspurger 1986).

By the Carboniferous, disseminules were specialized in several major plant groups: *Lepidocarpon* had female megaspores that were probably water dispersed and perhaps fertilized on the water surface (Phillips 1979); some cordaites had winged seeds; and some seed ferns probably had fleshy coverings on seeds. The absence of embryos in nearly all Paleozoic dispersed seeds indicates either delayed fertilization (a characteristic of living *Ginkgo*) or rapid fertilization without a period of dormancy (Rothwell 1988). The earliest evidence of such dormancy is found in Paleozoic conifers (Mapes et al. 1989).

2.2   *Inferring Autecology from Living Relatives*

The conclusion that the ecological preferences of an extinct species were similar to those of its living relatives is predicated on three basic assumptions: (1) that there is in fact a close relationship between the extinct and extant species; (2) that there has been little evolutionary change in the ecological preferences of the lineage between the time of the fossil and the present; and (3) that the living species are occupying the full range of ecological situations open to them. In practice these assumptions are very difficult to verify, and the validity of the first two assumptions decreases with increasingly old fossils. As a consequence, floristic analogy (as this method is often called) has been used most extensively on Cenozoic floras. The great advantage of floristic analogy is that it can be applied to any kind of fossil, including microfossils, and this increases stratigraphic and geographic density of samples. The numerous conclusions about vegetational and climatic change during the late Pleistocene-Holocene interval are based almost entirely on floristic analogy.

Floristic analogy has also played an important, though controversial, role in reconstruction of Tertiary climates (Axelrod 1987) and vegetation. Wolfe (1978) has criticized floristic analogy on the grounds that many living relatives of common early Tertiary species are highly relictual and no longer represent the full range of climatic conditions under which the lineage is capable of growing. Reconstruction of early Tertiary climate through floristic analogy probably is most reliable when large numbers of species are used (Hickey 1977) and when the plant groups used in the reconstruction are characterized by anatomical or physiological features that limit their climatic tolerance. For instance, the universal presence of parenchymatous axes and dispersed vascular bundles in palms (Arecaceae) makes it very unlikely that any extinct members of the group could have tolerated temperatures far below freezing.

2.3 Inferring Autecology from Sedimentary Environment

The distribution of plant fossils in the sediment can be an important source of autecological data. Much work has focused on spectacular cases of in situ forests or stump fields (Wnuk and Pfefferkorn 1984, 1987; Gastaldo 1986, 1987; DiMichele and DeMaris 1987; DiMichele and Nelson 1989), where the position and diameter of tree trunks can be observed directly. Although the discovery of a well-preserved "fossil forest" is an unusual event, horizons containing poorly preserved casts or concretions that formed around tree trunks are more common in the fossil record than is generally realized (Kraus 1988). Where in situ stems can be identified, they yield information on the size of individuals and the preferred substrate of species. This information leads to inferences about the successional status of the species. For instance, species commonly found rooted in, and preserved by, deposits representing flood events were probably early successional. Species commonly found rooted in coal (originally peat) or carbonaceous shale were probably tolerant of low-pH and/or low-oxygen substrates. Relatively uniform size of conspecific trunks in a local area may indicate an even-age stand, a characteristic of species that require large gaps or disturbances in order to colonize successfully. Groups of stumps or trunks reveal the spacing of individuals in a local population. Random spacing of conspecific trunks implies the absence of strong competition for resources (DiMichele and DeMaris 1987), whereas distant spacing has been taken as an indication of competition for light in the paleo-high-latitudes where there was a low angle of solar radiation (Jefferson 1982).

Autochthonous or nearly autochthonous compression fossil assemblages generally have received less attention than more spectacular fossil forests, but they may preserve similar kinds of information (Krassilov 1975). The abundant occurrence of remains of a species in a restricted sedimentary environment may indicate it grew there. For example, in the Potomac Group in eastern North America, early flowering plant leaves are abundant exclusively in point-bar deposits, an indication that these plants were successful only in unstable habitats (Hickey and Doyle 1977). Abundant compressed seeds and seedlings of *Joffrea speirsii* (Cercidiphyllaceae) in Paleocene fluvial point-bar deposits strongly imply that this species colonized freshly deposited mineral substrates in stream channels (Crane 1984; Crane and Stockey 1985). The abundance of foliage and pollen of taxodiaceous conifers in early Tertiary carbonaceous shales, but their near absence in coarser-grained, near-channel deposits, suggests these species did not do well in disturbed habitats but were very tolerant of the flooded and probably somewhat anoxic conditions represented by carbonaceous shale deposition (Wing 1984; Farley 1989). Fossil plants with morphological characters of floating aquatics (e.g., peltate leaves), or belonging to taxonomic groups that have many extant floating aquatic species (e.g., Salviniaceae, Nymphaeaceae), generally are abundant in deposits having the sedimentary features of small lakes or abandoned channels (Wing 1984); this is an example of how taxonomic, morphological, and sedimentological data can corroborate one another.

Recent studies of the relationships between forest floor litter and standing forests suggest the potential for inferring even more autecological details from compression fossil assemblages than the above examples indicate. Undisturbed leaf deposits may contain information on the size of individual tree canopies and the spacing between individuals, although recovering this information will require much greater sampling density than is generally used by paleobotanists (Burnham et al. 1992).

3   METHODS OF SAMPLING AND ANALYSIS AT THE COMMUNITY LEVEL

How and how well does the fossil record sample and represent the original vegetation? Taphonomic studies of plant dispersal, deposition, and representation in modern environments have contributed to an increasing understanding of the sampling biases of the macrofossil record. Until recently it was believed that only in rare instances, such as in cases where a forest has been buried in situ by an event deposit such as an ash fall or a catastrophic flood, could original spatial heterogeneity be recovered from megafossil samples. Recent taphonomic studies (Ferguson 1985; Gastaldo 1988; Burnham et al. 1992) suggest, however, that in some types of low-energy environments, there may be little or no movement of litter across the forest floor. In such situations, properly sampled leaf assemblages could yield information on spatial variation in the original vegetation (Hickey 1980; Wing 1984; Burnham 1987, 1988). Even highly dispersible pollen may yield information on local vegetational heterogeneity. Studies of pollen rain, that is, deposition from the air, have shown that although a very small proportion of pollen grains are transported thousands of kilometers, the vast majority are deposited within a few tens or hundreds of meters from their source (Tauber 1965; Raynor et al. 1970; Tsukada 1982), particularly in forested landscapes. Patterns of pollen distribution in ancient fluvial sediments also indicate that some palynofloras are highly local (Farley 1988, 1989) and so can be a basis for detailed community reconstruction.

Underlying the analyses of ancient assemblages at all levels is sampling. The fossil record is itself a sample of once living organisms and some of their interactions. Thus, sampling regimes must be designed to make the most of the information available, while compensating for numerous confounding biases. Sampling design is often overlooked in the preparation of a paleoecological study; this is unfortunate, because sampling frequently limits the kinds of analyses that can be carried out subsequently. In practice the fossil record rarely allows the ideal sampling scheme, but the robustness of many analysis methods can correct for this.

Faced with the dilemma of what to count, or even whether to count at all, and how to go about it, paleobotanists have devised numerous schemes to recover quantitative and spatial data from the fossil record. Not all of the methods return comparable information (Lamboy and Lesnikowska 1988a, b), and many are most successful with angiosperm-dominated leaf floras, where the objects sampled all have roughly similar size, shape, and taphonomic properties. The simplest method, applicable to all forms of preservation and environments of deposition, is to produce a species list with semiquantitative notations of the relative abundances (e.g., abundant, common, rare). The limitations such data impose on further analysis restrict their use. Paleobotanical sampling has been reviewed recently by Spicer (1988), but specific methods developed for use with compression-impression preservation, permineralized preservation, and microfossils merit attention here.

3.1 Sampling Compression-Impression Assemblages

Analyses of compression-impression fossils have spawned a number of related methods (Spicer 1988). The simplest is a direct count of the number of plant parts belonging to each species (Wing 1981; Taggart 1988; Burnham 1987). This method is most readily applied to dicotyledonous angiosperms, where most species produce leaves of relatively similar size and shape. Burnham et al. (1992) have examined in detail the relationship between the abundance of species in the forest and in in situ leaf litter. They find that on a per species basis leaf biomass can be used as a proxy for basal stem diameter, a major means for assessing biomass in modern forests. The leaf biomass of a species in a fossil sample can be calculated roughly by multiplying the number of leaves by the average size of the leaves by an estimate of leaf density (the last determined by reference to analogous extant taxa). Within limits, simple leaf number and estimated leaf biomass will produce similar results.

A point-count method was developed by Scott (1977, 1978) for field sampling of bedding-plane surfaces in the Late Carboniferous coal measures of England. The surface is sampled with a $1/2$-m^2 plexiglass sheet in which 100 randomly placed holes have been drilled; pins are placed through the holes, or a mark is made through them, and the fossils or bare rock to be tabulated are noted. Leary and Mickle (pers. comm., 1988) applied this method to the study of Early Pennsylvanian fossils in Illinois. In general, however, it has received little application because it is labor intensive and requires extended periods of access to undisturbed exposures. Lamboy and Lesnikowska (1988b) report on a modification of this technique in which blocks of mudstone were returned to the laboratory, divided there into microstratigraphic intervals, and broken up. Specimens from each interval then were laid out together and 100 random points per $1/2$ m^2 were marked for counting. In the laboratory all identifications can be checked at the microscope, and the specimens can be saved indefinitely.

Late Paleozoic fossil assemblages present special problems. The original vegetation contained a wide variety of growth architectures, and hence of plant litter types. Lycopsids and calamites produced mainly stem debris, some of very large size. In contrast, tree ferns and pteridosperms produced compound leaves, which in most species fragmented into isolated pinnules, pinnae, and rachises; some of the leaves were many meters in length (Laveine 1986). The problem of what and how to count is accentuated in these deposits. To address this problem Pfefferkorn et al. (1975) developed a technique in which hand specimens are treated as quadrats. The taxa occurring on each quadrat are noted, and a quantitative estimate of biomass is made per taxon as the number of quadrats occupied out of the total number of quadrats sampled. The advantages of this method are rapidity, ease of use with museum collections or mine spoils, and reduction of the biases imparted by taxon-specific size differences in litter (e.g., should one lycopsid stem equal one pteridosperm pinnule?).

Lamboy and Lesnikowska (1988b) compared the Scott method with that of Pfefferkorn et al., using the same suite of fossils, and found statistically significant differences in the estimates of biomass. DiMichele et al. (1991) compared simple counts of numbers of individual plant parts per taxon with the quadrat method (Pfefferkorn et al. 1975) using the same set of fossils and obtained nearly the same pattern of rank-order abundance, although the numbers were not directly comparable.

3.2 Sampling Permineralized Peats

Preservation of permineralized peat beds is uncommon in the fossil record. However, where they occur such deposits are an invaluable source of information on the structure of extinct plants and can be sampled quantitatively. The three-dimensional preservation of the plants requires that determinations of biomass be made from volumetric estimates of taxonomic abundance. To date, methods have been developed only for Late Carboniferous coal balls (Phillips et al. 1976), which are found mostly as calcium carbonate or siliceous concretions within coal seams (see Scott and Rex 1985). The basic method of analysis, developed by Phillips et al. (1974), could be applied to permineralized peat deposits of any age.

When dealing with coal balls, peels (analogous to thin sections; see Phillips et al. 1976) are made of the cut and etched surface of the concretion, which is generally between 1 and 100 cm across. A peel from the center of each coal ball is attached to a clear acetate cover grid ruled in square centimeters. The plant organ type, its taxonomic affinities, and preservational information are recorded for each square centimeter. Total cross-sectional area of each taxon-organ type is taken as a direct estimate of biomass. The small size of the sample quadrat reflects the detrital nature of coal-ball litter.

This method has been used in several studies of Late Carboniferous coal-swamp vegetation (Phillips and DiMichele 1981; Phillips et al. 1985; DiMichele and Phillips 1988; DiMichele et al. 1991). Pryor (1988) tested the Phillips et al. (1985) method against point counting and against the use of millimeter-square quadrats and found no significant differences.

3.3 Sampling Microfossils

Pollen transport curves (Tauber 1965; Raynor et al. 1970; Tsukada 1982), influx and dispersal patterns of pollen in lakes (Davis 1973), and attempts to relate palynomorph abundances to abundances of parent plants (Davis 1965, 1973; Faegri and Iversen 1975; Parsons and Prentice 1981; Webb et al. 1981) have produced a sophisticated understanding of how fossil pollen assemblages reflect the vegetation from which they were derived. Virtually all of these studies have been carried out on Recent or late Pleistocene pollen floras, in which taxonomic similarity to the present is high; but the observations made on postglacial palynofloras have been applied to ancient examples as well (Willard in press; Grady and Eble 1990; Mahaffy 1985).

Aside from specific, well-understood biases (e.g., Medullosan pteridosperm pollen is too large to appear in most microfossil analyses of Carboniferous rocks; some angiosperm groups produce weak-walled pollen that is virtually never preserved), the pollen-spore record appears to provide an excellent measure of local-to-subregional vegetation. Pollen transport, particularly in dense forests with closed or largely closed canopies, is limited and can be expected to be confined mostly to the plant community of origin. Extraneous elements are most likely to be introduced by water transport, not by wind dispersal (Farley 1988) and thus should only rarely confuse overall patterns of dominance and diversity. The major biases, particularly in older rocks where direct reference to the present is not possible, result from differential pollen production. Only where pollen-spore patterns can be compared directly to megafossil patterns can an understanding of over- or underrepresentation be gained (Smith 1962; Grady and Eble 1990; Willard in press).

A wide variety of methods have been proposed for the sampling and analysis of plant microfossils, mostly pollen and spores (see Traverse 1988; Farley 1988). Because microfossils generally are more ubiquitous and abundant than megafossils, collections can be made at close stratigraphic intervals or from a variety of depositional environments. Pollen and spores (palynomorphs) are released from matrix by maceration and most commonly are quantified by simply counting the number of grains of each type on a microscope slide. It has become standard practice to count 300 grains from any given sample (Maher 1972), although estimates of the relative abundances of rare species can be assessed more reliably from 500-grain counts (Farley 1988). Regardless of the number of grains in the sample, count data can be difficult to interpret because the proportions of species are autocorrelated; as one species becomes more common, others must make up less of the sample. To avoid this problem, some workers calculate the concentration of pollen of each species in the sediment; however, concentration data are also influenced by a variety of factors not related to the original vegetation, including chemical environment and depositional rate (Farley 1988). In Holocene sediments depositional rate can be factored out by calculating pollen influx rates, expressed in terms of the number of pollen grains accumulating per cm^2 per year (Davis et al. 1973). Unfortunately this method is limited to sedimentary sequences that supply closely spaced, precise, radiometric dates from which short-term depositional rates can be calculated.

Palynofloras are the mainstay of Quaternary paleoecological studies. A great deal of research has been devoted to understanding spore and pollen taphonomy (Davis 1973; Birks and Birks 1980), which provides insights into the behavior of these fossils at any time in the past. Integration of micro- and macrofossil studies can yield greater temporal resolution of ecological processes in the distant past as well (Taggart and Cross 1980; Willard 1990; Farley and Wing 1989; Wing and Farley 1990). Such studies will also pay greater attention to small plant particles enclosed in sediment, termed "palynodebris" (Manum 1976), or "phytodebris" (Tiffney 1989), which provide another source of data on paleovegetation. Fossil cuticles are probably one of the most promising, though largely untapped, sources of paleoecological information (Litke 1967; Kovach and Dilcher 1984; Upchurch 1989; Kerp 1989). Comparisons of dispersed cuticle with leaf floras show substantial overlap in composition and similarity in relative abundances (Kovach and Dilcher 1984), implying that dispersed cuticle floras could serve as stand-ins for megafossil assemblages in studies that require sampling with higher stratigraphic resolution or across a wider range of depositional environments (Wolfe and Upchurch 1987a).

3.4   Taphonomy and Effective Sampling Radius

Numerous papers in the last ten years have given us a broad base of understanding from which to analyze fossil deposits (see chap. 2). Generalities that have been established include the following:

1. In authochthonous or nearly autochthonous assemblages, rank-order dominance patterns of ancient forests are likely to be preserved (Burnham 1987; Burnham et al. 1992). However, even under nearly ideal sampling conditions, litter samples will miss a substantial proportion of the species diversity, omitting principally the rarer elements of the community (Burnham et al. 1992).

2. Assemblages preserved in channel-fill facies (lateral accretion or abandoned channel) preserve almost exclusively the vegetation growing along the channel margin (Burnham 1987; Gastaldo 1987; Scheihing and Pfefferkorn 1984), although species not typically part of the riparian vegetation may occur on the cut bank side of the channel and thus become preserved in oxbow fills. Extraneous elements may appear in coarse-grained, active channel facies, but are usually recognizably abraded; these elements largely tend to bypass the system and be flushed through (Gastaldo et al. 1987). These observations are important, because many, if not most, nonswamp fossiliferous deposits are preserved in channel-fill sequences.

3. Organic deposits in peats or clastic swamps undergo limited transport, and have few allochthonous elements (Cohen et al. 1987). Domed, ombrotrophic peats in particular have little contamination from extrinsic litter, because floodwaters cannot reach the peat surface (Anderson 1964; McCabe 1984, 1987). Studies of peat diagenesis (Cohen et al. 1987; Esterle et al. 1989) provide a great deal of detailed information on the patterns of cellular-level alteration of plant materials, which is important in the interpretation of intraswamp physical conditions and plant functional morphology.

4. Most of the fossil record preserves lowland, basinal vegetation, where burial by fine-grained water-borne sediment is most likely. Upland, or extrabasinal, vegetation is rarely preserved but can occur under unusual circumstances, such as in sinkholes (Leary and Pfefferkorn 1977), or more commonly in association with volcanogenic deposits (Taggart

1988; Bateman and Scott 1990). Forests buried by ash falls may have complex dynamics that can be envisioned only by study of modern analogues (Burnham and Spicer 1986). Thus, much of the ancient landscape is inaccessible to paleoecological analysis, even by pollen-spore studies. Fossil soils provide a means to infer more broadly the regional habitats associated with fossiliferous deposits; although the potential of paleopedology has been demonstrated (e.g., Retallack 1985), this tool remains to be used more widely by plant paleoecologists.

5. Certain kinds of deposits, such as Holocene pack-rat middens, are exceptionally well understood because direct comparison to modern counterparts is possible (Van Devender and Spaulding 1979; Cole 1985; chapter 2). This particular example emphasizes the importance that apparently odd, but taphonomically well understood, deposits play in our understanding of past ecosystems.

Overall, taphonomic studies of macrofossils and microfossils present a hopeful picture. Transport of plant debris appears to be limited in many depositional settings. Where transport is extensive and mixing of elements from distinctive communities may have occurred, physical factors, both preservational and lithological, point to allochthony. Taphonomic analyses suggest clearly, however, that considerable effort must be devoted to understanding the sedimentology at both local and regional scales if an accurate reconstruction of the physical habitat is to be made. Functional morphological-anatomical analysis of fossil plants cannot by itself provide an acceptable paleoecological picture.

3.5   Analytical Methods

Paleoecologists generate and analyze data that are remarkably similar to those of ecology. Samples of a fossil assemblage are taken and characterized by species (or taxa at some rank). The presence or abundance of these taxa in each sample is the basic variable. In recent years paleoecologists have turned to various ordination techniques (Spicer and Hill 1979; Kovach 1988), to study the co-occurrences of species among samples and portray variability in the composition of samples on a few axes. Many of these analyses are directed at finding gradients in species composition that presumably reflect gradients in environmental conditions. Ecologists studying extant vegetation can compare the patterns shown in ordinations of species occurrence data to environmental variables that have been measured directly (e.g., soil pH, nutrient levels). Although paleoecologists cannot measure these variables, they can measure or score other independent variables, such as sedimentary features, geochemical characteristics of sediment, and anatomical or morphological features of species. Multivariate methods are the subject of numerous texts and a large scientific literature. Summaries can be found in Digby and Kempton 1987, Pielou 1984, and Grieg-Smith 1983, among many others. An interesting variant on ordination techniques was used by Taggart (1988), who plotted localities on triangle diagrams in which each vertex represented one of the dominant elements in the regional vegetation. These served as base diagrams for the summary of other kinds of data, in much the same way that ordinations do.

4 SYNECOLOGY OF EXTINCT VEGETATION

Reconstruction of an extinct plant community or vegetation type is analogous to reconstruction of an extinct plant—the individual parts must be described and then assembled in a logical way, consistent with the data and with broad theoretical expectations. Many aspects of past vegetation can be inferred, given the proper depositional and preservational setting; trunk-size distribution, distance between individuals, canopy spread, number of species per unit area, relative abundances of species, aspects of vertical stratification, and distribution of species among life forms. Very rarely can all of these be inferred for a single fossil assemblage.

Reconstruction of individual communities at 1-to-100-year time scales is the basic level for paleoecological analysis. Fossil assemblages that accumulated over such short time intervals are useful because their interpretation is not confounded by the effects of time averaging (chapter 2), and because they permit the most direct comparisons with short-term studies of living vegetation. The successional or disturbance dynamics of fossil vegetation may be understood from temporally closely spaced "snapshots," or from samples that accumulated over periods of about 10^3 years. Approaches for studying both "instantaneous" and "time-averaged" samples are discussed in §4.1.

4.1 Synchronic Features of Extinct Vegetation

Diversity

Species richness—the number of species found in a given area—is one of the most widely used and conceptually simple descriptors of a local flora. The concept of diversity usually includes not only the number of species but also some measure of the evenness of the relative abundances of the species in an area (for a recent review see Magurran 1988). Although studies of richness and/or diversity are common in ecology and paleoecology, both the measurement and interpretation of diversity data are complex.

Accurate estimation of the number of species in a community or area generally involves large samples, especially if there are many rare species. In the fossil record, where it may be difficult to obtain large or replicate samples, it is in many cases difficult or impossible to get precise estimates of richness, because many rare species will not be detected (Koch 1978). This presents problems for studies of change in species composition through time, as well as for studies of diversity.

The problem of estimating richness has been studied specifically for autochthonous leaf litter in living forests. Complete species lists and precise estimates of richness are extremely difficult to obtain unless the forest has very few species and is spatially homogeneous, and even under these circumstances multiple closely spaced samples are required (Burnham et al. 1990; Burnham et al. 1992). In contrast, the relative abundances of the common species in a local 1000–5000-m² area of forest, those comprising more than 10% of the stem area of the forest, are recorded quite faithfully in the relative abundances of their leaves in the litter. These findings in extant forests should apply to autochthonous and nearly autochthonous fossil assemblages. Transported assemblages may sample a much larger part of the regional landscape, and therefore record more of the regional diversity, but they are less comparable to one another because it is almost never possible to determine the actual size of the area that was sampled. These observations cast substantial doubt on biological interpretations of local diversity change through time (e.g., Knoll 1986).

Structure

"Vegetational structure" refers broadly to cross-sectional height, stratification, the degree to which the canopy is open or closed, diversity of life-forms, and the relative spacing and placement of canopy trees, along with similar characteristics of understory and ground cover. The relative abundance and diversity of vines and epiphytes also must be considered. There is a great deal of terminology that encapsulates in a shorthand fashion such structural aspects of vegetation—e.g., steppe, pinyon-juniper woodland or parkland, boreal forest, multistratal rain forest—and much of the later fossil record can be described broadly in such terms.

Most of the structural attributes noted above can be inferred for a fossil assemblage, given low levels of taphonomic bias. Such inference relies strongly on the types of information and methods discussed in §2. Commonly not all of the ecological attributes of the species can be determined from a single fossil assemblage. For example, the height, spacing, canopy architecture, understory, and ground-cover characteristics, and even the relative constructional expense and reproductive output of the major plants in Carboniferous swamp communities dominated by *Lepidodendron* are reasonably well known. However, few of the assemblages so dominated preserve all of the important features. Rather, independent analyses, some of tree height and spacing (DiMichele and DeMaris 1987; Gastaldo 1987; Thomas and Watson 1976), others of life-history patterns and growth dynamics (Phillips and DiMichele in press), and still others of morphology (Watson 1907; DiMichele 1983), combine to amplify the analysis of any individual assemblage. The degree to which paleoecological studies rely on information from related or independent analyses for inferential support may be considerably greater than that for neoecological studies, but the pattern is inherently the same.

Not all structural attributes of a community need to be described in detail in comparative studies. In many instances only a particular aspect of an extinct community will be of interest, say, tree spacing, diversity of ground cover components, diversity of canopy elements, or the relative average expense of swamp vs. nonswamp communities at different points in time. Consequently assemblages that can be described only partially are not necessarily uninformative.

Leaf size and shape (leaf physiognomy) have been studied as characteristics of local or regional floras as well as of individuals or species (Bailey and Sinnott 1915, 1916; Wolfe 1979; Dolph and Dilcher 1980). Leaf characteristics of individual species are clearly influenced by many factors: habit, local habitat, seral status, and phylogenetic history. However, the aggregate leaf characteristics of the dicotyledonous species in a local flora relate strongly to climatic variables and vegetational features. For example, there is a strong positive correlation between the mean annual temperature under which a flora grows and the percentage of the dicot species present that have entire-margined leaves (Wolfe 1979). There is also a relationship between leaf size and mean annual precipitation; and long, attenuated leaf tips may be a mechanism for shedding rain in humid climates (Wolfe 1990). The diversity of inferred vine species in fossil leaf assemblages also has been used to infer the degree of forest stratification and vertical complexity (Wolfe 1979). Correlations of leaf features with climatic parameters have been used extensively to infer the paleoclimates experienced by fossil angiosperm floras (Wolfe 1972, 1978) but have been used little for preangiosperm floras because of the strong dependence on a modern analogue.

Short-Term Vegetational Dynamics

The dynamic aspects of a community are varied and include a wide range of attributes related to the concept of stability: what contributes to the structure and composition of a community and how is that character retained for extended periods of time (tens to thousands of years)? These are questions of great interest to neoecologists (Rahel 1990), and paleoecology cannot hope to address the dynamic aspects of past terrestrial communities with the same degree of temporal resolution. However, many of the larger-scale dynamic attributes of communities can be studied directly or inferred broadly.

Disturbance is one of the most important factors structuring a community on short time scales. Both the nature of disturbances and the regime, or intensity and frequency, have major impact on species composition and diversity, the nature of the dominant life histories, and hence the structure. Disturbances like fire often leave characteristic traces, such as charcoal; the regularity of fires and their impact on the vegetation can be studied if a sampling strategy is adopted that considers small-scale change over time. Flooding is variably detectable; it may leave little trace except in the biotic attributes of the plant assemblage, or it may leave a clear mark, as in cases where plants can be seen to have recovered from repeated burial by flood-borne waters. Finally, important disturbance factors such as wind throw can be almost impossible to detect. Canopy gap dynamics, which has proven to be important in controlling spatial heterogeneity in many tropical lowland forests (Brokaw 1985a, b), will always be difficult to infer from paleoecological data.

The effects of disturbance on vegetation, particularly the aerial coverage, is also difficult to detect. Consequently documenting the nature of vegetational recovery from various kinds of disturbances must be based on indirect inferences, such as the relative proportion of assemblages dominated by weedy taxa, within a larger suite of forested vegetation types. Succession has been described in some fossil sequences, for example in coal swamps of Pennsylvanian age (Smith and Butterworth 1967; Mahaffy 1985), but in such cases the controlling factors were demonstrably abiotic.

There are numerous other attributes of a community that contribute to its dynamics. These include factors such as rainfall and temperature regimes and response of the plants to potential seasonality, the patterns of reproductive timing and likely colonization of available space, nutrient regimes in the broad sense of low vs. high availability, and levels of herbivory. All of these factors can be analyzed directly from study of the rocks or of the plant fossils themselves. Temperature and rainfall are reflected strongly in aspects of plant morphology: growth rings and morphology of wood, leaf marginal characteristics, and aspects of morphology that indicate evapotranspirational stress. In combination with sedimentological indicators, such as evidence of seasonal variation in stream discharge, soil indicators of seasonal dryness vs. lack of it, evidence of extensive peat deposits or molds of evaporite crystals, etc., it is often possible to build a strong case for local climate and its pattern of variation. Reproductive patterns can be determined from both morphology and sedimentary context of plant fossils. For example, monocarpy (once in a lifetime reproduction) in some arborescent lycopsids of the Carboniferous correlates directly with indicators of flooded substrates and few fires and suggests preference on the part of such plants for very stable habitats with little disturbance (DiMichele and Phillips 1985).

The access to time dimensions, both short and long term, is perhaps the greatest strength paleoecology has in evaluating the dynamics of extinct systems. It is possible to track assemblages through a rock unit or, controlling for depositional environment, over extended periods of time. Even with reduced access to many of the details of disturbance or climatic patterns, many general patterns can be followed through time. For example, taxonomic and structural composition of a fire-prone vegetation (associated with abundant mineral charcoal) can be contrasted with that of vegetation in a similar habitat that was not fire prone. Paleoecological studies must focus on those dynamic attributes of ecosystems that are both accessible and can be examined profitably through time, aiming at problems that complement the studies of neoecology, but that are uniquely paleontological in scope. In this sense the universe of ecology is expanded, and we gain a general understanding of the way ecosystems respond to biotic and abiotic factors over the long term.

5 COMPARISON OF COMMUNITIES ACROSS TIME

Section 4 focused on means to describe extinct vegetation at a single time, or its characteristic patterns of changes over short periods of $10^0 - 10^2$ years. These descriptions, or "snapshots," of past vegetation are the units that are compared in studies that trace vegetational and ecological change across geological time. If the entire time interval across which comparisons are being made is on the order of $10^5 - 10^6$ years, then change in species composition is a viable, if incomplete, method for assessing ecological change.

Examples of paleoecological studies that span periods of $10^3 - 10^4$ years include work on patterns of change within single Late Carboniferous coal swamps through time (Smith and Butterworth 1967; Habib 1968; Phillips and DiMichele 1981; Mahaffy 1985; DiMichele and Phillips 1988; Grady and Eble 1990), within a single major Jurassic compression flora (Spicer and Hill 1979), or at numerous Holocene sites where changes in vegetation through time and space have been studied with pollen or macrofossils (Delcourt and Delcourt 1987). Interecosystem patterns of change have been studied on time scales of 10^6 years for Euramerican coal-swamp vegetation throughout Late Carboniferous (Phillips et al. 1985), through 10^5 years for back swamps of Eocene age within a single depositional basin (Wing and Farley 1990), and for Miocene floras within a single volcanogenically influenced landscape (Taggart and Cross 1980, 1990). These studies have used mainly taxic (the data compared consist of species compositions) approaches and have focused on assemblages that are composed of fundamentally the same kinds of plants, even if not the same species in all lineages throughout the interval under study. The Late Carboniferous coal-swamp study covered the longest time interval and encountered sufficient taxonomic change to render direct species-level comparison of assemblages problematic, forcing much of the focus onto higher-level taxonomic groups.

Questions concerning the stability of biotic assemblages through evolutionary time, the responses of communities to large-scale environmental perturbations, the relative roles of physical and biotic factors in structuring communities, and the effect of ecological role on evolutionary rate may require comparisons that encompass time spans on the order of $10^6 - 10^8$ years. Across such long time intervals extinctions and originations may have altered the species composition of the assemblages of interest, perhaps entirely. Thus, comparisons of communities across long time spans have to be based on characteristics other than taxonomic composition. These ataxonomic characteristics include vegetational structure, short-term dynamic behavior, and the kind and relative importance of different life-forms and life histories, as well as the more widely used characters of richness and diversity.

That we see the diversity of life as organized ecologically around repeated themes reflects an implicit taxon-independent perception (Cusset 1982). A major challenge for plant paleoecology is to transform this intuitive conceptualization into an explicit and consistent method for describing and comparing ecosystems and species across geological time. We see two major ways this might be achieved. The first method, which can be called "ecomorphic," relies entirely on the ecological interpretation of morphology to develop a fingerprint of the extinct community. The second method, which we call "ecological categorization," uses sedimentary context and analogy to extant relatives, as well as functional morphology, to assign extinct species to a priori ecological types. The relative dominance and diversity of these types then provides a community-level ecological description. These two methods are discussed in more detail in §§5.1 and 5.2.

5.1 Ecological Interpretation of Morphology at the Community Level

Just as the morphological attributes of a particular species can be used to infer its ecological role, so the morphological attributes of a group of co-occurring species can be used to make inferences about, and comparisons of, communities. Table 3.1 lists some ecologically and functionally important morphological features. These ecologically significant features can be analyzed in a variety of ways: the morphological characters can be qualitative (e.g., type of support tissue) or quantitative (e.g., axis diameter); the overall analysis can be univariate (based on a single feature) or multivariate (based on many features); and the basic unit of study can be the individual specimen or the species. The types of community paleoecological analyses are outlined in table 3.2

One approach to characterizing a plant assemblage is to plot its component species or individuals on a single, continuous morphological axis, such as seed size or leaf size. These assemblage distributions can then be compared to one another. This kind of approach was taken by Tiffney (1984, 1986) in comparing diaspore size in Late Cretaceous and Tertiary angiosperm diaspore assemblages. Tiffney observed a dramatic increase in mean and maximum seed size of angiosperm floras from the Cretaceous to the early Tertiary and concluded that this reflected a change in the ecological role of the group. Tiffney characterized each assemblage on the basis of just its mean and range of seed sizes, but entire diaspore size distributions could be compared with one another, as is done with body size distributions in mammalian assemblages (Legendre 1987). Traditional leaf margin analysis (Wolfe 1978), which relies on the percentage of species in an assemblage that have entire leaves, is the

Table 3.1   Morphological Characters of Functional/Ecological Importance

Organ	Character	Values
Stem	Axis diameter	Continuous
	Support tissue type	Wood, bark, roots
	Wood type	Pycnoxylic, manoxylic
	Tracheid/vessel diameter	Continuous
	Tracheid/vessel cross-sectional density	Continuous
	Stem shape	Columnar, branching
	Branching	Sparsely, densely, intercalary meristem
Leaf	Leaves	Absent, scalelike, laminar
	Laminar area	Continuous or semiquantitative (log scale)
	Area/perimeter ratio of leaf	Continuous
	Laminar margin	Entire, nonentire
	Laminar thickness	Continuous
	Cuticle thickness	Continuous
	Stomatal size	Continuous
	Stomatal state	Sunken, flush, papillate
	Stomatal distribution	Amphistomatous, hyposto-matous, epistomatous, bands
	Leaf hairs	Glabrous, pubescent, densely pubescent
Any	Antipredator devices	Absent, chemical, mechanical
Root	Root diameter	Continuous
	Root shape	Regular, buttress, prop, tap, aerenchymous
Male/Female	General reproduction type	Clonal, sexual
Female	Endosperm volume	Continuous
	Packaging attributes	Sarcotesta/fruit, hooks/barbs, aerodynamic devices, float
	Disseminule size	Continuous
Male/Female	Sexual system	Homosporous, heterosporous, seeds
	Distribution of repro. organs	Imperfect, perfect
Male	Male dispersal unit size	Continuous
	Male dispersal unit attributes	Sculpture, bladders, elaters
All	Life form	Tree, shrub, vine, ground cover, epiphyte

Table 3.2   Approaches to Community Paleoecology

	Univariate	Multivariate
Continuous	Units individuals—seed size[a]	Units individuals—early Tertiary mammals[b]
variables	Units species—mammal body size[c]	Units species—extant *Pinus*[d]
Discrete-state	Units individuals?	Units individuals?
variables	Units species—leaf margin analysis[e]	Units species—foliar physiognomy[f]

Notes: Ecological Morphology: (1) ecological classification of species is a posteriori; (2) paleoecological inferences are based solely on morphology; (3) uniformitarian assumptions are weak; (4) can be applied only to taxonomically and morphologically similar species; (5) data and analysis types in table.

Ecological Categorization: (1) ecological classification of species is a priori and somewhat subjective; (2) paleoecological inferences can be drawn from morphology, sedimentary environment, and taxonomic analogy; (3) uniformitarian assumptions strong; (4) analyses can include disparate groups of organisms.

[a] Tiffney 1984.
[b] R. K. Stucky, pers. comm., 1989.
[c] Legendre 1987.
[d] McCune 1988; discrete-state variables too.
[e] Wolfe 1979.
[f] Wolfe 1990.

same kind of analysis performed with a qualitative, two-state variable rather than a continuous, quantitative variable. In this kind of analysis the morphological features can be scored or measured for individual specimens (as in Tiffney's diaspore size work), or for species (as in Wolfe's leaf margin studies).

The univariate ecomorphological approaches discussed above can be expanded by using more than one morphological trait. Multivariate studies also can use either continuous or discrete-state variables and can treat specimens individually or consider only mean values of species. This kind of approach has been applied to living pines (McCune 1988), to South African fynbos vegetation (Campbell and Werger 1988), and to leaf physiognomy of dicotyledonous floras (Wolfe 1990). McCune used a mixture of continuous and discrete-state characters; the others used only discrete-state characters. The goal of this approach is to define a multidimensional space on the basis of morphological traits that are thought to have ecological significance. The dimensionality of the space can be reduced through a variety of ordination procedures (§3). Generally the first few axes of the ordination correlate with environmental or directly measured ecological variables. The great advantage to this approach is its objectivity. Clusters of species (or specimens) in the multidimensional morphological space should define groups of ecologically similar plants a posteriori, without any preconceived notions of ecological similarity.

Preliminary attempts to apply this approach to very broad comparisons in the plant fossil record have been confounded by four problems: (1) the major groups of vascular plants each have fundamentally different body plans, and therefore they have evolved fundamentally different morphological solutions to the same environmental problems. For example, only ferns have solved the problem of constructing a trunk by growing adventitious roots. Palms use persistent leaf bases for much the same purpose, but the fern and palm morphologies are utterly different. The result is that they cannot be compared readily along the same morphological axis. (2) the potential for plants to adapt to similar environmental conditions through changes in different organs (e.g., leaves or roots, not necessarily both) allows what are thought to be ecologically similar species to be morphologically divergent, especially if they belong to different higher taxa. (3) most plant species in the fossil record are based on only one or two organs, which means that only a limited set of morphological traits is usually observable for any one species. This interacts with (2) to make it impossible to characterize many fossil species. More complications can arise if ecologically (and morphologically) similar species are known from different fossilized organs and thus cannot be scored in the same categories. (4) the multivariate approach weights equally all the traits scored for each species, in spite of the likelihood that a small number of traits are primarily responsible for defining the role of the species in the ecosystem. The primacy of these traits is obscured in a multivariate analysis.

If the major groups of land plants have found fundamentally different morphological solutions to the same environmental-ecological problems, then direct morphological comparison probably is inappropriate. However, both univariate and multivariate analyses of ecologically significant morphological traits hold out great promise for taxonomically or temporally restricted ecological comparisons in the fossil record.

5.2 Ecological Categories

Ecological categorization uses information from morphology, sedimentary environment, or living relatives to infer the ecological roles of the species in a fossil assemblage. The species are then assigned to ecological categories that are determined a priori. These ecological categories are somewhat analogous to the guilds of Root (1967), in that they are an attempt to define similarities between species in their ecological functions. The advantage of this approach is that it is flexible; it can circumvent the noncomparability of morphology between major clades, and it incorporates ancillary information from sedimentological context or spatial and temporal distribution. However, this method is less objective than an analysis based on morphological traits (because ecological categories are defined a priori). The assignment of extinct species to ecological categories is also less explicit than strict morphological analysis (different species potentially being assigned to the same ecological category for different reasons), more tied to uniformitarian interpretation (ecological roles being influenced more by what we know of current roles), and more relativistic (because extinct species probably will be evaluated within the context of other species from the same time and place). Inferences about the different communities are compared rather than the measurable characteristics of the plant species that make up those communities. In spite of these defects, ecological categorization is the only method we know for comparing past plant communities that are preserved in very different ways or that are composed of species that are radically different in morphology. Preliminary application of an approach using ecological categories has produced encouraging results in an Eocene example (see below and Wing 1988).

An Eocene Example

The data for an Eocene example were taken from samples of compression floras collected from a variety of depositional environments and stratigraphic levels in the early Eocene Willwood Fm. of Wyoming (original data in Wing 1981). Census data had been taken using the count method outlined in §3.1, and a total of 52 species had been recovered from the 16 localities. Initial ordinations of the species abundances using reciprocal averaging (RA) did not reveal any obvious relationships between floral composition and other variables such as stratigraphic level or depositional environment. This lack of discernable order resulted from the large number of species that occurred only once and from the tendency for these unique occurrences to be concentrated at two or three localities that represented unusual depositional environments or were stratigraphically isolated. When the unusual samples were removed from the ordination, the remaining samples displayed a weak gradient related to stratigraphic level but no relation to local environment.

To reveal ecological groups better, the 52 taxa were assigned to 12 ecological types based on foliar features (such as the ones discussed in §2.1) or analogy with living relatives (§2.2). The ecological types, with example taxa and the criteria used in assigning taxa to ecological types, are given in table 3.3. The abundances of these ecotypes at the 16 localities were calculated by combining the values for the appropriate taxa; then the abundances of the ecological types were ordinated by reciprocal averaging (fig. 3.1). The ordination of ecological types revealed two major groupings on axis II: (1) the "forest group"—the two canopy tree types, understory trees and shrubs, and vines; and (2) a less-cohesive group including all the other types—aquatic plants, pioneer trees and shrubs, palms, large herbs, and ferns. Group 2 exhibited a gradient along axis I, with the emergent and floating aquatic forms at one end, the palms and pioneer trees and shrubs in the middle, and the scrambling ferns

Table 3.3   Ecological Categories for an Eocene Flora of 52 Species (Willwood Fm., Wyoming)

Abbreviation and Examples	Spp	Category	Criteria for Assignment to Category
CANOTR1 *Phoebe*, Annonaceae	4	Dicotyledonous canopy tree	Leaves simple, pinnate, micro- to notophyll size. Living relatives usually large canopy trees. Cuticle thickness may indicate deciduous or evergreen leaves.
CANOTR2 *Glyptostrobus*, *Metasequoia*	2	Coniferous canopy tree	Living species in these genera are large, deciduous conifers with fairly diffuse canopies and are tolerant of wet sites.
UNDETRS aff. *Hamamelis*, Theaceae	5	Understory tree/shrub	Noto- to mesophyll size leaves with thin cuticle. Usually members of families that have many understory spp.
SEEDVIN Menispermaceae, Cucurbitaceae	3	Dicotyledonous vine	Deeply cordate leaves, or membership in families that have many extant lianous species.
PIONTRS *Platycarya*, *Alnus*, *Populus*	22	Pioneer tree/shrub	Pinnately compound leaves, very small seeds, membership in families and/or genera dominated by early successional species.
LARGHRB Zingiberaceae, Masaceae	2	Large herbaceous plants	Large, thin, blade-shaped leaves. Living species in these families always herbaceous.
EMERAQU *Typha*, Alismataceae	2	Emergent aquatic plant	Living species in genus/family are emergent aquatics.
FLOTAQU *Salvinia*, *Spirodela*, *Azolla*	3	Floating aquatic plant	Peltate leaves. Roots and/or leaves morphologically very similar to those in living species of these genera that are specialized for floating aquatic existence.
FERNVIN *Lygodium*	2	Scrambling/climbing pteridophyte	Very similar to living species of *Lygodium*, which are scramblers.
FERNTRE *Cnemidaria*	1	Tree pteridophyte	Very similar to living tree fern *Cnemidaria tryonia*.
FERNOTH *Thelypteris*, *Equisetum*	5	Herbaceous pteridophyte	Small pinnae; living congenerics mostly herbaceous.
PALMTRE Palm leaf fragments	1	Palm tree	

Note: Abbreviations are used in fig. 3.1.

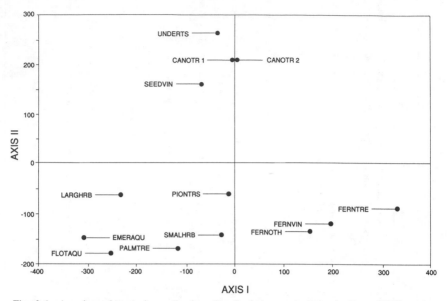

Fig. 3.1. A reciprocal averaging ordination of ecological types in the early Eocene Willwood flora, based on their abundances at 16 localities. Abbreviations for ecological types are explained in table 3.3. Ecological types typical of more stable sites, the "forest group," plot high on Axis II. Early successional types have low scores on Axis II and display a gradient on Axis I that may be related to moisture and/or disturbance. See text and table 3.3 for discussion of ecological types.

and tree ferns at the other end. Axis II appears to separate forest vegetation of more stable sites from a variety of kinds of seral vegetation. Axis I may represent a gradient from fully aquatic substrates in ponds to disturbed wetlands supporting ferns and some woody plants.

The interpretation of the axes of the RA analysis as representing ecological gradients is reinforced by the sedimentological features of the localities from which the collections were made. The RA plot of the sample sites is presented in figure 3.2, with the localities coded for environment of deposition. The two localities with features indicating deposition in oxbow ponds (Wing 1984) contained a flora dominated by floating and emergent aquatics, large herbs, and pioneer trees and shrubs. The remaining localities had sedimentary features indicative of deposition on a levee or crevasse splay near the channel or in floodplain swamps distant from the channel. One of the near-channel localities was dominated by pioneer trees and shrubs, whereas the sample at the other near-channel locality had species belonging to the deciduous canopy tree and understory trees and shrubs categories, with few pioneer elements. This second near-channel locality has finer-grained sediment than the first and was farther from the contemporaneous channel; these two factors may reflect

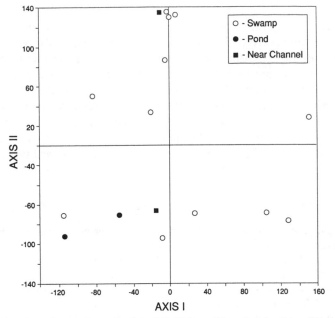

Fig. 3.2. A reciprocal averaging ordination of 16 Willwood Fm. plant fossil localities based on
the abundance of ecological types in the flora. Localities are coded for inferred environment of
deposition, which is based solely on sedimentological criteria. Both pond localities occur in the
same segment of the plot occupied by the emergent and floating aquatic species (see fig. 3.1),
showing concordance between paleoecological reconstruction based on plant morphology and
sedimentary features. One near-channel locality occurs in the area of the plot occupied by the
herbs and pioneer trees; the other is associated with the "forest group." Localities with sedimentary features typical of distal floodplain swamps are seen in all regions of the plot.

greater stability and explain the less successional vegetation. The distal swamp
localities were dominated either by the forest group or by ferns. The presence
of two taxonomically and ecologically distinct kinds of fossil plant assemblages in the distal swamp sedimentary environment suggests that these
groupings may result from conditions not reflected in the sediments, such as
local disturbance from tree falls or small fires.

6 CONCLUSION

This chapter has been concerned largely with the form-function correlations
that underlie ecological interpretations of fossil plants and with methods for
formalizing and combining these inferences to arrive at the paleocommunity
descriptions essential for evolutionary paleoecology. Although the plant fossil
record is rich in anatomical and morphological detail, the methods for combining autecological inferences into paleocommunity descriptions are not well

developed for pre-Pleistocene time. To some degree the development of these methods will rely on a closer integration of anatomical-morphological data with information derived from sedimentary context and spatial position of autochthonous fossils. Although methods of comparing pre-Pleistocene plant communities are not well developed, attempts based on one or a few morphological features have been made (Chaloner 1967; Tiffney 1984; Crepet 1984; Wing and Tiffney 1987). In the future we foresee comparisons of community characteristics across temporal or environmental gradients that use a greater array of morphological-ecological attributes and that will explicitly address the evolution and rise to dominance of new ecological types, as well as changes in species diversity.

REFERENCES

Anderson, J. A. R. 1964. The structure and development of the peat swamps of Sarawak and Brunei. *Journal of Tropical Geography* 18:7–16.

Augspurger, C. K. 1986. Morphology and dispersal potential of wind-dispersed diaspores of neotropical trees. *American Journal of Botany* 73:353–63.

Axelrod, D. I. 1987. The late Oligocene Creede flora, Colorado. *University of California Publications in Geological Sciences* 130:1–235.

Bailey, I. W., and E. W. Sinnott. 1915. A botanical index of Cretaceous and Tertiary climates. *Science,* n.s. 41:831–34.

———. 1916. The climatic distribution of certain types of angiosperm leaves. *American Journal of Botany* 3:24–39.

Baker, H. G. 1972. Seed mass in relation to environmental conditions in California. *Ecology* 53:997–1010.

Bateman, R. M., and A. C. Scott. 1990. A reappraisal of the Dinantian floras at Oxroad Bay, East Lothian, Scotland. 2. Volcanicity, palaeoenvironments, and palaeoecology. *Transactions of the Royal Society of Edinburgh* 81:161–94.

Batenburg, L. H. 1982. "Compression species" and "petrifaction species" of *Sphenophyllum* compared. *Review of Palaeobotany and Palynology* 36:335–59.

Beck, C. B. 1970. The appearance of gymnospermous structure. *Botanical Reviews* 45:379–400.

Birks, H. J. B., and H. H. Birks. 1980. *Quaternary Palaeoecology.* Baltimore: University Park Press.

Brokaw, N. V. L. 1985a. Gap-phase regeneration in a tropical forest. *Ecology* 66:682–87.

———. 1985b. Treefalls, regrowth, and community structure in tropical forests. 53–69 in S. T. A. Pickett and P. S. White, eds., *The Ecology of Natural Disturbance and Patch Dynamics.* Orlando, Fla: Academic Press.

Burnham, R. B. 1987. Inferring vegetation from plant-fossil assemblages: effects of depositional environment and heterogeneity in the source vegetation on assemblages from modern and ancient fluvial-deltaic environments. Ph.D. diss., University of Washington.

———. 1988. Paleoecological approaches to analyzing stratigraphic sequences. 105–25 in W. A. DiMichele and S. L. Wing, eds., *Methods and Applications of Plant Paleoecology.* Paleontological Society Special Publication no. 3.

Burnham, R. J., and R. A. Spicer. 1986. Forest litter preserved by volcanic activity at El Chichon, Mexico: A potentially accurate record of the preeruption vegetation. *Palaios* 1:158–61.

Burnham, R. J., S. L. Wing, and G. G. Parker 1990. Plant diversity and the fossil record: How reliable are the estimates? *Fourth International Congress of Systematics and Evolutionary Biology, Abstracts with Program,* 327.

———. 1992. Reflection of temperate forest composition and structure in the litter: Implications for the fossil record. *Paleobiology* 18:34–53.

Calkin, H. W., A. C. Gibson, and P. S. Nobel. 1985. Xylem water potentials and hydraulic conductances in eight species of ferns. *Canadian Journal of Botany* 63:632–37.

Campbell, B. M., and M. J. A. Werger. 1988. Plant form in the mountains of the Cape South Africa. *Journal of Ecology* 76:637–53.

Carlquist, S. 1975. *Ecological Strategies of Xylem Evolution.* Berkeley: University of California Press.

———. 1977. Ecological factors in wood evolution: A floristic approach. *American Journal of Botany* 64:887–96.

———. 1985. Vasicentric tracheids as a drought survival mechanism in the woody flora of southern California and similar regions: Review of vasicentric tracheids. *Aliso* 11:37–68.

Cerling, T. E., J. Quade, Y. Wang, and J. R. Bowman. 1989. Carbon isotopes in paleosol carbonates as paleoecological indicators. *Nature* 341:138–39.

Chaloner, W. G. 1967. Spores and land-plant evolution. *Review of Palaeobotany and Palynology* 1:83–89.

Cichan, M. A. 1986. Conductance in the wood of selected Carboniferous plants. *Paleobiology* 12:302–10.

Coddington, J. A. 1988. Cladistic tests of adaptational hypotheses. *Cladistics* 4:3–22.

Cohen, A. D., W. Spackman, and R. Raymond. 1987. Interpreting the characteristics of coal seams from chemical, physical, and petrographic studies of peat deposits. In A. C. Scott, ed., *Coal and Coal-Bearing Strata: Recent Advances.* Geological Society Special Publication 32:107–126.

Cole, K. 1985. Past rates of change, species richness, and a model of vegetational inertia in the Grand Canyon, Arizona. *American Naturalist* 125:289–303.

Crane, P. R. 1984. A re-evaluation of *Cercidiphyllum*-like plant fossils from the British early Tertiary. *Botanical Journal of the Linnaean Society* 89:199–230.

Crane, P. R. and R. A. Stockey. 1985. Growth and reproductive biology of *Joffrea speirsii* gen. et sp. nov., a *Cercidiphyllum*-like plant from the late Paleocene of Alberta, Canada. *Canadian Journal of Botany* 63:340–64.

Creber, G. T., and W. G. Chaloner. 1984. Climatic indications from growth rings in fossil woods. 49–77 in P. J. Brenchley, ed., *Fossils and Climate.* Chichester, England: John Wiley.

———. 1985. Tree growth in the Mesozoic and Early Tertiary and the reconstruction of palaeoclimates. *Palaeogeography, Palaeoclimatology, Palaeoecology* 52:35–60.

Crepet, W. L. 1984. Advanced (constant) insect pollination mechanisms: Pattern of evolution and implications vis-à-vis angiosperm diversity. *Annals of the Missouri Botanical Garden* 71:607–30.

Crepet, W. L., and E. M. Friis. 1987. The evolution of insect pollination in angiosperms. 181–201 in E. M. Friis, W. G. Chaloner, and P. R. Crane, eds., *The Ori-*

gins of Angiosperms and their Biological Consequences. New York: Cambridge University Press.

Cusset, G. 1982. The conceptual basis of plant morphology. 8–86 in R. Sattler, ed., *Axioms and Principles of Plant Construction*. The Hague: Martinus Nijhoff.

Davis, M. B. 1965. A method for determination of absolute pollen frequency. 674–85 in B. Kummel and D. M. Raup, eds., *Handbook of Paleontological Techniques*. San Francisco: W. H. Freeman.

———. 1973. Redeposition of pollen grains in lake sediments. *Limnology and Oceanography* 18:44–52.

Davis, M. B., L. B. Brubaker, and T. Webb III. 1973. Calibration of absolute pollen influx. 9–25 in H. J. B. Birks and R. G. West, eds., *Quaternary Plant Ecology*. London: Blackwell.

Delcourt, P. A., and H. R. Delcourt. 1987. *Long-Term Forest Dynamics of the Temperate Zone*. New York: Springer-Verlag.

Digby, P. G. N., and R. A. Kempton. 1987. *Multivariate Analysis of Ecological Communities*. London: Chapman and Hall.

DiMichele, W. A. 1983. *Lepidodendron hickii* and generic delimitation in Carboniferous lepidodendrid lycopods. *Systematic Botany* 8:317–33.

DiMichele, W. A., and P. J. DeMaris. 1987. Structure and dynamics of a Pennsylvanian-age *Lepidodendron* forest: Colonizers of a disturbed swamp habitat in the Herrin (no. 6) coal of Illinois. *Palaios* 2:146–57.

DiMichele, W. A., and W. J. Nelson. 1989. Small-scale spatial heterogeneity in Pennsylvanian-age vegetation from the roof shale of the Springfield Coal (Illinois Basin). *Palaios* 4:276–80.

DiMichele, W. A., and T. L. Phillips. 1985. Arborescent lycopod reproduction and paleoecology in a coal-swamp environment of late Middle Pennsylvanian age (Herrin Coal, Illinois, USA). *Review of Palaeobotany and Palynology* 44:1–26.

———. 1988. Paleoecology of the Middle Pennsylvanian-age Herrin coal swamp (Illinois) near a contemporaneous river system, the Walshville paleochannel. *Review of Palaeobotany and Palynology* 56:151–76.

DiMichele, W. A., T. L. Phillips, and G. E. McBrinn. 1991. Quantitative analysis and paleoecology of the Secor Coal and roof-shale floras (middle Pennsylvanian, Oklahoma). *Palaios* 6:390–409.

Dolph, G. E., and D. L. Dilcher. 1980. Variation in leaf size with respect to climate in the tropics of the Western Hemisphere. *Bulletin of the Torrey Botanical Club* 107:154–62.

Ehret, D. L., and T. L. Phillips. 1977. *Psaronius* root systems: Morphology and development. *Palaeontographica*, Abt. B 161:147–64.

Esterle, J. S., J. C. Ferm, and T. Yiu-Liong. 1989. A test for the analogy of tropical domed peat deposits to "dulling up" sequences in coal beds: Preliminary results. *Journal of Organic Chemistry* 14:333–42.

Ewers, F. W. 1985. Xylem structure and water conduction in conifer trees, dicot trees, and lianas. *IAWA Bulletin*, n.s. 6:309–17.

Faegri, K., and J. Iversen. 1975. *Textbook of Pollen Analysis*. 3d ed. New York: Hafner.

Farley, M. B. 1988. Environmental variation, palynofloras, and paleoecological interpretation. 126–46 in W. A. DiMichele and S. L. Wing, eds., *Methods and Applica-*

tions of Plant Paleoecology. Paleontological Society Special Publication no. 3.
————. 1989. Palynological facies fossils in nonmarine environments in the Paleogene of the Bighorn Basin. *Palaios* 4:565–73.

Farley, M. B., and S. L. Wing. 1989. Quantitative comparison of leaf and pollen samples from the same sites. Abstract. *American Journal of Botany* 76:164–65.

Feng, B.-C. 1989. Paleoecology of an upper Middle Pennsylvanian coal swamp from western Pennsylvania, USA. *Review of Palaeobotany and Palynology* 57:299–312.

Ferguson, D. K. 1985. The origin of leaf-assemblages: New light on an old problem. *Review of Palaeobotany and Palynology* 46:117–88.

Foster, S. A. 1986. On the adaptive value of large seeds for tropical moist forest trees: A review and synthesis. *Botanical Review* 52:260–99.

Foster, S. A., and C. H. Janson. 1985. The relationship between seed size and establishment conditions in tropical woody plants. *Ecology* 66:773–80.

Frankenburg, J. M., and D. A. Eggert. 1969. Petrified *Stigmaria* from North America, I. *Stigmaria ficoides,* the underground portions of Lepidodendraceae. *Palaeontographica* 128B:1–47.

Friis, E. M., and W. L. Crepet. 1987. Time of appearance of floral features. 145–79 in E. M. Friis, W. G. Chaloner, and P. R. Crane, eds., *The Origins of Angiosperms and Their Biological Consequences.* New York: Cambridge University Press.

Fritts, H. C. 1976. *Tree Rings and Climate.* New York: Academic Press.

Gartner, B. L., S. H. Bullock, H. A. Mooney, V. B. Brown, and J. L. Whitbeck. 1990. Water transport properties of vine and tree stems in a tropical deciduous forest. *American Journal of Botany* 77:742–49.

Gastaldo, R. A. 1986. Implications of the paleoecology of autochthonous lycopods in clastic sedimentary environments of the early Pennsylvanian of Alabama. *Palaeogeography, Palaeoclimatology, Palaeoecology* 53:191–212.

————. 1987. Confirmation of Carboniferous clastic swamp communities. *Nature* 326:869–71.

————. 1988. Conspectus of phytotaphonomy. 14–26 in W. A. DiMichele and S. L. Wing, eds., *Methods and Applications of Plant Paleoecology.* Paleontological Society Special Publication no. 3.

Gastaldo, R. A., D. P. Douglass, and S. M. McCarroll. 1987. Origin, characteristics, and provenance of plant macrodetritus in a Holocene crevasse splay, Mobile Delta, Alabama. *Palaios* 2:229–40.

Gibson, A. C., H. W. Calkin, and P. S. Nobel. 1985. Hydraulic conductance and xylem structure in tracheid-bearing plants. *IAWA Bulletin,* n.s. 6:293–302.

Givnish, T. 1979. On the adaptive significance of leaf form. 375–407 in O. T. Solbrig, S. Jain, G. B. Johnson, and P. Raven, eds., *Topics in Plant Population Biology.* New York: Columbia University Press.

Givnish, T., and G. Vermeij. 1976. Sizes and shapes of liane leaves. *American Naturalist* 110:743–46.

Gould, S. J., and R. C. Lewontin. 1979. The spandrels of San Marco and the Panglossian paradigm: A critique of the adaptationist programme. *Proceedings of the Royal Society of London, B* 205:581–98.

Grady, W. C., and C. F. Eble. 1990. Relationships among macerals, miospores, and paleoecology in a column of the Redstone coal (Upper Pennsylvanian) from north-central West Virginia (USA). *International Journal of Coal Geology* 15:1–26.

Grieg-Smith, P. 1983. *Quantitative Plant Ecology*, 3d ed. Berkeley: University of California Press.

Grime, J. P. 1977. Evidence for the existence of three primary strategies in plants and its relevance to ecological and evolutionary theory. *American Naturalist* 111: 1169–94.

Habib, D. 1968. Spore and pollen paleoecology of the Redstone Seam (Upper Pennsylvanian) of West Virginia. *Micropaleontology* 14:199–220.

Harper, J. L., K. G. Lovell, and P. H. Moore. 1970. The shapes and sizes of seeds. *Annual Review of Ecology and Systematics* 1:327–56.

Hickey, L. J. 1977. Stratigraphy and paleobotany of the Golden Valley Formation (early Tertiary) of western North Dakota. *Geological Society of America Memoir* 150:1–181.

―――. 1980. Paleocene stratigraphy and flora of the Clark's Fork Basin. 33–49 in P. D. Gingerich, ed., *Early Cenozoic Paleontology and Stratigraphy of the Bighorn Basin, Wyoming. University of Michigan Papers on Paleontology* 24.

Hickey, L. J., and J. A. Doyle. 1977. Early Cretaceous fossil evidence for angiosperm evolution. *Botanical Review* 43:3–104.

Horn, H. S. 1971. *The Adaptive Geometry of Trees*. Monographs in Population Biology 3:3–144.

Janzen, D. H. 1969. Seed eaters versus seed size, number, toxicity, and dispersal. *Evolution* 23:1–27.

Jefferson, T. H. 1982. The Early Cretaceous fossil forests of Alexander Island, Antarctica. *Palaeontology* 25:681–708.

Kerp, H. 1989. Cuticular analysis of gymnosperms. 36–63 in B. H. Tiffney, ed., Phytodebris, notes for a workshop on the study of fragmentary plant remains. Privately published for the Paleobotanical Section of the Botanical Society of America.

Knoll, A. H. 1986. Patterns of change in plant communities through time. 126–44 in J. Diamond and T. J. Case, eds., *Community Ecology*. New York: Harper and Row.

Koch, C. F. 1978. Bias in the published fossil record. *Paleobiology* 4:367–72.

Kortlandt, A. 1980. The Fayum primate forest: Did it exist? *Journal of Human Evolution* 9:27–297.

Kovach, W. L. 1988. Multivariate methods of analyzing paleoecological data. 72–104 in W. A. DiMichele and S. L. Wing, eds., *Methods and Applications of Plant Paleoecology*. Paleontological Society Special Publication no. 3.

Kovach, W. L., and D. L. Dilcher. 1984. Dispersed cuticles from the Eocene of North America. *Botanical Journal of the Linnean Society* 88:63–104.

Krassilov, V. A. 1975. *Paleoecology of Terrestrial Plants*. New York: John Wiley and Sons.

Kraus, M. J. 1988. Nodular remains of early Tertiary forests, Bighorn Basin, Wyoming. *Journal of Sedimentary Petrology* 58:888–93.

Lamboy, W. and A. Lesnikowska. 1988a. Some statistical methods useful in the analysis of plant paleoecological data. 52–71 in W. A. DiMichele and S. L. Wing, eds., *Methods and Applications of Plant Paleoecology*. Paleontological Society Special Publication no. 3.

―――. 1988b. Some statistical methods useful in the analysis of plant paleoecological data. *Palaios* 3:86–94.

Laveine, J. P. 1986. The size of the frond in the genus *Alethopteris* Sternberg (Pteridospermopsida, Carboniferous). *Geobios* 19:49–56.

Leary, R. L. and H. W. Pfefferkorn. 1977. An early Pennsylvanian flora with *Megalopteris* and Noeggerathiales from west-central Illinois. *Illinois State Geological Survey, circular 500.*

Lee, D. W., R. A. Boone, S. L. Tarsis, and D. Storch. 1990. Correlates of leaf optical properties in tropical forest sun and extreme-shade plants. *American Journal of Botany* 77:370–80.

Legendre, S. 1987. Analysis of mammalian communities from the late Eocene and Oligocene of southern France. *Palaeovertebrata* 16:191–212.

Lewis, A. M., and M. T. Tyree. 1985. The relative susceptibility to embolism of larger vs. smaller tracheids in *Thuja occidentalis*. Abstract. *IAWA Bulletin* n.s. 6:93.

Litke, R. 1967. Kutikularanalytischer Nachweis für einen Wechsel von warmgemässigtem zu warmen Klima im Jungtertiär. *Abhandlungen des Zentralen Geologischen Instituts* 10:123–27.

Loehle, C. 1988. Tree life history strategies: The role of defenses. *Canadian Journal of Forestry Research* 18:209–22.

Magurran, A. E. 1988. *Ecological Diversity and Its Measurement.* Princeton: Princeton University Press.

Mahaffy, J. F. 1985. Profile patterns of coal and peat palynology in the Herrin (No. 6) Coal Member, Carbondale Formation, Middle Pennsylvanian of southern Illinois. *Proceeding, Ninth International Congress of Carboniferous Stratigraphy and Geology* 5:25–34.

Maher, L. J., Jr. 1972. Nomograms for computing 0.95 confidence limits of pollen data. *Review of Paleobotany and Palynology* 13:85–93.

Manum, S. 1976. Dinocysts in Tertiary Norwegian-Greenland sea sediments (Deep Sea Drilling Project Leg 38), with observations on palynomorphs and palynodebris in relation to environment. 897–919 in M. Talwani and L. Udintsev, eds., *Initial Reports Deep Sea Drilling Project* 38.

Mapes, G., G. W. Rothwell, and M. T. Haworth. 1989. Evolution of seed dormancy. *Nature* 337:645–46.

Mazer, S. J. 1989. Ecological, taxonomic, and life history correlates of seed mass among Indiana dune angiosperms. *Ecological Monographs* 59:153–75.

———. 1990. Seed mass variation of Indiana Dune genera and families: Taxonomic and ecological correlates. Evolutionary Ecology 4:326–58.

McCabe, P. J. 1984. Depositional environments of coal-bearing strata. In R. A. Rahmani and R. M. Flores, eds., *Sedimentology of Coal and Coal-Bearing Sequences.* International Association of Sedimentologists Special Publication 7:13–42.

———. 1987. Facies studies of coal and coal-bearing strata. In A. C. Scott, ed., *Coal and Coal-Bearing Strata: Recent Advances. Geological Society Special Publication* 32:51–66.

McCune, B. 1988. Ecological diversity in North American pines. *American Journal of Botany* 75:353–68.

Nambudiri, E. M. V., W. D. Tidwell, B. N. Smith, and N. P. Hebert. 1978. A C_4 plant from the Pliocene. *Nature* 276:816–17.

Niklas, K. J. 1981. Airflow patterns around some early seed plant ovules and cupules: Implications concerning efficiency in wind pollination. *American Journal of Botany* 68:635–50.

———. 1984a. Size-related changes in the primary xylem anatomy of some early tracheophytes. *Paleobiology* 10:487–506.

———. 1984b. The motion of windborne pollen grains around conifer ovulate cones: Implications on wind pollination. *American Journal of Botany* 71:356–74.

O'Leary, M. H. 1988. Carbon isotopes in photosynthesis. *Bioscience* 38:328–36.

Oliver, F. W., and D. H. Scott. 1904. On the structure of the Paleozoic seed *Lagenostoma lomaxii*, with a statement of the evidence upon which it is referred to *Lyginodendron*. *Philosophical Transactions Royal Society London* B, 197:193–247.

Parkhurst, D. F., and O. L. Loucks. 1972. Optimal leaf size in relation to environment. *Journal of Ecology* 60:505–37.

Parrish, J. T., and R. A. Spicer. 1988. Late Cretaceous terrestrial vegetation: A near-polar temperature curve. *Geology* 16:22–25.

Parsons, R. W., and I. C. Prentice. 1981. Statistical approaches to *R*-values and the pollen-vegetation relationships. *Review of Palaeobotany and Palynology* 32: 127–52.

Pfefferkorn, H. W., and K. Fuchs. In press. A field classification of fossil plant substrate interactions. *Neues Jahrbuch für Geologie und Paläontologie, Abhandlungen*.

Pfefferkorn, H. W., H. Mustafa, and H. Hass. 1975. Quantitative Characterisierung Ober-Karboner Abdruckfloren. *Neues Jahrbuch für Geologie und Paläontologie, Abhandlungen* 150:253–69.

Phillips, T. L. 1979. Reproduction of heterosporous arborescent lycopods in the Mississippian-Pennsylvanian of Euramerica. *Review of Palaeobotany and Palynology* 27:239–89.

Phillips, T. L., M. J. Avcin and D. Berggren. 1976. *Fossil Peat from the Illinois Basin*. Illinois State Geological Survey, Educational Series 11.

Phillips, T. L. and W. A. DiMichele. 1981. Paleoecology of Middle Pennsylvanian age coal swamps in southern Illinois: Herrin Coal Member at Sahara Mine no. 6. 231–84 in K. J. Niklas, ed., *Paleobotany, Paleoecology, and Evolution*, vol. 1. New York: Praeger.

———. In press. Comparative ecology and life-history biology of arborescent lycopods in Late Carboniferous swamps of Euramerica. *Annals of the Missouri Botanical Gardens*.

Phillips, T. L., A. B. Kunz, and D. J. Mickish. 1977. Paleobotany of permineralized peat (coal balls) from the Herrin (No. 6) Coal Member of the Illinois Basin. 18–49 in P. N. Given and A. D. Cohen, eds., *Interdisciplinary Studies of Peat and Coal Origins*. Geological Society of America, Microform Publication 7.

Phillips, T. L., R. A. Peppers, M. J. Avcin, and P. J. Laughnan. 1974. Fossil plants and coal: Patterns of change in Pennsylvanian coal swamps of the Illinois basin. *Science* 184:1367–69.

Phillips, T. L., R. A. Peppers and W. A. DiMichele. 1985. Stratigraphic and inter-regional changes in Pennsylvanian coal-swamp vegetation: Environmental inferences. *International Journal of Coal Geology* 5:43–109.

Pielou, E. C. 1984. *The Interpretation of Ecological Data*. New York: John Wiley & Sons.

Pryor, J. S. 1988. Sampling methods for quantitative analysis of coal-ball plants. *Palaeogeography, Palaeoclimatology, Palaeoecology* 63:313–26.

Rahel, F. J. 1990. The hierarchical nature of community persistence: A problem of scale. *American Naturalist* 136:328–44.

Raynor, G. S., E. C. Ogden, and J. V. Hayes. 1970. Dispersion and deposition of ragweed pollen from experimental sources. *Journal of Applied Meteorology* 9: 885–95.

Remy, W., and R. Remy. 1980. Devonian gametophytes with anatomically preserved gametangia. *Science* 208:295–96.

Remy, W., R. Remy, H. Hass, St. Schultka, and F. Franzmeyer. 1980. *Sciadophyton* Steinmann: Ein Gametophyt aus dem Siegen. *Argumenta Palaeobotanica* 6:73–94.

Retallack, G. J. 1985. Fossil soils as grounds for interpreting the advent of large plants and animals on land. *Philosophical Transactions of the Royal Society of London* 309:105–42.

Ridley, H. N. 1930. *The Dispersal of Plants throughout the World*. Kent, England: L. Reeve.

Rockwood, L. L. 1985. Seed mass as a function of life form, elevation, and life form in neotropical forests. *Biotropica* 17:32–39.

Root, R. B. 1967. The niche exploitation pattern of the Blue Gray Gnatcatcher. *Ecological Monographs* 37:317–50.

Rothwell, G. W. 1977. Evidence of a pollination-drop mechanism in Paleozoic pteridosperms. *Science* 198:1251–52.

———. 1984. The apex of *Stigmaria* (Lycopsida), rooting organ of Lepidodendrales. *American Journal of Botany* 71:1031–34.

———. 1988. Cordaitales. 273–97 in C. B. Beck, ed., *Origin and Evolution of Gymnosperms*. New York: Columbia University Press.

Rothwell, G. W., and S. E. Scheckler. 1988. Biology of ancestral gymnosperms. 85–134 in C. B. Beck, ed., *Origin and Evolution of Gymnosperms*. New York: Columbia University Press.

Rothwell, G. W., and K. L. Whiteside. 1974. Rooting structures of the Carboniferous medullosan pteridosperms. *Canadian Journal of Botany* 52:97–102.

Rury, P. M., and W. C. Dickison. 1984. Structural correlations among wood, leaves, and plant habit. 495–540 in R. White and W. C. Dickison, eds., *Contemporary Problems in Plant Anatomy*. New York: Academic Press.

Salisbury, E. J. 1942. *The Reproductive Capacity of Plants*. London: Bell.

———. 1974. Seed size and mass in relation to environment. *Proceedings of the Royal Society of London B, Biological Sciences* 186:83–88.

Schabilion, J. T., and M. A. Reihman. 1985. Anatomy of petrified *Neuropteris scheuchzeri* pinnules from the Middle Pennsylvanian of Iowa: A paleoecological interpretation. *Proceedings, Ninth International Congress of Carboniferous Stratigraphy and Geology* 5:3–12.

Scheihing, M. H., and Pfefferkorn, H. W. 1984. The taphonomy of land plants in the Orinoco delta: A model for the incorporation of plant parts in clastic sediments of late Carboniferous age of Euramerica. *Review of Palaeobotany and Palynology* 41:205–40.

Schweitzer, H.-J. 1980. Die Gattungen *Taeniocrada* White und *Sciadophyton* Steinmann im Unterdevon des Rheinlandes. *Bonner Palaeobot. Mitteil.* no. 5.

Scott, A. C. 1977. A review of the ecology of Upper Carboniferous plant assemblages, with new data from Strathclyde. *Palaeontology* 20:447–73.

————. 1978. Sedimentological and ecological control of Westphalian B plant assemblages from West Yorkshire. *Proceedings of the Yorkshire Geological Society* 41:461–508.

Scott, A. C., and G. Rex. 1985. The formation and significance of Carboniferous coal balls. *Philosophical Transactions of the Royal Society of London, B* 311:123–37.

Silvertown, J. W. 1981. Seed size, lifespan, and germination date as coadapted features of plant life history. *American Naturalist* 118:860–64.

Smith, A. C. 1972. Buttressing of tropical trees: A descriptive model and new hypotheses. *American Naturalist* 106:32–46.

Smith, A. H. V. 1962. The palaeoecology of Carboniferous peats based on the miospores and petrography of bituminous coals. *Proceedings of the Yorkshire Geological Society* 33:423–74.

Smith, A. H. V., and M. A. Butterworth. 1967. *Miospores in the Coal Seams of the Carboniferous of Great Britain.* Special Papers in Palaeontology 1:1–324.

Sperry, J. S., M. T. Tyree, and J. R. Donnelly. 1988. Vulnerability of xylem to embolism in a mangrove vs. an inland species of Rhizophoraceae. *Physiology Plantarum* 74:276–83.

Spicer, R. A. 1988. Quantitative sampling of plant megafossils assemblages. 29–51 in W. A. DiMichele and S. L. Wing, eds., *Methods and Applications of Plant Paleoecology.* Paleontological Society Special Publication no. 3.

————. 1989. Physiological characteristics of land plants in relation to environment through time. *Transactions of the Royal Society of Edinburgh Earth Sciences* 80:321–29.

Spicer, R. A., and C. R. Hill. 1979. Principal component and correspondence analyses of quantitative data from a Jurassic plant bed. *Review of Palaeobotany and Palynology* 28:273–99.

Spicer, R. A., and J. T. Parrish. 1986. Paleobotanical evidence for cool north polar climates in middle Cretaceous (Albian-Cenomanian) time. *Geology* 14:703–6.

Taggart, R. E. 1988. The effect of vegetation heterogeneity on short stratigraphic sequences. 147–71 in W. A. DiMichele and S. L. Wing, eds., *Methods and Applications of Plant Paleoecology.* Paleontological Society Special Publication no. 3.

Taggart, R. E., and A. T. Cross. 1980. Vegetation change in the Miocene Sucker Creek Flora of Oregon and Idaho: A case study in paleosuccession. 185–210 in D. L. Dilcher and T. N. Taylor, eds., *Biostratigraphy of Fossil Plants.* Stroudsburg, Pa.: Dowden, Hutchinson, and Ross.

————. 1990. Plant successions and interruptions in Miocene volcanic deposits, Pacific Northwest. 57–68 in M. G. Lockley and A. Rice, eds., *Volcanism and Fossil Biotas.* Geological Society of America Special Paper 244.

Tauber, H. 1965. Differential pollen dispersion and the interpretation of pollen diagrams. *Danmarks Geologiske Undersogelse* 2:89.

Taylor, T. N. 1988. Pollen and pollen organs of fossil gymnosperms: Phylogeny and reproductive biology. 177–217 in C. B. Beck, ed., *Origin and Evolution of Gymnosperms.* New York: Columbia University Press.

Thomas, B. A. 1981. Structural adaptations shown by the Lepidocarpaceae. *Review of Palaeobotany and Palynology* 32:377–88.

Thomas, B. A. and J. Watson. 1976. A rediscovered 114-foot *Lepidodendron* from Bolton, Lancashire. *Geological Journal* 11:15–20.

Thomasson, J. R., M. E. Nelson, and R. J. Zakrzewski. 1986. A fossil grass (Gramineae: Chloridoideae) from the Miocene with Kranz anatomy. *Science* 233:876–78.

Tidwell, W. D., and E. M. V. Nambudiri. 1989. *Tomlinsonia thomassonii*, gen. et sp. nov., a permineralized grass from the upper Miocene Ricardo Formation, California. *Review of Palaeobotany and Palynology* 60:165–77.

Tiffney, B. H. 1984. Seed size, dispersal syndromes, and the rise of the angiosperms: Evidence and hypothesis. *Annals of the Missouri Botanical Garden* 71:551–76.

———. 1986. Evolution of seed dispersal syndromes according to the fossil record. 273–305 in D. R. Murray, ed., *Seed Dispersal*. North Ryde, N.S.W.: Academic Press Australia.

———. 1989. *Phytodebris: Notes for a workshop on the study of fragmentary plant remains*. Privately published for the Paleobotanical Section of the Botanical Society of America.

Tiffney, B. H., and K. J. Niklas. 1985. Clonal growth in land plants: A paleobotanical perspective. 35–66 in J. B. C. Jackson, L. Buss, and R. E. Cook, eds., *Population Biology and Evolution of Clonal Organisms*. New Haven: Yale University Press.

Traverse, A. 1988. *Paleopalynology*. Winchester, Mass.: Allen and Unwin.

Tsukada, M. 1982. *Pseudotsuga menziesii* (Mirb.) Franco: Its pollen dispersal and late Quaternary history in the Pacific Northwest. *Japanese Journal of Ecology* 32:159–87.

Upchurch, G. R., Jr. 1989. Dispersed angiosperm cuticles. 65–92 in B. H. Tiffney, ed., *Phytodebris, notes for a workshop on the study of fragmentary plant remains*. Privately published for the Paleobotanical Section of the Botanical Society of America.

Upchurch, G. R., Jr., and J. A. Wolfe. 1987. Mid-Cretaceous to Early Tertiary vegetation and climate: Evidence from fossil leaves and woods. 75–105 in E. M. Friis, W. G. Chaloner, and P. R. Crane, eds., *The Origin of Angiosperms and Their Biological Consequences*. Cambridge: Cambridge University Press.

van der Pijl, L. 1982. Principles of Dispersal in Higher Plants. 3d ed. Berlin: Springer-Verlag.

Van Devender, T. R., and W. G. Spaulding. 1979. Development of vegetation and climate in the southwestern United States. *Science* 204:701–10.

Watson, D. M. S. 1907. On a confusion of two species (*Lepidodendron harcourtii* Witham and *L. hickii* sp. nov.) under *Lepidodendron harcourtii* Witham, in Williamson's XIX Memoir, with a description of *L. hickii* sp nov. *Memoirs and Proceedings of the Manchester Literary Society* 49, *Memoir* 13.

Webb, T., III, S. E. Howe, R. H. W. Bradshaw, and K. M. Heide. 1981. Estimating plant abundances from pollen data: The use of regression analysis. *Review of Palaeobotany and Palynology* 34:269–300.

Willard, D. A. 1990. Palynology of the Springfield Coal of the Illinois Basin (Middle Pennsylvanian) with quantitative comparison of spore floras and coal-ball peats and implications for paleoecological studies. Ph.D. diss., University of Illinois, Urbana.

———. In press. Vegetational patterns in the Springfield Coal (Middle Pennsylvanian, Illinois Basin): Comparison of miospore and coal-ball records. *International Journal of Coal Geology*.

Wing, S. L. 1981. A study of paleoecology and paleobotany in the Willwood Formation (Early Eocene, Wyoming). Ph.D. diss., Yale University.

———. 1984. Relation of paleovegetation to geometry and cyclicity of some fluvial carbonaceous deposits. *Journal of Sedimentary Petrology* 54:52–66.

———. 1988. Taxon-free paleoecological analysis of Eocene megafloras from Wyoming. Abstract. *American Journal of Botany* 75.

Wing, S. L., and M. B. Farley. 1990. Stability in the composition of Paleogene forests. *Fourth International Congress of Systematic and Evolutionary Biology, Abstracts with Program.*

Wing, S. L., and B. H. Tiffney. 1987. The reciprocal interaction of angiosperm evolution and tetrapod herbivory. *Review of Palaeobotany and Palynology* 50:179–210.

Wnuk, C. 1985. The ontogeny and paleoecology of *Lepidodendron rimosum* and *Lepidodendron bretonense* trees from the middle Pennsylvanian of the Bernice Basin (Sullivan County, Pennsylvania). *Palaeontographica, B* 195:153–81.

———. 1989. Ontogeny and paleoecology of the Middle Pennsylvanian arborescent lycopod *Bothrodendron punctatum,* Bothrodendraceae (Western Middle Anthracite Field, Shomokin Quadrangle, Pennsylvania). *American Journal of Botany* 76:966–80.

Wnuk, C., and H. W. Pfefferkorn. 1984. The life habits and paleoecology of middle Pennsylvanian medullosan pteridosperms based on an in situ assemblage from the Bernice Basin (Sullivan County, Pennsylvania, USA). *Review of Palaeobotany and Palynology* 41:329–51.

———. 1987. A Pennsylvanian-age terrestrial storm deposit: Using plant fossils to characterize the history and process of sediment accumulation. *Journal of Sedimentary Petrology* 57:212–21.

Wolfe, J. A. 1972. An interpretation of Alaskan Tertiary floras. 201–33 in A. Graham, ed., *Floristics and Paleofloristics of Asia and Eastern North America.* Amsterdam: Elsevier.

———. 1978. A paleobotanical interpretation of Tertiary climates in the northern hemisphere. *American Scientist* 66:694–703.

———. 1979. Temperature parameters of humid to mesic forests of eastern Asia and relation to forests of other regions of the northern hemisphere and Australasia. *U.S. Geological Survey Professional Paper* 1106:1–37.

———. 1990. Palaeobotanical evidence for a marked temperature increase following the Cretaceous/Tertiary boundary. *Nature* 343:153–56.

Wolfe, J. A., and G. R. Upchurch, Jr. 1987a. Leaf assemblages across the Cretaceous-Tertiary boundary in the Raton Basin, New Mexico and Colorado. *Proceedings of the National Academy of Sciences USA* 84:5096–5100.

———. 1987b. North American nonmarine climates and vegetation during the Late Cretaceous. *Palaeogeography, Palaeoclimatology, Palaeoecology* 61:33–77.

Zimmermann, M. H. 1983. *Xylem Structure and the Ascent of Sap.* Berlin: Springer-Verlag.

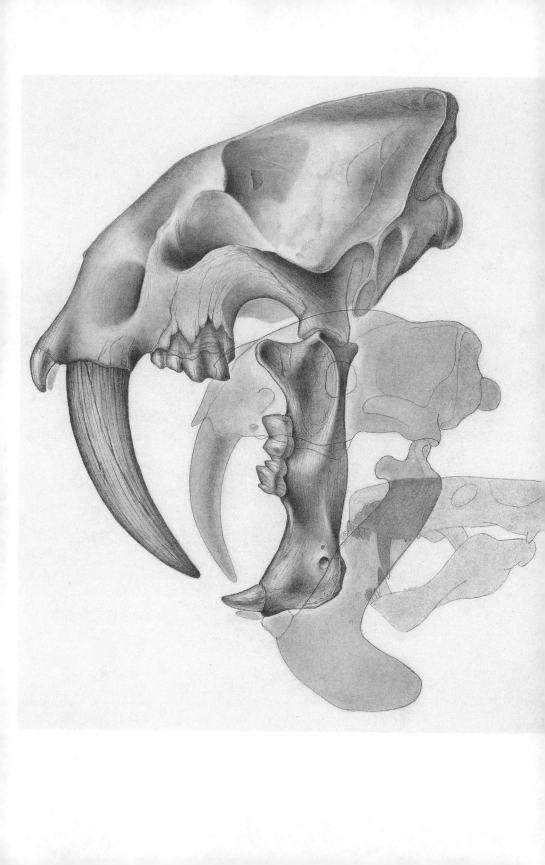

F O U R   Taxon-Free Characterization
of Animal Communities

John D. Damuth, RAPPORTEUR

IN COLLABORATION WITH DAVID JABLONSKI, JUDITH A. HARRIS, RICHARD POTTS,
RICHARD K. STUCKY, HANS-DIETER SUES, AND DAVID B. WEISHAMPEL

1 INTRODUCTION: TAXON-FREE CHARACTERIZATIONS

One important objective of paleoecological studies is the reconstruction of the
adaptations and modes of life of extinct organisms and the reconstruction of
the environments in which they lived. A second goal is to use our knowledge
of fossil faunas and paleoenvironments to investigate the general long-term
ecological or evolutionary processes that cause community and ecosystem
change over geological time (see chap. 1). The documentation of repeated historical patterns of community or ecosystem change should form the empirical
base upon which theories about such macroevolutionary processes may be
erected and evaluated.

This second goal can be approached only if we are able to make informative
comparisons of communities from different time horizons and geographic
areas. This requires that we compare communities in terms of general biological properties that are largely independent of specific historical circumstances
and the taxa that make up the fauna. In other words, we must be able to specify what properties of communities may be changing (or staying the same)
over time and across space other than their taxonomic composition per se.
These properties must be defined in ways that allow meaningful comparisons
across communities whose contained species may be only distantly related.

For example, there is reason to believe that the distribution of body sizes
among the vertebrates of a community is a significant feature of community
structure and indicates something about ecosystem function and environmental conditions. (For mammalian faunas, at least, see Fleming 1973; Andrews,
Lord, and Nesbit Evans 1979; Legendre 1984, 1989.) In such a case, the
distribution of body sizes (represented in some way, perhaps by a histogram)
becomes a *character* of the community—one that may change over time, and
one that also can be measured in any other community of animals. In this
sense it is a *taxon-free* characterization of that community. Taxon-free characterizations can consist of community-level characters such as species richness
or species diversity, or may consist of characters formed by aggregation of

species characteristics, such as counts or distributions of species according to locomotor type, food preference, habitat use, or body size. (Also, for continuous morphological measurements on species or specimens, or for ecological inferences expressed as continuous variables, the means, ranges, areas, or volumes defined by the values of the species of the community could be considered taxon-free characterizations.)

Of course a body-size distribution is only a partial characterization of the community, like that of characters contributing to the phenotype of an organism. Continuing this analogy, the organismal phenotype that we recognize is a combination of one or more characters that we have chosen out of the enormous array of potential characters that the organism possesses. Often the characters are further abstracted by being represented in terms of one of a limited number of character *states*. Yet, each organism is physically unique. What we have done in characterizing an organism by its phenotypic class is not to provide a complete description of that organism but rather to represent it in such a way that it can be meaningful compared with other organisms. We have ignored some of the organism's uniqueness and characteristics to accomplish a particular aim—perhaps to allow the assignment of fitness values to the phenotype, so that we can study some aspect of natural selection, or perhaps to allow phylogenetic reconstruction. Likewise, when we come up with a taxon-free characterization of a community, it is not intended to be a complete description of the community or the ecosystem. Rather we have chosen to ignore some unique features of the community to enable us to make particular comparisons across communities and across time.

Note also that it is the result (i.e., the characterization of the community) that is independent of explicit knowledge about taxa and their relationships, not necessarily the inferential pathways to that characterization. If we know, for example, that the body masses of species of different taxa are best estimated by different means, then of course we are allowed to use such knowledge. We want the best possible inferences about the ecosystem and its components—for use as building blocks for constructing characterizations of the community and developing interpretations of community change through time.

Various kinds of taxon-free analyses have proven useful in comparative community studies involving extant vertebrate faunas (Harrison 1962; Valverde 1964; Bourlière 1973; Fleming 1973; Eisenberg and McKay 1974; Eisenberg 1981; Eisenberg and Redford 1982; Glanz 1982; Emmons, Gautier-Hion, and Dubost 1983). Paleontologists are now able to make increasingly robust inferences concerning the ecological roles of fossil species (see below). Taxon-free analyses of vertebrate faunas have been widely employed in vertebrate paleoecology (Olson 1952; Fleagle 1978; Andrews, Lord, and Nesbit Evans 1979; Andrews and Nesbit Evans 1979; Van Couvering 1980; Webb 1983; Janis 1984; Legendre 1984, 1989; Van Valkenburgh 1985, 1987; Collinson and Hooker 1987; Stucky 1990). The utility of taxon-free characterizations has also been demonstrated in marine invertebrate paleontology, where use of this approach has yielded considerable insight on the ecological processes underlying major diversifications and extinctions (Bambach 1983, 1985; Bottjer and Ausich 1986).

2 ONE APPROACH TO TAXON-FREE CHARACTERIZATION

2.1 Ecomorphology and Ecological Types

From a purely biological point of view, the basic data on fossil communities come from the specimens found at a given locality. We thus would like to be able to turn species (or sometimes even specimen) lists into ecological characterizations of the community. Here we concentrate on one method of approach, the depiction of each species in terms of a set of ecological characteristics based primarily upon its morphology. (Rarely it may be possible to infer some aspects of an animal species' paleoecology directly from nonmorphological information, such as location or distribution within the sediment or other data associated with the fossil remains [Voorhies and Thomasson 1979; Horner 1982; Smith 1987]. Such nonmorphological inferences may be particularly important for fossil plants; see chap. 3.) For each ecological variable (e.g., diet) a species can be classified as belonging to one of a number of possible categories, each of which defines an ecological type (e.g., browser, grazer, etc.). The number or distribution of such ecological types can form the basic elements of some kinds of taxon-free characterizations (see below). This approach is essentially a version of the *ecomorphological* one employed by many ecologists, who infer ecological characteristics of extant species from their morphology and use these in community analysis (Karr and James 1975; Miles, Ricklefs, and Travis 1987). For this reason such ecological types, when based on inferences from morphology, have often been called ecomorphological types, or *ecomorphs* (Wing 1988; Martin 1989; Janis and Damuth 1990).

2.2 Functional Morphology and Autecological Inference

The functional morphology of vertebrates and related biomechanical principles are studied in an active body of research too extensive to be reviewed here (Alexander 1983; Hildebrand et al. 1985; Radinsky 1987). Nevertheless it is worthwhile to mention a number of major lines of evidence that allow inferences of autecology from data on fossil vertebrates.

Dietary inferences have traditionally relied upon studies of the design and function of vertebrate teeth and the functional morphology of skulls and jaws (Rensberger 1973, 1986, 1988; Kay and Hiiemae 1974; Kay 1975, 1978; Kay and Hylander 1978; Janis 1979; Lucas 1979; Krause 1982, 1986; Norman and Weishampel 1985; Hotton 1986; Maas, Krause, and Strait 1988; Fortelius

1982, 1985; Janis and Erhardt 1988; Janis and Fortelius 1988; Gordon and
Illius 1988; Solunias and Dawson-Saunders 1988; Van Valkenburgh 1988;
King, Oelofson, and Rubidge 1989). Additionally, studies of both gross dental wear (Rensberger 1973; Kay 1975, 1978; Fortelius 1987; Janis 1990) and
dental microwear (using scanning electron microscopy; see, e.g., Rensberger
1978; Young and Marty 1986; Young and Robson 1987; Teaford 1988; Solunias, Teaford, and Walker 1988; Van Valkenburgh, Teaford, and Walker
1990) allow interpretation of the kinds of foodstuffs actually eaten by the
organism.

The biomechanics of locomotion have also been extensively studied among
vertebrates (Gambaryan 1974; Childress 1977; Alexander 1982; Biewener
1900). Reconstructions of the locomotor habits of fossil vertebrates can reveal
much about their autecology (Aiello 1981; Jungers 1982; Jenkins and Krause
1983; Scott 1985, 1987; Van Valkenburgh 1985, 1987; Sues 1986; Kappelman 1988; Alexander 1989; Farlow 1990; Gatesy 1990). In addition, locomotor specializations and the functional morphology of limbs can reveal
features of the vegetation structure of the habitat (Emmons and Gentry 1983;
Reif and Silyn-Roberts 1987; Scott 1987; Kappelman 1988; Dudley and De
Vries 1990). Fossilized trackways also provide direct evidence of locomotor
behavior (Farlow 1981; Lockley 1986; Leakey and Harris 1987; Gillette and
Lockley 1989).

Animal body size correlates with a host of physiological, ecological, and
behavioral variables (Peters 1983; Calder 1984; Schmidt-Nielsen 1984; Damuth and MacFadden 1990). Fairly reliable estimates of vertebrate body size
can be obtained from even fragmentary skeletal and dental remains (Alexander 1989; Damuth and MacFadden 1990; Gingerich 1990).

A variety of aspects of behavioral ecology can be inferred from morphology and from other paleontological data (Boucot 1990). For example, diel
activity patterns in small mammals are related to the size of the orbit relative
to that of the skull (Kay and Cartmill 1977). Herding and other associative
behavior can be inferred from the occurrence of mass deaths (Voorhies and
Thomasson 1979; Turnbull and Martill 1988; Coombs 1990). Dinosaur nests
allow reconstruction of other aspects of social behavior, including parental
care (e.g., Horner and Makela 1979; Horner 1982; Coombs 1990; Forster
1990). Sexual dimorphism and morphological adaptations for sexual display
or combat can reveal features of social behavior, mating system, and population structure for extinct vertebrates (Barghusen 1975; Farlow and Dodson
1975; Hopson 1975; Sues 1978; Janis 1982; Fortelius 1983; Kitchner 1987;
Coombs 1990).

Morphologically based ecological inferences for fossil members of some
vertebrate groups (such as birds and small reptiles) are less common in the
literature than are those for the larger reptiles and mammals. However, useful
autecological and paleoenvironmental inferences involving these groups can

be made (Olson and Rasmussen 1986). Likewise, fossil terrestrial invertebrates have not yet been extensively studied from a functional perspective, but there is potential for further development of taxon-independent characterization of ecological roles (Manton 1977; Shear et al. 1984; Wilson 1987; Shear et al. 1989).

2.3 Continuous versus Category Variables

The ecological characteristics inferred for individual species can often be expressed in either of two forms: (1) as continuous measures along one or more axes, or (2) as membership in a discrete ecological category belonging to a particular ecological variable. For example, the proportion of plant fiber in the diet of a vertebrate species could be expressed by its position on an axis that represents a continuously varying tooth dimension that we know reflects relative dietary fiber. Alternatively, again based upon inferences from dental (and/or cranial) morphology, species could be placed in one of a series of discrete diet categories (e.g., high, medium, and low fiber; bark, grass, browse, fruit, etc.). Not all characteristics lend themselves equally well to representation in both ways. Choices of variable type and analytical strategy must be made that are appropriate to the question and data at hand.

An analysis using continuous variables assumes a close and continuous mapping of ecology onto particular morphological dimensions. Therefore continuous variables are most useful when the organisms being compared are closely related or are very similar in the general form-function relationship being examined (Van Valkenburgh 1985; Stucky 1990). Variables that take on continuous values are ideal for many kinds of studies of niche packing and guild structure. Indeed this is the way that ecomorphological studies are most often used in neoecology (Karr and James 1975; Miles, Ricklefs, and Travis 1987). Analyses using continuous-valued variables could in some cases be extended to comparisons among unlike organisms, if enough functional information were available to transform mathematically the morphologies of different species into a common scale of measurement.

However, category-valued variables have significant advantages when dissimilar organisms are being compared. For example, tooth dimensions differ consistently between browsers and grazers among both extant perissodactyls and artiodactyls, with grazers having narrower molars than browsers (Damuth 1990). However, because the teeth of members of the two orders differ significantly in basic plan, plotting measures on the same axes for both groups would be meaningless for direct dietary interpretation. As an extreme example, ants and heteromyid rodents are both granivores in the southwestern United States, yet they share no functionally relevant morphological characters that could be treated as part of a continuum. Because ecological categories allow assignment of rough ecological equivalence on the basis of different morphological criteria, they permit more general comparisons than do treatments using continuous variables. They will always be of value in making the most broad-ranging analyses.

The disadvantages of using categories include possible lack of precision and the possibly false suggestion of disjunction or discreteness, whereas in fact there may often be continuous gradation among the categories for a given variable. In addition, we tend to base our categories primarily upon our experience with extant and familiar organisms, and these categories may not be as applicable, useful, or illuminating when applied to very different types of organisms and communities in the remote past. Because categories have the power to channel our thinking in subtle ways, researchers should remain aware of the degree to which they themselves may have imposed a particular pattern upon nature. The variables and categories described below are probably widely useful but are not to be regarded as definitive.

2.4 A Basic Set of Ecological Variables and Categories

A set of ecological variables and their associated categories is listed in table 4.1. This particular scheme is only one of many possible ways to classify and express the ecological characteristics of fossil animals. We consider this scheme (or one similar to it) to incorporate the major ecological characteristics that can be reliably inferred for most well-represented fossil terrestrial animal species. We also consider it to provide a realistic and flexible basis for subsequent taxon-free community analyses of the type described more fully in §2.5. Readers should note that the variables and categories presented here are most directly applicable to fossil mammals and other large terrestrial vertebrates. This reflects the extensive fossil record of these organisms and the current state of research on their functional morphology. Other categories (and presumably variables) would have to be added or some of the present ones modified to characterize appropriately fossil birds or terrestrial invertebrates.

The categories for a number of the variables are presented in a hierarchical scheme to allow for different levels of resolution in comparison. They progress from the crudest or most general comparisons at level 1 to more fine-scaled comparisons at level 3. Although it may be possible to infer ecological characteristics of Neogene fossil vertebrates with some precision, it may not be possible to achieve the same level of resolution for more ancient creatures. In some cases the nature of the fossil record, or the current state of our knowledge of functional morphology, precludes any inference for certain variables.

2.5 Analyses Using Ecological Types

A mere listing of the ecological characteristics of the members of a paleocommunity (or a counting of ecological types) does not constitute an analysis or characterization of community structure or of any processes of organic change. Rather these basic ecological types can serve as the basis for a large number of different analyses, each relevant to a particular question we might be asking about similarities and differences among communities. Different analyses will focus on different categories and combinations of categories (or different variables and combinations of variables).

The results of such analyses allow interpretations, or further inferences, to be made concerning community structure, environment, or qualitative aspects of ecosystem function (such as this was an ecosystem of low productivity, or this community lived in a seasonal environment). These interpretations can be ends in themselves or may suggest new questions to ask. Many such higher-level inferences about and interpretations of ecosystems can be made only in view of sound empirical knowledge of extant ecosystems and through adoption of explicit theoretical perspectives concerning the way that ecosystems work and the factors that are important in organismal evolution and ecosystem change. The ultimate goal of paleocommunity and paleoecosystem studies is not the listing of ecological types but rather the use of these basic data (among others) to examine questions of empirical and theoretical interest (see chap. 1). Obviously we must have the questions before designing a particular data analysis to answer them.

In some cases the mere presence or absence of a particular type of organism can allow an interpretation of some aspect of a paleocommunity. Usually, however, more complex analyses involving many or all species and more than one variable will be revealing. A simple hypothetical example is presented in figures 4.1–4.3, involving only mammalian diet. Figure 4.1 shows the basic morphologically derived data for two imaginary paleocommunities in terms of counts of species in each ecological category. Figures 4.2–4.3 show two analyses and interpretations using these data.

The given interpretations are for illustration only and are not meant as serious proposals about valid ecological inferences. In particular, although all the interpretations of figure 4.3 are consistent with the analysis presented, we would require more and different kinds of independent information before we could consider them to be firmly established. Specifically, to support the given assertions about competition among predators, we would want reliable information about the relative abundances of species in the living community, and this may be difficult to obtain (Damuth 1982; chap. 2). Likewise, statements about competition imply a knowledge of resource levels, which might come from independent knowledge of paleoclimates. Ideally, all fossil localities to be compared should be *isotaphonomic* with respect to the preservation of the biological information required by a given analysis (chap. 2). Even the simplest analyses described here require that the species lists for the localities be comparable in completeness.

Figures 4.1–4.3 show how each analysis uses the same morphologically derived data to construct different taxon-free characterizations of the commu-

Table 4.1 Ecological Variables and Categories for Animals

Variable	Category		
	First Level	Second Level	Third Level
Food Preference	Carnivore	Carnivore	Carnivore Meat[1] Meat and Bone
	Herbivore	Invertebrate consumer	Insectivore (*sensu lato*) Soft[2] Mixed[2] Hard[2]
		Omnivore	Omnivore Animal dominated Plant dominated
		Herbivore	Herbivore Low relative Fiber[3] Frugivore Resin/gum Soft browser Granivore Medium Fiber[4] Selective Nonselective High Fiber Mixed browser-grazer Root specialist Corticivore (bark) Grazer Selective Nonselective
Food processing[5]	Oral processor[6] Gut processor[7]		
Locomotion[8]	Terrestrial	Terrestrial[9] Arboreal Scansorial Terrestrial Semifossorial Fossorial	(Terrestrial category can be broken down further according to a variety of classifications, e.g., cursorial, saltatory, ambush predator, etc.)
	Semiaquatic Aquatic Aerial		

Table 4.1 (*continued*)

| | Category | | |
Variable	First Level	Second Level	Third Level
Feeding habitat[8]	Terrestrial	Terrestrial[9] Canopy[10] Arboreal/surface Ground surface Subterranean Semiaquatic Aquatic Aerial	(Not broken down further)
Shelter habitat[8]	(Categories as for feeding habitat)		
Body mass	(Numerical value, or expressed in size classes)		
Encephalization quotient (EQ)[11]	(Numerical value)		
Activity period	Diurnal Crepuscular Nocturnal		
Miscellaneous[12]			

[1] Includes muscle, bone, fat.

[2] Refers to hardness of exoskeleton.

[3] I.e., high concentration of easily digested nutrients.

[4] Or Browser/Folivore.

[5] For ungulate mammals, a related category of "digestion" might also be included, with *hindgut fermenter, foregut fermenter (nonruminant), true ruminant,* etc., as entries.

[6] E.g., mammals, some dinosaurs.

[7] E.g., birds, some dinosaurs.

[8] The three categories *Locomotion, Feeding habitat,* and *Shelter habitat* are intended to be used together to characterize the part of the environment that is used by the species and to indicate to some degree how the animal uses the environment. Thus a giraffe is a ground-surface/canopy/ground-surface species, whereas a monkey might be an arboreal/canopy/canopy species. In this case the two species share the feeding habitat, but not the others.

[9] Only the terrestrial category is broken down further here.

[10] Could be further divided into upper and lower canopy.

[11] A measure of relative brain size. Other indices might be used, but this is the current standard.

[12] There are numerous aspects of social behavior and ecology that might be inferred from morphology or from other kinds of evidence (often of an extraordinary nature). Two examples: Sexual dimorphism (indicative of aspects of social organization), and evidence for group living (inferred from the occurrence of mass deaths). On the basis of body size, diet, and degree of sexual dimorphism, many mammals can be classified as probably solitary or social to varying degrees (see, e.g., Jarman 1974).

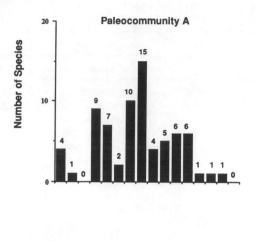

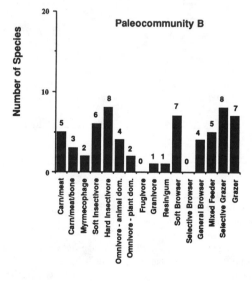

Food Type

Fig. 4.1. Basic ecomorphic dietary data for two hypothetical mammalian paleocommunities.

nities, directed toward the investigation of how different properties may differ between the two communities. In figure 4.2 the question being asked concerns the physical environment, and in figure 4.3 the question concerns ecological or evolutionary interactions among the members of a particular guild. Note that any valid, appropriate analytical techniques may be used, and that different analyses can allow us to arrive at the same interpretations. Also, a given analysis may support a number of different interpretations, as in figure 4.3.

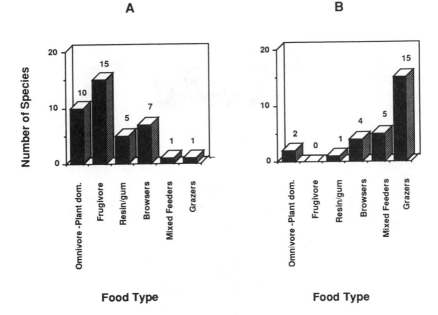

Fig. 4.2. An example of an analysis of the distribution of herbivore diets in two hypothetical paleocommunities (A and B). Data are from fig. 4.1. A possible ecological interpretation is that paleocommunity B is in a more open habitat with more grass.

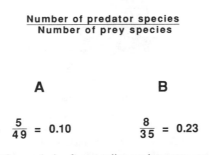

Fig. 4.3. An example of an analysis of mammalian predator-prey species ratios in two hypothetical paleocommunities (A and B). Data are from fig. 4.1. Possible interpretations of the results are: (1) Competition among predator species may be more intense in paleocommunity B; (2) Competition among prey species may be more intense in paleocommunity A; (3) Average niche breadth of predators is greater in paleocommunity A and less in paleocommunity B. These simplistic interpretations are presented for illustration purposes only. These interpretations are not necessarily either mutually exclusive nor compatible, nor are they proven on the basis of this analysis. Other interpretations are also consistent with this analysis (see text).

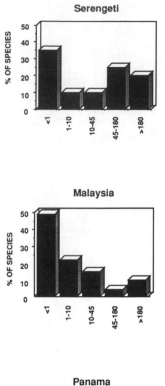

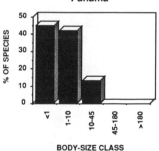

BODY-SIZE CLASS

Fig. 4.4. Body-mass distributions for the faunas of a tropical short-grass plains environment (Serengeti: Andrews, Lord, and Nesbit Evans 1979) and two lowland tropical rain-forest habitats (Malaysia: Harrison 1962; and Balboa, Panama: Fleming 1973). Values for the Serengeti were estimated from figures in Andrews, Lord, and Nesbit Evans 1979. Body-size classes are those used by Andrews, Lord, and Nesbit Evans 1979.

For comparison, figures 4.4 and 4.5 present body-size and food-type distributions for the primary consumers of three extant mammal communities (Harrison 1962; Fleming 1973; Andrews, Lord, and Nesbit Evans 1979). The Serengeti community is that of the short-grass plains, whereas the Malaysia and Panama communities are found in lowland tropical forest habitats. In both

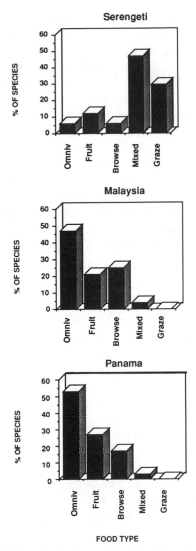

Fig. 4.5. Food-type distributions for the faunas of a tropical short-grass plains environment (Serengeti: Andrews, Lord, and Nesbit Evans 1979) and two lowland tropical rain-forest habitats (Malaysia: Harrison 1962; and Balboa, Panama: Fleming 1973). Values for the Serengeti were estimated from figures in Andrews, Lord, and Nesbit Evans 1979. "Omniv" refers to omnivores—plant dominated.

figures the Malaysian and Panamanian distributions are similar (in spite of having no species in common) and are obviously different from those of the open Serengeti environment. The pattern is especially marked for food type. From our knowledge of the extant ecosystems, the reason for the difference in food-type distributions is obvious: the dominance of grass as a resource for herbivores in the short-grass plains environment versus the lack of grass growing on the forest floor. Such differences in distributions across modern communities have been used to infer habitat for Miocene fossil assemblages (Andrews, Lord, and Nesbit Evans 1979; Andrews and Nesbit Evans 1979).

Note how all the interpretations for fossil data rely implicitly upon independently supported empirical or theoretical generalizations about community ecology for their validity. These generalizations, rules, laws, or processes are the results of research on extant ecosystems and communities. To judge the extent to which a particular ecological generalization based upon the Recent is applicable to the remote past, the causal mechanisms giving rise to present-day ecological patterns must be understood. Merely observing empirical patterns in the present is not sufficient. More work needs to be done with extant communities from a variety of habitats before our understanding of observed distributions of ecological types will be robust enough for widespread, routine interpretation of the fossil record. Ideally the paleoecologist is at least a bit of a neoecologist as well.

3   COMPLEMENTARY APPROACHES
AND HIGHER-LEVEL INFERENCES

Community characterization in terms of these morphologically based categories or variables is by no means the whole story in ecosystem and community paleoecology. Many of the things we would like to know about ancient ecosystems are only indirectly reflected in the ecological roles of members of the biotic community. From the point of view of contemporary evolutionary ecology, important features of both the physical setting of the ecosystem and its contained organisms include the quality of the resource base, density (or productivity of resources), spatial dispersion (patchiness) in three dimensions, and temporal patterning (i.e., seasonality and predictability). Working upward from the organism to these higher-level attributes of communities and ecosystems may involve tenuous chains of inference—if in a given case it is possible at all. Independent information from the record of the physical environment may often serve as a more direct indication of these features (see chap. 2). Such independent information is also frequently of crucial importance in allowing tests of our ecomorphic inferences.

Communities or ecosystems could be characterized in terms of such higher-level attributes. Similar kinds of qualities can also characterize particular animal or plant species, from the point of view of either their own ecology or the way in which they may provide resources for or interact with other species. However, for most fossil organisms it will be difficult to infer directly such higher-level attributes, since detailed records of resource use, short-term temporal fluctuations, and spatial distribution are generally not preserved. Rather, we would usually still base inferences about these upon more direct information about organisms' ecological roles combined with independent information about biotic interactants and the physical environment. The fact that sources of inferences about animals, plants, and the physical environment are largely independent provides a potentially rich basis for higher-level characterizations of communities.

The higher-level attributes described above make possible additional taxon-free characterizations of communities. In this case they are directed at answering questions posed within the context of the theories of phyletic evolution prevalent in the field of evolutionary ecology. The theoretical perspective from which the basic research questions derive determines not only the analytical methods to use but also whether organismic ecology as described by the ecological variables and categories of tables 4.1–4.3, by other continuous or qualitative variables, or by more general ecosystem characteristics is the appropriate starting place for taxon-free characterization of paleocommunities in a given study.

REFERENCES

Aiello, L. C. 1981. Locomotion in the Miocene Hominoidea. 63–97 in C. B. Stringer, ed., *Aspects of Human Evolution*. London: Taylor and Francis.

Alexander, R. M. 1982. *Locomotion of Animals*. New York: Chapman and Hall.

———. 1983. *Animal Mechanics*. 2d ed. Oxford: Blackwell Scientific Publications.

———. 1989. *Dynamics of Dinosaurs and Other Extinct Giants*. New York: Columbia University Press.

Andrews, P., and E. Nesbit Evans. 1979. The environment of *Ramapithecus* in Africa. *Paleobiology* 5:22–30.

Andrews, P., J. Lord, and E. M. Nesbit Evans. 1979. Patterns of ecological diversity in fossil and modern mammalian faunas. *Biological Journal of the Linnean Society* 11:177–205.

Bambach, R. K. 1983. Ecospace utilization and guilds in marine communities through the Phanerozoic. 719–46 in M. J. S. Tevesz and P. L. McCall, eds., *Biotic Interactions in Recent and Fossil Benthic Communities*. New York: Plenum.

———. 1985. Classes and adaptive variety: The ecology of diversification in marine faunas through the Phanerozoic. 191–253 in ed., J. W. Valentine, ed., *Phanerozoic Diversity Patterns*. Princeton: Princeton University Press.

Barghusen, H. R. 1975. A review of fighting adaptations in dinocephalians (Reptilia, Therapsida). *Paleobiology* 1:295–311.

Biewener, A. A. 1990. Biomechanics of mammalian terrestrial locomotion. *Science* 250:1097–1103.

Bottjer, D. J., and W. I. Ausich. 1986. Phanerozoic development of tiering in soft substrata suspension-feeding communities. *Paleobiology* 12:400–20.

Boucot, A. J. 1990. *Evolutionary Paleobiology of Behavior and Coevolution*. Amsterdam: Elsevier.

Bourlière, F. 1973. The comparative ecology of rainforest mammals in Africa and tropical America: Some introductory remarks. 279–92 in R. C. Meggers, E. S. Ayensu, and W. D. Duckworth, eds., *Tropical Forest Ecosystems in Africa and South America: A Comparative Review*. Washington, D. C.: Smithsonian Institution Press.

Calder, W. A., III. 1984. *Size, Function, and Life History.* Cambridge: Harvard University Press.

Childress, S. 1977. *Mechanics of Swimming and Flying.* New York: Courant Institute of Mathematical Sciences, New York University.

Collinson, M. E., and J. J. Hooker. 1987. Vegetational and mammalian faunal changes in the Early Tertiary of southern England. 259–304 in E. M. Friis, W. G. Chaloner, and P. R. Crane, eds., *The Origins of Angiosperms and Their Biological Consequences.* Cambridge: Cambridge University Press.

Coombs, W. P., Jr. 1990. Behavior patterns of dinosaurs. 32–42 in D. B. Weishampel, P. Dodson, and H. Osmólska, eds., *The Dinosauria.* Berkeley: University of California Press.

Damuth, J. 1982. Analysis of the preservation of community structure in assemblages of fossil mammals. *Paleobiology* 8:434–46.

————. 1990. Problems in estimating body masses of archaic ungulates using dental measurements. 229–53 in J. Damuth and B. J. MacFadden, eds., *Body Size in Mammalian Paleobiology: Estimation and Biological Implications.* New York: Cambridge University Press.

Damuth, J., and B. J. MacFadden, eds. 1990. *Body Size in Mammalian Paleobiology: Estimation and Biological Implications.* New York: Cambridge University Press.

Dudley, R., and P. DeVries. 1990. Tropical rain forest structure and the geographical distribution of gliding vertebrates. *Biotropica* 22:432–34.

Eisenberg, J. F. 1981. *The Mammalian Radiations.* Chicago: University of Chicago Press.

Eisenberg, J. F., and G. M. McKay. 1974. Comparison of ungulate adaptations in the New World and Old World tropical forests with special reference to Ceylon and the rainforests of Central America. 585–602 in V. Geist and F. Walther, eds., *The Behaviour of Ungulates and Its Relation to Management.* Morges, Switzerland: IUCN Publications, n.s. 24. International Union for the Conservation of Nature.

Eisenberg, J. F., and K. H. Redford. 1982. Comparative niche structure and evolution of mammals of the Nearctic and southern South America. 77–84, in M. A. Mares and H. H. Genoways, eds., *Mammalian Biology in South America.* Pymatuning Laboratory of Ecology, spec. pub. 6, University of Pittsburgh.

Emmons, L. H., A. Gautier-Hion, and G. Dubost. 1983. Community structure of the frugivorous-folivorous forest mammals of Gabon. *Journal of Zoology, London* 199:209–22.

Emmons, L. H., and A. H. Gentry. 1983. Tropical forest structure and the distribution of gliding and prehensile-tailed vertebrates. *American Naturalist* 121:513–24.

Farlow, J. O. 1981. Estimates of dinosaur speeds from a new trackway site in Texas. *Nature* 294:747–48.

————. 1990. Dynamic dinosaurs. *Paleobiology* 16:234–41.

Farlow, J. O., and P. Dodson. 1975. The behavioral significance of frill and horn morphology in ceratopsian dinosaurs. *Evolution* 29:353–61.

Fleagle, J. G. 1978. Size distribution of living and fossil primate faunas. *Paleobiology* 4:67–76.

Fleming, T. H. 1973. Numbers of mammal species in North and Central American forest communities. *Ecology* 54:555–63.

Forster, C. A. 1990. Evidence for juvenile groups in the ornithopod dinosaur *Tenontosaurus tilletti* Ostrom. *Journal of Paleontology* 64:164–75.

Fortelius, M. 1982. Ecological aspects of dental functional morphology in the Plio-Pleistocene rhinoceroses of Europe. 163–81 in B. Kurtén, ed., *Teeth: Form, Function, and Evolution.* New York: Columbia University Press.

———. 1983. The morphology and paleobiological significance of horns in *Coelodonta antiquitatis* (Mammalia: Rhinocerotidae). *Journal of Vertebrate Paleontology* 3:125–35.

———. 1985. Ungulate cheek teeth: Developmental, functional, and evolutionary interrelations. *Acta Zoologica Fennica* 180:1–76.

———. 1987. A note on the scaling of dental wear. *Evolutionary Theory* 8:73–75.

Gambaryan, P. P. 1974. *How Mammals Run: Anatomical Adaptations.* Trans. Hilary Hardin, Israel Program for Scientific Translations. New York: Wiley and Sons.

Gatesy, S. M. 1990. Caudofemoral musculature and the evolution of theropod locomotion. *Paleobiology* 16:170–86.

Gillette, D. D., and M. G. Lockley, eds. 1989. *Dinosaur Tracks and Traces.* Cambridge: Cambridge University Press.

Gingerich, P. D. 1990. Prediction of body mass in mammalian species from long bone lengths and diameters. *Contributions of the Museum of Paleontology, University of Michigan* 28:79–92.

Glanz, W. E. 1982. Adaptive zones of Neotropical mammals: A comparison of some temperate and tropical patterns. 95–110 in M. A. Mares and H. H. Genoways, eds., *Mammalian Biology in South America.* Pymatuning Laboratory of Ecology, spec. pub. 6, University of Pittsburgh.

Gordon, I. J., and A. W. Illius. 1988. Incisor arcade structure and diet selection in ruminants. *Functional Ecology* 2:15–22.

Harrison, J. L. 1962. The distribution of feeding habits among animals in a tropical rain forest. *Journal of Animal Ecology* 31:53–64.

Hildebrand, M., D. M. Bramble, K. F. Liem, and D. B. Wake, eds. 1985. *Functional Vertebrate Morphology.* Cambridge: Harvard University Press.

Hopson, J. A. 1975. The evolution of cranial display structures in hadrosaurian dinosaurs. *Paleobiology* 1:21–43.

Horner, J. R. 1982. Evidence for colonial nesting and "site fidelity" among ornithischian dinosaurs. *Nature* 297:675–76.

Horner, J. R., and R. Makela. 1979. Nest of juveniles provides evidence of family structure among dinosaurs. *Nature* 282:296–98.

Hotton, N., III. 1986. Dicynodonts and their role as primary consumers. 71–82 in N. Hotton III, P. D. MacLean, J. J. Roth, and E. C. Roth, eds., *The Ecology and Biology of Mammal-like Reptiles.* Washington, D.C.: Smithsonian Institution Press.

Janis, C. M. 1979. Mastication in the hyrax and its relevance to ungulate dental evolution. *Paleobiology* 5:50–59.

———. 1982. Evolution of horns in ungulates: Ecology and paleoecology. *Biological Reviews* 57:261–318.

———. 1984. The use of fossil ungulate communities as indicators of climate and environment. 85–104 in P. Benchley, ed., *Fossils and Climate.* London: Wiley and Sons.

———. 1990. The correlation between diet and dental wear in herbivorous mammals, and its relationship to the determination of diets of extinct species. 241–59 in A. J. Boucot, *Evolutionary Paleobiology of Behavior and Coevolution.* Amsterdam: Elsevier.

Janis, C. M., and J. Damuth. 1990. Mammals. 301–45 in K. J. McNamara, ed., *Evolutionary Trends*. London: Belhaven Press.

Janis, C. M., and D. Ehrhardt. 1988. Correlation of relative muzzle width and relative incisor width with dietary preference in ungulates. *Zoological Journal of the Linnean Society* 92:267–84.

Janis, C. M., and M. Fortelius. 1988. On the means whereby mammals achieve increased functional durability of their dentitions with special reference to limiting actors. *Biological Reviews* 63:197–230.

Jarman, P. J. 1974. The social organisation of antelope in relation to their ecology. *Behaviour* 58:215–67.

Jenkins, F. A., Jr., and D. W. Krause. 1983. Adaptations for climbing in North American multituberculates (Mammalia). *Science* 220:712–15.

Jungers, W. L. 1982. Lucy's limbs: Skeletal allometry and locomotion in *Australopithecus afarensis*. *Nature* 297:676–78.

Kappelman, J. 1988. Morphology and locomotor adaptations of the bovid femur in relation to habitat. *Journal of Morphology* 198:119–30.

Karr, J. R., and F. C. James. 1975. Ecomorphological configurations and convergent evolution. 258–91 in M. L. Cody and J. M. Diamond, eds., *Ecology and Evolution of Communities*. Cambridge, Mass.: Belknap Press.

Kay, R. F. 1975. The functional adaptations of primate molar teeth. *American Journal of Physical Anthropology* 43:195–216.

——. 1978. Molar structure and diet in extant Cercopithecidae. 309–39 in K. Joysey and P. Butler, eds., *Development, Function, and Evolution of Teeth*. London: Academic Press.

Kay, R. F., and M. Cartmill. 1977. Cranial morphology and adaptations of *Palaechthon nacimienti* and other Paromomyidae (Plesiadapoidea, ?Primates) with a description of a new genus and species. *Journal of Human Evolution* 6:19–53.

Kay, R. F., and K. M. Hiiemae. 1974. Jaw movement and tooth use in recent and fossil primates. *American Journal of Physical Anthropology* 40:227–56.

Kay, R. F., and W. L. Hylander. 1978. The dental structure of mammalian folivores with special reference to Primates and Phalangeroidea (Marsupialia). 173–91 in G. G. Montgomery, ed., *The Ecology of Arboreal Folivores*. Washington, D.C.: Smithsonian Institution Press.

King, G. M., B. W. Oelofsen, and B. S. Rubidge. 1989. The evolution of the dicynodont feeding system. *Zoological Journal of the Linnean Society* 96:185–211.

Kitchner, A. 1987. Fighting behavior of the extinct Irish elk. *Modern Geology* 11:1–28.

Krause, D. W. 1982. Jaw movement, dental function, and diet in the Paleocene multituberculate *Ptilodus*. *Paleobiology* 8:265–81.

——. 1986. Competitive exclusion and taxonomic displacement in the fossil record: The case of rodents and multituberculates in North America. In K. M. Flanagan and J. Lillegraven, eds., *Vertebrates, Phylogeny, and Philosophy*. Contrib. Geol., University of Wyoming Spec. Pap. 3:95–117.

Leakey, M. D., and J. M. Harris, eds. 1987. *Laetoli: A Pliocene Site in Northern Tanzania*. Oxford: Clarendon Press.

Legendre, S. 1984. Analysis of mammalian communities from the late Eocene and Oligocene of southern France. *Palaeovertebrata* 16:191–212.

————. 1989. Les communautés de mammifères du Paléogène (Eocène supérieur et Oligocène) d'Europe occidentale: Structures, milieux et évolution. *Münchner Geowissenschaftliche Abhandlungen* A 16:1–110.

Lockley, M. G. 1986. The paleobiological and paleoenvironmental importance of dinosaur footprints. *Palaios* 1:37–47.

Lucas, P. W. 1979. The dental-dietary adaptations of mammals. *Neues Jahrbuch für Geologie und Paläontologie Monatshefte* 1979:486–512.

Maas, M. C., D. W. Krause, and S. G. Strait. 1988. The decline and extinction of Plesiadapiformes (Mammalia: ?Primates) in North America: Displacement or replacement? *Paleobiology* 14:410–31.

Manton, S. M. 1977. *The Arthropoda: Habits, Functional Morphology, and Evolution.* Oxford: Clarendon Press.

Martin, L. D. 1989. Fossil history of the terrestrial Carnivora. 536–68 in J. L. Gittleman, ed., *Carnivore Behavior, Ecology, and Evolution.* Ithaca: Cornell University Press.

Miles, D. B., R. E. Rickelfs, and J. Travis. 1987. Concordance of ecomorphological relationships in three assemblages of passerine birds. *American Naturalist* 129: 347–64.

Norman, D. B., and D. B. Weishampel. 1985. Ornithopod feeding mechanisms: Their bearing on the evolution of herbivory. *American Naturalist* 126:151–64.

Olson, E. C. 1952. The evolution of a Permian vertebrate chronofauna. *Evolution* 6:181–96.

Olson, S. L., and D. T. Rasmussen. 1986. Paleoenvironment of the earliest hominoids: New evidence from the Oligocene avifauna of Egypt. *Science* 233:1202–4.

Peters, R. H. 1983. *The Ecological Implications of Body Size.* Cambridge: Cambridge University Press.

Radinsky, L. B. 1987. *The Evolution of Vertebrate Design.* Chicago: University of Chicago Press.

Reif, W.-E., and H. Silyn-Roberts. 1987. On the robustness of Moa's leg bones: An exercise in functional morphology of extinct organisms. *Neues Jahrbuch für Geologie und Paläontologie Monatshefte* 1987:155–60.

Rensberger, J. M. 1973. An occlusion model for mastication and dental wear in herbivorous mammals. *Journal of Paleontology* 47:515–28.

————. 1978. Scanning electron microscopy of wear and occlusal events in some small herbivores. 415–38 in P. M. Butler and K. A. Joysey, eds., *Development, Function, and Evolution of Teeth.* London: Academic Press.

————. 1986. Early chewing mechanisms in mammalian herbivores. *Paleobiology* 12:474–94.

————. 1988. The transition from insectivory to herbivory in mammalian teeth. *Mémoires, Muséum National d'Histoire Naturelle,* ser. C 53:351–65.

Schmidt-Nielsen, K. 1984. *Scaling: Why Is Animal Size So Important?* Cambridge: Cambridge University Press.

Scott, K. M. 1985. Allometric trends and locomotor adaptations in the Bovidae. *Bulletin of the American Museum of Natural History* 179:197–288.

————. 1987. Allometry and habitat-related adaptations in the postcranial skeleton of Cervidae. 65–80 in C. Wemmer, ed., *The Biology and Management of the Cervidae.* Washington, D.C.: Smithsonian Institution Press.

Shear, W. A., P. M. Bonamo, J. D. Grierson, W. D. I. Rolfe, E. L. Smith, and R. A.

Norton. 1984. Early land animals in North America: Evidence from Devonian age arthropods from Gilboa, New York. *Science* 224:492–94.

Shear, W. A., J. M. Palmer, J. A. Coddington, and P. M. Bonamo. 1989. A Devonian spinneret: Early evidence of spiders and silk use. *Science* 246:479–81.

Smith, R. H. 1987. Helical burrow casts of therapsid origin from the Beaufort Group (Permian) of South Africa. *Palaeogeography, Palaeoclimatology, Palaeoecology* 60:155–70.

Solunias, N., and B. Dawson-Saunders. 1988. Dietary adaptations and paleoecology of the Late Miocene ruminants from Pikermi and Samos in Greece. *Palaeogeography, Palaeoclimatology, Palaeoecology* 65:149–72.

Solunias, N., M. F. Teaford, and A. Walker. 1988. Interpreting the diet of extinct ruminants: The case of a non-browsing giraffid. *Paleobiology* 14:287–300.

Stucky, R. K. 1990. Evolution of land mammal diversity in North America during the Cenozoic. 375–432 in H. H. Genoways, ed., *Current Mammalogy*, vol. 2. New York: Plenum.

Sues, H.-D. 1978. Functional morphology of the dome in pachycephalosaurid dinosaurs. *Neues Jahrbuch für Geologie und Paläontologie Monatshefte* 1978:459–72.

―――. 1986. Locomotion and body form in early therapsids (Dinocephalia, Gorgonopsia, and Therocephalia). 61–70 in N. Hotton III, P. D. MacLean, J. J. Roth, and E. C. Roth, eds., *The Ecology and Biology of Mammal-like Reptiles*. Washington, D.C.: Smithsonian Institution Press.

Teaford, M. F. 1988. A review of dental microwear and diet in modern mammals. *Scanning Microscopy* 2:1149–66.

Turnbull, W. D., and D. M. Martill. 1988. Taphonomy and preservation of a monospecific titanothere assemblage from the Washakie Formation (Late Eocene), southern Wyoming: An ecological accident in the fossil record. *Palaeogeography, Palaeoclimatology, Palaeoecology* 63:91–108.

Valverde, J. A. 1964. Remarques sur la structure et l'évolution des communautés de vertébrés terrestres. *Revue d'Ecologie (La Terre et La Vie)* 111:121–54.

Van Couvering, J. A. H. 1980. Community evolution in East Africa in the Late Cenozoic. 272–98 in A. K. Behrensmeyer and A. P. Hill, eds., *Fossils in the Making*. Chicago: University of Chicago Press.

Van Valkenburgh, B. 1985. Locomotor diversity within past and present guilds of large predatory mammals. *Paleobiology* 11:406–28.

―――. 1987. Skeletal indicators of locomotor behavior in living and extinct carnivores. *Journal of Vertebrate Paleontology* 7:162–82.

―――. 1988. Trophic diversity in past and present guilds of large predatory mammals. *Paleobiology* 14:155–73.

Van Valkenburgh, B., M. F. Teaford, and A. Walker. 1990. Molar microwear and diet in large carnivores: Inferences concerning diet in the sabretoothed cat, *Smilodon fatalis*. *Journal of Zoology, London* 222:319–40.

Voorhies, M. R., and J. R. Thomasson. 1979. Fossil grass anthoecia within Miocene rhinoceros skeletons: Diet in an extinct species. *Science* 206:331–33.

Webb, S. D. 1983. The rise and fall of the Late Miocene ungulate fauna in North America. 267–306 in M. H. Nitecki, ed., *Coevolution*. Chicago: University of Chicago Press.

Wilson, E. O. 1987. The earliest known ants: An analysis of the Cretaceous species and an inference concerning their social organization. *Paleobiology* 13:44–53.

Wing, S. L. 1988. Taxon-free paleoecological analysis of Eocene megaflora from Wyoming. Abstracts, BSA Annual Meeting, 1988. *American Journal of Botany* 75 (suppl.): 120.

Young, W. G., and T. M. Marty. 1986. Wear and microwear on the teeth of a moose (*Alces alces*) population in Manitoba, Canada. *Canadian Journal of Zoology* 64: 2467–79.

Young, W. G., and S. K. Robson. 1987. Jaw movements from microwear on the molar teeth of the koala *Phascolarctos cinereus*. *Journal of Zoology, London* 213: 51–61.

F I V E   Paleozoic Terrestrial Ecosystems

William A. DiMichele and Robert W. Hook, RAPPORTEURS

IN COLLABORATION WITH
RICHARD BEERBOWER, JÜRGEN A. BOY, ROBERT A. GASTALDO, NICHOLAS HOTTON
III, TOM L. PHILLIPS, STEPHEN E. SCHECKLER, WILLIAM A. SHEAR, AND
HANS-DIETER SUES

1 INTRODUCTION

The Paleozoic records the origin, assembly, and modernization of terrestrial ecosystems. Significant ecological and evolutionary events included the colonization of land by nonvascular plants, the appearance and diversification of vascular plants, the radiation of terrestrial arthropods, the origin of air-breathing vertebrates, the rise of amniotes, and the development of tetrapod herbivory. These events, occurring within ecological frameworks fundamentally different from any that came later, appear almost directional when viewed in retrospect. Paleozoic ecosystems progressed, although not continuously or necessarily gradually, toward a modern system of interactions among primary producers, arthropod and tetrapod detritivores and herbivores, and predatory animals during a period of approximately 200 million years. By the Late Permian, ecosystem structure and dynamics were not substantially different from those of the present day, even though all the species were different.

The major events that occurred during the Paleozoic dictated the future course of ecological and evolutionary history in the terrestrial biosphere. Consequently the Paleozoic can be visualized as a time of "canalization" or entrenchment of the major kinds of interactions among organisms and environments at the ecosystem level. As ecosystems became more complex, the existing spectra of interactions and ecological opportunities confined the course of subsequent events, both ecological and evolutionary. It also was a time of establishment of the basic patterns of physiological ecology of terrestrial organisms, including water and nutrient acquisition and transport in plants, means of acquiring and processing food in animals, and other biological attributes still found in the present-day biota. Yet the apparent directionality of these patterns is belied by detailed examination. The Paleozoic illustrates, perhaps better than any other time, how the ecosystems and the species that comprised them interacted, placing increasingly narrow limits on the course of subsequent evolutionary events.

Our objective is to integrate plant and animal fossil records. Differences in the nature of these records quickly become obvious, and there are few fossil deposits where both abundant plant and animal remains co-occur. Within much of the late Paleozoic, plant localities are abundant, and both macrofossils and microfossils can be sampled. Plants frequently are transported only limited distances (Gastaldo 1988; chap. 2), thereby permitting certain parts of the original landscape to be reconstructed. Paleozoic animal occurrences, however, are usually sufficiently rare and widely spaced stratigraphically to provide only glimpses of animal assemblages, often rendering community reconstruction problematic.

Inferences of function, on the other hand, can be drawn far more readily for individual animals than for many plants. The preservation of even fragmentary animal hard parts may offer information on feeding mechanisms, food preference, and motility. In contrast, fragmentary plant material can be difficult to place in a whole-plant context, which is necessary for inference about life history and ecological strategy. Even whole plants may be uninformative in the lack of quantitative data on population dynamics and habitat preferences. This is particularly true of Paleozoic plants for which closely related modern descendants, or even potential modern ecological analogues, are lacking.

Because co-occurrences are limited, and because the interactions between plants and animals and their environments (herbivory, detritivory, structural and chemical defenses, habitat and space requirements, etc.) are difficult or impossible to observe in the fossil record, integration relies on our ability to build scenarios consistent with the data. We do not hope to reconstruct what happened point by point but rather focus on broad time intervals and describe the components and their possible interactions. Through comparison of inferred structural and dynamic features of ecosystems from different time intervals, it is possible to evaluate these systems for properties that are independent of taxonomic composition.

Prior to the Late Silurian, all inferences for terrestrial ecosystems are made from indirect evidence. This evidence comes primarily from analysis of paleosols and from studies of the microbial biota of time-equivalent marine facies. From the Late Silurian onward, however, there is unequivocal megafossil evidence of terrestrially adapted plants and animals, as well as rhizome or rootlike traces and invertebrate burrows in the upper horizons of paleosols. Such paleosols also provide evidence of chemical zonation and organic enrichment attributable to a terrestrial community. Prior to the Late Silurian, from rocks of Late Ordovician to mid-Silurian age, there are abundant remains of plants or plantlike organisms that share many of the organizational properties of "true" land plants (here meaning embryophytes). These constitute compelling evidence for a terrestrial biota rich in primary producers, but of uncertain taxonomic affinities.

2 THE ORDOVICIAN

Evidence for an Ordovician terrestrial biota is conclusive, although the exact nature of its constituents remains conjectural. Paleosols from the Dunn Point Formation of Nova Scotia show reduction mottles and surficial erosion patterns that suggest soil formation and stabilization by clumps of nonvascular plants; rooting structures, however, are not preserved (Retallack 1981, 1986a).

Studies of paleosols from the Juniata Formation of Pennsylvania led Retallack (1985a, 1985b) and Feakes and Retallack (1988) to describe fossil soils from two contrasting environments. One series comprised immature soils formed along streamsides, whereas the other formed on low terraces and was more mature. Both show reduction mottles indicative of bacterial modification of organic matter, chemical profiles suggestive of plant colonization, and burrows of presumed arthropod origin (Retallack and Feakes 1987). Overall the Ordovician landscape of tropical paleolatitudes was relatively barren, with a low-diversity and sparsely clumped low-growing vegetation and a few invertebrates (Feakes and Retallack 1988).

Further evidence of the vegetation of this time comes from analyses of the microfossil record from near-shore marine sediments (Gray 1985). These include obligate spore tetrads, reminiscent of some modern liverworts, as well as sheets of tissue from enigmatic nematophytalean plants (nonvascular plants of uncertain affinity). These data suggest that the first major adaptive radiation by plants onto land occurred in Middle to Late Ordovician and involved plants at or near the nonvascular grade of vegetative organization. These might have included bryophytic plants, lichens, and fungi, in addition to the microbial mats that probably existed on land since the Late Precambrian. The Middle to Late Ordovician may have been an interval of rapid colonization by founding populations with limited genetic diversity (indicated by the monotony of the obligate spore tetrads), with life histories that included ecophysiological tolerance to desiccation, and with short vegetative life cycles (Gray 1985). Judged from their ubiquitous occurrence in deposits of this age, such forms occurred worldwide.

3 THE SILURIAN

The evidence for an Early Silurian terrestrial biota consists mainly of the same varieties of microfossils reported by Gray (1985) in the Ordovician. Beginning in mid–late Early Silurian, however, obligate spore tetrads of uniform morphology were replaced by trilete spores (single spores with a mark on one face, formed by being pressed together in a tetrahedral tetrad following meiosis) (Pratt et al. 1978). These diversified rapidly in ornamentation and laesural morphology (shape of the trilete mark), perhaps indicating a second major radiation as plants approached and attained the vascular plant grade of

vegetative organization (Gray 1985). The diversity of spore types that appeared during this second radiation suggests establishment of large populations of plants and greater intertaxon variation (Gray 1985).

The microfossil record agrees well with the appearance of megafossils
(Edwards 1980; Edwards and Fanning 1985) in the middle Silurian (Wenlock-
ian) (see fig. 5.1 for stratigraphic chart) and the gradual diversification into
the latest Silurian (Pridolian) of a flora consisting largely of *Cooksonia-* or
Salopella-like rhyniophytes (very simple, possibly nonvascular, sticklike,

PERIOD	EPOCH		AGE	DATE (Ma)
PERMIAN	LATE			245 / 255
PERMIAN	EARLY		LEONARDIAN	
PERMIAN	EARLY		WOLFCAMPIAN	
CARBONIFEROUS	PENNSYLVANIAN	LATE	STEPHANIAN	290 (300) / 303 (306)
CARBONIFEROUS	PENNSYLVANIAN	LATE	WESTPHALIAN	317
CARBONIFEROUS	PENNSYLVANIAN	LATE	NAMURIAN	323 (319) / 333
CARBONIFEROUS	MISSISSIPPIAN	EARLY	VISEAN	
CARBONIFEROUS	MISSISSIPPIAN	EARLY	TOURNAISIAN	349 / 363
DEVONIAN	LATE		FAMMENIAN	
DEVONIAN	LATE		FRASNIAN	
DEVONIAN	MIDDLE		GIVETIAN	377
DEVONIAN	MIDDLE		EIFELIAN	386
DEVONIAN	EARLY		EMSIAN	
DEVONIAN	EARLY		SIGENIAN (PRAGIAN)	
DEVONIAN	EARLY		GEDINNIAN (LOCHKOVIAN)	409
SILURIAN	LATE		PRIDOLEAN	
SILURIAN	LATE		LUOLDVIAN	
SILURIAN	EARLY		WENLOCKIAN	424
SILURIAN	EARLY		LLANDOVERIAN	439
OROVICIAN	LATE		ASHGILLIAN	
OROVICIAN	LATE		CARADOCIAN	
OROVICIAN	MIDDLE			464
OROVICIAN	EARLY			510

Fig. 5.1. Chronostratigraphy for the Ordovician through the Late Permian. Numbers in paren-
theses are alternative dates for age boundaries.

rootless and leafless plants with terminal sporangia). Whereas the early rhyniophyte floras are typical of Euramerican deposits, an approximately coeval flora has been found in Australia that includes a complex vascular plant, *Baragwanathia* (Timms and Chambers 1984). Because *Baragwanathia* is a lycopsid with true roots and leaves, it is more advanced morphologically than any plant of comparable age from Euramerica, which has led to doubts regarding its age (Bateman 1991; Hueber 1992), determined from associated graptolites (Garratt and Rickards 1984). Should future studies support the Silurian determination, *Baragwanathia* will place the evolution of Gondwanan vegetation considerably ahead of that of Euramerica.

Most Silurian floras occur in marine facies. When found in terrestrial beds, the fossils occur most often in fluvial sandstones (Edwards 1980; Edwards and Fanning 1985). The few plant axes found in laterally equivalent floodplain deposits also are allochthonous, and there is no direct evidence for plant communities (Edwards and Fanning 1985) except for those inferred from analysis of Silurian paleosols.

Retallack (1985a, 1986) described a Late Silurian (Ludlovian) paleosol from Pennsylvania that has structures similar to rhizome traces of an early vascular plant, such as a rhyniophyte. This paleosol is reported to be much better developed than the Ordovician examples noted above (Retallack 1986). Although the evidence is sparse, Retallack concluded that a variety of poorly known, nonvascular plants preceded primitive vascular plants on land, and that large soil animals of uncertain affinities were present. The discovery of Late Silurian terrestrial predatory forms, trigonotarbid arachnids and centipedes (Jeram et al. 1990), lends weight to this speculation, in that predatory invertebrates indicate the existence of simple food webs that would have included nonpredatory forms. Based on present knowledge of Early Devonian biotas, the Silurian food webs probably were based on as yet unknown detritivores or grazers that fed on microorganisms.

Ecological reconstruction of Silurian plant communities can be derived from the architecture of the fragments and their sedimentological context; further environmental inferences can be based on paleosol data. Axes are small and fork at wide angles, and all terminal parts end in sporangia. Floodplain deposits show fewer, but larger axes, whereas channel sandstones show abundant, fragmentary small axes, often almost completely covering ripple surfaces, along with rounded pieces of silicified *Prototaxites*. Many paleosols are reported to contain calcareous nodules that may represent pedogenic carbonates indicative of low sedimentation rates and possibly long-term subaerial exposure (see chap. 2).

Edwards (1980) concludes that *Cooksonia* and similar plants probably produced little cover of any height because every aerial axis tip was determinate and ended in a sporangium. Edwards and Fanning (1985) speculate that in such an ecosystem, in which active river systems were flanked by broad, well-drained floodplains, plants probably grew adjacent to the channels and also on bar tops. They further suggest the possibility that the simplicity of these early plants may not be primitive but may reflect reduction associated with adaptation to either swamp or arid, edaphically stressed environments. Mummified axes and sporangia, for example, show extensive sclerenchyma (thick-walled cells) in the outer layers of axes, an anatomical modification often found in plants from either kind of habitat. These plants, however, may have been nonvascular; in this case the cortical cells would have provided the principal support, confounding a uniformitarian comparison and inference.

Using paleogeographic reconstructions (see figure 5.2) and sedimentological analyses, Edwards and Fanning (1985) suggest that the plants of relatively unstable floodplain environments in warm and seasonally dry climates may have been opportunists with rapid sexual maturation of gametophytes in the wet season; this resulted in early establishment of more drought-tolerant sporophytes. Plants like *Cooksonia* and *Salopella* expended minimal energy in vegetative growth and quickly reached simultaneous spore production from determinate aerial branches. A single plant may have covered a large area by rhizomatous growth resulting in a patchy vegetation dominated locally by just one species.

Sherwood-Pike and Gray (1985) describe probable ascomycete hyphae from Late Silurian (Ludlow) strata. They suggest that these were terrestrial fungi at least contemporaneous with the earliest land plants. Some hyphae occurred inside small rounded pellets of apparent invertebrate fecal material; if this is correct, it suggests a fauna that included mycophagous (fungus-eating) microarthropods. Rolfe (1985b) accepts this suggestion and identifies millipeds from Late Silurian (Ludlow-Pridoli), now complemented by the discovery of predatory animals (Jeram et al. 1990). The importance of these observations is that they confirm at least a minimal terrestrial food web consisting of primary producers, decomposers, secondary consumers, and predators.

4   THE EARLY TO MIDDLE DEVONIAN

Lower to lower Middle Devonian rocks provide the first adequate glimpse of terrestrial ecosystems from numerous sites around the world. Most of the continental land surface was in the southern hemisphere (Boucot and Gray 1983; Scotese in Gensel and Andrews 1984; Livermore et al. 1985; Scotese et al. 1985) (Figure 5.2 illustrates inferred continental positions in the Late Silurian). The known fossiliferous deposits are mostly from wetland habitats in tropical and subtropical regions, although some are known from higher latitudes (Gensel and Andrews 1984; Raymond et al. 1985a).

Floras appear to have been relatively cosmopolitan, with some distinctions apparently a consequence of physical barriers and latitudinal, possibly climatic, factors (Timms and Chambers 1984; Raymond et al. 1985a). Edwards and Fanning (1985), however, conclude that such generalizations on the early biogeography of vascular plants are premature because assemblages from elsewhere in the world generally are poorly preserved and are not independently dated, making detection of provincialism in the Late Silurian and Early Devonian difficult. They disagree specifically with Ziegler et al. (1981), who identify three floral units. Raymond et al. (1985a) find a pattern similar to that of Ziegler et al. (1981) for the late Early Devonian (middle Siegenian-Emsian), including (1) a large equatorial and low latitude unit that includes floral assemblages from China, Kazakhstan, Siberia, Laurussia, and northern Gondwana; (2) a small southern low-latitude unit that includes assemblages from Australia; and (3) a small middle-to-high latitude unit that includes floral assemblages from South Africa and Australia. For early Early Devonian (Gedinnian–Early Siegenian), in contrast, Raymond et al. (1985a) find just two poorly differentiated units: (1) an Australian unit characterized by *Yarravia* and *Baragwanathia;* and (2) a widespread "northern hemisphere" unit characterized by *Zosterophyllum, Taeniocrada, Cooksonia,* and *Drepanophycus.* Note, however, that floras from China, which are only now being studied in full, contain many plants not seen elsewhere in the Northern Hemisphere.

Animals are known from only a few localities, and faunas may have been morphologically simple and of rather low diversity. At present no biogeographic patterns can be detected.

4.1 Plants

Diversification of vascular plants accelerated in the Early Devonian. During Gedinnian time (fig. 5.1) a new clade, the zosterophylls, had appeared; by Siegenian (Pragian) time, two more, the lycopsids and trimerophytes, were present. The new morphologies and morphological combinations represented by these clades provided opportunities for still-further diversification and added new architectures to terrestrial landscapes, making community structure and dynamics more complex. Examples of these architectural innovations include pseudomonopodial growth (a "main trunk") secondary vascular tissues that transported nutrients and added support, thus permitting taller plants, and root systems that supplemented and amplified the anchorage and absorptive capacities needed by larger plants. Reconstructions of many of these plants are provided by Gensel and Andrews (1984).

The Late Silurian and Early Devonian rhyniophytes and zosterophylls were small plants that formed local, monospecific patches as a result of clonal growth. Most spread vegetatively by the growth of prostrate axes from which aerial axes arose (Niklas and O'Rourke 1982; Tiffney and Niklas 1985). Where known, these all had only tufts of rhizoids on the bottoms of their horizontal axes, which provided anchorage and permitted absorption (Kidston and Lang 1917, 1920a, 1921a; Gensel et al. 1975; Hueber and Banks 1979;

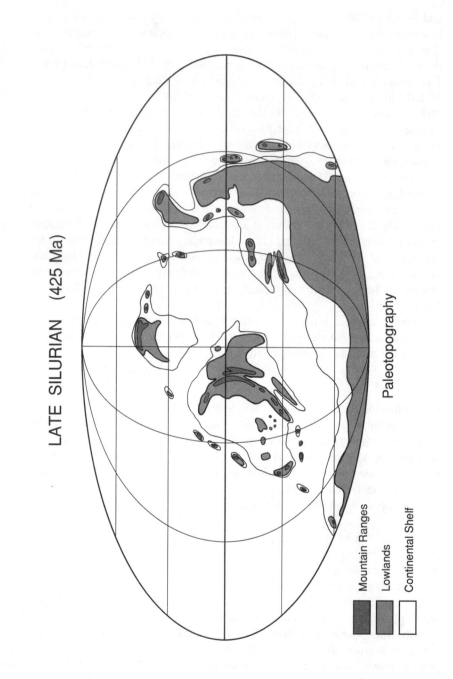

LATE SILURIAN (425 Ma)

Paleotopography

Mountain Ranges

Lowlands

Continental Shelf

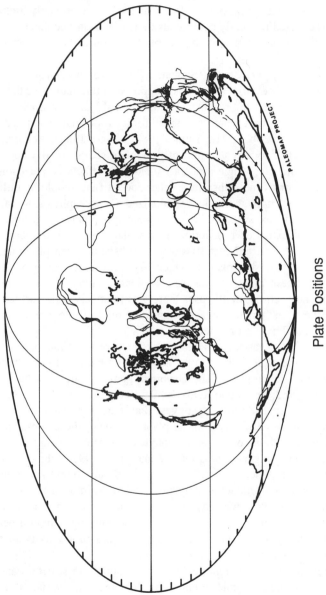

Plate Positions

Fig. 5.2. Generalized continental positions and orography in the Late Silurian. (Courtesy of C. R. Scotese and the PaleoMap Project, University of Texas, Arlington.)

Rayner 1983; Banks 1985). Rhizoids had only limited ability to penetrate the substrate to acquire water and nutrients, a problem possibly ameliorated by mycorrhizal associations (Pirozynski and Malloch 1975; Malloch et al. 1980). Thus, they were able to utilize only surface water and were probably limited to habitats characterized by nearly continuous moisture. Water conductance was limited by reliance on rhizoids as well as by the presence of few stomata on the aerial parts.

The middle Early Devonian (Siegenian) trimerophytes had complex branching patterns. The aerial shoot was composed of tall main axes that bore highly dichotomized lateral branches. Lycopsids had leafy aerial axes borne on rhizomes that also produced sparse, small forked roots that penetrated soils (Kidston and Lang 1920b, 1921a; Tanner 1983; Rayner 1984). According to Remy et al. (1980b), root penetration of floodplain sediments did not occur until the Siegenian, after which root traces and their effects on paleosols are common. For the first time, subsurface water and nutrients could be utilized, so that slightly drier habitats might be colonized. Trimerophytes are not yet known to have had root systems, although their xylem was thicker and evidences somewhat extended production of vascular tissues. Their large size (some *Pertica* species are 2 m tall) suggests substantial transpirational demands (compared to earlier plants), which may have required roots. Aerial axes in most of these plants varied from leafless to spiny to being densely covered with short leaves; in all cases there was relatively little surface area for photosynthesis and small biomass, suggesting relatively low requirements for water, CO_2, and mineral nutrients.

All Early Devonian vascular plants were homosporous. This entails a life history in which the diploid sporophyte alternates with a free-living, haploid, sexual phase, the gametophyte. The Devonian gametophytes that have been reported (Remy 1982; Remy and Remy 1980a; Remy et al. 1980a, b; Remy et al. 1991; Schweitzer 1980, 1981) suggest that water-mediated fertilization was the rule, as in modern, free-sporing plants. Some of these gametophytes are quite large, and those of the Rhyniophytes in particular suggest an isomorphic alternation of generations (Remy and Hass 1991 a, b, c; Kenrick and Crane 1991). In general, spore production and subsequent gametophytically mediated sexual reproduction may have been most important in dispersal and location of new sites, whereas clonal growth was the major means by which established plants expanded within local habitats. Sporangium-bearing axes are relatively uncommon in many species, and productivity in these plants probably was very low (Beerbower 1985).

Early terrestrial plant communities were quite simple in structure and dynamics, regardless of taxonomic composition and geographic variation. Landscape diversity was low; at most 10 to 15 genera (many represented by a single species) comprised a flora drawn generally from a range of habitats (Gensel and Andrews 1984; Knoll et al. 1984; Niklas et al. 1985). Exposure-scale or single-bed collections generally have still lower diversity, 1 to 5 species (Bateman, in press), suggesting limited interhabitat interactions. Both vascular and nonvascular plants made up the vegetation. The time of appearance of true tracheophytes is equivocal. Depending in part on how tracheary cells are defined (Kenrick, Edwards, and Dales 1991; Kenrick, Remy, and Crane 1991; Kenrick and Crane 1991), many plants originally considered to have been vascular plants in fact may be nonvascular but share other derived features with early tracheophytes. Included are well-known plants such as *Cooksonia* (Edwards and Edwards 1986) and *Aglaophyton (Rhynia) major* (Edwards 1986). Kenrick and Crane (1991) also argue for a group of early tracheophytes with water-conducting cells distinct histologically from other tracheophytes. This diversity of vascular and nonvascular plants—of which many were distinct morphologically from tracheophytes but contributed significantly to early land vegetation (Gray 1985; Taylor 1982)—gives floras from this time an unusual degree of heterogeneity at the highest taxonomic levels.

Plants that form the record of this time occupied predominantly wetland areas and the margins of watercourses where substrates were wet much of the time (Diane Edwards 1980; Schweitzer 1983; Beerbower 1985). Canopy heights were low (less than 2 m) and evidence of stratification—more than one layer of vegetation—is equivocal.

Early Devonian plant communities usually consisted of an array of patches, each patch dominated by a single species (Andrews et al. 1977; Diane Edwards 1980; Edwards and Fanning 1985). An example is the Trout Valley flora of Maine (Andrews et al. 1977), which Gensel and Andrews (1984) considered typical of many Early Devonian assemblages. Vegetation consisted of low-diversity patches; patch size varied but usually was composed of one to three or four species. It is probable that patches were occupied by large clones, within which sexual reproduction was relatively rare (Edwards and Fanning 1985). Knoll et al. (1979) refer to this pattern of space occupation as "turfing in." Because these early land plants had limited capacity to obtain nutrients from the substrate and limited potential for vertical growth (Niklas 1982), a space-occupation strategy permitted them to control access to water and nutrients, as well as to compensate for limited photosynthetic area. At some localities, certain species appear to occur most often in specific lithologies, suggesting habitat or microhabitat specificity among groups of species. This raises the issue of how to delimit communities in the Early Devonian. Interactions between clones at the margins of patches and competition during the location and initial exploitation of new sites may have been the major types of plant-plant dynamics; this suggests landscape-level processes similar to those found today in disturbed settings.

The expansion and diversification of plants in the Early Devonian also had an effect on the dynamics of floodplains and on edaphic properties of eco-

systems. Secondary succession, manifested as vegetational recovery following ecosystem disturbance, became dominant over the primary successional
colonization of previously unoccupied surfaces. Bank-dwelling plants may
have acted to stabilize stream courses, thereby lessening the effects or frequency of flooding. In general, soil formation was less often and less critically
interrupted by clastic sedimentation (Retallack 1981), although flood disturbance remained a major modifier of ecosystem patterns (Andrews et al. 1977).
With the development of large stands of plants with tree habit in the Middle
and Late Devonian, even greater portions of the landscape would have been
stabilized.

Our considerable morphological knowledge of Devonian plants enables us
to infer how their architecture or reproduction affected their ecology. Many
occurrences of Early Devonian zosterophylls, lycopsids, or trimerophytes are
preserved in near-channel, overbank deposits in which parallel alignment of
numerous axes of a single species suggest that "the plants grew in dense
monotypic water-side stands that were flattened by sudden floods" (Edwards
and Fanning 1985; other examples in Andrews et al. 1977; Diane Edwards
1980; Gensel 1982; Gensel and Andrews 1987). The rhizomatous habit from
which aerial axes arose and propagated is now firmly established in all the
major clades: rhyniophytes (Kidston and Lang 1917, 1920a, 1921a), trimerophytes (Doran 1980), zosterophylls (Lang 1927; Lele and Walton 1961;
Gensel et al. 1975; Hueber and Banks 1979; Rayner 1983), and lycopsids
(Kidston and Lang 1920b, 1921a; Gensel et al. 1969; Rayner 1984).

In situ preservation of monocultures of *Rhynia, Horneophyton, Aglaophyton,* and *Lyonophyton* occurs in the permineralized peat of the Rhynie
Chert (Kidston and Lang 1917, 1920a, 1921b; Remy 1982; Remy and Remy
1980a, 1980b; Bateman in press). *Sciadophyton,* a probable branched gametophyte, is preserved in situ in floodplain sediments (Remy et al. 1980a,
1980b).

A consistent image emerges from studies of Devonian plants and their habi-
tats, one of evolving terrestrial ecosystems, driven by plant evolution itself.
Summaries are found in Diane Edwards (1980), Scott (1980), Edwards and
Fanning (1985), and Collinson and Scott (1987a).

4.2 Animals

Virtually all terrestrial animal fossils from the Early Devonian are arthropods.
A quadrupedal trackway has been reported from Australia (Warren et al.
1986), but its affinities are unclear, and actual remains of terrestrial vertebrates are unknown in rocks of this age. Arthropod assemblages are reconstructed from a few well-studied localities and several isolated occurrences.
The arthropod communities were diverse and included detritivores and predators, though surprisingly few, if any, herbivores. This suggests that terrestrial
arthropod assemblages were diversifying and increasing in complexity at the

same time as terrestrial plants. The best-known lower Middle Devonian locality is the Rynie Chert of Scotland (Siegenian age; Rolfe 1980), which is complemented by the one-stage-younger Alken an der Mosel locality of Germany (Emsian age; Stormer 1970), and an upper Middle Devonian site from Gilboa of New York state (Givetian age; Shear et al. 1984).

The major interaction between plants and animals at this time was through detritus feeding, as deduced from animal morphologies, coprolite distributions, and the nature of plant damage. Although limited evidence has been adduced from damage repair to plants to support the existence of arthropod herbivory in the Early Devonian (Banks 1981; Kevan et al. 1975), it is not clear that such damage reflects systematic herbivory rather than misplaced detritivory. A number of detritus feeders are known, although they are not the dominants in species diversity or numbers of individuals at any of the best-studied sites. Millipeds appear to have been abundant throughout the Devonian (Almond 1985), although it is difficult to convert this abundance to a measure of ecological importance. Small arthropleurids are known from Alken an der Mosel (Størmer 1976) and Gilboa (Shear, unpublished data), whereas collembolans (springtails), which are very important soil and detritus feeders today, are reported in the Devonian only from the Rhynie Chert (Greenslade and Whalley 1986). Mites occur at Rhynie and Gilboa (Kethley et al. 1989); six species represent two major groups (orbatids and endeostigmatids) in the latter fauna. The earliest insects are archeognathans (machilid-like insects) from the Early Devonian of the Gaspé peninsula, Canada (Labandeira et al. 1988). Insect fossils are rare throughout the Devonian (Shear et al. 1984).

Numerically the most abundant animals at Rhynie and Gilboa are predators. This excess of predators, recorded consistently from the few Devonian localities known, persisted until the Late Carboniferous. The most important of these predators are the trigonotarbid arachnids, which are dominant numerically at Rhynie and are both dominant and diverse (9 species, more than 3 genera; Shear et al. 1987) at Gilboa. The earliest evidence of spiders (1 species), and silk production, comes from the Gilboa assemblage (Shear et al. 1989), which also includes other predators, such as centipeds (possibly 2 species), scorpions, and pseudoscorpions.

The earlier Devonian animal community is distinct from all post-Devonian assemblages. The dominance of predatory forms raises the question of how energy entered animal food webs and suggests a substantial bias. The pattern is noteworthy, however, because a predominance of predators similarly characterizes Carboniferous and Early Permian terrestrial tetrapod assemblages.

4.3 Synthesis

There were no directly reciprocal interactions between early plant and animal assemblages. Plant primary productivity clearly supported the animal food webs, but only through a detritus chain and not through herbivory. Detritivores contributed to soil development and nutrient cycling, which would have provided feedback to the plant communities. Plants, through their effects on resource accessibility and by moderating thermal extremes and providing protection from desiccation and predation, contributed to the patchiness and diversity of habitats. Through these effects plants created microhabitat variation and opportunities for specialization and evolutionary diversification of the animals (Rolfe 1985b). In general the system was integrated weakly, with animals as "epibionts" on the plant community. Dynamic interactions appear to have been confined mostly to plant-plant or animal-animal subsystems.

Plant communities of the Early and Middle Devonian had relatively high guild depth (the number of different species with similar ecological strategies and resource requirements). Plant species were morphologically stereotyped, mostly clonal space occupiers, lacking leaves and roots or with very small leaves and limited root systems. Although microhabitat specialization may have existed, entire groups of species probably had similar preferences. However, morphology alone suggests some differences in strategy among these plants. For example, *Rhynia gwynne-vaughnii* was probably a colonizing species (Diane Edwards 1980); deciduous branches may have allowed rapid dispersal across exposed or disturbed substrates. In contrast, *Aglaophyton* (previously *Rhynia*) *major,* a site-occupying form (Edwards 1986), had a typical clonal growth habit and probably "turfed in" in more physically stable environments. Both species occur in the Rhynie Chert.

The known aspects of morphology and environment suggest that the structure of early plant communities was controlled largely by the ability of plants to locate patches opened for colonization by disturbance or aggradation and by the ability of clones to exclude competitors through rapid space occupation or to escape competition or disturbance by rapid completion of the life cycle. On an ecological time scale, such ecosystems probably had very high structural stability, since many species had similar growth forms and life histories and could substitute for each other. If we use the terms defined by Pimm (1984), early Devonian plant communities were stable, low in species richness, and in a general sense, low in connectance (there were few interspecific interactions). The basic structural organization and dynamics should have remained the same in the face of all but the most severe and widespread perturbations, even if certain taxa disappeared or dominance-diversity patterns changed.

On a geological time scale (millions of years) such systems probably had low resistance, in the sense that the low connectance among the component taxa permitted incursions of new, more structurally complex plants. During the Devonian there were global directional increases in the structural complexity of plant communities. This is reflected both by increasing species richness (Niklas et al. 1980; Knoll 1986) and by the continuous morphological modernization of the flora (Chaloner and Sheerin 1979). At current levels of stratigraphic resolution, these increases appear to be continuous and relatively gradual. The early assemblages, composed of plants with primitive morphologies, were far from the biomechanical limits that could have been attained, given the constraints of tissues and developmental biology. The appearance of "biomechanically superior" plants (Knoll and Niklas 1987), which were taller and which had some combination of central root systems, leaves, seeds, and other less conspicuous features, initiated the displacement of the older species and the communities they formed, possibly through interspecific competition (Knoll 1984). Comparable generalizations can be drawn from the terrestrial invertebrate record of the Devonian, in which species richness and diversity of morphological specializations increase through time.

5 THE LATE DEVONIAN TO EARLY CARBONIFEROUS

Diversity, structural complexity, and biogeographic variability of ecosystems increased throughout the Devonian, and ecosystems developed some modern aspects by the Late Devonian and Early Carboniferous (Raymond 1985; Scheckler 1986a). During this time interval, modernization of animal-plant interaction and of animal trophic relationships appears to have lagged behind that of plant community structure (Beerbower 1985).

Fossil deposits of the Late Devonian and Early Carboniferous range from tropical to temperate paleolatitudes (Raymond 1985; Ziegler et al. 1981) (figure 5.3). On the basis of the degree of phytogeographic differentiation and the inferred changes in continental configuration during the Early Carboniferous, Raymond et al. (1985b) and Raymond (1985) infer climatic amelioration, trending toward warmer and wetter conditions throughout the period. In general, climate during the later Devonian and earliest Carboniferous in the Old Red Sandstone portion of Euramerica appears to have been warm and seasonally dry (Woodrow et al. 1973; Allen 1979). Studies of plant megafossils (Rowley et al. 1985; Raymond 1985; Raymond et al. 1985b) and microfossils (Zwan 1981; Clayton 1985) suggest three major phytogeographic subunits during this period, with the largest—the equatorial midlatitudes—further subdivided latitudinally.

An extinction in the Late Devonian had a major impact on plant communities and provides the earliest opportunity for study of plant community dynamics following a major species turnover. The first evidence of this extinction was found in the record of marine invertebrates at the Frasnian-Famennian boundary (McLaren 1982, 1983). Allowing for the vagaries of correlating marine and continental sections, biotic changes appear to occur earlier in terrestrial deposits, first among plants (Scheckler 1986a) and somewhat later among freshwater invertebrates (McGhee 1982).

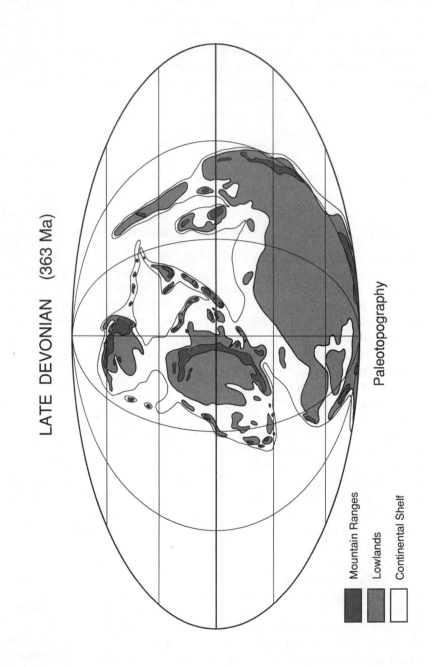

LATE DEVONIAN (363 Ma)

Paleotopography

Mountain Ranges

Lowlands

Continental Shelf

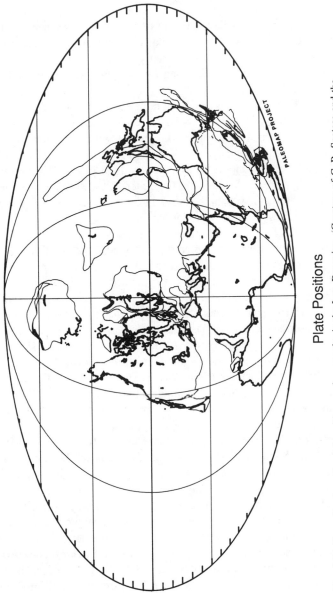

Plate Positions

Fig. 5.3. Generalized continental positions and orography in the Late Devonian. (Courtesy of C. R. Scotese and the PaleoMap Project, University of Texas, Arlington.)

5.1 Plants

Late Devonian

Early Frasnian plant assemblages from New York state (Scheckler 1986c) are diverse and well known, but their paleoecology has not been assessed in detail. These plants are described briefly in order to provide a counterpoint to the postextinction late Frasnian floras, which are better understood paleoecologically.

The oldest Frasnian floras were dominated in terms of biomass by aneurophytalean progymnosperms, bushy shrubs to small trees. A few species of the progymnosperm *Archaeopteris,* large trees with trunk diameters up to 1 m, made their first appearance at this time, but in low abundances (Banks 1966, 1980; Scheckler 1986a). Many other kinds of plants were present, including early fernlike cladoxylaleans, Iridopteridales, and both herbaceous-trailing and arborescent lycopsids. In many constituents, these floras reflect ties to Middle and even Early Devonian floras.

During the middle of the Frasnian, significant floristic changes appear to have occurred, but the extent to which these changes reflect differences in the depositional settings is unclear. Aneurophytalean progymnosperms declined precipitously in abundance, and diversity drops from six to two genera in the New York section. Archaeopterid progymnosperms increased in abundance and diversity. Herbaceous "ferns" and lycopsids became rare to absent, although lycopsid trees remained. The landscape was altered drastically by the subsequent rise of *Archaeopteris* forests. These dominance-diversity changes have been attributed to the same factors that ultimately caused the Frasnian-Famennian marine extinctions (Scheckler 1986a), possibly climatic change associated with the onset of Gondwanan glaciation (Veevers and Powell 1987; Crowley and North 1988). The decline in diversity and major shifts in dominance patterns continued into the late Frasnian.

Late Frasnian plant communities also are known from localities in New York (Scheckler 1986a). The most characteristic plants are *Archaeopteris* species, some of which produced growth rings indicating some kind of seasonality (Creber and Chaloner 1984). Because nearly all species had small laminar leaves arrayed on large, flattened branch systems, a shaded forest was likely. These trees appear to have produced many deciduous branches with attached laminate foliage, which resulted in a much greater yield of litter than that produced by earlier plants; the branches may have been shed seasonally, perhaps at the beginning of the dry season (Scheckler 1978). Increased amounts of litter may have accentuated the role of fire in these communities as a factor in disturbance and succession (Cope and Chaloner 1980; Chaloner and Cope 1982; but see also Beck et al. 1982). *Archaeopteris* trees formed low-diversity gallery forests in waterlogged soils along streamsides or on wet floodplains, as suggested by abundant *Archaeopteris* remains in channel and organic-rich sediments (Beck 1964; Retallack 1985a, b). Late Frasnian calcareous paleosols have been interpreted as well-drained areas that may have supported shrubby and herbaceous vegetation (Retallack 1985a). Such landscapes probably had considerable spatial heterogeneity, which is more detectable with paleosols than with the limited spectrum of environments that preserve megafossils.

Forested areas had become common by the late Frasnian and there were clear habitat distinctions among the plants. Structural and taxonomic diversity, although low, were increased relative to earlier Devonian plant communities. However, guild depths of trees in many of these environments were low, with limited overlap in any one assemblage. This should have predisposed these ecosystems to extensive structural change in the face of extinctions.

Throughout the Famennian, diversity and structural complexity of plant communities increased, both within and between habitats. Lowland nearshore vegetation appears to have been differentiated into a number of discrete types composed of shrubby colonizers, forest trees, ground cover, vines, and stress-tolerant swamp specialists. This is demonstrated by studies of Famennian deposits from the eastern United States, southwestern Ireland, and Arctic Norway. In the Famennian of Virginia and West Virginia (Scheckler 1986b), the first peat swamps have been identified and are clearly segregated taxonomically from floodplain forests, which contain the earliest evidence of seed plants (Scheckler 1986b; Gillespie et al. 1981). Peat swamps from this area were nearly monodominant stands of the fern *Rhacophyton*. These floras occur at four spatially and temporally distinct localities in fluvial channel-fill and coastal peat deposits. In addition to *Rhacophyton*, they contain minor amounts of apparently allochthonous debris from aborescent lycopsids that grew in swampy areas fringing the peat deposits (Beeler and Scheckler 1983). By the Early Carboniferous, these lycopsids were the dominant elements of peat-forming swamps, a position they occupied until the Stephanian in Euramerica (Phillips et al. 1974; Scheckler 1986c).

Famennian forests in non-peat-accumulating, lowland environments of Virginia and West Virginia were dominated by a mixture of *Rhacophyton ceratangium* and three species of *Archaeopteris* (*A. halliana, A. macilenta, A. hibernica*). Although these sites appear to have been better drained than the coexisting swamps (Scheckler 1986b), floodplain paleosols suggest relatively wet conditions (Retallack 1985a).

A still-greater diversity in lowland habitats is suggested by storm deposits that also contain linguloid brachiopods and fish (Scheckler 1986b). The dominant elements in these deposits are *Rhacophyton ceratangium* (40% of the vegetation) and up to five species of *Archaeopteris* (three above, as well as *A. obtusa* and *A. sphenophyllifolia;* an additional 40% of the biomass). The remaining 20% of the biomass is divided among lycopsids, seed plants, and ground cover, such as *Sphenophyllum subtenerrimum*. Although these plants are allochthonous, they record a structurally and taxonomically diverse vegetation that existed in wetlands surrounding the peat swamps. Seed plants in these deposits may have been colonizers, because they appear to have been small and produced small seeds (4 to 5 mm in diameter).

A general picture of the latest Devonian plant communities comes from deposits in Arctic Norway that demonstrate the control of floristic heterogeneity largely by physical factors (Scheckler 1985, 1986a; Fairon-Demaret 1986; Kaiser 1970). Work on this flora is in progress, but four major habitats appear to be represented. Peat-forming wetlands, preserved as coals or organic-rich shales, were dominated by a lycopsid tree, *Cyclostigma*. This genus had a stigmarian-like root system and is the earliest evidence of such lycopsid trees in peat swamps. Streamside habitats, represented by clastic rocks overlying the coals and in overbank deposits, were dominated by the clonal *Pseudobornea*, a putative calamite precursor. Poorly drained soils supported a low-diversity forest of large *Archaeopteris hibernica* trees. In better-drained areas of floodplains, the dominant tree was *Archaeopteris fimbriata*, a tree smaller than *A. hibernica;* this forest was patchy, with *Rhacophyton ceratangium* being either a local dominant or part of an understory along with *Sphenophyllum*.

A very different picture is presented by latest Devonian floras from southwestern Ireland (Matten et al. 1984; Matten et al. 1980a; Matten et al. 1980b). Integration of the vegetational patterns with depositional environments (Bridge et al. 1980) indicates that the flora was dominated by seed plants, probably of the "seed fern" groups Calymopityaceae and Lyginopteridaceae, which makes this the oldest-known seed-plant dominated vegetation. These assemblages occur in floodplain sequences that include soils with well-developed nodular carbonate beds indicative of well-drained conditions and low sedimentation rates on the floodplain. The small stature of the vegetation and its reproductive attributes suggest r-selected plants adapted to periodic environmental disturbance, probably by flooding. Such a habitat would have been more conducive to seed-plant dominance than to dominance by lower vascular plants; the larger lower vascular plants would have required more extended periods of undisturbed, predictably wet conditions.

These three floras demonstrate that vegetational patterns strongly reflected physical habitat variability by the end of the Devonian. A wide range of ecological strategies is seen in the plants; most of them also have distinct preferences for specific environmental conditions. At opposite ends of the ecological spectrum are well-drained habitats, newly accessible to seed plants because of their ability to reproduce without the need of water, and the long-inhabited wetlands. Vegetation of Late Devonian (Famennian) peat-accumulating wetlands (*Rhacophyton*-dominated) and similar environments of the Early Carboniferous (Tournaisian) (*Lepidodendropsis*-dominated) contain just one species each (Scheckler and Beeler 1984; Scheckler 1985, 1986a, 1986b, 1986c). Similar monospecific dominance, by an early gymnosperm, is found in pioneer dry habitats in the Late Devonian (Famennian) (Rothwell and Scheckler 1988). Wetlands, however, seem to differ from other settings in that they were apparently not areas of taxonomic innovation. Rather, as demonstrated by the examples discussed below, the wetlands became refugia for relict plants (Knoll, 1985; DiMichele et al. 1987).

Diversification of early gymnosperms may have occurred in the absence of interspecific competition for habitat space (Rothwell and Scheckler 1988; DiMichele et al. 1989). Early gymnosperms prospered less by successful competition than by rapid exploitation of a previously unused habitat, such as primary successional areas on a prograding delta margin. This pattern of seed-plant dominance in nonswamp habitats was to characterize the Carboniferous. However, many small lower vascular plants accompanied seed plants in the drier floodplain environments of the Devonian.

Early Carboniferous

Early Carboniferous plant assemblages, although known from a relatively small number of localities, occur worldwide (Raymond et al. 1985b; Raymond 1985). There are major differences between Late Devonian and Early Carboniferous floras and their vegetational structures, and these are reflected in major changes in dominance-diversity patterns. Because both *Archaeopteris* and *Rhacophyton* were extinct by the Early Carboniferous, Tournaisian (basal Carboniferous; see fig. 5.1) landscapes were markedly different. Areas that were forest dominated for the previous 15 million years gave way to an entirely new plant-community organization. No tree-sized counterparts of *Archaeopteris* capable of forming dense, shady forests are known prior to the end of the Tournaisian and beginning of the Visean, when the gymnosperm *Pitus* appears (Scott 1988; Retallack and Dilcher 1988). The remaining vegetation, including newly evolved forms growing in floodplains and other lowland environments, appears to have been enriched in species more to the r-selected end of the life-history spectrum; these were largely shrubby pteridosperms, less than 2 m in height. Swamp floras continued to be taxonomically distinct from contemporaneous floodplain vegetation, and some swamps were dominated by trees, mostly midsized tree lycopsids (Scheckler 1986a). The physicochemical attributes of swamp environments may have buffered these habitats from many of the perturbations that affected other lowland vegetation, imparting a degree of taxonomic and vegetational conservatism (Knoll 1986; DiMichele et al. 1985, 1987).

Reports of Early Carboniferous plant assemblages from the Tournaisian and Visean indicate significant regional diversification and further development of the ecological distinctions suggested by the latest Devonian floras (see summaries in Scott et al. 1984; Scheckler 1986a; Scott and Rex 1987; Bateman and Rothwell 1990). Among Early Carboniferous floras, those from the late Tournaisian have received the most detailed paleoecological study.

Within a deltaic, coal-bearing sequence of Virginia (Scheckler and Beeler 1984; Scheckler 1986a, c), the lycopsid tree *Lepidodendropsis* dominated coastal swamps of lagoonal or back-barrier settings, as well as floodplain swamps. Calamite trees or shrubs (*Archaeocalamites*) were common in stream-margin environments, whereas better-drained habitats were dominated by ferns and pteridosperms, such as *Neurocardiocarpus* on levees and *Triphyllopteris* and *Rhodeopteridium* in drier interdistributary areas. This basic subdivision of lycopsid abundance in peat and some clastic swamps, pteridosperm dominance of better-drained habitats (presumably levees and floodplains), and local sphenopsid abundance in disturbed or stream-margin environments is typical of the Late Carboniferous (Oshurkova 1974; Scott 1978; Pfefferkorn and Thomson 1982) and appears to have been established by the late Tournaisian. Notwithstanding these generalizations, distributions of the major plant groups overlapped broadly within lowland areas.

A different suite of environments within volcanigenic terrains is known from numerous late Tournaisian and Visean exposures in western Europe (Scott et al. 1984; Rowe and Galtier 1989). Despite the volcanic overprint, the basic partitioning of ecological resources and habitats by the major plant groups appears to be the same as the partitioning identified in the roughly co-eval coal-bearing sequence of Virginia (Scheckler 1986a). The floras are preserved in a variety of coastal-lowland paleoenvironments, ranging from lakes and small swamps in which muddy peat accumulated to disturbed areas that were subject to periodic volcanic ash falls and debris flows (Rex 1986; Bateman and Scott 1990). The general climate of the regions in which the plant-bearing deposits were formed appears to have been seasonally wet, perhaps within an area of monsoonal circulation and occasionally heavy rainfall (Bateman and Scott 1990). These environments present a strong contrast to those that characterize Upper Carboniferous coal-bearing sequences, where poor drainage, increased equability of rainfall, and reduced levels of disturbance permitted development of regionally extensive, relatively homogeneous floras of tree-sized plants (Pfefferkorn and Thomson 1982). The volcanigenic floras of the Early Carboniferous suggest that disturbed, moisture-limited environments existed during times of increased lowland floral homogeneity, such as in the Late Carboniferous, but are not preserved. The floras of Lower Carboniferous deltaic lowlands (Scheckler 1986a, c) appear to be linked more closely to Late Carboniferous coastal and deltaic lowland floras than do Lower Carboniferous floras from volcanigenic lowlands.

The most completely studied Tournaisian plant assemblage is that of Oxroad Bay in Scotland (Scott et al. 1984, 1985b; Bateman and Scott 1990; Bateman and Rothwell 1990). This assemblage is drawn from many plant-bearing horizons from a variety of depositional environments. It is a microcosm of the physical heterogeneity that appears to characterize all the Lower Carboniferous sites of western Europe. A general picture emerges of local variability in floristic composition that was controlled largely by physical conditions. In the wettest parts of floodplains, the lycopsid tree *Lepidodendron calamopsoides* and the pteridosperm *Stenomyelon tuedianum* formed the dominant canopy vegetation (Scott and Rex 1987), while the more disturbed parts of the wetter habitats were dominated by the scrambling, ground-cover lycopsid *Oxroadia* and other lower vascular plants, including calamites and some ferns. The major plant groups that dominated these wetter environments, even within the disturbed areas, were basically the same as those from more uniformly wet, deltaic lowlands, but the species and generic composition was generally very different. Pteridosperms, particularly those with r-selected life histories (small stature, small seeds, high reproductive output), formed an open vegetation in areas that were more heavily disturbed by volcanic activity and that were subject to water deficits (Rothwell and Scott 1985). This type of flora varies greatly from site to site and is distinct from floras of coastal wetlands.

Another putatively Tournaisian flora (possibly early Visean; R. Bateman pers. comm. 1991) is the "lower flora" from Loch Humphrey Burn, Scotland, which is dominated by ferns. A large proportion of these ferns are fusinized (preserved as charcoal), which leads Scott et al. (1985b) to suggest that ferns were colonizing volcanigenic areas that were subject to fires; the early radiation of ferns may have occurred in these environments during the Lower Carboniferous (Scott and Galtier 1985; Galtier and Scott 1985). Possibly the most significant observation about these floras is that they contain numerous taxa not recorded from any Tournaisian nonvolcanic settings, implying further habitat and biogeogragraphic subdivision among the pteridosperms and ferns of nonswamp environments.

Late Visean floras from volcanigenic terrains show similar vegetational patterns to those of the Tournaisian, despite the addition of many new species (Scott 1990). Floras and depositional environments have been described from a number of sites, including Kingswood (Scott et al. 1986), Pettycur (Rex and Scott 1987; Scott and Rex 1987), Loch Humphrey Burn (upper flora: Scott et al. 1984), and East Kirkton (Rolfe et al. 1990), in Scotland, and Esnost (Rex 1986) in France. Although the range of depositional environments represented at some of these localities is similar (Kingswood, Pettycur, and the upper Loch Humphrey Burn beds are disturbed volcanigenic landscapes with local lakes or peaty mud deposits), they have very few taxa in common (Scott et al. 1986; Bateman and Rothwell 1990). A comparison of Kingswood with the several localities at Pettycur (Scott et al. 1986) suggests a complex vegetational mosaic. Open lakes were fringed by a vegetation of herbaceous lycopsids with some gymnosperms of indeterminate affinities. Surrounding these lakes were better-drained areas subject to frequent fires. Such disturbances favored weedy seed plants, particularly pteridosperms, which were diverse at

the Kingswood locality. On poorly drained, weathered lava surfaces, ferns occupied areas that were periodically burned. In these relatively disturbed volcanic areas, the most diverse vegetation is found in swamp deposits from
Pettycur, which formed in depressions on lava-flow surfaces. Like most other
Carboniferous localities, these swamps were dominated by lycopsid trees of
several kinds, along with sphenopsids, filicalean ferns, and small pteridosperms (e.g., *Heterangium*).

These Tournaisian and Visean deposits demonstrate that the major plant
groups had partitioned the land surface along distinct ecological lines. In particular there was a division between plant groups common in wetlands and
those of better-drained habitats. The division was largely complete by the late
Tournaisian. Within better-drained areas, pteridosperms and ferns were further subdivided, so that a vegetation composed largely of r-selected colonists
dominated the most marginal habitats. The basic vegetational patterns seen in
these deposits and the associations of higher taxa with particular habitats persisted until the end of the Carboniferous.

Diversification of species and plant architectures continued through the
Early Carboniferous (Knoll et al. 1984; Niklas 1986). During this time all
major taxonomic groups of vascular plants evolved: seed plants, ferns,
sphenopsids, and several lycopsid groups. Relative to subsequent Phanerozoic
plant radiations, it was an explosive, high-level taxonomic radiation (the angiosperms, although diverse, are still only a variation on the seed plant organizational theme). Accompanying this taxonomic diversification was an
architectural diversification. Among the trees, major taxonomic groups appear to have had fundamentally different architectures, different basic ecological habitat preferences, and different life-history patterns. This led to a
significant congruence between phylogenetic lineages of plants and ecological
strategies, such that taxa of the same clade tended to have similar ecologies.
This taxonomic and ecological overlap, which is unique to the Carboniferous,
reached its zenith in the Westphalian (DiMichele et al. 1987).

Despite the taxonomic differences between the plants of the Carboniferous
and those of later times, vegetation became structurally modern during this
time, probably by the Visean (Scheckler 1986a). Canopy stratification developed. Lianas and possibly epiphytes became part of the vegetation. Spatial
heterogeneity was great, and plants occupied and extended the diversity of
basinal lowland habitats. Viewed from afar, the forests may not have looked
too different from those of today, although taxonomic composition was totally
different.

5.2 Animals

There are very few terrestrial animal assemblages known from the Late Devonian through the Early Carboniferous (Frasnian-Visean). Because better-
known Late Carboniferous assemblages suggest continued importance of

arthropods as the base of animal trophic relationships, such relationships are presumed to have prevailed in the Late Devonian and Early Carboniferous. Terrestrial vertebrates probably underwent major radiations during this time, but there is little direct evidence for this.

Arthropods

Detritivorous arthropods appear to have remained the major link between plants and animal food webs. Millipeds are present in Devonian to Upper Carboniferous rocks (Almond 1985; Hannibal and Feldmann 1981), minute arthropleurids first appear in the Middle Devonian (Shear 1986), and gigantic arthropleurids up to 2 m long are present in Upper Carboniferous deposits (Rolfe and Ingham 1967; Hahn et al. 1986). Mites and collembolans (springtails) first appear in the fossil record in the Early Devonian, and although they are unrecorded by body fossils in the Late Devonian and throughout the Carboniferous, they undoubtedly were present. In addition to these major detritivorous groups, many early litter-reducing insects (e.g., blattoids), though not found as fossils in Upper Devonian and Lower Carboniferous rocks, were probably in existence.

Scorpions, centipeds, and several orders of arachnids were, by inference, major predators during the Early Carboniferous, although body fossils of these groups are rare. Documentation of their presence may soon be forthcoming from the Lower Carboniferous of Scotland (Wood et al. 1985).

A diversity of winged insects is known from Namurian rocks (Wooten 1981), suggesting an earlier diversification. Certain protorthopteroids had biting mouthparts and mantislike front legs and may have been predators of other insects (Carpenter 1971; Burnham 1983). Grasshopper-like forms with biting mouthparts (Sharov 1968) also were present and may have been herbivorous. The majority of Late Carboniferous winged insects had mouthparts apparently adapted for pulling apart plant reproductive organs and sucking up the contents, suggesting that herbivory may have been established in the less-well-known Namurian record (see §6.2). At present it is not possible to estimate the extent of arthropod biomass supported by herbivory. However, changes in plant structure, particularly hardened, sclerotic seed coats and increased abundance of tough sclerenchyma in pollen organs and stems, suggest that by the beginning of the Namurian, herbivorous insects had become a significant selective factor.

Vertebrates

Evidence of terrestial vertebrates in the Devonian is sparse but geographically widespread. In addition to amphibian trackways from Australia (Warren and Wakefield 1972) and questionable footprints from Brazil (Leonardi 1983), the first amphibian fossils come from the uppermost Devonian of East Greenland and Russia (Jarvik 1952; Milner et al. 1986; Clack 1988, 1989; Coates and

Clack 1990). These include ichthyostegid and anthracosaurian amphibians, which probably preyed on arthropods or aquatic animals.

There are about ten Lower Carboniferous (mostly Visean) tetrapod-bearing deposits known from Scotland and eastern North America. With few exceptions, these deposits also contain freshwater fishes and may be regarded, in a colloquial sense, as lake assemblages (Smithson 1985a; Milner et al. 1986). These assemblages are dominated by a variety of aquatic to semiaquatic tetrapods. A recently rediscovered Visean-age deposit near East Kirkton, Scotland (Wood et al. 1985; Rolfe et al. 1990) has a fully terrestrial tetrapod biota that indicates a greater degree of tetrapod adaptation to the terrestrial environment than previously known.

Despite a predominance of amphibious-to-fully-aquatic amphibians in Devonian and Lower Carboniferous deposits, a diversity of terrestrial forms has been recognized, all of which were insectivorous or carnivorous (table 5.1).

1. Ichthyostegalians. *Ichthyostega,* known only from the Devonian of East Greenland, was approximately a meter in length. Its stout forelimbs were certainly capable of terrestrial locomotion, but other characteristics suggest a more amphibious habit (Jarvik 1952, 1980). The appendicular skeleton of *Acanthostega* indicates an aquatic mode of life (Coates and Clack 1990).

2. Anthracosaurs. Postcranial remains from the Devonian of Russia have been ascribed to a new proterogyrinid anthracosaur (Lebedev 1984). Complete proterogyrinid skeletons from the Visean of Iowa, Scotland, and West Virginia represent amphibians about 0.5 to 1.5 m in length (Holmes 1984; Milner et al. 1986; Bolt et al. 1988). Although these short-bodied forms had substantial limbs (Holmes 1980), their somewhat reduced forelimbs, which lacked well-ossified carpals, and dorsoventrally expanded tail suggest that they were amphibious predators. Proterogyrinids are accompanied by comparably sized eoherpetontid anthracosaurs in Lower Carboniferous deposits. With jaw architectures for feeding on land and massive limbs, *Eoherpeton* is interpreted as a terrestrial predator (Smithson 1985b).

3. Temnospondyls. Discoveries from East Kirkton include at least two types of intermediate-sized (30 to 50 cm in length) terrestrial amphibians that resemble the lower Westphalian temnospondyl *Dendrerpeton* (Wood et al.

Table 5.1   Lower Carboniferous (Visean) Terrestrial Amphibians

	Small	Large
Carnivores		Proterogyrinid anthracosaurs
		Eoherpetontid anthracosaurs
Insectivores	Aïstopods?	
	"Dendrerpetontid" temnospondyls	

Note: Size categories are based on estimated total body lengths of largest specimens (not applicable to aïstopods), small < 0.5 m, large ≥ 0.5 (but generally less than 1.5 m).

1985; Milner et al. 1986). These forms were somewhat froglike, with large orbits, short snouts, relatively open palates, short presacral columns, and well-ossified limbs. Large posteriorly facing emarginations at the skull table–cheek junction, generally termed otic notches, may have supported tympanic membranes capable of transmitting air-borne sounds to the inner ear. Although further study is necessary to confirm details of otic structures, such sensory adaptations would have facilitated an active terrestrial existence.

Isolated remains of rhachitomous amphibians and trackways bearing forefoot imprints with four digits indicate the existence of other, presumably terrestrial, Early Carboniferous temnospondyls that exceeded a meter in length.

4. Aïstopods. Because of their aberrant morphology and widespread occurrence, the life habits of aïstopod amphibians are difficult to assess. These snakelike amphibians lacked limbs and girdles and occur in fish-dominated as well as non-fish-bearing assemblages throughout the Carboniferous (Wellstead 1982; Milner et al. 1986). Cranial and dental specializations within the group indicate dietary preferences for either soft- or hard-bodied prey. A complete, fully ossified specimen measuring approximately 30 mm in length suggests further that aïstopods may have been characterized by direct development without an obviously aquatic larval stage (Baird 1965). Although most authors regard the group as aquatic, some aïstopods may have thrived in non-aquatic settings.

5. Amniotes. A virtually complete amniote skeleton in the Visean-age East Kirkton deposit provides further evidence of a taxonomically diverse terrestrial fauna in the Early Carboniferous. Authoritatively identified as an amniote on the basis of the skull-roof pattern and tarsal configuration (Smithson 1989), this specimen also has the long, lightly built limbs of a highly terrestrial form. Further ecological inference and taxonomic assignment await detailed description.

Because most tetrapod-bearing deposits represent wetland or lowland areas, tetrapods of this time have been considered to be strongly linked to the aquatic environment for food and reproduction. However, glimpses of a fully terrestrial fauna such as that provided by the East Kirkton site and the appearance of terrestrial forms within nearly all major tetrapod groups by the end of the Westphalian belie the traditional picture of a solely water-based Early Carboniferous terrestrial fauna. It appears likely that just as plants and arthropods were moving rapidly and adaptively into a broad spectrum of habitats, so were vertebrates.

5.3 Synthesis

Throughout the Late Devonian and most of the Early Carboniferous, the major linkage between animal food webs and primary plant productivity appears to have been the detritivorous arthropods. Insect herbivory, which may have appeared before the Namurian, altered considerably the selective environment for plants and favored structural and biochemical defenses. It is difficult to demonstrate a direct response to insect predation pressures because of the complex interaction of plants with many aspects of their environment and the possibility of the random fixation of traits during the early radiation of plant lineages. As indirect, and certainly equivocal, evidence of the effects of predation, stony seed coats and other potentially protective seed-coat complexities appeared; seeds with open integuments were less common by the Namurian. Successful pteridosperm (Medullosaceae and Lyginopteridaceae) radiations in drier habitats may in part reflect the presence of tissues enriched in resinlike substances and cortical sclerenchyma, both of which could have deterred insect predation. The expansion of insect herbivory would have introduced a strong modifier on plant-plant interspecific and intraspecific interaction and would have expanded the resource base for insectivores. Insect herbivory provides the potential for significantly more integration of the plant and animal components of the ecosystem, including the potential for a negative feedback between plants and animals due to the direct effects of natural selection. Extensive herbivory should have had a strong influence on plant structure and life history and may have altered ecosystems in ways that we may be unable to detect at a cause-effect level.

Plant community structure and dynamics during this time had increased in complexity over the earlier Devonian. During the Late Devonian, habitat or niche partitioning appears to have been well established. Guild depth, however, was low. The ecomorphotype-to-species ratio would have been very close to one. As a consequence the options for successional recovery were few. Over short time spans (thousands of years) such systems may have been relatively stable with higher levels of species-species connectedness because of greater species resource specificity than in earlier Devonian ecosystems. Local and regional patterns of disturbance and recovery may have been comparable in their degree of regularity to modern systems in similar physical settings. However, over millions of years the low guild depth should have made these communities very susceptible to dramatic changes in both dominance-diversity hierarchy and structure. This may be reflected best by the disappearance of *Archaeopteris* forest trees at the end of the Devonian. With this change, the canopy in the wetter parts of many floodplain areas went from many meters in height to perhaps 2 m or less. Structural and ecomorphic recovery from these kinds of events appears to have occurred on a time scale of millions of years.

In contrast to earlier assemblages, Late Devonian and especially Early Carboniferous plant communities may have been more resistent to invasion by "biomechanically superior" forms (*sensu* Knoll and Niklas 1987). High levels of resource partitioning are indicated by the environmental restrictions of many plants. For example, the abundance and diversity of lycopsid trees was centered in wetlands, particularly swamps, from near the beginning of the Early Carboniferous to the later part of the Late Carboniferous. The lycopsid trees were replaced as dominants in wetlands by ferns and pteridosperms only after extrinsically induced extinctions (Phillips and Peppers 1984). Less dramatic examples can be provided for the ecological roles and physical distribution of sphenopsids, ferns, and progymnosperms. In essence, once an ecological resource space was occupied by biologically complex plant groups the likelihood of further invasion or partitioning of that resource space by other groups declined; these data appear to support the tessera model of Valentine (1980). Changes in the habitat breadth of major lineages, and hence the ecological status quo, appear to be linked to extinction of earlier occupants and the creation of broad resource voids by extrinsic physical disruption rather than by biotic competition (see theoretical discussions in Valentine 1980; Vermeij 1987; Erwin et al. 1987; Valentine and Walker 1987).

The Late Devonian and Early Carboniferous saw the establishment of ecosystems with potential for persistence over time intervals of millions of years. What can be thought of as "ecological constraints" developed on the dynamics of the botanical part of terrestrial ecosystems. The established vegetational structure reflected a relatively high degree of convergence between ecomorphotypic and phylogenetic groups. This convergence means that most major clades had relatively narrow ranges of ecological tolerance; this restricted available habitat space for each clade and channeled subsequent plant evolution by strongly limiting the range of ecological opportunities open to novel phenotypes. The likelihood of occurrence and especially of the survival of new phenotypes was diminished as long as extinctions did not eliminate major clades in this ecotaxonomic landscape. Although this picture is simplified, it encapsulates the main threads of a pattern that was established in the Late Devonian and Early Carboniferous and persisted into the Permian.

The small number of Upper Devonian and Lower Carboniferous animal deposits limits our ability to gauge their temporal dynamics. The differences between Late Devonian and Late Carboniferous assemblages suggest that animal communities of the Early Carboniferous were developing rapidly. The basic ecomorphic structure of carnivore domination seen in Late Carboniferous vertebrate and arthropod assemblages seems to have been in place by the end of the Early Carboniferous. Only insect herbivory may have been a primary agent altering plant community structure and dynamics, which appear relatively modern well before modernization of the animal assemblages. Terrestrial tetrapods of the Early Carboniferous radiated into a landscape devoid of preexisting vertebrate competitors. The early vertebrate communities were unlike any that have existed since. In the absence of herbivores, and with the strong aquatic linkages of most of the known species, early tetrapods appear to have had little impact on plants during a time when many of the most basic attributes of plant communities were evolving.

6 THE LATE CARBONIFEROUS: WESTPHALIAN
AND STEPHANIAN

The Westphalian was the zenith of peat formation in Euramerica, which was located in tropical latitudes during the Late Carboniferous. The beginning of the agglomeration of the continents into Pangaea during this time is thought to have affected patterns of air circulation and climate (Parrish 1982), which also may have been influenced by glaciation in southern Gondwana during the Westphalian and Stephanian (Crowell and Frakes 1975; Visser 1987, 1989; Veevers and Powell 1987). Although the details of continental configuration (fig. 5.4) and specific climatic inferences are debated, patterns of air circulation and associated rainfall and changes in latitudinal positions of the continents may have had a major impact on plant and animal distribution (DiMichele et al. 1987; Knoll 1984; Olson 1985b).

During this time, ecological dominance of lowland habitats was shared by more orders and classes of plants than at any time in earth history. Global vegetational provinces had developed: Euramerican vegetation, the best known, was distinct from that of Gondwana (southern hemisphere) and Angara (Siberia), and to some extent from that of Cathaysia (China) (Chaloner and Lacey 1973; Chaloner and Meyen 1973; Zhao and Wu 1979; Raymond et al. 1985b; Zhang et al. 1987; Laveine et al. 1987). In Euramerica, Angara, and Cathaysia, the dominant plant groups were lycopsids, cordaites, and pteridosperms, although there were marked differences in generic and specific composition. Gondwanan floras were distinct and apparently had little taxonomic exchange with the other regions (Archangelsky 1986).

Westphalian and Stephanian plants are probably the best known of the Paleozoic, if not the entire pre-Cretaceous, largely because of their preservation in coal-bearing strata. Similarly, Westphalian arthropods and tetrapods from peat-accumulating wetland environments are better known than those of the Devonian and Lower Carboniferous. Plants from better-drained habitats, or from lowland environments that were not sites of major peat accumulation, especially in western parts of North America, also are known from a few occurrences (e.g., Pepperburg 1910; Read 1934; Read and Merriam 1940; Arnold 1941; Mamay and Read 1956; Tidwell 1967, 1988; Tidwell et. al. 1974; Leary and Pfefferkorn 1977; Leary 1981; Ash and Tidwell 1982; Winston 1983; Rothwell and Mapes 1988), and demonstrate the existence of xeromorphic vegetation in areas proximate to the wetlands (Havlena 1970).

There were major changes in the taxonomic composition of lowland Euramerican vegetation from the Westphalian to the Stephanian. The arborescent lycopsids *Lepidodendron, Lepidophloios, Asolanus, Diaphorodendron, Anabathra,* and *Bothrodendron* have not been found in Stephanian-equivalent rocks of North America and are presumed to have been extirpated throughout most of this part of Euramerica (Phillips et al. 1974). *Sigillaria* survived in North America, and several other genera, notably *Lepidodendron, Lepi-*

dophloios, and *Asolanus,* persisted in parts of southern and north-central Europe and in the Donets Basin of eastern Europe into the Stephanian (Stschegolev 1975; Boersma 1978; Lorenzo 1979; Kerp and Fichter 1985). Pteridosperms and tree ferns became the dominant elements of Stephanian lowland-wetland assemblages throughout Euramerica. As monographic research proceeds on ferns and pteridosperms, it appears that major taxonomic changes in these groups coincide with the changes in lycopsids (Mickle 1984; Lesnikowska 1989) throughout the wet lowlands (Phillips and Peppers 1984; Pfefferkorn and Thomas 1982), possibly because of increasingly dry or seasonal climate (Phillips and Peppers 1984; Cecil et al. 1984; Durden 1984a; Cecil and England 1989).

The Westphalian was the last Phanerozoic epoch to lack tetrapod herbivores. Although the diversity of amphibious and terrestrial tetrapods had increased, they remained insectivorous or carnivorous. The first tetrapod herbivores are thought to have evolved in the Stephanian within habitats that supported mesomorphic-to-xeromorphic plants. Arthropods, including flying insects, were diversified by the Late Carboniferous and appear to have had nearly as large a spectrum of ecological strategies as they do today: predators, detritivores, and herbivores were present, some with large body sizes. The substantial fossil record of arthropods suggests they had an important role in shaping the structure and dynamics of Westphalian and Stephanian plant communities (Labandeira 1990).

6.1 Plants

Upper Carboniferous plant assemblages are known from hundreds of localities in the Euramerican lowland wetlands. In contrast, comparatively little is known of floras from true upland environments (*sensu* Leary 1981), and there even has been speculation that higher elevations were not vegetated (Remy 1975). Westphalian vegetation can be divided broadly into that of wetland areas where peat formation occurred, and that of typically better-drained environments in which little or no peat accumulated. Within deltaic and coastal wetland environments, a more specific subdivision into peat-forming and non-peat-forming habitats is possible (Havlena 1970; Mapes and Gastaldo 1986). Transitional clastic swamp habitats supported vegetation intermediate between peat swamps and better-drained clastic soils. In overview it is clear that peat-forming and non-peat-forming habitats were edaphically distinct physical settings that supported different kinds of plant assemblages. During the Westphalian, when these distinctions were greatest, peat-forming habitats were dominated by several genera of lycopsid trees and, in some settings, by cordaitean gymnosperms (Phillips et al. 1985). Wetlands were otherwise dominated by pteridosperms, with increasing abundances of ferns late in the Westphalian (Pfefferkorn and Thomson 1982; see Collinson and Scott 1987b for a review of plant biologies and environments).

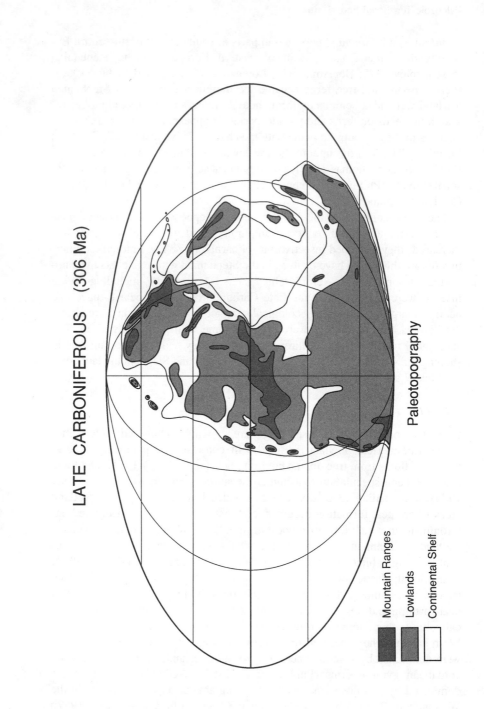

LATE CARBONIFEROUS   (306 Ma)

Paleotopography

Mountain Ranges

Lowlands

Continental Shelf

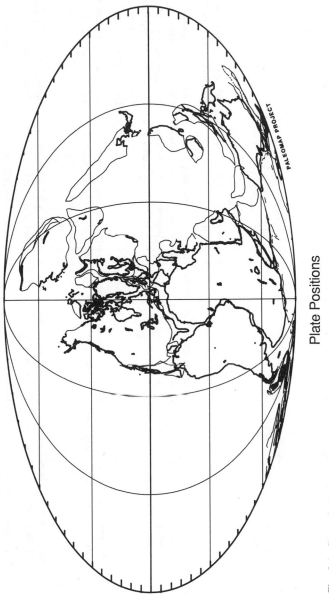

Plate Positions

Fig. 5.4. Generalized continental positions and orography in the Late Carboniferous. (Courtesy of C. R. Scotese and the PaleoMap Project, University of Texas, Arlington.)

Table 5.2 Upper Carboniferous Plants

Plant Group	Environmental Distribution	Habit	Reproductive Biology	Major Taxa
Lycopods	Peat and mineral substrate swamps	Mostly trees, some shrubs or herbs	Heterosporous, some with seedlike reproduction; many forms monocarpic, K-selected	Lepidodendron, Lepidophloios, Diaphorodendron, Sigillaria, Anabathra (Paralycopodites), Asolanus
Tree ferns	Flood basin habitats, including peat swamps in late Westphalian and Stephanian	Large and small trees	Homosporous; required some form of substrate exposure to reproduce; r-selected, weedy	Psaronius (Pecopteris); many species
Small ferns	Wide distribution; in most kinds of environments	Sprawling ground cover and vines	Mostly homosporous; extensive vegetative reproduction; span r/K spectrum	Botryopteris, Anachoropteris, Ankyropteris, Zygopteris, Psalixochlaena, Stauropteris
Pteridosperms	Better-drained parts of flood basins; in nutrient-rich parts of swamps	Large and small trees, ground cover vines	Seed plants; seeds ranging from small to very large; some possibly insect pollinated; many K-selected trees; some r-selected weedy forms	Medullosa and Sutcliffia (Common foliage types include Sphenopteris, Alethopteris, Neuropteris, Mariopteris, Odontopteris, Eusphenopteris), Callistophyton, Heterangium, Lyginopteris, other small forms
Sphenopsids	Wet parts of flood basins; high-energy aggradational environments	Trees (calamites) and shrubs (sphenophyllum)	Mostly homosporous; some heterosporous; most reproduction vegetative; more on K-selected end of spectrum	Arthropitys, Calamodendron, Calamites (Foliage types: Annularia, Asterophyllites), Sphenophyllum
Cordaites	Peat and mineral substrate swamps; undefined parts of wet floodbasins	Trees and shrubs, possibly mangroves	Seed plants; wind pollinated and dispersed; K-selected	Mesoxylon, Cordaixylon (Pennsylvanioxylon), Cordaites
Conifers	Well-drained areas, perhaps marginal-to-wet lowlands	Trees	Seed plants; wind pollinated and dispersed; K-selected	Walchia, Ernestiodendron, Hermitia

Although the same basic physical environments existed in the Stephanian as in the Westphalian, the distinction between peat-forming and other lowland vegetation types is blurred taxonomically (DiMichele et al. 1985). Peat swamps were dominated by ferns and pteridosperms (Phillips et al. 1985), as were surrounding flood-basin habitats (Pfefferkorn and Thomson 1982). A flora dominated by conifers and other distinct elements is clearly identifiable in more mesic, well-drained habitats by the Stephanian (Winston 1983; Mapes and Gastaldo 1986; Rothwell and Mapes 1988; Broutin et al. 1990; Mamay and Mapes in press). Its earlier existence in the Westphalian is suspected from allochthonous debris of conifers discovered in otherwise-typical lowland assemblages (Scott and Chaloner 1983; Lyons and Darrah 1989). The hints of a largely unsampled upland or inland flora suggest a large-scale bias in the tropical Late Carboniferous record (see chap. 2), possibly reflective of a low-relief landscape with a wide rainy belt (Parrish 1982; Ziegler 1990).

During the Westphalian, before major extinctions altered the basic flora that had evolved in the early Early Carboniferous, plant ecological dominance was distributed among lycopsids, ferns, sphenopsids, and seed plants (table 5.2). Within seed plants there were two major phylogenetic lineages, cordaites and pteridosperms, with environmentally distinct distributions. During the Stephanian, lycopsid trees ceased to be ecologically significant in all but a few habitats (Phillips et al. 1974; Phillips et al. 1985). Opportunistic ferns expanded to a dominant position in the wetter habitats, and seed plants, particularly cordaites and conifers, became important in a wider array of habitats. This proceeded as a stepwise loss of dominance by lower vascular (spore-producing) plant groups that were ecologically significant during the Early Carboniferous and Westphalian and continued through the Permian and into the Mesozoic (see chap. 6). Seed plants diversified and increased their ranges as other groups declined; this resulted ultimately in terrestrial domination by this one major clade of plants by the end of the Paleozoic. A similar pattern appears to have continued among lineages of seed plants.

Lycopsids

The lycopsid trees are better understood ecologically than any other group of Carboniferous plants. At least eight genera (DiMichele 1983, 1985; Wnuk 1989; Bateman and DiMichele 1991) and numerous species populated mainly peat and clastic swamp habitats. They ranged from dominants in physically stable habitats to colonizers of disturbed environments (DiMichele and Phillips 1985; DiMichele and DeMaris 1987; Gastaldo 1987). All lycopsids reflect a basic "pole-tree" architecture, with two variants. Some species had deciduous lateral branches on which cones were borne over an extended per.od of time (for reconstructions see Wnuk 1985, 1989; DiMichele and Phillips 1985). In many species individual trees spent much of their lives in a prereproductive phase, as an unbranched pole. In these species the development of a determinate crown signaled the beginning of reproduction, rendering the trees functionally monocarpic; the crown was the final phase of growth and can be envisioned as a reproductive adaptation rather than a light-capture structure. Many lycopsids produced massive, seed-like female reproduction units that appear to have been aquatically dispersed (Phillips 1979). Plant densities varied, but some lycopsid forests evidently had very high densities and high standing biomass per hectare (Gastaldo 1986b; DiMichele and DeMaris 1987; Wnuk and Pfefferkorn 1987). Some species may have formed small, high-density, monospecific patches within a more complex vegetation (DiMichele and Nelson 1989). The pole-tree forests were probably very open to light penetration, with random spacing between trees.

Lycopsid construction was relatively "cheap" compared to that of woody trees. The major support tissue was thick bark, which apparently was waterproof and decay resistant. The small amount of wood was very efficient at water conduction (Cichan 1986). There was no secondary phloem (Arnold 1960; Eggert and Kanemoto 1977); the lack of phloem presents a dilemma regarding the means by which the root systems of these trees received the sugars generated by photosynthesis. Anatomical aspects of the root system and their shallow burial suggest the possibility that leaflike rootlets borne on the stemlike stigmarian rhizomorphs were photosynthetic, and that roots and shoots independently generated their photosynthetic carbon (Phillips and DiMichele 1992). Leaves were sclerenchymatous and also lacked phloem or had little of it, suggesting very local use of photosynthate. Because the vegetative parts were composed of tough, lignified tissues containing few available nutrients, they offered little high-quality forage for insect of other arthropod herbivores, in spite of high total biomass. In contrast, large cones, both male and female, would have been likely targets for herbivory. Cones were produced in great numbers in monocarpic species, but only once in the life of a tree. Although it is not clear that whole stands reproduced synchronously, sedimentological factors associated with the preservation of in situ stumps (DiMichele and DeMaris 1987) suggest that some lycopsid stands may have been even aged.

Sigillaria was the only abundant lycopsid tree in Stephanian wetlands of Euramerica. Peat-forming wetlands of the early Stephanian were extremely variable in composition and included marshlands as well as forests (Phillips et al. 1974). When conditions favorable to the accumulation of thick, widespread peats returned, the remaining lycopsid trees did not regain their prior ecological prominence. Rather, tree fern and pteridosperm dominance of peat swamps ensued.

Pteridosperms

The taxonomic and structural diversities of pteridosperms were very high by the Late Carboniferous (Phillips 1981). Most medullosan pteridosperms were monoaxial trees (Pfefferkorn et al. 1984; Wnuk and Pfefferkorn 1984), 6 to

10 m in maximum height. Leaves were compound and could be several meters long (e.g., *Alethopteris;* Laveine 1986). These fronds were the dominant organs of the plant and had thick, leathery pinnules and robust construction. After death, frond remains may have folded back against the stem forming a "skirt" around the axis below the live crown; this is suggested by the recurved leaf bases found on many stems (Pfefferkorn et al. 1984). Such skirts may have provided important habitats for arthropods and small scansorial tetrapods.

Medullosans may have been poor quality food for herbivores. Stems had several bundles of vascular tissue, surrounded by secondary phloem (Delevoryas 1955; Stewart and Delevoryas 1956). However, these were buried in a dense ground tissue, including periderm, and stems retained leaf bases, further burying vascular tissue. In the frond axes vascular tissue was dispersed in small, isolated bundles (Ramanujam et al. 1974). Surrounding the vascular bundles and subepidermally in the stem and leaf axes were dense sclerenchymatous bundles. In addition, medullosan stems and leaves had abundant small canals filled with resinlike material (Lyons et al. 1982) that may have deterred insect predation. On the basis of the evidence, most arthropod damage to medullosan foliage (Scott and Taylor 1983) is not clearly due to herbivory, as opposed to detritivory. Lesnikowska (1989) notes no evidence of herbivory in an examination of hundreds of anatomically preserved coalball specimens, although Labandeira (1990) suggests possible leaf mining and folivory on the basis of damage to late Westphalian compression-impression medullosan leaves. In general, these plants appear to have been of tough construction and to have offered limited targets for either sucking or chewing on vegetative plant parts.

Reproductive organs, in contrast to vegetative parts, may have been major targets for predation. Although medullosan seeds were large, they were protected by complex seed coats, typically with hardened middle layers and fleshy outer layers (Taylor 1965). Seed-coat specializations may have protected the seeds against crushing or perhaps may have facilitated passage through the gut of herbivorous tetrapods or arthropods. In some cases, fleshy outer layers may have attracted animals. The great diversity in the size, shape, and anatomical structure of medullosan seeds suggests ecological specialization, possibly including close life-history associations with animals. Pollen organs (Millay and Taylor 1979; Stidd 1980) were very large in some species and contained large pollen grains, a potential energy source for animals that could tear through the tough outer layers of the reproductive structure (Kukalova-Peck 1983). Most medullosan pollen exceeded 200 μm in diameter, a size so large that animal—probably insect—pollination vectors are considered likely (Dilcher 1979).

A scrambling-to-viney habit was common among a wide diversity of small pteridosperms (Stidd and Phillips 1973; Rothwell 1981; Pigg et al. 1987). These plants generally produced small seeds and pollen organs, had scle-

renchymatous plates or fibers in the stems and fronds, and had small compound leaves with relatively thin laminate foliage. Their vegetative parts and
reproductive organs should have been accessible to low-browsing omnivorous
or herbivorous tetrapods and arthropods. Most were very patchy in distribution, although some, such as *Callistophyton boysettii* of the late Westphalian (Phillips and DiMichele 1981; DiMichele and Phillips 1988) or
Lyginopteris species of the early Westphalian (Phillips et al. 1985), were
common and relatively abundant for such small plants. Fortuitous preservation of pollen drops and the morphology of some pollen grains indicate wind
pollination in many of these taxa (Rothwell 1972, 1977).

Throughout the Late Carboniferous, pteridosperms were dominant elements on clastic soils of floodplain habitats (Peppers and Pfefferkorn 1970;
Scott 1977, 1978, 1979; Pfefferkorn and Thomson 1982). Their numbers increased in peat swamps during the Westphalian, although only a few genera
from clastic-compression floras were able to grow in peat-forming environments. No species were unique to peat swamps (DiMichele et al. 1985).

Pteridosperms produced relatively large amounts of litter. In some trees
much of the litter may have remained in the crowns as fronds died and decayed
while still attached to stems; in others, such as many neuropterid species, pinnules abscised and the recurved rachises may have added support to the
slender stems (Pfefferkorn et al. 1984).

Ferns

Ferns were important in most lowland environments during the Westphalian
and Stephanian. They can be divided into two major groups: herbaceous
ground cover, including vines; and trees or subtrees. Epiphytic habit has been
suggested for some ferns, although the evidence is equivocal (Rothwell
1989). Small ferns were present throughout the Westphalian (Phillips 1979) at
a low biomass (1% to 5%) but in considerable taxonomic and structural diversity. Tree ferns, some of which were of low stature, specifically the marattialean *Psaronius,* were present in low but increasing numbers up to the late
Westphalian, when tree-fern abundances rose rapidly throughout the Euramerican lowlands (Phillips et al. 1974; Pfefferkorn and Thomson 1982;
Phillips and Peppers 1984). Marattiales appeared first in peat-forming environments during the middle Westphalian, perhaps in the nutrient-poor central areas of raised peat bodies (Eble 1990). It is not clear that these
populations were ancestral to later peat-swamp or floodplain marattialean
ferns. During the Stephanian, tree ferns were the dominant trees in most peat
swamps (Phillips and Peppers 1984) and in many other habitats (Pfefferkorn
and Thomson 1982).

Small ferns were among the most important ground-cover elements during
the Late Carboniferous (e.g., *Psalixochlaena, Botryopteris, Anachoropteris,
Ankyropteris, Zygopteris, Stauropteris*). Most were of sprawling habit, many

perhaps were facultative climbers similar to poison ivy (J. Mickle, pers. comm. to DiMichele 1987), and some were true vines (Millay and Taylor 1980, Trivett and Rothwell 1988). Small but abundant fronds, thin laminate foliage, and little or no sclerenchymatous or woody tissue characterize these plants as a group (Phillips 1974). Reproductive organs varied in complexity, but some, such as those of *Zygopteris* and some *Botryopteris* species, were produced in large, armored aggregations (Murdy and Andrews 1957; Phillips and Andrews 1968; Galtier 1971) that would have been accessible to foraging animals. Most reproductive organs were foliar borne, which may have discouraged predation, because the nutrient-rich tissues were dispersed and would have caused animals to ingest less-digestible vegetative tissue along with energy-rich sporangia.

Marattialean ferns were uncommon and of low taxonomic diversity throughout much of the early Westphalian (Corsin 1948; Millay 1979; Pfefferkorn and Thomson 1982; Gastaldo 1984; Phillips et al. 1985). Beginning in the middle Westphalian and accelerating during the late Westphalian, this group of plants diversified and became a dominant element of many kinds of vegetation. Even more so than lycopsids, marattialeans of tree habit were "cheaply" constructed. They attained aborescence by mantling a relatively fleshy stem of small diameter in a thick cover of supporting, adventitious roots (Ehret and Phillips 1977). The roots were mostly air spaces and provided strength in aggregate. Fronds several meters long were borne in an umbrella-like crown. *Psaronius* foliage was thin but often bore reproductive organs in high density on the undersides of pinnules (Millay 1979). Many species contained sclerenchyma bundles and sheaths in leaves and stems and, less commonly, very abundant mucilage ducts or cavities throughout stem and leaves (Morgan 1959; Mickle 1984). "Cheap" construction, high dispersal abilities, and tolerance of all but deeply flooded physical environments led to the ecological expansion of this plant group.

Psaronius plants probably were major targets for herbivores, particularly insects. At present there is little supporting evidence for this hypothesis except for a report on wound repair associated with tissue damage and fecal pellets in Stephanian-age *Psaronius* fronds (Lesnikowska 1990). Lesnikowska (1989) also has found a major size difference between Westphalian and Stephanian *Psaronius* species in peat swamps; whereas Westphalian forms generally were small subtrees, larger, heavily root-mantled stems appear in the later Westphalian and become characteristic in the Stephanian. Such increased size would have kept all but climbing and flying animals from the fertile fronds of *Psaronius*.

Sphenopsids

There were two major groups of sphenopsids during the Carboniferous, both clonal. The sphenophylls were ground-cover plants and the calamites were arborescent forms. *Sphenophyllum* encompassed many species and occurred in almost all lowland habitats. In general the plants were sprawling, with small woody stems and delicate heterophyllous leaves borne in whorls. Batenburg (1982) differentiates peat-swamp species from those of other environments. In floodplain and streamside settings, many *Sphenophyllum* species apparently formed dense thickets of highly branched stems linked by barbs or hooks on the leaves. These thickets may have been almost impenetrable to large animals and were possibly favored by smaller, relatively defenseless forms. Densities of *Sphenophyllum* in shales frequently are high, with stems crossing bedding planes vertically, suggesting partial burial in place and an ability to survive such burial. Although abundances of *Sphenophyllum* were patchy in peat swamps, local densities could be high; there is no evidence of interlocking of the stems and leaves in such assemblages. Cones were of small to moderate size, and sexual reproduction may have been secondary to clonal growth as a means of local spread in such environments.

Calamites were the only trees of the Late Carboniferous that were clonal (Tiffney and Niklas 1985). They were variable in size and form (Hirmer 1927) but could be quite large trees. The largest woody stems, almost 0.5 m in diameter, have been found in Stephanian-age coal-ball deposits (Galtier and Phillips 1985). Westphalian calamites appear to have been of smaller girth and presumably were shorter than the largest Stephanian forms. They were most abundant in streamside habitats, such as channel bars, at lake margins, and in other near-water settings (Oshurkova 1974; Brzyski et al. 1976; Scott 1978). In these environments, vegetative clonal growth was probably significantly more important than sexual reproduction. Although they were occasionally the dominant element of physically disturbed wetland habitats, calamites were minor components of the vegetation for the most part. Woody stems, thin leaves, and relatively low reproductive allocation suggest that they were not significant targets for herbivores.

Cordaites and Conifers

Cordaites were a diverse group of woody plants related closely to the modern conifers (Mapes 1987). They ranged from large forest-forming trees, encountered almost exclusively in deposits associated with well-drained, clastic-soil habitats, to small, thicket-forming shrubs that were possibly mangrovelike within peat-swamp environments (Raymond and Phillips 1983; Rothwell and Warner 1984; Costanza 1985; Trivett and Rothwell 1985; Raymond 1988). The biology and ecology of peat-swamp cordaites is much better known than that of nonswamp taxa.

Cordaites had dense wood. Leaves were typically strap shaped and in most species were coriaceous, with thick cuticle and abundant sclerenchyma. Winged seeds with large nutrient reserves often were produced in abundance in exposed positions on small conelike shoots. In general these seeds, of which several different types are known, had a very sclerenchymatous seed coat that may have deterred predation by animals (for discussion of cordaite seed structure, see Roth 1955; Leisman 1961; Taylor and Stewart 1964). In most instances papery wings suggest wind dispersal, although in *Cardiocarpus magnicellularis* a fleshy outer seed coat (Baxter and Roth 1954) may have attracted certain animals. Both attractant and protective aspects of these seeds could have functioned after, as well as before, dispersal.

Cordaites were abundant in peat swamps during the middle and late Westphalian but were minor elements of Stephanian swamps. In lowland areas peripheral to peat formation (traditionally "uplands" or "extrabasinal lowlands"; Havlena 1970; Pfefferkorn 1980), they appear to have been important trees from the earliest Late Carboniferous (Leary 1981). Occurrences of cordaitean trees continued into the Stephanian and Early Permian (Winston 1983; Lyons and Darrah 1989) on better-drained, clastic soils.

Conifers occurred first in the middle Westphalian (Scott and Chaloner 1983; Lyons and Darrah 1989) as fragmentary allochthonous material and as microfossils in areas adjacent to contemporaneous uplands. They are encountered rarely in Westphalian rocks. Several occurrences in the late Westphalian are discussed by Lyons and Darrah (1989), again largely from areas proximal to contemporaneous "uplands," or areas of erosion. The oldest well-preserved conifers in the Appalachian basin appear at or near the Westphalian-Stephanian boundary (McComas 1988). Conifers are more commonly encountered in Stephanian assemblages (Cridland and Morris 1963; Winston 1983; Mapes and Gastaldo 1986; Rothwell and Mapes 1988; Mapes and Rothwell 1988), mostly in habitats peripheral to peat formation and, rarely, from strata associated with coals (Canright and Blazey 1974; McComas 1988).

The primitive conifers *Walchia* and *Ernestiodendron* (*Hermitia* and *Culmitzchia* if not differentiable to one of the other genera; Visscher et al. 1986) were similar to many modern conifers vegetatively, with needlelike leaves, wood, tree habit, and moderate-to-large size (Florin 1951; Visscher et al. 1986). The cones were not as compact or as well protected against predation as those of modern conifers (Mapes 1987), but they may not have been accessible to large arthropods or low-browsing tetrapods. These trees apparently tolerated drier, better-drained conditions, as suggested by both their structure and the sedimentary contexts in which they are most often preserved (Winston 1983; Mapes and Gastaldo 1986; Rothwell and Mapes 1988; Lyons and Darrah 1989; but see McComas 1988). Woody stems, evergreen leaves, seed dormancy (Mapes et al. 1989), and co-occurrence with large cordaites in a seed-plant-dominated vegetation suggest growth in stable habitats with xeric conditions at least part of the year. At some sites, such as the earliest conifer from England or the Stephanian-age deposits at Garnett and Hamilton Quarry,

Kansas, conifer remains are often preserved as fusain, suggesting forest fires and periodically fire-prone or water-stressed habitats (Scott and Chaloner 1983; Rothwell and Mapes 1988). These conifers probably would have been relatively low-quality forage; as with cordaites, early herbivores may have been interested only in their seeds or pollen.

6.2 Invertebrates

The record of Westphalian and Stephanian terrestrial invertebrates indicates a broad ecological spectrum, approximating that of post-Paleozoic to modern faunas, though diversity is not comparable. Nearly all information comes from lowland, equatorial areas of peat accumulation (Wootten 1981; Durden 1984a). The record includes the first appearance of terrestrial molluscs (Solem and Yochelson 1979). Most of the information on arthropods comes from collecting sites that have exceptional preservation, such as Mazon Creek (late Westphalian, Illinois), or Commentry and Montceau-les-Mines (Stephanian, France), which present opportunities to study diversity and anatomy. Supplementing these deposits are other, less prolific assemblages, such as those in roof shales above coals (Jarzembowski 1987) or from coals (Bartram et al. 1987), which provide an important indication of distribution. The most common arthropods are myriapods, arachnids, and insects, particularly blattoids. The greatest diversity is recorded at Mazon Creek, the most intensively studied assemblage. Ecological inferences are based in large part on functional analyses of mouthparts in living relatives and on the nature of damage to plants.

Detritivores

A large proportion of primary plant productivity in the Westphalian and Stephanian continued to reach animal food webs through arthropod detritivores. A great variety of detritivores are known or suspected. Among the most abundant terrestrial arthropods were millipeds and arthropleurids. Millipeds appear to have had the same ecological role as today, namely, as detritus feeders and soil formers in forested regions. The Late Carboniferous record included cylindrical forms, which burrowed or pushed through soft substrates, and flat "litter splitters" (Hoffman 1969; Rolfe 1985a). Spiny millipeds, both flat and cylindrical, were more common in the Late Carboniferous than today (Hannibal and Feldman 1981), and some probably were too large to hide in litter. Evidence of damage to spines (Rolfe 1985a) suggests predation by amphibians and reptiles, some of which swallowed prey whole. Rolfe (1985a) and others have suggested further that some Carboniferous myriapods may have had the protection of repugnatorial glands, similar to those in extant forms.

Arthropleurids were of enormous size by the Late Carboniferous, perhaps greater than 2 m in length (Rolfe and Ingham 1967; Rolfe 1980; Hahn et al. 1986). Evidence from gut contents indicates that their diet included wood as well as higher nutrient materials. The large size of archipolypod millipeds and arthropleurids and the presence of long tergal spines in the former suggest that these animals filled a niche not yet shared with tetrapods. Whether the demise of archipolypods and giant arthropleurids was due to increasingly effective predators, competition from predators, extensive environmental disruption, or some combination of these is unclear. Though not giant, large myriapods live in tropical to temperate regions today.

Among chelicerates, mites may have been detritivores during the Late Carboniferous, but coprolites provide the only evidence of these animals (Scott and Taylor 1983). Mites may not have radiated by this time into the diversity of ecological roles they have today, and most mites probably fed on fungi and the wood these fungi invaded.

Hexapods included several important detritivorous groups. Among insects, blattoids ("roaches") were the most widespread and abundant arthropods of the Carboniferous (Durden 1969, 1984a, 1984b, 1988; Wootten 1981; Scott and Taylor 1983). The extinct order Protelytroptera, ancestors of modern earwigs, were beetle-like. Collembolans also are inferred to have been important detritivores, despite a lack of fossils in the Carboniferous, because of occurrences in both Devonian and Permian strata (Riek 1976; Greenslade and Whalley 1986).

Herbivores

Insect herbivory appears to have been well developed by the Late Carboniferous. Chewed leaves, bored seeds, and pollen in coprolites and guts (Scott and Taylor 1983) all suggest herbivory, although such damage could be post mortem. Inferences drawn from insect functional morphology (e.g., Labandeira 1990; Labandeira and Beall 1990), however, reinforce the herbivory interpretation, as does evidence of plant-tissue repair and characteristic feeding patterns, such as leaf mining (Labandeira 1990). The rise of insect herbivory may have been related indirectly to the evolution of wings in the Early Carboniferous. With flying abilities, insects no longer were limited to a cryptic habit (Durden 1984a) and could exploit new resources. Such herbivory would have enhanced the entry of plant primary productivity into animal food webs and greatly strengthened links between animal and plant communities.

The paleopteran orders Megasecoptera and Paleodictyoptera included a variety of forms, with wingspans ranging from 9 mm to 43 cm. Wings of the adult insects were permanently outstretched horizontally. This configuration affects abilities to perch under windy conditions, makes walking difficult, constrains flying to areas of open vegetation, and limits hiding places (Carpenter 1971). Complex color patterns on the wings of Paleodictyoptera may have served as disruptive concealment, apostematic warnings, or perhaps even communication in mating or mate competition. The animals may have preferred open vegetation beneath moderately dense canopies, which could be found in pteridosperm or tree-fern-dominated areas. Both nymphs and adults of many species were armored or spiny, suggesting antipredator adaptations.

Mouthparts of megasecopterans and paleodictyopterans were modified for piercing and sucking; some paleodictyopteran species had a pronounced cibarial pump. In *Monsteropterum moravicum,* the beak was 20 cm long and contained two mandibular and two maxillary stylets and one hypopharyngeal stylet; all rested in a Z-shaped labial trough (Kukalova-Peck 1972, 1985). Such insects may have torn apart fructifications with claws and imbibed the spore or pollen contents (Kukalova-Peck 1983); spores have been found filling the guts of larger specimens of several species. Bored holes in seeds and megaspores (Sharov 1973; Scott and Taylor 1983) suggest predation by sucking insects. Sap feeding from vegetative tissues, specifically phloem, also may have occurred (Scott and Taylor 1983), although Kukalova-Peck (pers. comm. to Shear 1987) argues that sap feeding requires mouthparts that are not found in these insects and that only the small species fed on sap, probably from ovules rather than stem or leaf phloem. Certainly most Late Carboniferous lowland plants were not optimal targets for vegetative herbivory (thick bark, little or highly dispersed phloem) but did present easily accessible megaspores, ovules, and microspores or pollen. Nymphs of these orders also were winged, terrestrial forms and apparently were herbivorous, feeding mostly on fructifications (Kukalova-Peck 1978; Shear and Kukalova-Peck 1990).

Among the Neoptera, herbivores included the Protorthoptera (Burnham 1983) and several other related groups in the Hemipteroidea (Rasnitsyn 1980), and the Orthoptera (Sharov 1968). Hemipteroidea in general were diverse and abundant in the Carboniferous, with a variety of mouthparts, from those for chewing or sucking with a domed postclypeus to those with triangular or elongated stylets. The Protorthoptera had chewing mouthparts, although some had raptorial, mantislike front legs (Carpenter 1971). Among Orthoptera, most of the modern ecomorphs were present in the Permo-Carboniferous. As in modern forms, hind legs were adapted for jumping, and biting mouthparts were present, suggesting herbivory.

Predators

Predatory chelicerates and insects were common and diverse in Westphalian and Stephanian communities. Some of the forms were probably in resource competition with tetrapods, and the Late Carboniferous may have been a transitional time in the relationship between arthropod and tetrapod predators. Among chelicerates, scorpions and arachnids were the major terrestrial predators. Scorpion cuticle has been recovered in large quantities from macerations of British coals (Bartram et al. 1987), indicating the presence of scorpions in peat-forming habitats, as well as in better-drained settings (Kjellesvig-Waering 1986). Predatory arachnids included large, heavily armored trigonotarbids (aphanototarbids) with reduced eyes, and the closely related anthracomartids, which were eyeless; both lacked web-building or poison glands. These groups are rare but diverse at Mazon Creek (Richardson and Johnson 1971). All Late Carboniferous spiders belong to the suborder Mesothelae. They were small and very much like their modern descendants, except *Megaranea* (Hünicken 1980), which had a leg span of more than 40 cm. Present-day relatives of these spiders build silk-lined burrows. Evidence for extensive use of silk in aerial webs in the Upper Carboniferous is scant and ambiguous. The earliest clear indication of web-building capabilities is from the Triassic in fossils that appear modern morphologically. Other arachnid groups of the Late Carboniferous were the phalangiotarbids, a group of sit-and-wait, spiderlike predators (Beall, pers. comm. to Shear 1987), the Uropygi and Amblypygi, also sit-and-wait predators, and the Opiliones and Solpugida, which were voracious, active hunters. The Ricinulei were larger and more diverse predators in litter during the Carboniferous than today (Selden 1986).

Predatory insects of note are the Protodonata (dragonflies) and Ephemerata (mayflies). Protodonata are common as fossils because they frequented lake and pond margins and had aquatic larvae (Kukalova-Peck, pers. comm. to Shear 1987). The protodonata often reached large sizes, with wing spans up to 63 cm reported for Stephanian forms (Carpenter 1960). However, wingspans of as little as 30 mm have been reported (Wootten 1981), which is small even by modern standards. The legs of the larger animals were proportionately more robust than those of extant dragonflies, suggesting that they caught large insects in flight. Ephemerata are rare in the Late Carboniferous, but giant forms with wingspans up to 42 cm are known from Germany and Illinois (Kukalova-Peck 1983). Adults differed from modern mayflies in having functional biting mouthparts. Their aquatic larvae may have preyed upon small aquatic vertebrates.

In summary, the spectrum of Paleozoic predatory arthropods was quite different from that of today. They tended to be much larger, and species diversity may have been lower. Most of the arachnids were cursorial or sit-and-wait predators. There probably were no web-building spiders. Predaceous arthropods may have preyed upon herbivorous and detritivorous arthropods, which had increased greatly in abundance and diversity over earlier times, or on small tetrapods. Herbivorous forms were more numerous and more diverse than predators by the Late Carboniferous, a reversal of previous conditions.

6.3 Vertebrates

Late Carboniferous terrestrial amphibians and amniotes are known mainly from eleven extensively collected deposits that are dominated by aquatic-to-amphibious taxa and from five localities that are distinguishable as terrestrial assemblages. The records are from southern Euramerica. Nearly all Westphalian-age examples occur in coal-bearing, fluviodeltaic sequences and represent autochthonous samples of tropical wetland faunas (Gersib and McCabe 1981; Rust et al. 1984; Hook and Ferm 1988; Hook and Baird 1988). In contrast, a great range of paleoenvironmental circumstances is presented by Stephanian tetrapod-bearing deposits; these include, but are not limited to, sapropelic fillings of small lakes and clastic fillings of small-to-large lakes (Boy 1977), carbonaceous and calcareous channel fills within well-drained coastal settings (Reisz et al. 1982; French et al. 1988), and braided stream deposits within semiarid alluvial plains (Berman et al. 1987a; Eberth 1987). Although a few Stephanian examples are taphonomically comparable to those of the Westphalian, no Westphalian deposit has the better-drained edaphic aspects of certain Stephanian localities.

The exceedingly narrow paleoenvironmental scope of Westphalian tetrapod occurrences has confounded evolutionary and ecological interpretations. Some of the more terrestrial elements of the fish-dominated assemblages, for example, have been regarded as "erratics" from an otherwise unsampled "uplands" biota (A. R. Milner 1980a; Boyd 1984; Olson 1985a), a notion that appears to be supported by upright tree stumps that contain terrestrial forms (Carroll et al. 1972). Sedimentological and petrographic data (Hook and Hower 1988), however, reveal that such distinctions are artificial and that aquatic, amphibious, and terrestrial animals coexisted at a landscape scale within a mosaic of lowland habitats. Thus, like most of the autochthonous plant deposits, the present Westphalian tetrapod sample is of use in discussing only one type of Late Carboniferous ecosystem.

A small number of Stephanian vertebrate localities also offer plant assemblages that suggest well-drained substrates or dry climatic conditions. Taxonomically, however, none of the Stephanian assemblages appears to be the lineal descendant of faunas from the Westphalian wetlands; rather the assemblages are composed of amphibian and amniote groups that persist well into the Early Permian, a pattern paralleled to a large extent by the associated plants (Mapes and Gastaldo 1986).

In both ecological and morphological terms, most Paleozoic tetrapods cannot be portrayed fairly by simple analogy to modern forms. Because many of the major groups of Late Carboniferous land-dwelling amphibians and reptiles are extinct, only ecomorphic generalizations, based mainly on postcranial morphologies and to a lesser extent on sensory adaptations, are possible. Interpretation of several amphibian groups is hindered further by the likelihood that individual taxa had strong life-history ties to aquatic environments and may have become active in nonaquatic setting only as adults.

Anthracosaurs

Two types of anthracosaur amphibians, embolomeres and gephyrostegids, are known from Westphalian coal deposits. Several embolomere genera, in particular *Anthracosaurus* (Panchen 1977, 1981) and similar North American taxa, have been interpreted as possible terrestrial forms. Although this is difficult to assess without more complete representation of postcranial skeletons, their massive dentitions and deep skulls indicate a predatory habit. On the other hand, gephyrostegids were short-trunked, generally lightly built tetrapods that averaged approximately 0.4 m in total length (Carroll 1970) and probably were insectivorous. They are known first from a nearly complete skeleton from the Namurian of Germany (Boy and Bandel 1973). The largest gephyrostegid is the enigmatic reptiliomorph *Solenodonsaurus,* a more robust form from the late Westphalian (Carroll 1970; Panchen 1972).

Fragmentary remains of archeriid embolomeres occur in both Westphalian and Stephanian rocks of North America. On the basis of their similarity to the well-known genus *Archeria* from the Lower Permian (Romer 1957; Clack and Holmes 1988; Holmes 1989), these Upper Carboniferous records probably represent an intermediate-sized gavial-like form that was primarily aquatic.

Nectrideans

Despite an abundance of nectrideans in Upper Carboniferous vertebrate deposits, only *Scincosaurus* appears to have been suited for terrestrial existence. This genus is known from both Westphalian and Stephanian coal-bearing facies. A small, short-bodied form with a long, relatively inflexible tail, *Scincosaurus* exhibited robust girdles and limbs indicative of a nonaquatic habit. A disproportionately small skull and small pedicellate teeth suggest a diet of soft-bodied prey (A. C. Milner 1980).

Microsaurs

Microsaurs were small, predominantly terrestrial amphibians. Although microsaurs are present at several Westphalian localities, the Stephanian record of land-dwelling taxa is poor. *Tuditanus,* a well-known late Westphalian genus (Carroll and Baird 1968), was a small (approximately 15 cm total length), superficially salamander-like form that probably fed on insects and other small arthropods; other tuditanomorphs reached a length of approximately 30 cm (*Asaphestera,* Carroll and Gaskill 1978; *Trihecaton,* Vaughn 1972). Other Late Carboniferous microsaurs, such as *Elfridia* (Thayer 1985), *Leiocephalikon,* and *Sparodus* (Carroll 1966), possessed crushing dentitions of large, blunt, conical teeth set in massive jaws and in some cases sets of large accessory teeth. Such species probably had an omnivorous diet of hard-shelled invertebrates and possibly nutrient-rich, hard-covered seeds.

Temnospondyls

The greatest variety of terrestrial amphibians during the Late Carboniferous is found in the Temnospondyli. As a group, they ranged from small to intermediate-sized insectivores and carnivores that were highly terrestrial, to large carnivores that probably favored marginal aquatic settings and a main diet of freshwater fish, sharks, and amphibians. *Dendrerpeton,* known from the early Westphalian of Ireland and Nova Scotia, represents a generalized design with a boxlike skull and well-ossified, modestly scaled postcranial skeleton; dentition consisted of marginal tooth rows of many simple conical teeth accompanied by palatal tusk pairs and finely denticulated palatal surfaces (Carroll 1967a; A. R. Milner 1980b). A primitive impedance-matching system for perception of air-borne vibrations may have been present (Clack 1983; Godfrey et al. 1987). Although no complete skeletons have been described, ample specimens indicate a small terrestrial form that probably fed on invertebrates.

By the late Westphalian, the major amphibious-to-terrestrial temnospondyl groups were differentiated. Westphalian edopoids are known best from *Cochleosaurus,* a long-skulled, crocodile-like form (Rieppel 1980; A. R. Milner 1980a). The preponderance of small to intermediate-sized specimens in a large sample of over 50 individuals from Nýřany, Czechoslovakia, suggests increased terrestriality with growth. Large individuals exceeded 1.5 m in length and appear to have been sit-and-wait predators. More primitive edopoid genera are known from the early Westphalian, and others are present in late Westphalian assemblages (Milner 1987).

Amphibamus, a salamander-like form of late Westphalian age, is the oldest representative of the Dissorophoidea, a group of terrestrial Permo-Carboniferous temnospondyls that is characterized by, among other features, a froglike stapes and, presumably, a tympanic ear (Bolt and Lombard 1985). Growth stages of *Amphibamus* range from larval forms with external gills to fully terrestrial adults with estimated body lengths up to 25 cm (Bolt 1979; Milner 1982; Hook and Baird 1984). A similar genus is recorded in the Stephanian (Daly 1988, in press). The trematopids are a second group of Late Carboniferous dissorophoids that have a conventional body design. They are interpreted as land-dwelling forms on the basis of marginal dentitions, which consisted of small but acuminate, moderately recurved teeth, and general skull morphologies. The posterior part of the elongated narial opening of trematopids has been interpreted as the site of a salt gland comparable to that in some modern lizards (Bolt 1974b). Trematopids are known from the Westphalian (*Mordex;* Milner 1986) and the Stephanian; postcranial remains from the Stephanian of Kansas and New Mexico suggest body lengths of up to 50 cm (Milner 1985; Berman et al. 1987a).

Two groups of aberrant dissorophoids that are regarded traditionally as Permian forms are recorded in middle to late Stephanian deposits. Genera from Ohio and the southwest United States had bizzare pelycosaur-like "sails" composed of greatly extended and somewhat expanded neural spines (*Astreptorhachis,* Vaughn 1971; *Platyhystrix,* Berman et al. 1981). They reached a meter or more in body length and appear to have been fully terrestrial carnivores. The second group, the so-called armored dissorophoids, were somewhat smaller and are discussed in §7.3, Temnospondyls.

Remains of eryopoids are recorded in early Stephanian deposits. The Appalachian species of *Eryops* exceeded a meter (Vaughn 1958; Murphy 1971) in length and was probably an amphibious-to-terrestrial carnivore as an adult. Actinodontids and other eryopoids from the Permo-Carboniferous of Europe (Boy 1988, 1990) generally were smaller and even less terrestrial.

Aïstopods

Two distinct aïstopod amphibian families occurred in the Late Carboniferous. Ophiderpetontids, which reached a length of nearly 2 m, are regarded as the more aquatic group (A. R. Milner 1980a), whereas the generally smaller and less heavily scaled phlegethontids show several specializations that may represent terrestrial adaptations (McGinnis 1967). For example, a radically modified skull may have allowed certain phlegethontid genera to ingest oversized prey in a manner similar to that of some modern-day snakes (Lund 1978).

Cotylosaurs

Cotylosaurs (*sensu* Heaton 1980) have long attracted attention because they offer a blend of amphibian and reptilian characters in what has been regarded classically as *the* primitive terrestrial morphotype. Limnoscelids are represented by fragmentary remains from three North America localities that range from late Westphalian to early Stephanian age (Carroll 1967b, 1984; Berman and Sumida 1990). Recognition of these materials, however, relies on a virtually complete skeleton of *Limnoscelis* from the upper Stephanian or possibly lowermost Permian of New Mexico (Romer 1946; Fracasso 1980, 1987). Despite the size of this specimen (nearly 2 m in total length), a poorly ossified postcranial skeleton implies an amphibious, perhaps crocodile-like habit.

Diadectid cotylosaurs, the earliest possible herbivorous tetrapods known at present, first appear in the early Stephanian (Carroll 1984) and range from the southwestern United States through the Ohio Valley to central Europe. The larger Stephanian diadectids equal the dimensions of small *Diadectes* specimens from the Lower Permian (Vaughn 1969, 1972) and are characterized similarly by chisel-shaped "incisors" and transversely expanded cheek teeth that show wear facets; however, both the degree of expansion and amount of wear on the teeth of Stephanian forms are significantly less than those exhibited by Permian diadectids. Short, broad limbs indicate limited locomotor abilities and suggest a diet of only slow-moving, if not stationary, low-to-the-ground objects. Shelled invertebrates, seeds, and pollen-bearing organs that required crushing are plausible food items; living plant materials, such as roots and low-to-the-ground shoots or foliage, cannot be excluded, but lack of extreme tooth wear such as that seen in later much larger diadectids suggests that Stephanian representatives were omnivorous. The increased body size and greater dental specialization of Permian diadectids suggest a change in feeding habits during the Permo-Carboniferous.

Amniotes

The Late Carboniferous vertebrate record is distinguished by the first representation of a diverse amniote stock. Two major groups, the captorhinomorphs and the synapsids, appear in Westphalian coal-bearing facies, and a third clade, the diapsids, is known mainly from a wealth of specimens from a single Stephanian locality. Whereas Carboniferous captorhinomorphs and synapsids appear as generalized forms transitional to the far greater array of amniote ecomorphotypes seen in the Early Permian, diapsids show comparatively little change through this time period. The unheralded appearance of well-differentiated amniote clades in the late Westphalian, along with insights provided by the recent East Kirkton discovery (Smithson 1989), underscore the inadequacies of our present sample of early tetrapod faunas.

Protorothyridid captorhinomorphs are well known as small (average body length of approximately 20 cm), lizardlike forms that appear first in Westphalian coal-bearing facies (Carroll and Baird 1972). The closely spaced, bluntly pointed, conical teeth of the marginal series indicate that the protorothyridid captorhinomorphs were insectivorous, and elongation of the limbs suggests an active existence that may have included tree climbing (Reisz and Baird 1983). Protorothyridid remains are very poorly represented in the Stephanian record, but trackways corresponding to their pedal morphologies and body proportions are known.

The pelycosaur *Archaeothyris* is the only Westphalian-age synapsid represented by reasonably good fossil material (Reisz 1972, 1975). Over 50 cm long, this form exhibits a high skull with sharp, slightly recurved and compressed teeth that bear modest serrations. These attributes, along with those of its poorly known postcranial skeleton, indicate that *Archaeothyris* was a terrestrial predator.

A rich diversity of pelycosaurs is recorded at Garnett, Kansas, in a Stephanian-aged channel fill that occurs within a carbonate-dominated coastal plain setting (Reisz et al. 1982); as noted above, the Garnett plant assemblage is characterized by taxa that suggest well-drained substrates and perhaps seasonally dry climate. The Garnett edaphosaurid, *Ianthasaurus,* is particularly noteworthy because of its relatively small size (50 cm estimated length) and insectivorous dentition (Reisz and Berman 1986; Modesto and Reisz 1990); later edaphosaurids were much larger, with more specialized dentitions, bulky trunks, and small heads.

The sphenacodontid *Haptodus* occurs first at Garnett and is the earliest representative of a very successful group of dominant terrestrial predators. Also known from excellent European materials, the postcranial skeleton represents an agile but powerful form, and the marginal dentition features well-differentiated caniniform teeth and modifications for slicing rapidly through prey (Currie 1977, 1979). Another sphenacodontid from Garnett has a more massive, bulbous dentition that suggests a somewhat different diet (Reisz 1990). A much larger (over 3 m in estimated length) sphenacodontid occurs in Stephanian coal deposits of Czechoslovakia, though it is represented only by incomplete material (Romer 1945; Zajíc and Stamberg 1985).

Upper Stephanian to possibly lowest Permian deposits in New Mexico, representing better-drained or drier habitats, have furnished fragmentary remains of large varanopseid pelycosaurs that exceeded one meter in length (Romer and Price 1940; Eberth and Brinkman 1983). By comparison with slightly younger varanopseids from the same region, these early synapsids were dominant predators, although more lightly built than contemporaneous sphenacodontines.

A lizardlike habitus typifies the Stephanian diapsid *Petrolacosaurus,* which occurs in abundance at Garnett (Reisz 1981). In comparison with protorothyridids, the longer limbs, neck, and tail impart a more gracile appearance to this larger but also insectivorous form.

Characteristics of Terrestrial Vertebrate Faunas in the Late Carboniferous

Table 5.3 summarizes the size ranges and feeding habits inferred for terrestrial tetrapods of the Westphalian and Stephanian. Although these gross ecological groupings suggest a dichotomy between small insectivores and medium-sized carnivores, there was almost certainly overlap; insectivores preyed upon a variety of small animals, including other tetrapods, and predators of intermediate size probably also consumed a mixed diet of smaller tetrapods and some arthropods. Large predators are represented by several anthracosaurs and temnospondyls, which may have been more amphibious and adapted to passive sit-and-wait strategies, and by intermediate- to large-sized sphenacodontid pelycosaurs, which were undoubtedly more active and probably attacked amphibious-to-terrestrial tetrapods of any size with success.

Increased size in diadectids through the Permo-Carboniferous interval may have conferred certain physiological advantages, such as an ability to utilize high-fiber plant tissues, but also may have been a response to predator pressures. Given the abundance of vertebrate insectivores and predatory insects, a similar interpretation may account for the appearance of very large representatives of several arthropod groups in the Upper Carboniferous.

In all Westphalian assemblages, associated with peat formation and presumably very wet conditions, terrestrial carnivores were outnumbered by insectivores. This relationship is difficult to evaluate in the Stephanian because assemblages are few and samples are small, but it is clearly reversed in a majority of Early Permian assemblages. There is no unequivocal evidence of terrestrial herbivores in the Upper Carboniferous, but some forms, the om-

Table 5.3 Upper Carboniferous Terrestrial Amphibians and Reptiles

	Small	Intermediate	Large
Carnivores	Phlegethontid aïstopods	Ophiacodontid pelycosaurs Sphenacodontid pelycosaurs Varanopseid pelycosaurs Solenodonsaurid anthracosaurs "Platyhystrixid" and trematopid temnospondyls	Sphenacodontid pelycosaurs Anthracosaurid anthracosaurs? Edopoid temnospondyls Eryopoid temnospondyls Limnoscelid cotylosaurs?
Insectivores	Protorothyridid captorhinomorphs Gephyrostegid anthracosaurs Scincosaurid nectrideans Tuditanid microsaurs "Dendrerpetontid" temnospondyls Dissorophid temnospondyls	Petrolacosaurid diapsids Edaphosaurid pelycosaurs	
Omnivores	Gymnarthrid microsaurs Pantylid microsaurs		Diadectid cotylosaurs

Note: Size categories are based on estimated total body lengths of largest specimens, small < 0.5 m, intermediate 0.5–1.5 m, large ≥ 1.5 m.

nivores of table 5.3, at least had some abilities to exploit low-fiber plant tissues, such as reproductive organs, megaspores, seeds, and shoot buds, in addition to animal prey. Because the nutritional value of these structures would have been high, it is conceivable that some animals may have survived on them alone.

As a whole the Stephanian records the emergence and diversification of amniotes as the dominant predators and the possible beginnings of tetrapod herbivory among diadectid cotylosaurs. Early amniotes and possibly cotylosaurs appear to have been tolerant of a wide range of environmental conditions. For example, some of the Stephanian amniotes preserved with xeromorphic plants in North America are found also in coal-bearing limnic basins of central Europe. That these same tetrapod groups persisted and were widespread well into the Early Permian suggests that a cosmopolitan terrestrial vertebrate fauna was established before the end of the Carboniferous.

6.4   Synthesis

The Late Carboniferous landscape was complex and spatially heterogeneous, encompassing a variety of ecosystems, each with distinctive attributes. Despite levels of complexity that rival those of later times, the dynamics of Late Carboniferous ecosystems are unlike those of the post-Paleozoic. Most notable is the continuation of detritivory as the major entry point of plant productivity into animal food webs, with expanding, but still limited, herbivory. Plant communities continued to be composed of a mixture of different major groups of vascular plants. The taxonomic structure of the landscape was not randomly distributed across habitats; rather classes were distributionally centered in specific habitats, an ecologically primitive pattern remaining from the much older radiation of plants in the Late Devonian and Early Carboniferous. Post-Carboniferous ecosystems became increasingly dominated by seed plants in all habitats. Tetrapod assemblages also retained patterns established earlier and continued to be dominated by insectivores and carnivores. Clearly terrestrial vertebrates depended heavily on an invertebrate food source, mainly insects, and in this sense there appear to have been much stronger linkages between these animal groups than between the vertebrates and plants.

The Late Carboniferous, particularly the Stephanian, records glimpses of the change that was to occur in the terrestrial biota during the Permian. Traces of floras with xeromorphic elements from peripheral extrabasinal habitats are detectable beginning in the early Westphalian throughout Euramerica. Likewise, there is evidence of distinct vertebrate assemblages within various lowland environments during the Stephanian. Unfortunately the record is confounded by limited exposure of drier tropical habitats and of paratropical areas in general. It is clear, however, that the wetland terrestrial biotas of the Westphalian are minor elements, missing from most areas during the Early Permian.

Vegetational Patterns

Most Late Carboniferous fossil plant assemblages were deposited in or near coastal wetlands (distributory channels, estuaries, interdistributary bays, peat swamps; see chap. 2). These wetlands were physically complex areas that included peat and mineral-soil swamps, floodbasins, and levees; within these areas many microhabitats and associated floras have been detailed (peat-forming habitats: Smith and Butterworth 1967; Peppers 1979; Phillips and Di-Michele 1981; DiMichele and DeMaris 1987; DiMichele and Phillips 1988; DiMichele and Nelson 1989; Eble et al. 1989; Feng 1989; Pryor 1988; Raymond 1987, 1988; Raymond and Phillips 1983; Winston 1986, 1988, 1989, 1990; Winston and Stanton 1989; Mahaffy 1985, 1988; Eble 1990; Eble and Grady 1990; Grady and Eble 1990; Bateman in press; Willard in press; clastic-soil habitats: Oshurkova 1974, 1978; Besly and Fielding 1989; Gastaldo 1982, 1986a, 1986b, 1987, 1988; Scott 1977, 1978, 1979, 1984; Wnuk 1986; Wnuk and Pfefferkorn 1987). The vegetational structure and patterns of taxonomic dominance and diversity differed spatially among these habitats (Oshurkova 1974; Scott 1978; DiMichele et al. 1985; Gastaldo 1988), and within any one of them through time (Phillips et al. 1974, 1985; Phillips 1980; Pfefferkorn and Thomson 1982). Enough is known about the ecologies of individual species and the communities they comprised to suggest that long-term persistence of assemblages over millions of years within any one kind of habitat was the rule rather than the exception. Temporal changes between different dominance-diversity patterns appear to have been relatively abrupt rather than continuous and gradual, and were possibly forced by changes in the physical environment rather than by biotic competition (Phillips and Peppers 1984; DiMichele et al. 1987).

Coexisting with wetland habitats as far back as the earliest Carboniferous were peripheral areas that were well drained or, in some cases, seasonally dry. In the Late Carboniferous record, these areas appear to have been part of a continuum of environmental variation, although floristic overlap was limited. Floras from such habitats are relatively uncommon in the fossil record. Included are the large Namurian-age Rock Island floras (Leary and Pfefferkorn 1977; Leary 1981), and tropical-to-paratropical floras from the Namurian through Stephanian of the western United States (Manning Canyon Shale and Moab floras, Utah: Tidwell 1967, 1988; Spotted Ridge flora, Oregon: Mamay and Read 1956; several floras in New Mexico: Ash and Tidwell 1982; Mamay and Mapes in press; Garnett, Hamilton, and Baldwin floras of Kansas: Cridland and Morris 1963; Winston 1983; Rothwell and Mapes 1988) and several areas of Europe (Kerp and Fichter 1985; Broutin et al. 1990). In comparison to contemporaneous wetland floras, these assemblages often contain or are dominated by genera and species that are unknown from peat-forming wetland assemblages. In general they suggest that the plant composition of the more distal wetlands, of better-drained areas, or of areas with seasonal dryness was distinct from that of basinal wetland habitats in the Euramerican tropical belt. Many of these floras were dominated by seed plants and later ones, of Stephanian age, were enriched in conifers, taeniopterids, and other genera that were typically absent from coeval wetland assemblages.

Within the wetlands, Westphalian plant communities had reached a level of structural complexity comparable to that of extant plant communities. This inference is based on the number and kinds of canopy trees, the distribution of ground cover, and the occurrence of epiphytes and vines. Subsequent Carboniferous landscapes were not as structurally complex; for example, Stephanian peat swamps had considerably fewer structural components than those of the Westphalian. In addition there was a strong taxonomic component to the local variability of Westphalian landscapes. At no subsequent time have so many higher taxonomic groups of plants been dominant at the landscape level or partitioned the landscape so strongly along taxonomically specific, ecological lines. This taxonomic-ecologic partitioning began to break down during the Westphalian-Stephanian floral transition.

The signal of change to come began in the late Westphalian with the rise of tree-fern dominance in some compression-impression (floodbasin) assemblages (Pfefferkorn and Thomson 1982). It was approximately at this same time that the earliest floras rich in conifers began to appear sporadically within lowland settings (McComas 1988; Lyons and Darrah 1989). Conifers are known earlier only from transported materials, supposedly derived from "distant" areas (Scott and Chaloner 1983). In peat-forming swamps, which appear to have been more buffered against extrinsic changes than most mineral-soil habitats (DiMichele et al. 1987; Knoll 1985), major floristic changes also began in the latest Westphalian. Westphalian peat swamps were taxonomically distinct from the surrounding lowland vegetation of mineral soils (DiMichele et al. 1985), but during the floral transition to the Stephanian, major extinctions occurred in the lycopsids (Phillips et al. 1974) and tree ferns (Lesnikowska 1989) of peat swamps, and probably in pteridosperms as well (Taylor 1965). The combination of a rise in tree-fern abundance and the sporadic appearance of conifers in mineral-soil habitats, along with a dramatic change in peat-swamp vegetation from lycopsid to tree-fern dominance, strongly suggest physical changes in lowland habitats. Throughout the lowlands, the rise of tree ferns brought in the first potentially closed canopy forests. Tree ferns, as major litter producers and as targets of insect herbivory, may have increased substantially the resources available to both detritivorous and herbivorous arthropods.

Tree-fern dominance of the broader wetlands was a short-lived phenomenon geologically, lasting perhaps as much as 6 My. By the late Stephanian, these plants appear to have been increasingly confined to narrow parts of the lowlands as coniferophytes and new kinds of "cycadophytic" seed plants began to appear more frequently in lowland settings (Read and Mamay 1964; Kerp and Fichter 1985; Kerp et al. 1989). By the Early Permian in both Europe and the United States, tree ferns and medullosan pteridosperms remained inhabitants of peat swamps and wetlands but were surrounded by an increasingly different, xeromorphic vegetation (Blazey 1974; Canright and Blazey 1974; Galtier and Phillips 1985; Broutin et al. 1990). This reflects a broad trend that continued into the Permian, a trend in which xeromorphic forms replaced more hygromorphic to mesomorphic elements of the vegetation (Ziegler 1990).

Animal Assemblages and Plant-Animal Interactions

The animal component of Late Carboniferous ecosystems can be divided broadly into four groups: detritivores, omnivores and herbivores, insectivores, and carnivores. Arthropods comprised almost all of the detritivores and probably accounted for most of the herbivory. Predatory arthropods also were abundant. Tetrapods were mostly insectivorous and carnivorous. By inference, the interactions among these animals, expressed in the likely trophic relationships among them, were very highly developed and complex and cannot be viewed as a simple, or even necessarily a primitive, system. In comparison with later systems that include tetrapod herbivores, the fundamentally different insectivore-carnivore tetrapod assemblages represent an alternative form of energy transfer and trophic organization within highly vegetated ecosystems.

Arthropod detritivory remained the major entry point for plant productivity to animal food webs. A variety of litter feeders is known, some very large or with what appears to be armor; either large size or armor would have deterred many predators. Some forms of insect herbivory were probably well established in the Westphalian, particularly sap feeding, predation on pollen-or-spore-producing organs and seeds, and perhaps leaf mining. Evidence for folivory is less clear; damaged foliage is scarce, and it usually cannot be determined if the damage was done during the life of the leaf or after it became part of the litter. Insect mouthparts suitable for eating leaves also are suitable for consuming litter and other insects. The most convincing evidence is the wide variety of features developed by plants that could serve, even if by happenstance, to deter insect predation. To the extent that detritivores depended on decomposers rather than plant detritus itself, the advent of insect herbivory served to simplify food webs. Modern trophic structure, however, did not appear prior to significant foliar herbivory by tetrapods in the Permian.

Throughout the Westphalian and through much, if not all, of the Stephanian, tetrapod faunas lacked herbivores and were composed almost entirely of insectivores and carnivores with some potential omnivores. None of the known Westphalian tetrapods appear to have been capable of foliar herbivory.

Omnivory may have existed; small amphibians with jaws capable of puncture-crushing invertebrates could have dealt with tough, brittle seed coats or cones, and the low "metabolic size" of these animals would have permitted them to survive on such a diet. However, the amount of energy transferred to animal food webs through such means would have been small. In the Stephanian, diadectids may have had some impact as omnivores or herbivores. Diadectid remains are almost universal in Stephanian fossil assemblages of Euramerica, but they are not particularly abundant in any one assemblage, and their diversity is low. Even if they were herbivorous, diadectids probably effected only low rates of transfer of energy and protein into tetrapod food webs.

The interactions between animals and plants, and hence their mutual linkages and reciprocal effects as selective factors, were greater in Late Carboniferous than in Early Carboniferous ecosystems. Although there is evidence of insects using the aerial parts of plants for sites to lay eggs and develop larvae, the major linkage is through insect herbivory, which, with the development of insect flight, brought a whole new range of selective factors to the plants. Most predation appears to have been on reproductive organs, particularly those bearing pollen and spores. The extent of such herbivory, that is, the volume of plant material consumed relative to that available, is impossible to estimate, but the evidence for general foliar herbivory is limited and equivocal. The possible existence of herbivory in drier habitats, where plant tissues could have provided both nutrients and water, requires critical consideration.

Although detritivory may have had some selective effect on plants (e.g., hardened seed coats or chemical defenses of dispersed parts), it points to rather limited interactions between the plant and animal subcomponents of the larger wetland ecosystem. Because nearly all tetrapods relied entirely upon the secondary productivity of arthropods or on other vertebrates for food resources, plants were immediately important only in defining habitat spaces. For plants, except for any disturbance created, vertebrates had limited impact as direct selective vectors, although we cannot rule out even occasional seed predation as shaping plant reproductive biologies.

By the beginning of the Westphalian, many plants evidenced sophisticated mechanical and possible chemical defenses, presumably against pre- and postdispersal arthropod predation. Potential chemical defenses included resins, which occur abundantly in the tissues of medullosan pteridosperms, potentially glandular trichomes of lyginopterid pteridosperms, and mucilages in marattialean tree ferns. Stephanian tree ferns show evidence of damage to live tissue, whereas medullosans do not (Lesnikowska 1989). Mechanical protection may have included hardened seed coats and highly fibrous or sclerenchymatous pollen and spore-bearing organs. Hardened seed coats may have protected against post-dispersal consumption by detritivores. A modern analogue may be the extant *Ginkgo biloba,* which retains many apparently primitive reproductive attributes; pollination occurs prior to dispersal, but fertilization and embryo formation may occur after separation of the ovule from the parent plant, resulting in a very long period between dispersal and germination. Dispersed seeds or ovules would have been a prime target for detritivores, particularly if dispersal occurred in large numbers, as was apparently the case for many Late Carboniferous plants, such as cordaites and many small pteridosperms. Very large seed size, such as that found in the giant pachytesta seeds of some medullosans (Taylor 1965), also may have discouraged ingestion by all animals except large omnivores with crushing dentitions.

It is problematic to attribute the origin of sclerotic nests in vegetative tissues (such as in many pteridosperms or tree ferns), infolded leaves or hairy leaf surfaces, or "armored" stem surfaces (as in some lycopsids) to adaptation resulting from animal predation. These ultimately may have served such a function, but their origins may reflect quite different selective agents or none at all. Many of these basic features appeared in the earliest members of their respective higher taxonomic groups, well before any evidence of insect or vertebrate herbivory.

Ecosystem Complexity and Persistence

The modernization of plant-community dynamics occurred before that of the arthropods or vertebrates. A high level of complexity was reached before the end of the Carboniferous when animal dynamics were far from "modern," which suggests strongly that animals were not necessary for the expression of the basic structural and dynamic aspects of plant communities. Subsequent to the evolution of significant herbivory, animals probably influenced specific plant-plant interactions, although the specific effects of herbivory on modern plant communities are subject to considerable debate.

The wetland ecosystems of the Late Carboniferous, viewed from a shorter "ecological" time scale, do appear to have been persistent for remarkably long periods (2 to 3 My) in terms of taxonomic composition and dominance-diversity characteristics. Overall the persistence of the vegetation is even longer when viewed in ecomorphic terms. This suggests that such communities were resistant to invasion by new taxa with biomechanically superior traits, or that long-term ecosystem persistence suppressed the evolution of biomechanically superior forms by limiting safe sites for their ecological establishment. Biotic change appears to coincide with changes in prevailing climate, which accompanied the formation of Pangaea during the Late Carboniferous and Permian (Stanley 1988; Parrish et al. 1986, Ziegler 1990). Although interpretation of temporal changes within floras and faunas is somewhat confounded (at million-year time scales) by taphonomic factors, a basic signal can be detected: long intervals of persistence punctuated by short intervals of change. These patterns of relative persistence of ecosystems, resistance to biotically induced change, and major response to marked change in extrinsic conditions appear to characterize most of post-Carboniferous time to the present (see chapter 6).

Because of the potentials for plant-animal coevolution, the entire ecosystem must still be seen as impersistent on a time scale of tens of millions of years. Such ecosystems remained susceptible to invasion or disruption by certain major evolutionary innovations, particularly in the animal component. The plant elements throughout the later Paleozoic appear to have responded most to physical disruption of the environment, and taxonomic assemblages persisted for much longer than older assemblages of plants (Knoll 1984). More importantly, plant community dynamics, as far as we can discern them, were more or less modern by the Westphalian. Taxonomic replacements do not seem to have altered the spectrum of dynamics significantly, although they may have changed the specific pattern that was most common. Furthermore, the demise of the wetland floras does not appear to have resulted from invasion and competitive displacement but from wholesale habitat replacement, bringing in a new biota. The evolution of vertebrate herbivory and the diversification of insect herbivores must have selected against certain dynamics, especially patterns of reproduction and energy allocation to reproductive versus vegetative tissue, although there is presently little evidence to suggest that herbivores caused any major, sudden change in the spectrum of possible interactions among the plants. The evolution of vertebrate herbivory in particular appears to have expanded in the drier, more xeromorphic vegetation that moved into lowlands as they dried out.

7 THE PERMIAN

During the Permian, terrestrial biotic patterns characteristic of the Carboniferous gave way to those typical of Mesozoic and younger biotas (Frederiksen 1972). The transition from Carboniferous to Permian floras and faunas was globally diachronous (Knoll 1984) and occurred over an extended interval of time, spanning the late Stephanian and Early Permian. Although the time involved was great, the change was not gradual or geographically uniform within or between basins; the typical Carboniferous and Permian biotas remained largely at the ends of an edaphic continuum during the transition. Increasing spatial heterogeneity and increasing predominance of drier or better-drained conditions in lowland regions brought "Permian-type" assemblages to prominence in the fossil record. The nature of this biotic transition and the complex intrabasinal facies changes have obscured the Carboniferous-Permian boundary in continental rocks (Barlow 1975). Perhaps the best-documented transitional sequences in Euramerica are the Dunkard (Pittsburgh) Basin of the Appalachians (Barlow 1975), those of the southwestern United States (Read and Mamay 1964), southern Spain (Broutin 1981), and the Rotliegendes of central Europe (Boy and Fichter 1982; Kerp and Fichter

1985; Boy et al. 1990). Floristic changes also occurred over an extended period of time in the Angaran region (Meyen 1982) and in Cathaysia (Li and Yao 1982). During this transitional period, there was a reduction in the frequency of occurrence of floras rich in hygrophilous taxa, particularly medullosan pteridosperms, marattialean tree ferns (pecopterids), and sphenopsids; at the same time, floras dominated by conifers, callipterids (some of which had affinities with the peltasperms; Kerp 1982a), and "pteridospermous" seed plants of many new kinds increased. Faunal changes continued in directions first evidenced in the Late Carboniferous, including the ongoing diversification of insects and insectivores, and increasingly specialized feeding adaptations, such as herbivory, among tetrapods.

Accompanying the temporal changes in biotas was an increase in provinciality, particularly detectable in the vegetation (Read and Mamay 1964; Havlena 1970; Chaloner and Meyen 1973; Chaloner and Lacey 1973; Asama 1985; Ziegler 1990). Four major floristic realms have been recognized (Chaloner and Lacey 1973); they represent possibly the greatest amount of global floristic differentiation prior to the late Tertiary. In broad terms these floristic provinces were (1) the Euramerican, which can be subdivided into more typically European and western North American components; (2) the Cathaysian, divisible into northern and southern subprovinces, both closely linked to Euramerican floras; (3) the Angaran (largely Asian USSR), which may have been subdivided into frost-free sub-Angaran and temperate subprovinces (Meyen 1982); and (4) the Gondwanan, in the Southern Hemisphere. Ziegler (1990) has refined this pattern considerably and recognizes ten biomes, the same number and kind found in extant vegetation, varying from tropical rainforest through tundra. Although these biomes can be grouped generally into the larger four floristic provinces, they are better defined floristically than the "provinces" of earlier authors and are related closely to inferred global climatic patterns. Inferred continental positions during the Late Permian are illustrated in figure 5.5.

The vertebrate record, although excellent in some areas, does not provide the same biogeographic resolution as the plant record. To the extent that faunal composition can be determined, terrestrial tetrapod assemblages do not appear to have been subdivided biogeographically, and many genera have remarkably widespread distributions. Differences between European and North American freshwater aquatic records, however, suggest some provinciality.

Increased provinciality and the areal expansion of more mesomorphic to xeromorphic floras have been related to increasing continentality associated with the formation of Pangaea and to waning polar glaciation (Olson 1985b; Ziegler et al. 1981). It has been suggested that the northward movement of many of the major sampling areas of the Northern Hemisphere may have altered regional climates even without the effects of continental agglomeration (Cecil 1990).

Throughout the Early Permian, only loose links existed between plant pro-

ductivity and animal food webs, continuing primarily through insect herbivory and arthropod detritivory (Olson 1966, 1967, 1975). Thus, the most
conspicuous aspects of ecosystem dynamics also continued from the Late Carboniferous into the Early Permian. The expansion of vertebrate herbivory in
the latest Early Permian heralded major changes to come during the Late Permian, when vertebrate herbivory became well established and food webs became essentially modern.

7.1 Plants

Permian vegetation was generally more variable, particularly at the landscape
level, than Upper Carboniferous vegetation. At a local level, floristic heterogeneity was very great in some assemblages, with many co-occurring
distinctive ecomorphs, representative of different life habits. At a higher level,
interregional variability in dominance patterns and ecomorphic diversity became greater during the Early Permian, and appears to reflect greater edaphic
variability, and possibly climatic zonation, across the Pangaean continent
(Parrish et al. 1986; Ziegler 1990). Tropical and paratropical areas of Euramerica and Cathaysia experienced a great increase in spatial variability of
floras, an increase in seed-plant numbers, and a decline in hygromorphic
ferns, lepidodendrids, and sphenopsids. Major floristic change occurred earlier in more northern areas, such as the Angaran subcontinent, where a distinct, cordaitean-rich, seed-plant-dominated flora rose to prominence during
the Carboniferous, replacing lycopsid-dominated vegetation (Meyen 1982).
In some places very wet conditions, associated with peat formation, persisted
into the Early Permian and provided sites for the continued dominance of
Stephanian-type plant associations (e.g., Appalachian Dunkard Basin); in
others, such as southern Cathaysia, Westphalian-type lepidodendrid-domi-
nated peat swamps persisted through the end of the Permian. At the same
time, extremely seasonal moisture availability in other areas favored the establishment of a flora strongly dominated by xeromorphic seed plants (e.g.,
Hermit Shale flora of Arizona).

The earliest Permian floras appear to have retained a strong Late Carboniferous aspect only in areas of high moisture availability. Because many of
the European deposits represent disjunct intermontane basins (Havlena 1970,
1975), a relatively high degree of taxonomic variation developed and resulted
in floristic changes (compare the floras discussed by Kerp and Fichter [1985]
to those discussed by Gillespie et al. [1975]; see Bode [1975] and subsequent
discussion). Within peat-forming, basinal areas, floras from the wetter mineral-soil environments were dominated largely by pecopterid ferns, with a
large admixture of medullosan pteridosperms such as *Neuropteris* and *Odon-
topteris,* and few callipterid taxa. Floras dominated by more typically Permian
elements, such as numerous callipterids, *Taeniopteris,* and coniferophytes,
appeared only sporadically in peat-forming wetlands. The Dunkard Basin, for

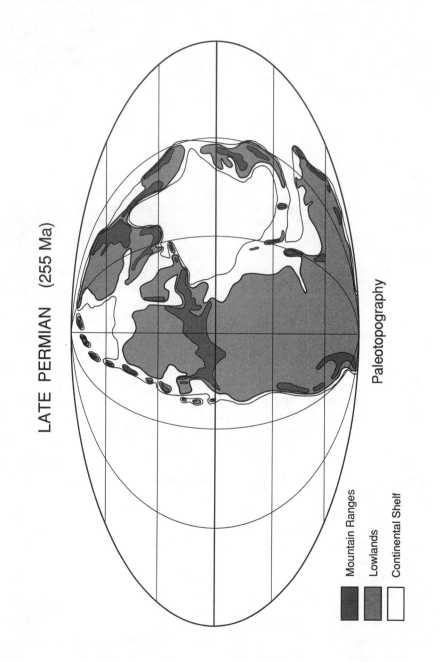

LATE PERMIAN (255 Ma)

Paleotopography

Mountain Ranges

Lowlands

Continental Shelf

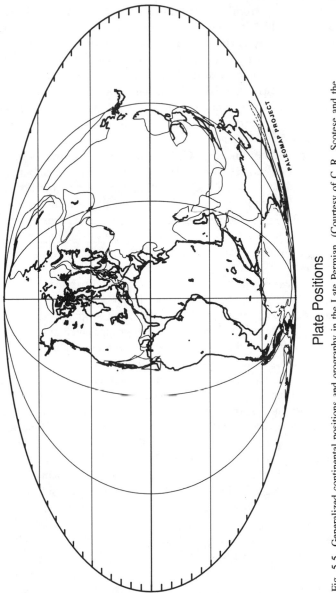

Plate Positions

Fig. 5.5. Generalized continental positions and orography in the Late Permian. (Courtesy of C. R. Scotese and the PaleoMap Project, University of Texas, Arlington.)

example, was sufficiently wet that conifers, most of the callipterids, and many associated plants were rare and were confined to isolated deposits (White 1933; Gillespie et al. 1975).

Early Permian floras from western Euramerica (western United States) were conspicuously enriched in conifers and other mesomorphic-to-xeromorphic plants. During the early Early Permian (Wolfcampian), these western floras retained many taxa typical of the Late Carboniferous, although these became less abundant through time as western Euramerica became increasingly dry.

During the late Early Permian (Leonardian), western Euramerican floras became distinctly different from their antecedents in the region. The development of regional variation was reflected in three distinct floras (Read and Mamay 1964), which contained many plants previously unknown in North America, some possibly linked to Cathaysian floras of China (White 1912; Mamay 1967, 1986; but see Asama 1985). All of these western floras had decidedly more mesomorphic to xeromorphic aspects than older floras of the region. Floristic distinctiveness may derive from regional orogenic activity that altered local climate and created barriers to species exchange within the area (White 1929; Read 1947; Read and Mamay 1964). Judging from current biostratigraphic estimates, no well-known floras of comparable age occur in eastern Euramerica.

The autecologies, life histories, growth forms, and habitats of many Permian plants are poorly known. In the cases of pecopterid ferns, sphenopsids, many medullosans, and lycopsids, analogy to Late Carboniferous antecedents provides a basis for further inference. For other groups, such as conifers, morphology and comparison with better-known younger descendents provide reasonable first approximations of habitat preference and are consistent with broader patterns of distribution. Nonetheless, there has been little attempt to reconstruct Early Permian plant communities (but see Kerp et al. 1989). To the degree that it is now known, the drier parts of the landscape were occupied by conifers and xeromorphic "pteridospermous" plants. Callipterids, taeniopterids, gigantopterids, and some medullosan pteridosperms dominated more mesic, but still seasonally dry areas. Pecopterid ferns, medullosans such as *Neuropteris,* and sphenopsids grew along watercourses or in wetter soils. In the wettest areas pecopterids and the few remaining lycopsids were dominant, with medullosans variably abundant. Thus, there remained a recognizable component of ecological segregation along major taxonomic lines. However, seed plants were clearly a much more conspicuous part of the entire flora than in similar Carboniferous habitats.

Wetlands Patterns

The most conservative Early Permian plant communities were swamps. Palynological analyses of Dunkard Basin coals (Clendening 1974, 1975) indicate a flora rich in ferns, pteridosperms, and sphenopsids, with only small numbers of coniferophytes. Floras of surrounding mineral-soil substrates also retained a slowly changing but strongly Stephanian aspect. Clastic-compression floras in the Dunkard Basin are dominated mainly by pecopterids, medullosan pteridosperms, and sphenopsids, with assemblages of conifers, callipterids, and other more typically Permian plants confined to isolated occurrences (White 1933). Southern China also remained very wet well into the Permian (Ziegler 1990), and included lycopsid, tree-fern, and cordaite-dominated swamps (T. L. Phillips unpublished data), as well as Carboniferous-type assemblages in the adjacent, clastic lowlands (Chaloner and Meyen 1973; Meyen 1982; Laveine et al. 1987). Such wetland floristic conservatism led Clendening to conclude that Dunkard rocks are entirely Upper Carboniferous.

Callipterids and conifers were often conspicuous to dominant in Early Permian intermontane basins of western Europe (Havlena 1975; Kerp 1982b; Kerp and Fichter 1985; Kerp et al. 1989; Broutin et al. 1990), preserved in a wide array of freshwater lowland settings, including lakes and channels, splays, and swamps (Boy and Fichter 1982). In the Dunkard Basin, however, the scattered occurrence of conifers, callipterids, and other "Permian" plants as early as the latest Westphalian or earliest Stephanian (McComas 1988) suggests that a conifer-*Callipteris* flora was established in parts of the lowlands where preservation generally was unlikely but occurred on occasion. The patterns of plant distribution in time and space clearly indicate the contemporaneous existence of a conifer-callipterid flora with a fern-and-pteridosperm-dominated lowland-wetland flora during the Stephanian. It is possible that Permian plants periodically invaded refugial wetlands from surrounding upland habitats, perhaps during brief intervals of increased moisture limitation, establishing ephemeral stands in temporarily suitable habitats (White 1933).

Southwestern United States: Emergence of Xeromorphic Floras

Permian floras of the southwestern United States have been studied intensively and are of considerable interest because they occur with abundant, well-studied vertebrate remains. The oldest plant assemblages of this region are of Wolfcampian age and are rich in pecopterid ferns, medullosan pteridosperms, and sphenopsids. Conifers, *Taeniopteris,* and callipterids also are present, with conifers dominant at some sites (Read and Mamay 1964; Ash and Tidwell 1982; Sander 1987). The morphological-ecomorphotypic heterogeneity among these assemblages suggests considerable variation in the regional physical habitats. At some exceptional sites, such as the Geraldine bone bed in north-central Texas (Sander 1987), there is very high species richness. The Geraldine deposit includes calamites, pecopterid ferns, medullosan pteridosperms, conifers, cordaites, callipterids, questionable lycopsids, and the cycad-like *Russelites* (Mamay 1968) in a mudstone deposited in a single standing-water body. This astounding diversity of ecomorphs seems to represent a mixture of plants drawn from an edaphically diverse sampling area that ranged from very wet to well-drained soils.

Leonardian floras of the southwestern United States have received considerable taxonomic and morphological study and continue to yield new genera and species. Three contemporaneous "floras" existed in the southwest during this time, described by Read and Mamay (1964) as the *Gigantopteris, Glenopteris,* and *Supaia* floras.

The *Glenopteris* flora is known only from the Leonardian of central Kansas. The characteristic and dominant plant, *Glenopteris* (Sellards 1908), had very thick, presumably fleshy, leaves. In general, the flora was relatively diverse and included pecopterid ferns, cordaites, lycopsids, and medullosan pteridosperms, among other elements. Mamay (personal communication 1988) suggests a coastal habitat, because the plant-bearing beds are intercalated regionally with marine limestones.

The *Gigantopteris* flora grew to the south, or approximately paleowest, of the *Glenopteris* flora. The *Gigantopteris* flora was highly diverse and included gigantopterids, a group of plants prominent in Cathaysian floras of Permian age, but no *Gigantopteris* itself (Mamay 1986). The flora also included pecopterids, medullosan pteridosperms, sphenopsids, cordaites, callipterids, conifers, and a large number of enigmatic forms, such as early cycads and peltasperms. The dominant plants of the flora vary and include pecopterid ferns, medullosans, conifers, and several different groups of seed plants, broadly characterized as "seed ferns" (Mamay 1967). As with older Permian floras, it is difficult to reconstruct the source communities from existing data. Diverse florules from several localities, such as those described briefly by Ash and Tidwell (1982) or in a series of papers by Mamay (the Emily Irish locality of north-central Texas; Mamay 1968, 1975, 1986, 1988), suggest considerable habitat heterogeneity surrounding the sites of deposition. This is supported further by site-to-site variation in dominance-diversity patterns within florules of the *Gigantopteris* flora. Most plant assemblages from north-central Texas are preserved in drab mudstones that infilled small water bodies within alluvial to coastal-plain settings and probably represent samples of local floras (Read 1943; Mamay 1967, 1968). A number of these deposits are from coastal areas, close to brackish or marine waters (Mamay et al. 1984). If such channel deposits favor the preservation of mesophilous to hygrophilous elements, those species that grew along watercourses will be overrepresented (Scheihing and Pfefferkorn 1984; Gastaldo et al. 1987; Burnham 1989), and the broader landscape may have supported a considerably more xeromorphic flora (see Burnham 1990 for a Tertiary parallel). Most of the plants that appear for the first time in this flora are seed plants, suggesting the increased importance and diversity of this group.

The *Supaia* flora, known from the Leonardian of Arizona, New Mexico and

Utah, was to the approximate paleonorth and paleonorthwest of the *Glenopteris* and *Gigantopteris* floras (White 1929; Mamay and Breed 1970; Ash and Tidwell 1982). It is characterized by a lower diversity: the total number of "whole-plant" species may be as low as 15 to 20. A lack of pre-Permian species is most notable, with no representatives of many of the common earlier Permian groups such as cordaites, sphenopsids, medullosans, and most pecopterids. Most of the *Supaia* flora plants were highly xeromorphic, with leathery leaves, hairy surfaces, spines, scales, sunken veins, and simple, rather than compound, leaves. Seed plants or presumed seed plants overwhelmingly dominate the flora. The best-known florule of the *Supaia* flora is from a channel-fill deposit in the Hermit Shale of the Grand Canyon of Arizona, where plant remains are associated with mud cracks and pseudomorphs of salt crystals, indicative of a seasonally dry climate (White 1929). Some elements of this flora appear in France during the Late Permian.

In summary, the floras of the southwestern United States record great local and regional variation comparable to that of a modern lowland modified by seasonal climatic fluctuations (Ziegler 1990). If this region had received high rainfall throughout the year, as may have been the case earlier, a much more homogeneous flora would have been present because extrinsic conditions would have dampened the effects of micro- and macrohabitat differences (Stebbins 1952; Axelrod 1967, 1972). The Leonardian manifestation of floristic heterogeneity probably is related to climatic conditions that accentuated physical differences within habitats.

Southwestern floras of the United States are the focal point for several inferences about Early Permian biogeography. Most notable are taxonomic overlap between floras of the southwestern United States (western Euramerica) and China (eastern Euramerica), a pattern first recognized by White (1912). Although there are no current paleocontinental reconstructions (e.g., Chaloner and Lacey 1973; Ziegler et al. 1981; Asama 1985) that provide possible migration routes, several Cathaysian taxa, such as *Russelites* (Mamay 1968), and the gigantopterids *Cathaysiopteris, Zeilleropteris,* and *Gigantonoclea* (Mamay 1986, 1988), are known from the Permian of the western United States; other plants of this region that may have phylogenetic links to Cathaysian groups include *Delnortea* (Mamay et al. 1986, 1988), *Gigantopteridium* (Mamay 1988), and *Tinsleya* (Mamay 1966). Asama (1985), however, believes that the morphological similarities among the plants from the two regions reflect parallel evolution. Links between the southwestern United States and western Europe are indicated by some plants in the Rotliegend flora of Germany, including *Autunia* (*Sandrewia*) (Kerp 1986) and *Sobernheimia* (Kerp 1983), an early cycad possibly related to the cycads described by Mamay (1973, 1976).

Discussion of the Cathaysian flora and references to literature can be found in Li and Yao (1982). Meyen (1982) provides a synthesis of Angaran Permo-

Carboniferous floras. Archangelsky (1986) discusses Gondwanan floras of Permian age, which are distinct from those of Euramerica, Angara, and Cathaysia.

7.2 Arthropods

The record of Permian arthropods reveals continued diversification of herbivorous insects and evolution or expansion of a number of major groups. Diverse insect assemblages are known from Euramerica (Durden 1984a, 1988), the Kuznetz basin of Siberia, within the temperate portion of Angaraland (Meyen 1982), and from the Late Permian of Australia and South Africa (Wootten 1981). The largest and best known faunas of the Late Permian are from the USSR. Only insects and scorpions have been examined in detail from most of these assemblages (Schneider 1978, 1980, 1982; Wootten 1981; Durden 1984a; Kjellesvig-Waering 1986).

Notable Permian events are the evolution or expansion of hemipterans, sucking forms that fed from phloem or xylem (Evans 1954, 1956; Rasnitsyn 1980), and beetles (Ponomarenko 1969). The beetles (Coleoptera) and bugs (Homoptera) may have evolved in cold temperate regions, as suggested by their distribution in Angaran and Gondwanan deposits, where the floras and sedimentary environments suggest temperature seasonality. As detritivores, herbivores, or pollinators, the beetles opened a spectrum of possibilities for plant-animal coevolution. Also appearing in the Permian were small Diaphanopterodea, which resembled mosquitos in body size, suctorial mouthparts, and structure (Kukalova-Peck 1974), and which may have been bloodsucking forms (Pruvost 1919).

Herbivorous protorthopterans (Carpenter 1971; Burnham 1983) and orthopterans (Sharov 1968) with modern patterns of structural organization, increase in abundance in Permian deposits, as do predaceous ephemeropterans (mayflies) (Kukalova-Peck 1985). Odonates also continue to be numerically abundant as predators (Wootten 1981). Collembolans are documented from South Africa (Riek 1976).

The low number of well-known Permian arthropod assemblages offers an incomplete picture of the ecological roles played by these animals in the Permian. In combination with the Carboniferous record, however, there is enough evidence to indicate that herbivory was global, and that many new groups of herbivores were evolving and dispersing widely, including beetles. The record does not, however, reveal large detritivores like those of the Carboniferous. Many earlier groups, such as millipeds, existed through the Permian, which suggests that detritivores remained important in some, probably most, Permian habitats. Most arthropod-bearing deposits of the Permian are not paleoenvironmentally similar to the peat-forming wetlands of the Late Carboniferous and thus, do not support direct comparison. Permian assemblages generally derive from drier and possibly seasonally cooler settings. A tropical, everwet setting of Permian age may have been dominated by detritivores in much the same manner as Carboniferous wetlands, or modern, equatorial, peat-forming swamps.

7.3 Vertebrates

Early Permian

Our understanding of Early Permian terrestrial vertebrates is derived largely from numerous sites in north-central Texas and adjacent parts of Oklahoma and from several deposits within the nearby Four Corners region of north-central New Mexico, southwestern Colorado, and southeastern Utah. Smaller assemblages are known from northern Oklahoma, Kansas, the Dunkard Basin of the Appalachians, Prince Edward Island (Canada), the English Midlands, and continental Europe. These reptile-dominated faunas appear to have been geographically widespread, with little taxonomic variation among the major land-dwelling forms. Contemporaneous aquatic assemblages from small and large lakes are known, both in North America and continental Europe (Olson 1977; Boy 1977, 1987); the composition of these freshwater faunas varies significantly between the classic Rotliegend and Texas-Oklahoma collections. Although a relatively small sample of terrestrial tetrapods is known from Europe, associated trackway-bearing beds in Germany record a diverse terrestrial fauna (Haubold and Katzung 1978; Boy and Fichter 1982; Fichter 1983; Fichter and Kowalczyk 1983).

The most productive of the more commonly occurring Lower Permian vertebrate deposits in North America represent mudstone infillings of abandoned channels, sometimes carbonate-rich and often termed ponds by various workers (Read 1943; Olson 1958; Berman 1978; Olson and Mead 1982; Eberth 1985, 1987). Characterized by sedimentological evidence indicative of reducing conditions and by channel-like geometries, these deposits yield exceptional specimens of aquatic-to-terrestrial animals and in some cases well-preserved plants and invertebrates. Associated or discrete channel-lag conglomerates often contain a great assortment of generally fragmentary vertebrate remains (Read 1943; Olson 1958; Hlavin 1972; Murry and Johnson 1987). Complete articulated skeletons of terrestrial tetrapods are exceedingly rare and occur most commonly in isolation or in monospecific or very low diversity assemblages within floodplain sediments (Williston 1911; White 1939; Romer 1969; Berman et al. 1987b, 1988). Disarticulated remains of small and almost exclusively terrestrial vertebrates of Early Permian age have been collected in large quantities from fissure fills in southern Oklahoma (Peabody 1961; Olson 1967, 1991; Bolt 1980). Table 5.4 summarizes size categories and dietary groups of Early Permian tetrapods.

MICROSAURS   The record of terrestrial microsaurs is sparse, and they have a suite of features that suggests secondarily acquired aquatic adaptations. *Pan-*

Table 5.4 Lower Permian Terrestrial Amphibians and Reptiles

	Small	Intermediate	Large
Carnivores	Eothyridid pelycosaurs Dissorophid temnospondyls Phlegethontid aïstopods	Sphenacodontid pelycosaurs Varanopseid pelycosaurs Seymouriid cotylosaurs Tseajaiid cotylosaurs Parioxid eryopoid temnospondyls Dissorophoid temnospondyls Ostodolepid microsaurs	Sphenacodontid pelycosaurs Limnoscelid cotylosaurs?

Insectivores	Single–tooth row captorhinid
	  captorhinomorphs
	Protorothyridid captorhinomorphs
	Dissorophid temnospondyls
	Hapsidopareiontid microsaurs
	Goniorhynchid microsaurs
	Brachystelechid microsaurs
	Ostodolepid microsaurs
		Single–tooth row captorhinid
		  captorhinomorphs
		Araeosceloid diapsids	

Omnivores	Bolosaurid reptiles Gymnarthrid microsaurs Pantylid microsaurs	Primitive multiple–tooth row captorhinid captorhinomorphs	Edaphosaurid pelycosaurs? Diadectid cotylosaurs?

Herbivores		Caseid pelycosaurs
		Advanced multiple–tooth row
		  captorhinid captorhinomorphs
			Caseid pelycosaurs
			Advanced multiple–tooth row
			  captorhinid captorhinomorphs
			Edaphosaurid pelycosaurs?
			Diadectid cotylosaurs?

Note: Size categories are based on estimated total body lengths of largest specimens, small < 0.5 m, intermediate $0.5 – 1.5$ m, large ≥ 1.5 m.

tylus, which is well represented in Wolfcampian and Leonardian deposits of Texas, is an exception. This genus had a disproportionately large, massive skull and a dentition that consisted of relatively few, stout, blunt marginal teeth and complexes of palatal and coronoid teeth that resembled the marginal series both in size range and shape. All teeth show wear facets, with some palatal tooth crowns reduced to mere stubs. An unequivocal lack of direct occlusion indicates that the attrition arose from grinding resistant food materials, if not also an admixture of silt and sand grains (Carroll 1968). Whereas the slightly overhanging snout with posteroventrally directed premaxillary teeth suggests a grubbing habit, the entire dentition and powerfully built jaws indicate that food was crushed. The postcranial skeleton includes an expanded rib cage and stout but short limbs that would have held the body close to the ground. *Pantylus* probably obtained food from the soil, and this food may well have included plant roots and rhizomes (Romer 1969), among other resistant plant parts, such as stems and tough-coated seeds. A monospecific suite of more primitive pantylid microsaurs from New Mexico may represent a failed aestivation assemblage (Berman et al. 1988).

The majority of other Permian terrestrial microsaurs are from the Leonardian of Texas and Oklahoma. The marginal dentition of gymnarthrids superficially resembled that of *Pantylus* in consisting of few large, conical teeth, but accessory series were not elaborated greatly, and wear was less extensive (Carroll and Gaskill 1978; Bolt and DeMar 1983). In general, gymnarthrid teeth also were more acuminate, indicative of a piercing, rather than a crushing bite, suited for dealing swiftly with small prey (Gregory et al. 1956). An elongated presacral vertebral column, only moderately developed limbs, and preservation in aquatic as well as terrestrial assemblages suggest that gymnarthrids were less terrestrial than *Pantylus* (Olson 1939; Gregory et al. 1956).

The ostodolepidids were the largest known microsaurs, attaining lengths of greater than 50 cm. Their skulls were strongly wedge shaped with an expanded occipital surface that presumably accommodated a well-developed dorsal axial musculature (Carroll and Gaskill 1978). These features, along with simple, closely spaced, peglike teeth, suggest that ostodolepidids may have sought burrowing prey.

TEMNOSPONDYLS As in the Carboniferous, the greatest variety of terrestrial amphibians during the Permian is found among the dissorophoid temnospondyls. At least ten valid taxa of dissorophoids are known from Lower Permian deposits in Texas, Oklahoma, and New Mexico; although many taxonomic problems exist within this group, as many as four distinct genera are known to occur within single assemblages. The family Dissorophidae shows trends toward increased skull size and the development of extensive dermal ossifications on top of, and in some cases attached to, presacral neural spines. Inappropriately termed armor, these ossifications probably enhanced terrestrial locomotor abilities by strengthening the axial skeleton (DeMar 1966; Bolt 1974a). As a group, armored dissorophoids were of intermediate size or smaller and probably fed on insects and small tetrapods. *Tersomius*, however, lacked dermal armor and was somewhat smaller than other dissorophoids. The remarkable abundance of juvenile *Tersomius* specimens in assemblages that include a significant aquatic component (Carroll 1964a; Olson 1970; Daly 1973; Bolt 1974a, 1977) suggests a change from an aquatic to a terrestrial habit with growth, a pattern that may pertain to other dissorophoid taxa (Olson 1985a).

Trematopid dissorophoids, which are known first from scant records in the Upper Carboniferous, are represented in the Lower Permian by two genera. These small to intermediate-sized forms are interpreted as agile, land-dwelling predators on the basis of marginal dentitions, which consisted of small-to-large acuminate and moderately recurved teeth, and well-ossified but lightly built postcranial skeletons (Olson 1941; Dilkes and Reisz 1987; Dilkes 1990).

The second major group of terrestrial Permian amphibians, eryopoid temnospondyls, is recognized most commonly as a semiaquatic predator. The flat head, dentition, and postcranial anatomy of *Eryops* suggests that this widespread and very long-lived form was crocodile-like in its habits, practicing lie-and-wait predation in or at the margins of water bodies (Miner 1925; Olson 1936; Sawin 1941; Moulton 1974). Although not regarded as a terrestrial predator, *Eryops* was obviously capable of terrestrial locomotion and was certainly preyed upon by large land-dwelling carnivores (Bakker 1982; Olson 1983). *Parioxys*, an intermediate-sized temnospondyl from Texas that is commonly regarded as an eryopoid (Moustafa 1955; Carroll 1964b), was probably a more terrestrial form. In contrast, the various eryopoids known from the Early Permian of Europe appear to have been aquatic ecomorphs (Werneberg 1989; Milner 1989).

COTYLOSAURS (*SENSU* HEATON 1980)   The genus *Seymouria* comprises virtually all materials regarded as seymouriamorph cotylosaurs from the Early Permian. First represented in middle Wolfcampian rocks of New Mexico, Utah, and Texas (Berman et al. 1987b) and particularly well known from Leonardian deposits in Texas, *Seymouria* epitomizes a robust terrestrial reptiliomorph of intermediate size (Romer 1928). A generalized dentition of small, fairly uniform, conelike teeth set in modestly constructed jaws suggests a grab-and-gulp feeding style and a diet of soft-bodied invertebrates and small vertebrates (White 1939). A similar form occurs in the late Early Permian of Europe (Martens 1989).

Another fully terrestrial cotylosaur, *Tseajaia*, is known from the Wolfcampian of Utah and New Mexico. Although the described material is larger and more stocky than contemporary *Seymouria* species, these genera share many postcranial similarities (Moss 1972). The more heavily built skull and greater development of premaxillary and caniniform maxillary teeth in *Tseajaia* suggest modest dietary differences between it and *Seymouria.*

Some of the morphologic attributes of diadectid cotylosaurs have been considered previously in the discussion of Late Carboniferous tetrapods. Compared to Stephanian forms, the teeth of Permian diadectids are more specialized and more extensively worn, and their bodies much larger, in some cases exceeding 3 m in length. Their considerable mass and expanded torso indicate a large gut, which together with a low rate of ingestion dictated by a small head, suggests they had the capacity for prolonged chemical processing of a large food mass. Such capacities imply further that a large proportion of the diadectid diet during the Permian may have consisted of large pieces of fibrous plant tissues. As ectotherms, however, they would not have had to process the amount of food required by mammalian herbivores, and such processing may have been quite inefficient by endothermic standards. The basic morphological suite is comparable to that of large, extant, herbivorous tortoises.

The well-documented record of *Diadectes* in the Texas sequence shows an increase in size through Lower Permian deposits (Romer 1944; Olson 1947). Despite intensive collecting efforts in Texas, no diadectid remains have been found in late Leonardian deposits (Olson and Mead 1982; Murry and Johnson 1987).

Fragmentary remains from the Wolfcampian of West Virginia and Colorado have been described as limnoscelid cotylosaurs (Romer 1952; Lewis and Vaughn 1965). More complete materials from slightly younger Early Permian rocks in Germany await description (Martens 1989). These intermediate-to-large carnivorous forms appear to have been capable of movement on land but, like *Limnoscelis,* may have existed mainly as amphibious predators.

AMNIOTES   The differentiation of amniotes chartered upon meager occurrences in the Carboniferous is proven in full by the Early Permian record. At least twenty-five amniote genera are known from the Wolfcampian-Leonardian interval of north-central Texas. More than half of these are pelycosaurs, which collectively encompass a very wide range of ecomorphotypic variation. Though there are fewer captorhinomorphs in the Texas-Oklahoma section, they too offer a broad spectrum of ecologically significant adaptations. This array of amniote diversity is expanded further by several curious taxa that cannot be assigned readily to any major group.

CAPTORHINOMORPHS (*SENSU* HEATON AND REISZ 1986).   Early Permian protorothyridids from early Wolfcampian deposits of Texas and West Virginia differ little from earlier members of this family of small, agile insectivores (Carroll and Baird 1972; Clark and Carroll 1973). Protorothyridids are last recorded in North America in fissure fillings near Fort Sill, Oklahoma, estimated to be middle Leonardian in age on the basis of the entire tetrapod assemblage (Bolt 1980; Reisz 1980; Olson 1991). A possible younger occurrence of a protorothyridid-captorhinid intermediate is known from a recent discovery in the Upper Rotliegendes of Germany (Martens 1989).

Relatively abundant remains of captorhinid captorhinomorphs from the Early Permian of Texas and Oklahoma record the diversification of a group characterized by massive skulls that had sharply downturned snout tips. Wolfcampian captorhinids were somewhat larger and more heavily built than contemporaneous protorothyridids (Clark and Carroll 1973). Unlike later captorhinids, which had more elongate, rounded snouts and enlarged cheek areas for powerful jaw adductor musculature, these primitive captorhinid genera had relatively high skulls that are triangular in plan view. The maxillary dentition of early captorhinids resembled that of most protorothyridids in consisting of a single tooth row of simple cones spaced between mandibular teeth. More advanced single-tooth-row captorhinids from lower Leonardian deposits had proportionately larger, more closely spaced maxillary and dentary teeth that exhibit distinct occlusal facets (Heaton 1979). The enlarged anterior premaxillary and dentary teeth also show pronounced wear surfaces indicative of shearing bite.

A greater degree of specialization, doubtless related to enhanced food processing, is found in Leonardian captorhinids that have multiple maxillary and dentary tooth rows. Although these superficially resemble the tooth batteries of the microsaur *Pantylus,* detailed study of *Captorhinus* has shown that teeth were variably chisel shaped and that wear was less severe (Bolt and DeMar 1975; Ricqles and Bolt 1983). Rather than crushing and grinding abrasive materials, *Captorhinus* appears to have used its massive jaws to shear food. This multiple-tooth-row plan is elaborated in late Leonardian and Guadalupian (early Late Permian) captorhinids, most of which are large and are regarded as herbivores (Stovall 1950; Olson 1962a, 1962b).

Captorhinids as a group appear to have been fully terrestrial (Heaton and Reisz 1980, Sumida 1989a, 1990), notwithstanding an interpretation of the intermediate-sized, single-tooth-row form *Labidosaurus* as semiaquatic (Olson 1952, 1983). The postcranial skeleton of small captorhinids conforms to a generalized lizard morphotype, and functional analysis of the pectoral limb and girdle indicates a sprawling gait common to primitive tetrapods (Holmes 1977; Heaton and Reisz 1980; Dilkes and Reisz 1986). Whereas the most primitive captorhinids were probably insectivorous, the feeding habits of later Early Permian taxa, particularly genera with multiple-tooth rows, are open to speculation and are described most often as omnivorous. The large, recumbent premaxillary teeth of *Captorhinus* consistently show wear facets on the anterior surface that does not occlude any other teeth (Hotton unpublished data); these large teeth and the hooked, beaklike snout may have been used in some sort of grubbing activity. In foraging, these reptiles may have sought invertebrates among coarse plant litter, or perhaps even in rotten logs, as suggested by the occurrence of galleried wood preserved in captorhinid-bearing sequences. One of the very few records of trophic interaction from the Early Permian is a *Captorhinus* skull that has a partially articulated skeleton of another individual in its mouth (Eaton 1964). "As the two reptiles appear to be conspecific, the specimen constitutes an object lesson on the evils of cannibalism" (Baird and Carroll 1967).

SYNAPSIDS.   Nearly all Early Permian terrestrial vertebrate assemblages are dominated by pelycosaurs. From robust forms with small heads and short faces, resembling *Heloderma,* to gracile forms with large heads and long faces, resembling *Varanus,* pelycosaurs ranged from approximately 0.5 m to over 3 m in length, of which half or more consisted of tail. In pose of limbs, configuration of feet, and sprawled gait, Early Permian pelycosaurs were identical to their Late Carboniferous antecedents.

Sphenacodontid pelycosaurs occur in abundance throughout the Early Permian and are known in North America and Europe (Romer and Price 1940; Bakker 1975a, 1980; Olson 1976, 1983). During this time they continued their role as top predators and by the Late Permian attained estimated body weights greater than 150 kg (Romer and Price 1940; Olson 1983). *Dimetrodon, Sphenacodon,* and *Ctenospondylus* are characterized by large, laterally compressed, serrate, and recurved teeth set along a well-differentiated marginal tooth row, an enormous gape, and powerful jaws, and were presumably adapted to attacking prey of approximately their own size (Vaughn 1964, 1970; Barghusen 1973; Berman 1978; Eberth 1985). In limb proportions and pedal configurations, these forms were comparable to the largest of the modern lizards (Romer 1922, 1927), and there were no structural features to suggest aquatic habits. *Secodontosaurus,* another Texas sphenacodontid, co-existed with and resembles *Dimetrodon* and *Ctenospondylus* in the development of greatly heightened neural spines that form a "sail" (Romer and Price 1940). Such sails, which also occur in the contemporaneous pelycosaurs *Edaphosaurus* and *Lupeosaurus* (Sumida 1989b), may have served as thermoregulatory organs (Romer 1948; Bramwell and Fellgett 1973).

Varanopseids are the only other reasonably well known group of Early Permian pelycosaurs that can be interpreted confidently as terrestrial predators. Found in the Wolfcampian of New Mexico and the Leonardian of Texas, varanopseids were more gracile than sphenacodontids, reached lengths of approximately 1 m, and conformed to a general lizard morphotype. Teeth were not differentiated to the degree seen in sphenacodontids but were laterally compressed and dramatically recurved in a manner that suggests a very wide gape (Langston and Reisz 1981; Berman and Reisz 1982). Although varanopseid remains are exceedingly rare, taphonomic bias is shown by *Varanops,* which is known solely from a "herd" of some twenty juveniles collected from a single Leonardian-age deposit (Williston 1911; Romer and Price 1940). Taphonomic biases also may be shown by the near nonoccurrence of eothyridid pelycosaurs, a poorly represented but distinct group of small carnivores.

The best known ophiacodontid pelycosaur, *Ophiacodon*, has been regarded traditionally as an aquatic piscivore on the basis of comparatively poor ossification, reduced forelimbs, an absence of claws otherwise characteristic of pelycosaurs, and general skull construction (Romer and Price 1940; Brinkman 1988). Primitive ophiacodontids, such as *Stereophallodon* from the early Wolfcampian of Texas (Brinkman and Eberth 1986), are inadequately represented at present and may include terrestrial forms.

The Early Permian pelycosaur *Edaphosaurus*, like its cotylosaur contemporary *Diadectes*, has the characteristic barrel-shaped trunk and small head of a terrestrial herbivore capable of consuming high-fiber plant tissue. This interpretation of *Edaphosaurus* is weakened, however, by the dentition, which consisted of numerous teeth packed closely along the margins of the jaws and on special palatal expansions that opposed similar batteries on an elaborated medial surface of the mandible. The marginal teeth were slightly larger than the medial teeth, and all were small and bluntly pointed (Romer and Price 1940). Unlike *Diadectes* and advanced captorhinids, wear facets are absent in virtually all specimens. This specialized crushing, not grinding, dentition may have enabled *Edaphosaurus* to feed on small shelly mollusks and arthropods, as well as, or instead of, plant material. If edaphosaurids sought food in shallow, vegetation-choked water bodies or along watercourses, perhaps enough plant material was ingested in the process to supplement a primary diet of animal protein.

The postcranial anatomy of *Edaphosaurus* provides further cause for conjecture. Despite massive limbs unquestionably used for terrestrial locomotion, *Edaphosaurus* is commonly held to be a semiaquatic swamp dweller. This interpretation appears to be predicated largely on the occurrence of well-preserved plants with exceptional and abundant *Edaphosaurus* specimens at a single Texas locality, the remarkable Geraldine bone bed (Sander 1987); in fact most Permian plant-bearing beds of Texas do not represent swamps and do not yield pelycosaurs. Moreover, movement among dense vegetation would have been made difficult, if not impossible, by the full-sized "sail" that was studded with protuberant lateral tubercles. Examples of broken-in-life neural spines are not uncommon, but these in no way define the habit of *Edaphosaurus* or any other "fin-back" pelycosaur. Successive Texas species of *Edaphosaurus*, like several other lineages of large tetrapods in the Early Permian, show progressive size increases during late Wolfcampian and early Leonardian times (Romer and Price 1940).

A close evolutionary relationship between edaphosaurid and caseid pelycosaurs, which was once held to be likely and of no small ecological consequence (Romer and Price 1940; Stovall et al. 1966), has been refuted on the basis of new materials and phylogenetic analysis (Langston 1965; Brinkman and Eberth 1983; Reisz and Berman 1986). Known first from the late Leonardian of Texas, caseids stand alone as the earliest unequivocal "conventional" terrestrial herbivores. The oldest caseid taxon was of intermediate size, but *Cotylorhynchus,* from the Early Permian of Oklahoma, was of gargantuan size, nearly 3 m in length and exceeding an estimated body weight of 300 kg (Stovall et al. 1966). Much of the increased mass was accounted for by an enormous torso that was held close to the ground (contrary to an often-reproduced restoration of Stovall et al. 1966). The entire postcranial skeleton was massive, with particularly robust forelimbs and strong, clawed feet. The rotund body carried a disproportionately small head distinguished by an abbreviated but pronounced snout that bore enormous external narial openings (Olson 1968). In general, the caseid dentition consisted of relatively few, high-crowned, spatulate marginal teeth that became more conical anteriorly. The more laterally compressed teeth were coarsely serrate, like those of herbivorous iguanid lizards, and would have served to cut or slice vegetation. In comparison with *Edaphosaurus,* the development of palatal and mandibular accessory teeth was slight.

The size of the caseid body cavity suggests that large amounts of high-fiber plant tissues were processed through microbial fermentation. Well-preserved hyoid structures known in some genera indicate a large, mobile tongue, and the enlarged external nares indicate an uncertain sensory specialization possibly related to feeding (Olson 1968). Because the rewards of a labor-intensive, digging search for food probably would not have sustained such large animals, the powerful, claw-bearing forelimbs are more likely to have been used in tearing apart aerial plant organs.

DIAPSIDS AND OTHER REPTILES.   Remains of Early Permian diapsid reptiles occur in North America and Europe. With general lizardlike body proportions and posture, the best-known genus, *Araeoscelis,* resembled earlier diapsids (Vaughn 1955). This agile form had a modestly differentiated dentition and probably captured hard-bodied invertebrates such as insects (Reisz et al. 1984).

Several small, problematic reptiles from the Early Permian expand the range of ecomorphotypic diversity. *Bolosaurus,* from the Wolfcampian of Texas and New Mexico, and related genera from the Late Permian of the USSR, possessed extremely specialized marginal dentitions (Watson 1954; Ivakhnenko and Tverdokhlebova 1987). The relatively few teeth were highly differentiated and heavily enameled, and the most anterior were somewhat like those of *Diadectes* in being directed anteriorly and chisel shaped. Posteriorly along both upper and lower jaws, the teeth became larger, more bulbous, and variously modified by ridges that pass obliquely from the crown to the base of the tooth. Wear facets were present on most teeth, and those of the cheek series resulted from a shearing pattern of occlusion that "self-sharpened" opposing teeth. The skull and jaw construction suggests a complex musculature that probably was unique among early tetrapods. The bolosaurid postcranial skeleton is virtually unknown.

The front teeth of bolosaurids appear to have collected food materials, which were then sliced, perhaps quite finely, by the cheek teeth. If a specialized diet of "hard vegetation" is accepted (Ivakhnenko and Tverdokhlebova 1987), bolosaurids represent the only small herbivores known from the Early Permian. In *Bolosaurus,* however, the delicate anterior teeth may have been unsuited for removing attached plant structures, and a diet of small invertebrates is perhaps more plausible (Watson 1954).

A single skull from the middle Leonardian of the north-central Texas has recently been declared the oldest therapsid (Laurin and Reisz 1990). Long known as *Tetraceratops,* this unique and exceedingly odd specimen was so named because of numerous cranial rugosities that may have supported hornlike structures similar to those found on later amniote herbivores. The dentition, however, is that of a specialized carnivore, complete with a diastema anterior to an enlarged caniniform tooth. This most unusual form, like later oddities from the otherwise relictual Leonardian assemblage of Texas and Oklahoma, demonstrates that rare and perhaps progressive faunal elements may go uncollected or unrecognized even within the most intensively sampled regions.

Characteristics of Terrestrial Vertebrate Faunas in the Early Permian   A summary of Early Permian terrestrial tetrapods indicates a variety of feeding habits and considerable diversity within some ecomorphotypic groups (table 5.4). Intensive collecting of deposits in Texas and Oklahoma has demonstrated that two or more taxa of one particular adaptive strategy often coexisted; in particular, the record shows that the depth of various predator guilds was increased significantly during this time. The most diverse assemblages are known from the most productive beds, which are probably early Leonardian in age (Hook 1989). A subsequent decline in small-to-intermediate-sized predators has been interpreted as an ecological response to environmental change (Olson 1952, 1983), but this decline may be linked to variations in conditions suitable for the preservation of smaller animals (Olson 1962a; Murry and Johnson 1987).

On the basis of extensive work in north-central Texas and Oklahoma, Olson (1952, 1961, 1966, 1975, 1983) concluded that Early Permian terrestrial vertebrate communities probably relied heavily upon the productivity of aquatic systems. Rather than diminishing through time, this archaic trophic linkage is said to have increased with the apparent decline of insectivores as prey in the latest Early Permian. The recently enhanced record of Carboniferous tetrapods, however, shows that most terrestrial groups of the Early Permian are known first from Stephanian deposits and that a diverse, fully terrestrial tetrapod community existed in the Early Carboniferous. Concurrently, species diversity within aquatic-to-amphibious systems appears to have declined steadily through the Stephanian and into the Permian. These data are difficult to reconcile with the anomalous abundance of top-level carnivores in the Texas record and their postulated dependence on contemporaneous freshwater faunas.

Within the carnivorous and omnivorous feeding categories, several phylogenetic lineages show increases in body size through the Leonardian. This pattern may be indirect evidence of the evolution of herbivores from predaceous ancestors in edaphosaurids, diadectids, and captorhinids, all three of which are represented earlier by smaller insectivorous-to-omnivorous species. Large captorhinids with multiple-tooth rows continued into the Later Permian and occur together with caseids, but edaphosaurids and diadectids apparently were extinct by the end of the Early Permian. Although an immediate phylogenetic antecedent of caseids is unknown, they appear to be most closely related to eothyridids, a poorly known group of small, predaceous pelycosaurs. Thus, adaptations for or at least approaching herbivory evolved independently in several distinct tetrapod groups.

In the later Early Permian, when large herbivorous caseids first appeared, they were not associated with large carnivores (Stovall et al. 1966; Olson 1970); in younger caseid-bearing Guadalupian deposits of Texas and Oklahoma, carnivores are exceedingly rare (Olson 1962a, 1965). Because these caseids and contemporaneous captorhinids are regarded almost universally as herbivores, the late Leonardian assemblages in which they occur may be seen as the earliest record of an herbivore-dominated terrestrial fauna.

Late Permian

Late Permian tetrapod faunas were dramatically different from those of the Early Permian and hence from the rest of the Paleozoic. Herbivorous tetrapods were abundant and diverse, and faunal similarities between northern and southern parts of Pangaea can be recognized (Sues and Boy 1988). Most assemblages were composed largely of mammal-like reptiles. Both terrestrial and aquatic amphibians declined sharply in numbers and diversity. Comparisons between aquatic faunas (Boy 1977, 1987) suggest substantial taxonomic changes and an almost total change between the Early and Late Permian. The abrupt and widely encompassing changes between the Early and Late Permian tetrapod assemblages, terrestrial and aquatic, suggest a global hiatus in the terrestrial fossil record (Boy 1987). In eastern Europe several groups of terrestrial amphibians (eryopoids, dissorophoids) are accompanied by very few aquatic amphibians (e.g., gavial-like archegosaurids). In addition some more or less aquatic families appeared for the first time (melosaurids, ianthanosaurids, dvinosaurids). The first stereospondylous amphibians (alligator-like amphibious-to-aquatic forms) appear in Gondwana (South Africa, India, Australia) and in European and Asian parts of the USSR.

Late Permian tetrapod assemblages are characterized by a high diversity of fully terrestrial diapsids and synapsids (Romer 1973) and a significant reduction in terrestrial amphibians. Included are unquestionable herbivores in large numbers, as well as omnivorous, insectivorous, and carnivorous forms, all abundant and diverse. Tetrapod herbivores occur phylogenetically mainly among the advanced mammal-like reptiles (therapsids), and probably among procolophonids and pareiasaurids. In some South African assemblages, dicynodont therapsids outnumber predatory taxa 10 to 1 (Bakker 1975b), a marked contrast with Early Permian assemblages. Dicynodont head-body size ratios suggest high ingestion rates and an ability to comminute food finely, implying high rates of digestion. Dicynodonts also show evidence of complex behavior, including burrowing (Smith 1987), and of an increased scope of activities. The radiation in herbivores is accompanied by increasing diversity in predators and in reptile ecological diversity. Terrestrial predators such as the gorgonopsids were very vagile, slender forms with high levels of activity (Hotton 1980; Kemp 1982; Sues 1986). The first gliding reptile, *Coelurosauravus*, is recorded from near-shore marine deposits of Europe and Madagascar (Evans and Haubold 1987). These changes indicate the establishment of complex interactions among tetrapods, and between tetrapods and plants. As tetrapods joined insects as herbivores with sizable impact on plant communities, ecosystems reached the spectrum of dynamic interactions that have characterized them since that time.

The dicynodont therapsids formed the first major radiation of herbivorous tetrapods and were by far the most abundant terrestrial vertebrates during the Late Permian (King 1981, 1988; Hotton 1986). They first appeared in lower Upper Permian rocks of South Africa. Dicynodonts left an especially rich fossil record in the Karoo Supergroup of southern and eastern Africa but also have been found in the Upper Permian of Brazil, China, India, Scotland, and Russia (King 1988). They survived the end-Permian extinction, but their early Mesozoic diversity was greatly reduced, and they became extinct in the latter half of the Triassic. The Permian genera range in basal skull length from 50 mm (*Emydops*) to greater than 50 cm (*Rhachiocephalus*) (King 1988). Dicynodonts were quadrupedal animals with short, robust limbs and probably foraged within one meter of the ground, and all shared a stereotyped complex masticatory apparatus that presumably enabled them to be effective herbivores. The skull was rigidly constructed, and the mandibular symphysis is fused. Except for upper canines, the snout was typically edentuous. The rostral ends of both the upper and lower jaws were covered with a horny beak with sharp margins between which food was sliced when the mandible was retracted. Mandibular retraction was facilitated by the laterally widely exposed external adductor mandibulae muscle, and a highly modified jaw joint permitted extensive anteroposterior translation of the mandible relative to the cranium (Crompton and Hotton 1967; King 1981). An elongate trunk, which was posteriorly wide in some forms, and broad pelvic region presumably reflect development of a capacious gut.

Recent work has provided evidence for a considerable range of foraging modes among dicynodonts (Hotton 1986; Smith 1987). *Cistecephalus* and *Kawingasaurus* had distinctive, boxlike skulls and highly modified limbs with structural specializations that are very similar to those in fossorial mammals such as chrysochlorids (golden moles). These small dicynodonts were probably mole-like burrowers (Cluver 1978). Hotton (1986) interprets *Diictodon* as a terrestrial grubber with access to subterranean parts of plants because of its long, clawlike ungual phalanges and the shape of the rostral end of the snout, which he relates to shearing roots and rhizomes; this genus has been found curled up within helical burrows (Smith 1987). Larger dicynodonts, such as *Oudenodon,* may have been browsers that fed on shrublike vegetation, including equisetaleans, pteridosperms with *Glossopteris*-type foliage, ferns, and perhaps low-statured conifers (Hotton 1986).

The Dinocephalia are a group of primitive therapsids of early Late Permian age that includes both herbivorous and carnivorous forms. They were up to 3 or 4 m long, massively built animals, often with considerably thickened bones; they are characterized especially by their peculiar, interlocking incisor teeth (Chudinov 1983; King 1988). Dinocephalians are known from the Dinocephalian Assemblage Zone of the Beaufort Group in South Africa and from the Upper Permian Zones I and II of European Russia (Chudinov 1983). On the basis of body form and taphonomic data, it is inferred that the Tapinocephalidae had a hippopotamus-like, semiaquatic mode of life (Boonstra 1955). They are generally considered to have been herbivores. The dentition consists of numerous chisel-like teeth without clearly differentiated canines. The Anteosauridae, Titanosuchidae, and Brithopodidae were more gracile in build, and judging from the marginal dentition with long canines and robust, conical postcanine teeth, they were predatory forms (Kemp 1982).

Late Permian herbivores also included the Pareiasauria and Procolophonia, both primitive amniote groups of uncertain affinities. Pareiasaurs were large (up to 3 m long), ponderous forms with sprawling limbs; they are known mostly from the Upper Permian of South Africa and Russia. The marginal dentition is composed of relatively simple, laterally compressed, leaf-shaped teeth (Ivakhnenko 1987). The massive trunk region indicates the development of a voluminous gut. Procolophonians were relatively small animals with solidly built skulls and vaguely lizardlike appearance, with short but robust limbs. The orbital margin is embayed posteriorly to accommodate the anterior portion of the extensive jaw adductor musculature. In adult specimens, the more posterior marginal teeth became distinctly expanded transversely. Tooth wear was extensive, producing extensive occlusal surfaces on each tooth in larger species (Ivakhnenko 1979; Carroll and Lindsay 1985). Gow (1978) has argued that only the larger forms had an exclusively herbivorous mode of life and considered the smaller species and perhaps younger individuals of the larger forms mostly insectivorous.

The Gorgonopsida were the top predators in terrestrial environments during the Late Permian (Russell-Sigogneau 1989). The rigidly built, large skulls of these primitive therapsids bore enormous bladelike canines with serrated edges and robust, interlocking incisor teeth (Kemp 1969). The postcanine dentition was greatly reduced. The temporal area for the origin of the jaw adductor musculature was considerably expanded. The limbs are relatively slender and carried the trunk well above the ground (Kemp 1982). A dicynodont jaw has been found between the ribs of at least one gorgonopsian skeleton (Huene 1950).

Another group of carnivorous therapsids is represented by the Therocephalia, which include both robust, short-limbed forms such as the large *Pristerognathus* and smaller, possibly insectivorous forms such as *Ictidosuchoides* (Kemp 1982).

7.4 Synthesis

The Permian terrestrial fossil record differs from that of the Carboniferous in ways that complicate direct comparisons. Two major kinds of biotas overlap in tropical latitudes: that of poorly drained wetlands, characterized by high levels of moisture availability throughout the year, and that of better-drained, probably seasonally dry habitats. These biotas were part of a continuum; the end members manifest dramatic floristic differences, but there are few distinctions among the tetrapod faunas. The Early Permian fossil record documents increasing prominence of the drier-site plant assemblages in lowland depositional environments, a probable indication of shrinking habitat areas for biotas restricted to wetter conditions. As a consequence of this rise in habitat and biotic variability and a proportional increase of difficulty in reconstructing patterns across the landscape, we rely heavily upon sequences that were either poorly drained or that were at least seasonally dry in order to isolate the end members. A combination of complex ecosystem patterns, paleobiogeographic variation, and in some cases inadequate data complicates ecological inferences.

Havlena (1970) subdivides Permo-Carboniferous tropical floras into a hierarchy of progressively more restricted ecological units. The most geographically encompassing units are the "close-to-the-sea" and "intermontane" biomes or ecotopes. The close-to-the-sea biome is far better known and accounts for nearly all of the plant fossil record. Within each of these biomes are floras from predominantly aggradational ("basinal") environments, including swamps, and from predominantly erosional, better-drained ("extrabasinal") environments; the former are much better preserved and hence provide most of the lowland plant fossil record. The character of these aggradational lowlands changed as seasonal dryness in the lowlands became more widespread; in effect, the lowlands of the Carboniferous were also "wetlands," whereas the Permian lowlands became more seasonally dry in many areas.

The general pattern that began in the Stephanian and continued into the Permian is the movement of floras from predominantly erosional or better-drained portions of the landscape into the more mesic-to-hydric aggradational portions of lowland areas. Whether this pattern of species migration can be extended to the larger biomes, with movement from intermontane areas into the close-to- the-sea habitats, is not certain but is suspected for groups such as conifers, which have a scant but definitive record prior to the Stephanian, out of range of all but the most exceptional preservation (Scott and Chaloner 1983).

Throughout most of the Permo-Carboniferous, the vertebrate record does not offer clear evidence for such a mechanism of faunal change. The unexpected appearance of several amniote groups in the late Leonardian and early Guadalupian, however, has been attributed to a pattern of community evolution similar to that described for the plants (Olson 1962a).

Plant-Animal Interactions

The complexity, regional variation, and temporal fluctuation in environmental conditions within Early Permian lowlands suggests a mosaic of ecosystem patterns. Included was the persistence of Carboniferous-type plant-animal interactions and animal food-chain organization in the most stable wetland areas, particularly those of the earliest Permian. In general, however, there were major changes in ecosystem dynamics that began in the Early Permian, most notably the rise of tetrapod herbivory, that led to nearly complete modernization of basic terrestrial ecosystem dynamics by the Late Permian.

Sometime during the late Wolfcampian (middle Early Permian), seed plants became dominant in all areas but those with moderately wet soils. The high diversity of seed plants, particularly a great array of "cycadophytic" forms along with the conifers, suggests further floristic subdivision of better-drained lowland areas than was common in earlier Permo-Carboniferous times. Most of the plants were xeromorphic, even armored in the seasonally driest areas, and would have presented animals with a formidable problem in processing and digestion. As long-lived trees, woody conifers and many of the cycadophytes were much more K-selected than dominants of earlier floras, their more r-selected ancestors. They had greater investment in relatively expensive woody tissues and thick evergreen foliage than did lower vascular plants such as tree ferns. For early herbivores, with limited capacity to reach elevated plant parts, much of the primary productivity of these trees would have been inaccessible. It is unlikely that tetrapod herbivory was a significant selective force in shaping the morphologies of Early Permian trees.

In contrast, plant-insect interactions appear to have modernized considerably during the Early Permian. Little direct evidence has been presented, but the appearance of major groups such as beetles suggests expanded herbivory. The main impact of insects may have been as small, highly specialized herbivores that had a selective effect on plant micromorphology, and possibly even on population structure. The degree to which arthropod detritivory remained dominant in the Early Permian requires investigation, particularly in drier habitats. Detritivory undoubtedly remained important, although insect herbivores may have begun to displace detritivores as the main entry point of terrestrial-plant productivity into animal food webs.

Adaptations for tetrapod herbivory and omnivory were conspicuous in the Early Permian. The impact of herbivores and omnivores, given their limited reach, must have been mostly on ground-cover ferns, sphenopsids and small pteridosperms, shrubby seed plants, and young plants of all types. Folivory may have evolved independently in at least four tetrapod lineages during this time, and in at least another four between the mid-Permian and the Triassic. This pattern suggests responses, probably in the Leonardian, to new ecological opportunities, possibly the spread of xeromorphic, shrubby vegetation with energy-rich leaves, a general increase in nutrient availability, and perhaps distinct growing seasons.

Plant-animal integration within Early Permian ecosystems was increasing, but still reflected mostly plant-insect interaction, through herbivory and possibly pollination. Arthropods still appear to have been the link between plants and vertebrates within the larger ecosystem. This is reflected in the continued dominance of tetrapod assemblages by carnivores and a variety of insectivores into the middle Leonardian, despite the emergence of vertebrate herbivory. The available evidence for most animal groups indicates increased overlap and specialization within ecological roles first defined in the Late Carboniferous. To the extent that tetrapods depended on insectivory and carnivory rather than herbivory throughout most of the Early Permian, tetrapod food chains retained a primitive character.

Early Permian ecosystem structure and dynamics were clearly intermediate between those of earlier Paleozoic and Mesozoic ecosystems. Floras show increased physiognomic and taxonomic affinities with the Mesozoic, whereas general trophic interactions remain Paleozoic in aspect. The evolution of tetrapod herbivory in the late Early Permian opened the possibility for radical changes in the pattern of energy flow to higher trophic levels. The major effect of vertebrate herbivory was to shorten the length of food chains and thus to increase the amount of energy from plants available to higher-level consumers. As the sophistication of herbivory increased, herbivores undoubtedly became major selective forces that shaped plant phenotypes and community structure. Judging from the extent of vertebrate herbivory in Guadalupian deposits (Late Permian), it appears that ecosystems had undergone significant modernization before the Late Permian, and that plants and animals were linked in a more singular set of interdependent dynamics based on plant primary productivity.

8   PALEOZOIC SUMMARY AND DISCUSSION

The Paleozoic was the time during which terrestrial ecosystems were organized and assembled. The plants, invertebrates, and tetrapods can be thought of as three separate subsystems of the larger terrestrial ecosystem. Dominance-diversity patterns, structure, and interactions within these subsystems modernized more or less independently and at different rates throughout most of the Paleozoic. Concomitantly, these three separate subsystems became integrated through trophic interactions and the creation or modification of habitat conditions. This partitioning of ecosystems into "phylogenetic" subunits differs radically from the organization of highly integrated modern systems.

The appearance of structurally and ecomorphically modern plant communities in some parts of the landscape primarily during the Late Devonian and Early Carboniferous included the evolution of life histories, particularly seed habit, that permitted the exploitation of a wide variety of terrestrial environments. The independent evolution of arborescence in nearly all major groups of plants was key to the development of multistratal forests in some habitats at this time. By the Late Carboniferous, complex forest structure and a wide variety of forest profiles and dynamics were clearly in existence across all major land masses.

Global provincialization of vegetation began during the Early Carboniferous and increased more or less continuously until the Permian, when it reached its Paleozoic maximum; this was perhaps the greatest provincialization known prior to the later Tertiary. Such provincialism was made possible by several factors. The physical opportunities increased in diversity and abundance during the Paleozoic as environments changed globally. Because the plants were at the base of the major terrestrial radiation, they evolved increasing tolerances to demanding physical conditions and expanded the extent of the vegetated land surface. This expansion into unoccupied parts of the terrestrial environment was accompanied by an increase in local and regional vegetational heterogeneity, a consequence of local variation in both physical conditions and evolutionary dynamics.

The first animal groups to exploit plant productivity were the arthropods, and then primarily as detritivores. Detritivory was the main entry point of plant primary productivity into animal food webs until the end of the Early Permian. Herbivory, largely by insects, is known from the Early Carboniferous and may have existed in the Devonian. Herbivory expanded rapidly during the Late Carboniferous and was an established part of ecosystem dynamics by the Stephanian or possibly late Westphalian and into the Early Permian. As suggested by the large size of medullosan pollen, insect pollination probably had evolved by the Westphalian. The selective effects of arthropods on plants may be seen in the evolution of sclerotic seed coats and other structural features that would have served to deter post- and predispersal predation. Clear evidence of damage to living foliage does not appear until the later Westphalian.

Tetrapods were the final element to enter the web of interactions in terrestrial communities. They began, and continued, as carnivores and insectivores through the Carboniferous and Early Permian, situated at the ends of long food chains that were mediated primarily by arthropod detritivory. Although predation on insect herbivores would have shortened some of these pathways and added trophic links to the system, it was not until the expansion of tetrapod herbivory in the late Early Permian that a full integration of the three subsystems was established. The integration had progressed greatly by the Late Permian. However, even by the end of the Paleozoic, the tetrapod component of ecosystems was distinctly different in the sizes, modes of feeding, and range of resources exploited from Mesozoic and younger ecosystems.

8.1   Ecosystem Stability and Persistence

The fossil record affords an opportunity to examine ecological patterns over extended periods of time, periods much greater than those accessible to any neoecological studies. Although there are flaws in this record that limit resolution and the kinds of questions that can be addressed, basic patterns are evident.

The evolution of terrestrial ecosystems during the last half of the Paleozoic encompasses 150 to perhaps as much as 200 My. The modernization of terrestrial communities that occurred during this time was neither rapid nor an unbroken continuum. Evolutionary innovation played a major role in the process, modified by global changes in environmental conditions and by dramatic changes in the interactions among organisms on the land surface. In addition, as ecosystems became increasingly complex, the complexity itself appears to have had a role in constraining the kinds of structural changes that could be further accommodated, and hence the direction of evolutionary events.

The times of greatest change in ecosystem structure were the Late Devonian and the earliest Early Carboniferous, and the transition from the Carboniferous to the Permian. During the first of these, most major architectural and life-history patterns evolved within plants. These patterns were expressed strongly along taxonomic lines and generally reflect what have been described as classes; as a consequence ecological resources appear to have been partitioned mainly along taxonomic lines. This ecotaxonomic partitioning established a framework for subsequent evolutionary events throughout the Carboniferous and into the Permian. There unquestionably were changes in the taxonomic composition of terrestrial communities during the Carboniferous, most notably the rise of a peat-forming, wetlands vegetation, the evolution of insect diversity, and the appearance of amniotes. Despite these major changes in the constituent plant and animal groups, the basic biotic partitioning of the landscape and its resources remained largely the same throughout the interval.

At a detailed level, the Euramerican tropical biota of the Late Carboniferous graphically illustrates that ecosystems are fundamentally persistent entities if extrinsic conditions remain relatively unchanged. Peat-swamp floras underwent four major changes during this time interval, all on a basic theme that reflects changes in dominance-diversity structure more than fundamental taxonomic change in the vegetation. Profound change in the basic biotic composition of the wet lowlands appears to be driven by climatic change. The effect of evolutionary novelties on the structure of communities, at least as a primary driving force of change, appears to have been minimal. However, by all abiotic indicators, changes in physical conditions were much more gradual than were resultant changes in ecosystems, suggesting that the systems may have had organizational properties that engendered persistence and maintenance of biotic interactions and dynamics until thresholds were crossed. At such times biotic change occurred rapidly. The record indicates that most responses were migrational, but they also may have involved extinction and the establishment of novel phenotypes.

The changes in ecosystem dynamics that occurred during the Permian appear, at least superficially, to be more complex than those of the Devonian-Early Carboniferous. Complex ecosystems were already established, global biotic provincialism was developing or had developed in many groups of organisms, and global climates were influenced by the formation of Pangaea. Changes in plant communities during this time reflect a combination of regional extinction, the loss of many of the dominant lower vascular plants of wetlands, and the rise to dominance of seed plants in most kinds of habitats. This was the final breakdown of a strong, high-level taxonomic overprint on ecological partitioning. Despite the complexity of the global pattern, physical factors appear again to have had preeminence over biotic factors as the ultimate driving force of most ecological change.

During the floristic changes that accompanied the Permo-Carboniferous transition and the changes in faunal dominance-diversity patterns that accompanied them, tetrapod herbivory appeared. As a profound change in the means of energy transfer from plants to animal food webs, tetrapod herbivory may be one of the best examples of how an evolutionary innovation can modify some aspects of otherwise stable ecosystems. The rarity of such events, the restriction of their rise and ecological fixation to relatively narrow intervals of time, and more importantly, their nearly simultaneous occurrence in more than one phylogenetic lineage comprise a pattern repeated for critical adaptations in both plants and animals and one that characterizes both major times of ecosystem change in the Paleozoic.

8.2 Conclusions

1. The assembly of terrestrial ecosystems during the Paleozoic to a large extent dictated the direction of subsequent ecological and evolutionary events and the kinds of interactions possible between plants, invertebrates, and vertebrates. As with species-level evolution, historical constraints play an important role in understanding patterns of ecosystem evolution. The Markovian nature of evolutionary processes insures the influence of Paleozoic events on all younger systems.
2. The Recent is a poor analogue for nearly all aspects of Paleozoic ecosystems except for those of the latest Permian. The lack of tetrapod herbivory, the narrow spectrum of plant-insect interactions, the importance of detritivory as the base of the food chain, and the strong partitioning of ecological resource space along widely divergent phylogenetic lines in plants are themes that run throughout most of the Paleozoic. Because of these and other fundamental differences, ecological models based on the present cannot be applied to Paleozoic examples in a uniformitarian manner.
3. The generality of ecological and evolutionary principles can be tested through attempts to apply them to Paleozoic ecosystems. These principles include persistence, the relationship between complexity and stability, the responses of ecosystems to large-scale extrinsic stresses, and the effects of local disturbances on such processes as succession or persistence. The great differences between the Paleozoic and the Recent in taxonomic composition and in the basic biotic interactions within the ecosystems present us with a unique source of data to use in refining ecological concepts and testing their generality.

REFERENCES

Allen, J. R. L. 1979. Old Red Sandstone facies in external basins, with particular reference to southern Britain. In M. R. House, C. T. Scrutton, and M. G. Bassett, eds, *The Devonian System*. Palaeontological Association, Special Papers in Palaeontology 23:65–80.

Almond, J. E. 1985. The Silurian-Devonian fossil record of the Myriapoda. *Philosophical Transactions of the Royal Society*, B 309: 227–37.

Andrews, H. N., A. E. Kasper, W. H. Forbes, P. G. Gensel, and W. G. Chaloner. 1977. Early Devonian flora of the Trout Valley Formation of northern Maine. *Review of Palaeobotany and Palynology* 23:255–85.

Archangelsky, S. 1986. Late Paleozoic floras of the Southern Hemisphere: Distribution, composition, and paleoecology. *University of Tennessee, Department of Geological Sciences, Studies in Geology* 15:128–42.

Arnold, C. A. 1941. Some Paleozoic plants from central Colorado and their stratigraphic significance. *University of Michigan, Museum of Paleontology, Contributions* 6:59–70.

———. 1960. A lepidodendrid stem from Kansas and its bearing on the problem of cambium and phloem in Paleozoic lycopods. *Contributions, Museum of Paleontology, University of Michigan* 15:249–67.

Asama, K. 1985. Permian to Triassic floral change and some problems of the paleobiogeography, parallelism, mixed floras, and the origin of the angiosperms. 199–218 in K. Nakazawa and J. M. Dickens, eds., *The Tethys*. Tokyo University Press.

Ash, S., and W. D. Tidwell. 1982. Notes on the Paleozoic plants of central New Mexico. 245–48 in *New Mexico Geological Survey Guidebook, Thirty-third Field Conference, Albuquerque Country II*.

Axelrod, D. I. 1967. Drought, diastophism, and quantum evolution. *Evolution* 21:201–9.

———. 1972. Edaphic aridity as a factor in angiosperm evolution. *American Naturalist* 106:311–20.

Baird, D. 1965. Paleozoic lepospondyl amphibians. *American Zoologist* 5:287–94.

Baird, D., and R. L. Carroll. 1967. *Captorhinus* vs. *Hypopnous* (Reptilia, Captorhinomorpha). *Journal of Paleontology* 24:264–65.

Bakker, R. T. 1975a. Dinosaur renaissance. *Scientific American* 232:58–78.

———. 1975b. Experimental and fossil evidence for the evolution of tetrapod bioenergetics. 365–99 in D. M. Gates and R. B. Schmerl, eds., *Perspectives of Biophysical Ecology*. New York: Springer-Verlag.

———. 1980. Dinosaur heresy—dinosaur renaissance: Why we need endothermic archosaurs for a comprehensive theory of bioenergetic evolution. 351–462 in R. D. K. Thomas and E. C. Olson, eds., *A Cold Look at the Warm-Blooded Dinosaurs*. Boulder, Colo.: Westview Press.

———. 1982. Juvenile-adult habitat shift in Permian fossil reptiles and amphibians. *Science* 217:53–55.

Banks, H. P. 1966. Devonian floras of New York State. *Empire State Geogram, New York Geological Survey* 4:10–24.

———. 1980. Floral assemblages in the Siluro-Devonian. 1–24 in D. L. Dilcher and T. N. Taylor, eds., *Biostratigraphy of Fossil Plants*. Stroudsburg, Pa.: Dowden, Hutchinson and Ross.

———. 1981. Peridermal activity (wound repair) in an Early Devonian (Emsian) trimerophyte from the Gaspé Peninsula, Canada. *Palaeobotanist* 28/29:20–25.

———. 1985. Early land plants. *Philosophical Transactions of the Royal Society of London* B 309:197–200.

Barghusen, H. R. 1973. The adductor jaw musculature of *Dimetrodon* (Reptilia, Pelycosauria). *Journal of Paleontology* 47:823–34.

Barlow, J. A., ed. 1975. *The Age of the Dunkard*. Morgantown: West Virginia Geological and Economic Survey.

Bartram, K. M., A. J. Jeram, and P. A. Selden. 1987. Arthropod cuticles in coal. *Journal of the Geological Society* 144:513–17.

Bateman, R. M. 1991. Paleoecology. 34–116 in C. J. Cleal, ed., *Plant Fossils in Geological Investigation: The Paleozoic*. Chichester: Ellis Horwood.

Bateman, R. M., and W. A. DiMichele. 1991. *Hizemodendron*, gen. nov., a pseudo-herbaceous segregate of *Lepidodendron* (Pennsylvanian): Phylogenetic context for evolutionary changes in lycopsid growth architecture. *Systematic Botany* 16:195–205.

Bateman, R. M., and G. W. Rothwell. 1990. A reappraisal of the Dinantian floras at Oxroad Bay, East Lothian, Scotland. 1. Floristics and development of whole-plant concepts. *Transactions of the Royal Society of Edinburgh* 81:127–59.

Bateman, R. M., and A. C. Scott. 1990. A reappraisal of the Dinantian floras at Oxroad Bay, East Lothian, Scotland. 2. Volcanicity, palaeoenvironments, and palaeoecology. *Transactions of the Royal Society of Edinburgh* 81:161–94.

Batenburg, L. H. 1982. "Compression species" and "petrifaction species" of *Sphenophyllum* compared. *Review of Palaeobotany and Palynology* 36:335–59.

Baxter, R. W., and E. A. Roth. 1954. *Cardiocarpus magnicellularis* sp. nov., a preliminary report. *Transactions of the Kansas Academy of Science* 57:458–60.

Beck, C. B. 1964. Predominance of *Archaeopteris* in Upper Devonian flora of western Catskills and adjacent Pennsylvania. *Botanical Gazette* 125:126–28.

Beck, C. B., K. Coy and R. Schmid. 1982. Observations on the fine structure of *Callixylon* wood. *American Journal of Botany* 69:54–76.

Beeler, H. E., and S. E. Scheckler. 1983. A new Upper Devonian tree lycopod. *American Journal of Botany* 70(5/2):67–68.

Beerbower, J. R. 1985. Early development of continental ecosystems. 47–91 in B. H. Tiffney, ed., *Geological Factors and the Evolution of Plants*. New Haven: Yale University Press.

Berman, D. S 1978. *Ctenospondylus ninevehensis,* a new species (Reptilia, Pelycosauria) from the Lower Permian Dunkard Group of Ohio. *Annals of the Carnegie Museum* 47:493–514.

Berman, D. S, and R. R. Reisz. 1982. Restudy of *Mycterosaurus longiceps* (Reptilia, Pelycosauria) from the Lower Permian of Texas. *Annals of the Carnegie Museum* 51:423–53.

Berman, D. S, and S. S. Sumida. 1990. A new species of *Limnoscelis* (Amphibia, Diadectomorpha) from the Late Pennsylvanian Sangre de Cristo Formation of Central Colorado. *Annals of the Carnegie Museum* 59:303–41.

Berman, D. S, D. A. Eberth, and D. B. Brinkman. 1988. *Stegotretus agyrus,* a new genus and species of microsaur (amphibian) from the Permo-Pennsylvanian of New Mexico. *Annals of the Carnegie Museum* 57:293–323.

Berman, D. S, R. R. Reisz, and M. A. Fracasso. 1981. Skull of the Lower Permian dissorophid amphibian *Platyhystrix rugosus. Annals of the Carnegie Museum* 50:391–416.

Berman, D. S, R. R. Reisz, and D. A. Eberth. 1987a. A new genus and species of trematopid amphibian from the Late Pennsylvanian of north-central New Mexico. *Journal of Vertebrate Paleontology* 7:252–69.

———. 1987b. *Seymouria sanjuanensis* (Amphibia, Batrachosauria) from the Lower Permian Cutler Formation of north-central New Mexico and the occurrence of sexual dimorphism in that genus questioned. *Canadian Journal of Earth Sciences* 24:1769–84.

Besly, B. M., and C. R. Fielding. 1989. Paleosols in Westphalian coal-bearing and red-bed sequences, central and northern England. *Palaeogeography, Palaeoclimatology, and Palaeoecology* 70:303–30.

Blazey, E. B. 1974. Fossil flora of the Mogollon Rim, central Arizona. *Palaeontographica*, Abt. B. 146:1–20.

Bode, H. 1975. The stratigraphic position of the Dunkard. 143–54 in J. A. Barlow, ed., *The Age of the Dunkard*. Morgantown: West Virginia Geological and Economic Survey.

Boersma, M. 1978. A survey of the fossil floras of the "Illinger Flözzone" ("Huesweiler Schichten," Lower Stephanian, Saar, German Federal Republic). *Review of Palaeobotany and Palynology* 26:41–92.

Bolt, J. R. 1974a. Armor of dissorophids (Amphibia: Labyrinthodontia): An examination of its taxonomic use and report of a new occurrence. *Journal of Paleontology* 48:135–42.

———. 1974b. Osteology, function, and evolution of the trematopsid (Amphibia: Labyrinthodontia) nasal region. *Fieldiana, Geology* 33:11–30.

———. 1977. Dissorophoid relationships in ontogeny, and the origin of the Lissamphibia. *Journal of Paleontology* 51:235–49.

———. 1979. *Amphibamus grandiceps* as a juvenile dissorophid: Evidence and implications. 529–63 in M. H. Nitecki, ed., *Mazon Creek Fossils*. New York: Academic Press.

———. 1980. New tetrapods with bicuspid teeth from the Fort Sill locality (Lower Permian, Oklahoma). *Neues Jahrbuch für Geologie und Palaeontologie, Monatshefte* 1980:449–59.

Bolt, J. R., and R. E. DeMar. 1975. An explanatory model of the evolution of multiple rows of teeth in *Captorhinus aguti*. *Journal of Paleontology* 49:814–32.

———. 1983. Simultaneous tooth replacement in *Euryodus* and *Cardiocephalus* (Amphibia: Microsauria). *Journal of Paleontology* 57:911–23.

Bolt, J. R., and R. E. Lombard. 1985. Evolution of the amphibian tympanic ear and the origin of frogs. *Biological Journal of the Linnean Society* 24:83–99.

Bolt, J. R., R. M. McKay, B. J. Witzke, and M. P. McAdams. 1988. A new Lower Carboniferous tetrapod locality in Iowa. *Nature* 333:768–70.

Boonstra, L. D. 1955. The girdles and limbs of the South African Deinocephalia. *Annals of the South African Museum* 42:185–326.

Boucot, A. J., and J. Gray. 1983. A Paleozoic Pangaea. *Science* 222:571–81.

Boy, J. A. 1977. Typen und Genese jungpaläozoischer Tetrapoden-Lagerstätten. *Palaeontographica*, Abt. A, 156:111–67.

———. 1987. Die Tetrapoden-Lokalitäten des saarpfälzischen Rotliegenden (?Ober-Karbon–Unter-Perm; SW-Deutschland) and die Biostratigraphie der Rotliegend-Tetrapoden. *Mainzer Geowissenschaftliche Mitteilungen* 16:31–65.

———. 1988. Über einige Vertreter der Eryopoidea (Amphibia: Temnospondyli) aus dem europäischen Rotliegend (?höchstes Karbon-Perm). 1. *Sclerocephalus*. *Paläontologische Zeitschrift* 62:107–32.

———. 1990. Über einige Vertreter der Eryopoidea (Amphibia: Temnospondyli) aus dem europäischen Rotliegend (?höchstes Karbon-Perm). 3. *Onchiodon*. *Paläontologische Zeitschrift* 64:287–312.

Boy, J. A., and K. Bandel. 1973. *Bruktererpeton fiebigi* n. gen. n. sp. (Amphibia, Gephyrostegida): Der erste Tetrapode aus dem Rheinisch-Westfälischen Karbon (Namur B; W.-Deutschland). *Palaeontographica* A145: 39–77.

Boy, J. A., and J. Fichter. 1982. Zur Stratigraphie des saarpfälzischen Rotliegenden

(?Ober-Karbon–Unter-Perm; SW Deutschland). *Zeitschrift der deutschen geologischen Gesellschaft* 133:607–42.

———. 1988. Zur Stratigraphie des höheren Rotliegend im Saar-Nahe-Becken (Unter-Perm; SW Deutschland) und seiner Korrelation mit anderen Gebieten. *Neues Jahrbuch für Geologie und Paläontologie Abhandlungen* 176:331–94.

Boy, J. A., D. Meckert and T. Schindler. 1990. Probleme der lithostratigraphischen Gliederung im unteren Rotliegend des Saar-Nahe-Beckens (?Ober-Karbon–Unter-Perm; SW-Deutschland). *Mainzer geowissenschaftliche Mitteilungen* 19:99–116.

Boyd, M. J. 1984. The Upper Carboniferous tetrapod assemblage from Newsham, Northumberland. *Palaeontology* 27:367–92.

Bramwell, C. D., and P. B. Fellgett. 1973. Thermal regulation in sail lizards. *Nature* 242:203–5.

Bridge, J. S., P. M. Van Veen, and L. C. Matten. 1980. Aspects of sedimentology, palynology, and palaeobotany of the Upper Devonian of southern Kerry Head, Co. Kerry, Ireland. *Geological Journal* 15:143–70.

Brinkman, D. 1988. Size-independent criteria for estimating relative age in *Ophiacodon* and *Dimetrodon* (Reptilia, Pelycosauria) from the Admiral and lower Belle Plains Formations of west-central Texas. *Journal of Vertebrate Paleontology* 8:172–80.

Brinkman, D., and D. A. Eberth. 1983. The interrelationships of pelycosaurs. *Breviora* 473:1–35.

———. 1986. The anatomy and relationships of *Stereophallodon* and *Baldwinonus* (Reptilia, Pelycosauria). *Breviora*. 485:1–34.

Broutin, J. 1981. Etude paléobotanique et palynologique du passage Carbonifère-Permien dans les bassins continentaux du Sud-Est de la zone d'Ossa-Morena (environs de Guadalcanal, Espagne du Sud). Implications paléogeographiques et stratigraphiques. Thesis, Université Pierre et Marie Curie, Paris.

Broutin, J., J. Doubinger, G. Farjanel, P. Freytet, H. Kerp, J. Langiaux, M.-L. Lebreton, S. Sebban, and S. Satta. 1990. Le renouvellement des flores au passage Carbonifère Permien: Approaches stratigraphique, biologique, sédimentologique. *Comptes Rendues, Académie des Sciences Paris* 311:1563–69.

Brzyski, B., R. Gradzinski, and R. Krzanowska. 1976. Upright calamite stems from Brynow and conditions of their burial (in Polish with an English summary). *Annales de la Société Géologique de Pologne* 46:159–82.

Burnham, L. 1983. Studies on Upper Carboniferous insects: 1. The Geraridae (Order Protorthoptera). *Psyche* 90:1–57.

Burnham, R. J. 1989. Relationships between standing vegetation and leaf litter in a paratropical forest: Implications for paleobotany. *Review of Palaeobotany and Palynology* 58:5–32.

———. 1990. Some Late Eocene depositional environments of the coal-bearing Puget Group of western Washington State, USA. *International Journal of Coal Geology* 15:27–57.

Canright, J. E., and E. B. Blazey. 1974. A Lower Permian flora from Promontory Butte, central Arizona. 57–62 in S. R. Ash, ed., *Guidebook to Devonian, Permian, and Triassic plant localities, east-central Arizona*. Paleobotanical Section of the Botanical Society of America, Field Trip in conjunction with twenty-fifth Annual A.I.B.S. Meeting, Tempe, Arizona.

Carpenter, F. M. 1960. Studies on North American Carboniferous insects. 1. The Protodonata. *Psyche* 67:98–110.

———. 1971. Adaptations among Palaeozoic insects. *Proceedings of the North American Palaeontological Convention (1969)* 1:1236–51.

Carroll, R. L. 1964a. Early evolution of the dissorophid amphibians. *Bulletin of the Museum of Comparative Zoology, Harvard University* 131:161–250.

———. 1964b. The relationships of the rhachitomous amphibian *Parioxys*. *American Museum Novitates* 2167:1–11.

———. 1966. Microsaurs from the Westphalian B of Joggins, Nova Scotia. *Proceedings of the Linnean Society* 177:63–97.

———. 1967a. Labyrinthodonts from the Joggins Formation. *Journal of Paleontology* 41:111–42.

———. 1967b. A limnoscelid reptile from the Middle Pennsylvanian. *Journal of Paleontology* 41:1256–61.

———. 1968. The post-cranial skeleton of the Permian microsaur *Pantylus*. *Canadian Journal of Zoology* 46:1175–92.

———. 1970. The ancestry of reptiles. *Philosophical Transactions of the Royal Society of London* B, 257:267–308.

———. 1984. Problems in the use of terrestrial vertebrates for zoning of the Carboniferous. *Ninth International Congress, Carboniferous Stratigraphy and Geology, C.R.* 2:135–47.

Carroll, R. L., and D. Baird. 1968. The Carboniferous amphibian *Tuditanus* (*Eosauravus*) and the distinction between microsaurs and reptiles. *American Museum Novitates* 2337:1–50.

———. 1972. Carboniferous stem-reptiles of the family Romeriidae. *Bulletin of the Museum of Comparative Zoology, Harvard University* 143:321–64.

Carroll, R. L., and P. Gaskill. 1978. The order Microsauria. *American Philosophical Society Memoirs* 126:1–211.

Carroll, R. L., and W. Lindsay. 1985. The cranial anatomy of the primitive reptile *Procolophon*. *Canadian Journal of Earth Sciences* 22:1571–87.

Carroll, R. L., E. S. Belt, D. L. Dineley, D. Baird, and D. C. McGregor. 1972. Vertebrate paleontology of eastern Canada. *Excursion A59, Twenty-fourth International Geological Congress, Montreal*, 1–113.

Cecil, C. B. 1990. Paleoclimate controls on stratigraphic repetition of chemical and siliciclastic rocks. *Geology* 18:533–36.

Cecil, C. B., and K. J. England. 1989. Origin of coal deposits and associated rocks in the Carboniferous of the Appalachian Basin. In C. B. Cecil, et al., eds., *Carboniferous Geology of the Eastern United States. Twenty-Eighth International Geological Congress, Field Trip Guidebook T143:* 84–104.

Cecil, C. B., R. W. Stanton, S. G. Neuzil, F. T. Dulong, L. F. Ruppert, and B. S. Pierce. 1985. Paleoclimatic controls on Late Paleozoic sedimentation and peat formation in the central Appalachian Basin (USA). *International Journal of Coal Geology* 5:195–230.

Chaloner, W. G., and M. J. Cope. 1982. Interaction of plant evolution, wildfire, atmospheric composition, and climate. *Third North American Paleontological Convention, Proceedings* 1:83–85.

Chaloner, W. G., and W. S. Lacey. 1973. The distribution of Late Paleozoic floras. In

N. F. Hughes, ed., Organisms and Continents through time. *Palaeontological Association Special Paper* 12:271–89.

Chaloner, W. G., and S. V. Meyen. 1973. Carboniferous and Permian floras of the northern continents. 169–86 in A. Hallam, ed., *Atlas of Palaeogeography*. Amsterdam: Elsevier Scientific.

Chaloner, W. G., and A. Sheerin. 1979. Devonian macrofloras. *Special Papers in Palaeontology* 23:145–61.

Chudinov, P. K. 1983. Early Therapsids (in Russian). *Trudy Palaeontologicheskogo Instituta, Akademiya Nauk SSSR* 202:1–227.

Cichan, M. A. 1986. Conductance in the wood of selected Carboniferous plants. *Paleobiology* 12:302–10.

Clack, J. A. 1983. The stapes of the Coal Measures embolomere *Pholiderpeton scutigerum* Huxley (Amphibia: Anthracosauria) and otic evolution in early tetrapods. *Zoological Journal of the Linnean Society* 79:121–48.

———. 1988. New material of the early tetrapod *Acanthostega* from the Upper Devonian of East Greenland. *Palaeontology* 31:699–724.

———. 1989. Discovery of the earliest known tetrapod stapes. *Nature* 342:425–30.

Clack, J. A., and R. Holmes. 1988. The braincase of the anthracosaur *Archeria crassidisca* with comments on the inter-relationships of primitive tetrapods. *Palaeontology* 31:85–107.

Clark, J., and R. L. Carroll. 1973. Romeriid reptiles from the Lower Permian. *Bulletin of the Museum of Comparative Zoology at Harvard University* 144:353–407.

Clayton, G. 1985. Dinantian miospores and inter-continental correlation. In J. L. Escobedo, L. F. Granados, B. Mendez, R. Pignatelli, R. Rey, and R. H. Wagner, eds., *C.R. Tenth International Congress of Carboniferous Stratigraphy and Geology* 4:9–23.

Clendening, J. A. 1974. Palynological evidence for a Pennsylvanian age assignment of the Dunkard Group in the Appalachian Basin: Part 2. *West Virginia Geological and Economic Survey, Coal Geology Bulletin* 3:1–107.

———. 1975. Palynological evidence for a Pennsylvanian age assignment of the Dunkard Group in the Appalachian basin: Part 1. 195–216 in J. A. Barlow, ed., *The Age of the Dunkard*. West Virginia Geological and Economic Survey.

Cluver, M. A. 1978. The skeleton of the mammal-like reptile *Cistecephalus* with evidence for a fossorial mode of life. *Annals of the South African Museum* 76:213–46.

Coates, M. I., and J. A. Clack. 1990. Polydactyly in the earliest known tetrapod limbs. *Nature* 347:66–69.

Collinson, M. E., and A. C. Scott. 1987a. Factors controlling the organization and evolution of ancient plant communities. 399–420 in J. H. R. Gee and P. S. Giller, eds. *Organization of Communities Past and Present*. Oxford: Blackwell Scientific.

———. 1987b. Implications of vegetational change through the geological record on models for coal-forming environments. 67–85 in A. C. Scott, ed., *Coal and Coal-Bearing Strata: Recent Advances*. Geological Society Special Publication no. 32.

Cope, M. J., and W. G. Chaloner. 1980. Fossil charcoal as evidence of past atmospheric composition. *Nature* 283:647–49.

Corsin, P. 1948. Reconstitutions de Pecopteridées: Genres *Caulopteris* Lindley and Hutton, *Megaphyton* Artis et *Hagiophyton* nov. gen. *Annales de la Sociéte Géologique du Nord* 67:6–25.

Costanza, S. H. 1985. *Pennsylvanioxylon* of Middle and Upper Pennsylvanian coals from the Illinois basin, and its comparison with *Mesoxylon*. *Palaeontographica, Abt. B.* 197:81–121.

Creber, G. T., and W. G. Chaloner. 1984. Influence of environmental factors on the wood structure of living and fossil trees. *Botanical Review* 50:357–448.

Cridland, A. A. 1964. *Amyelon* in American coal balls. *Palaeontology* 7:186–209.

Cridland, A. A., and J. E. Morris. 1963. *Taeniopteris, Walchia,* and *Dichophyllum* in the Pennsylvanian System of Kansas. *University of Kansas Science Bulletin* 44: 71–85.

Crompton, A. W., and N. Hotton III. 1967. Functional anatomy of the masticatory apparatus of two dicynodonts (Reptilia, Therapsida). *Postilla* 109:1–51.

Crowell, J. C., and L. A. Frakes. 1975. The Late Paleozoic glaciation. 313–31 in K. S. W. Campbell, ed., *Gondwana Geology.* Papers presented at the Third Gondwana Symposium, Canberra, Australia, 1973. Canberra: Australian National University Press.

Crowley, T. J., and G. R. North. 1988. Abrupt climate change and extinction events in Earth history. *Science* 240:996–1002.

Currie, P. J. 1977. A new haptodontine sphenacodont (Reptilia: Pelycosauria) from the Upper Pennsylvanian of North America. *Journal of Paleontology* 51:927–42.

———. 1979. The osteology of haptodontine sphenacodonts (Reptilia: Pelycosauria). *Palaeontographica, Abt. A* 163:130–68.

Daly, E. 1969. A new procolophonoid reptile from the Lower Permian of Oklahoma. *Journal of Paleontology* 43:676–87.

———. 1973. A Lower Permian vertebrate fauna from southern Oklahoma. *Journal of Paleontology* 47:562–89.

———. 1988. A note on the dissorophoid amphibian of Hamilton quarry. *Kansas Geological Survey, Guidebook Series* 6:185–87.

———. In press. The Amphibamidae (Amphibia, Temnospondyli), with a description of a new genus from the Pennsylvanian of Kansas. *University of Kansas, Museum of Natural History, Special Publications.*

Delevoryas, T. 1955. The Medullosae: Structure and relationships. *Palaeontographica, Abt. B.* 97:114–67.

DeMar, R. E. 1966. The phylogenetic and functional implications of the armor of the Dissorophidae. *Fieldiana, Geology* 16:55–88.

Dilcher, D. L. 1979. Early angiosperm reproduction: An introductory report. *Review of Palaeobotany and Palynology* 27:291–328.

Dilkes, D. W. 1990. A new trematopsid amphibian (Temnospondyli: Dissorophoidea) from the Lower Permian of Texas. *Journal of Vertebrate Paleontology* 10: 222–43.

Dilkes, D. W., and R. R. Reisz. 1986. The axial skeleton of the Early Permian reptile *Eocaptorhinus laticeps* (Williston). *Canadian Journal of Earth Sciences* 23:1288–96.

———. 1987. *Trematops milleri* Williston, 1909 identified as a junior synonym of *Acheloma cumminsi* Cope, 1882, with a revision of the genus. *American Museum Novitates* 2902:1–12.

DiMichele, W. A. 1983. *Lepidodendron hickii* and generic delimitation in Carboniferous lepidodendrid lycopods. *Systematic Botany* 8:317–33.

————. 1985. *Diaphorodendron* gen. nov., a segregate from *Lepidodendron* (Pennsylvanian age). *Systematic Botany* 10:453–58.

DiMichele, W. A., and P. J. DeMaris. 1987. Structure and dynamics of a Pennsylvanian-age *Lepidodendron* forest: Colonizers of a disturbed swamp habitat in the Herrin (No. 6) Coal of Illinois: *Palaios* 2:146–57.

DiMichele, W. A., and W. J. Nelson. 1989. Small-scale spatial heterogeneity in Pennsylvanian-age vegetation from the roof shale of the Springfield Coal (Illinois Basin). *Palaios* 4:276–80.

DiMichele, W. A., and T. L. Phillips. 1985. Arborescent lycopod reproduction and paleoecology in a coal-swamp environment of late Middle Pennsylvanian age (Herrin Coal, Illinois, USA). *Review of Palaeobotany and Palynology* 44:1–26.

————. 1988. Paleoecology of the Middle Pennsylvanian–age Herrin coal swamp (Illinois) near a contemporaneous river system, the Walshville paleochannel. *Review of Palaeobotany and Palynology* 56:151–76.

DiMichele, W. A., J. I. Davis, and R. G. Olmstead. 1989. Origins of heterospory and the seed habit: The role of heterochrony. *Taxon* 38:1–11.

DiMichele, W. A., T. L. Phillips, and R. G. Olmstead. 1987. Opportunistic evolution: Abiotic environmental stress and the fossil record of plants. *Review of Palaeobotany and Palynology* 50:151–78.

DiMichele, W. A., T. L. Phillips, and R. A. Peppers. 1985. The influence of climate and depositional environment on the distribution and evolution of Pennsylvanian coal-swamp plants. 223–56 in B. H. Tiffney, ed., *Geological Factors and the Evolution of Plants*. New Haven: Yale University Press.

Doran, J. B. 1980. A new species of *Psilophyton* from the Lower Devonian of northern New Brunswick, Canada. *Canadian Journal of Botany* 58:2241–62.

Durden, C. J. 1969. Pennsylvanian correlation using blattoid insects. *Canadian Journal of Earth Sciences* 6:1159–77.

————. 1984a. Carboniferous and Permian entomology of western North America. *Ninth International Congress of Carboniferous Stratigraphy and Geology*, C.R. 2:81–89.

————. 1984b. North American provincial insect ages for the continental last half of the Carboniferous and first half of the Permian. *Ninth International Congress of Carboniferous Stratigraphy and Geology*, C.R. 2:606–12.

————. 1988. Hamilton insect fauna. *Kansas Geological Survey, Guidebook Series* 6:117–24.

Eaton, T. H., Jr. 1964. A captorhinomorph predator and its prey (Cotylosauria). *American Museum Novitates* 2169:1–3.

Eberth, D. A. 1985. The skull of *Sphenacodon ferocior*, and comparisons with other sphenacodontines (Reptilia: Pelycosauria). *New Mexico Institute of Mining and Technology Circular* 190:1–39.

————. 1987. Stratigraphy, sedimentology, and paleoecology of Cutler Formation red-beds (Permo-Pennsylvanian) in north-central New Mexico, Ph.D. diss., University of Toronto.

Eberth, D. A., and D. Brinkman. 1983. *Ruthiromia elcobriensis*, a new pelycosaur from El Cobre Canyon, New Mexico. *Breviora* 474:1–26.

Eble, C. F. 1990. A palynological transect, swamp interior to swamp margin, in the Mary Lee coal bed, Warrior Basin, Alabama. 65–81 in R. A. Gastaldo, T. M.

Demko, and Yuejin Liu, eds., *Carboniferous Coastal Environments and Paleocommunities of the Mary Lee Coal Zone, Marion and Walker Counties, Alabama.* Alabama Geological Survey, Guidebook for Field Trip 6. Thirty-Ninth Annual Meeting, Southeastern Section, Geological Society of America.

Eble, C. F., and W. C. Grady. 1990. Paleoecological interpretation of a Middle Pennsylvanian coal bed in the central Appalachian Basin. *International Journal of Coal Geology* 16:255–86.

Eble, C. F., W. C. Grady, and W. H. Gillespie. 1989. Palynology, petrology, and paleoecology of the Hernshaw-Fire Clay coal bed in the central Appalachian Basin. 133–42 in C. B. Cecil et al., eds., *Carboniferous Geology of the Eastern United States.* Field Trip Guidebook T143.

Edwards, Diane 1980. Early land floras. 55–85 in A. L. Panchen, ed., *The Terrestrial Environment and the Origin of Land Vertebrates.* New York: Academic Press.

Edwards, Diane, and D. S. Edwards. 1986. A reconsideration of the Rhyniophytina. *Systematics Association Special Volume* 31:199–220.

Edwards, Diane, and U. Fanning. 1985. Evolution and environment in the Late Silurian–Early Devonian: The rise of the pteridophytes. *Philosophical Transactions of the Royal Society of London* B 309:147–65.

Edwards, D. S. 1980. Evidence for the sporophyte status of the Lower Devonian plant *Rhynia gwynne-vaughnii* Kidston and Lang. *Review of Palaeobotany and Palynology* 29:177–88.

———. 1986. *Aglaophyton major,* a non-vascular land-plant from the Devonian Rhynie Chert. *Botanical Journal of the Linnaean Society* 93:173–204.

Eggert, D. A., and N. Y. Kanemoto. 1977. Stem phloem of a Middle Pennsylvanian *Lepidodendron. Botanical Gazette* 138:102–11.

Ehret, D. L., and T. L. Phillips. 1977. *Psaronius* root systems: Morphology and development. *Palaeontographica,* Abt. B. 161:147–64.

Erwin, D. H., J. W. Valentine, and J. J. Sepkoski. 1987. A comparative study of diversification events: The Early Paleozoic versus the Mesozoic. *Evolution* 41:1177–86.

Evans, J. W. 1954. The periods of origin and diversification of the Superfamilies of the Homoptera-Auchenorhyncha (Insecta) as determined by a study of the wings of Paleozoic and Mesozoic fossils. *Proceedings of the Linnean Society of London* 175:171–81.

———. 1956. Paleozoic and Mesozoic Hemiptera (Insecta). *Australian Journal of Zoology* 4:165–258.

Evans, S. E., and H. Haubold. 1987. A review of the Upper Permian genera *Coelurosauravus, Weigeltisaurus,* and *Gracilisaurus* (Reptilia: Diapsida). *Zoological Journal of the Linnean Society* 90:275–303.

Fairon-Demaret, M. 1986. Some uppermost Devonian megafloras: A stratigraphic review. *Annales de la Société Géologique Belgique* 109:43–48.

Feakes, C. R., and G. J. Retallack. 1988. Recognition and chemical characterization of fossil soils developed on alluvium: A Late Ordovician example. *Geological Society of America Special Paper* 216:35–48.

Feng, B.-C. 1989. Paleoecology of an upper Middle Pennsylvanian coal swamp from western Pennsylvania, USA. *Review of Palaeobotany and Palynology* 57:299–312.

Fichter, J. 1983. Tetrapodenfährten aus dem saarpfälzischen Rotliegenden (?Ober-Karbon–Unter-Perm; SW-Deutschland), Teil II: Die Fährten der Gattungen *Foli-*

ipes, Varanopus, Ichniotherium, Dimetropus, Palmichnus, Phalangichnus, cf. *Cheilichnus,* cf. *Laoporus* und *Anhomoiichnium. Mainzer Naturwissenschaftliches Archiv* 21:125–86.

Fichter, J., and G. Kowalczyk. 1983. Tetrapodenfährten aus dem Rotliegenden der Wetterau und ihre stratigraphische Auswertung. *Mainzer Geowissenschaftliche Mitteilungen* 12:123–58.

Florin, R. 1951. Evolution in cordaites and conifers. *Acta Horti Bergiani* 15:285–388.

Fracasso, M. A. 1980. Age of the Permo-Carboniferous Cutler Formation vertebrate fauna from El Cobre Canyon, New Mexico. *Journal of Paleontology* 54:1237–44.

———. 1987. Braincase of *Limnoscelis paludis* Williston. *Postilla* 201:1–22.

Frakes, L. A., and J. C. Crowell. 1969. Late Paleozoic glaciation. 1. South America. *Geological Society of America, Bulletin* 80:1007–42.

Frederiksen, N. O. 1972. The rise of the Mesophytic flora. *Geoscience and Man* 4:17–28.

French, J. A., W. L. Watney, and J. E. Anderson. 1988. Stratigraphic and sedimentological considerations relating to the fossiliferous limestones at Hamilton quarry, Greenwood County, Kansas. *Kansas Geological Survey, Guidebook Series* 6:37–58.

Galtier, J. 1971. La fructification de *Botryopteris forensis* Renault (Coenopteridales de Stephanien Francais): Précisions sur la sporanges et les spores. *Naturalia Monspeliensia Botanique* 22:145–55.

Galtier, J., and T. L. Phillips. 1985. Swamp vegetation from Grand Croix (Stephanian) and Autun (Autunian), France, and comparisons with coal-ball peats of the Illinois basin. *C.R. Ninth International Congress of Carboniferous Stratigraphy and Geology* 4:13–24.

Galtier, J., and A. C. Scott. 1985. The diversification of early ferns. *Proceedings of the Royal Society of Edinburgh* 86B:289–301.

Garratt, M. J., and R. B. Rickards. 1984. Graptolite biostratigraphy of early land plants from Victoria, Australia. *Proceedings of the Yorkshire Geological Society* 44:377–84.

Gastaldo, R. A. 1982. A preliminary assessment of Early Pennsylvanian megafloral paleoecology in northeastern Alabama. *Third North American Paleontological Convention, Proceedings* 1:187–92.

———. 1984. Studies on North American pecopterids: 3. *Pecopteris buttsii* D. White, from the Early Pennsylvanian of Alabama. *Journal of Paleontology* 58:63–77.

———. 1986a. An explanation for lycopod configuration, "Fossil Grove" Victoria Park, Glasgow. *Scottish Journal of Geology* 22:77–83.

———. 1986b. Implications on the paleoecology of autochthonous lycopods in clastic sedimentary environments of the early Pennsylvanian of Alabama. *Palaeogeography, Palaeoclimatology, Palaeoecology* 53:191–212.

———. 1986c. Upper Carboniferous paleoecological reconstructions: Observations and reconsiderations. *Tenth International Congress of Carboniferous Stratigraphy and Geology, C.R.* 2:281–96.

———. 1987. Confirmation of Carboniferous clastic swamp communities. *Nature* 326:869–71.

———. 1988. Conspectus of phytotaphonomy. In *Methods and Applications of Plant Paleoecology.* Paleontological Society, Special Publication no. 3:14–28.

Gastaldo, R. A., D. P. Douglass, and S. M. McCarroll. 1987. Origin, characteristics, and provenance of plant macrodetritus in a Holocene crevasse splay, Mobile Delta, Alabama. *Palaios* 2:229–40.

Gensel, P. G. 1982. *Oricilla*, a new genus referable to the zosterophyllophytes from the late Early Devonian of northern New Brunswick. *Review of Palaeobotany and Palynology* 37:345–59.

Gensel, P. G., and H. N. Andrews. 1984. *Plant Life in the Devonian*. New York: Praeger.

———. 1987. The evolution of early land plants. *American Scientist* 75:478–89.

Gensel, P. G., H. N. Andrews, and W. H. Forbes. 1975. A new species of *Sawdonia* with notes on the origin of microphylls and lateral sporangia. *Botanical Gazette* 136:50–62.

Gensel, P. G., A. Kasper, and H. N. Andrews. 1969. *Kaulangiophyton*, a new genus of plants from the Devonian of Maine. *Bulletin of the Torrey Botanical Club* 96:265–76.

Gersib, G. A., and P. J. McCabe. 1981. Continental coal-bearing sediments of the Port Hood Formation (Carboniferous), Cape Linzee, Nova Scotia, Canada. *Society of Economic Paleontologists and Mineralogists Special Publications* 31:95–108.

Gillespie, W. H., G. J. Hennen, and C. Balasco. 1975. Plant megafossils from Dunkard strata in northwestern West Virginia and southwestern Pennsylvania. 223–44 in J. A. Barlow, ed., *The Age of the Dunkard*. Morgantown: West Virginia Geological and Economic Survey.

Gillespie, W. H., G. W. Rothwell, and S. E. Scheckler. 1981. The earliest seeds. *Nature* 293:462–64.

Godfrey, S. J., A. R. Fiorillo, and R. L. Carroll. 1987. A newly discovered skull of the temnospondyl amphibian *Dendrerpeton acadianum* Owen. *Canadian Journal of Earth Sciences* 24:796–805.

Gow, C. E. 1978. The advent of herbivory in certain reptilian lineages during the Triassic. *Palaeontologia Africana* 21:133–41.

Grady, W. C., and C. F. Eble. 1990. Relationships among macerals, miospores, and paleoecology in a column of the Redstone coal (Upper Pennsylvanian) from north-central West Virginia (USA). *International Journal of Coal Geology* 15:1–26.

Gray, J. 1985. The microfossil record of early land plants: Advances in understanding early terrestrialization. *Philosophical Transactions of the Royal Society of London* B 309:167–95.

Greenslade, P., and P. E. S. Whalley. 1986. The systematic position of *Rhyniella praecursor* Hirst and Maulik (Collembola), the earliest known hexapod. 319–323 in R. Dallai, ed., *Second International Symposium on Apterygota*. Siena, Italy: University of Siena.

Gregory, J. T., F. E. Peabody, and L. I. Price. 1956. Revision of the Gymnarthidae American Permian microsaurs. *Peabody Museum of Natural History, Yale University, Bulletin* 10:1–77.

Hahn, G., R. Hahn, and C. Brauckmann. 1986. Zur Kenntnis von *Arthropleura* (Myriapoda; Ober-Karbon). *Geologica et Palaeontologica* 20:125–37.

Hannibal, J. T., and R. M. Feldmann. 1981. Systematics and functional morphology of oniscomorph millipedes (Arthropoda: Diplopoda) from the Carboniferous of North America. *Journal of Paleontology* 55:730–46.

Haubold, H., and G. Katzung. 1978. Palaeoecology and palaeoenvironment of tetrapod footprints from the Rotliegend (Lower Permian) of central Europe. *Palaeogeography, Palaeoclimatology, Palaeoecology* 23:307–23.

Havlena, V. 1970. Einige Bemerkungen zur Phytogeographie und Geobotanik des Karbons und Perms. *Sixth International Congress of Carboniferous Stratigraphy and Geology (Sheffield, 1967)*, C.R. 3:901–12.

———. 1975. European upper Paleozoic, *Callipteris conferta*, and the Permo-Carboniferous boundary. 7–22 in J. A. Barlow, ed., *The Age of the Dunkard*. Morgantown: West Virginia Geological and Economic Survey.

Heaton, M. J. 1979. Cranial morphology of primitive captorhinid reptiles from the Late Pennsylvanian and Early Permian, Oklahoma and Texas. *Oklahoma Geological Survey Bulletin* 127:1–84.

———. 1980. The Cotylosauria: A reconsideration of a group of archaic tetrapods. 497–551 in A. L. Panchen, ed., *The Terrestrial Environment and the Origin of Land Vertebrates*. New York: Academic Press.

Heaton, M. J., and R. R. Reisz. 1980. A skeletal reconstruction of the Early Permian captorhinid reptile *Eocaptorhinus laticeps* (Williston). *Journal of Paleontology* 54:136–43.

———. 1986. Phylogenetic relationships of captorhinomorph reptiles. *Canadian Journal of Earth Sciences* 23:402–18.

Hirmer, M. 1927. *Handbuch der Paläobotanik*. Berlin: Oldenbourg.

Hlavin, W. J. 1972. Early Permian vertebrates from the upper Washington Formation at Belpre, Ohio. 30–31 in T. Arkle, Jr., director, *I. C. White Memorial Symposium, Field Trip Log*. Morgantown: West Virginia Geological and Economic Survey.

Hoffman, R. L. 1969. Myriapoda, exclusive of Insecta. R572–R606. in R. C. Moore, ed., *Treatise on Invertebrate Paleontology*. Lawrence: Geological Society of America and University of Kansas Press.

Holmes, R. 1977. The osteology and musculature of the pectoral limb of small captorhinids. *Journal of Morphology* 152:101–40.

———. 1980. *Proterogyrinus scheelei* and the early evolution of the labyrinthodont pectoral limb. 351–76 in A. L. Panchen, ed., *The Terrestrial Environment and the Origin of Land Vertebrates*. New York: Academic Press.

———. 1984. The Carboniferous amphibian *Proterogyrinus scheelei* Romer, and the early evolution of tetrapods. *Philosophical Transactions of the Royal Society of London* B 306: 431–527.

———. 1989. The skull and axial skeleton of the Lower Permian anthracosauroid amphibian *Archeria crassidisca* Cope. *Palaeontographica* A 207: 161–206.

Hook, R. W. 1989. Stratigraphic distribution of tetrapods in the Bowie and Wichita Groups, Permo-Carboniferous of North-Central Texas. 47–53 in R. W. Hook, ed., *Permo-Carboniferous Vertebrate Paleontology, Lithostratigraphy, and Depositional Environments of North-Central Texas*. Field Trip Guidebook no. 2, Forty-Ninth Annual Meeting of the Society of Vertebrate Paleontology.

Hook, R. W., and D. Baird. 1984. *Ichthycanthus platypus* Cope, 1877, reidentified as the dissorophoid amphibian *Amphibamus lyelli*. *Journal of Paleontology* 58: 697–702.

———. 1988. An overview of the Upper Carboniferous fossil deposit at Linton, Ohio. *Ohio Journal of Science* 88:55–60.

Hook, R. W., and J. C. Ferm. 1988. Paleoenvironmental controls on vertebrate-bearing abandoned channels in the Upper Carboniferous. *Palaeogeography, Palaeoclimatology, Palaeoecology* 63:159–81.

Hook, R. W., and J. C. Hower. 1988. Petrography and taphonomic significance of the vertebrate-bearing cannel coal of Linton, Ohio (Westphalian D, Upper Carboniferous). *Journal of Sedimentary Petrology* 58:72–80.

Hotton, N., III. 1980. An alternative to dinosaur endothermy: The happy wanderer. In R. D. K. Thomas and E. C. Olson, ed., *A Cold Look at the Warm-blooded Dinosaurs*. American Association for the Advancement of Science, Selected Symposia Series 28:311–50.

———. 1986. Dicynodonts and their role as primary consumers. 71–82 in N. Hotton III, P. D. MacLean, J. J. Roth, and C. E. Roth, eds., *The Ecology and Biology of Mammal-like Reptiles*. Washington, D.C.: Smithsonian Institution Press.

Hueber, F. M. 1992. Speculations on the early lycopsids and zosterophylls. *Annals of the Missouri Botanical Garden*.

Hueber, F. M., and H. P. Banks. 1979. *Serrulacaulis furcatus* gen. et sp. nov., a new zosterophyll from the lower Upper Devonian of New York State. *Review of Palaeobotany and Palynology* 28:169–89.

Huene, F. von. 1950. Die Theriodontier des ostafrikanischen Ruhuhu-Gebietes in der Tuebinger Sammlung. *Neues Jahrbuch für Geologie und Palaeontologie, Abhandlungen* 92:47–136.

Hünicken, M. A. 1980. A giant fossil spider (*Megaranea servinei*) from Bajo de Veliz, Upper Carboniferous, Argentina. *Boletino, Academia Nacional de Ciencias* (Cordoba) 53:317–28.

Ivakhnenko, M. F. 1979. Permian and Triassic procolophonians of the Russian Platform (in Russian). *Trudy Paleontologicheskogo Instituta, Akademiya Nauk SSSR* 164:1–80.

———. 1987. Permian parareptiles of the USSR (in Russian). *Trudy Paleontologicheskogo Instituta, Akademiya Nauk SSSR* 223:1–159.

Ivakhnenko, M. F., and G. I. Tverdokhlebova. 1987. A revision of the Permian bolosauromorphs of eastern Europe. *Paleontological Journal* 1987:93–100.

Jarvik, E. 1952. On the fish-like tail in the ichthyostegid stegocephalians with description of a new stegocephalian and a new crossopterygian from the Upper Devonian of East Greenland. *Meddelelser om Grønland* 114:1–90.

———. 1980. *Basic Structure and Evolution of Vertebrates*, vol. 1. New York: Academic Press.

Jarzembowski, E. A. 1987. The occurrence and diversity of Coal Measure insects. *Journal of Geological Society, London* 144:507–11.

Jennings, J. R. 1980. Fossil plants from the Fountain Formation (Pennsylvanian) of Colorado. *Journal of Paleontology* 54:149–58.

Jeram, A. J., P. A. Selden, and D. Edwards. 1990. Land animals in the Silurian: Arachnids and myriapods from Shropshire, England. *Science* 250:658–61.

Kaiser, K. 1970. Die Oberdevon-Flora des Bäreninsel. 3. Mikrofloras des höheren Oberdevons und des Unterkarbons. *Palaeontographica*, Abt. B 129:71–124.

Kemp, T. S. 1969. On the functional morphology of the gorgonopsid skull. *Philosophical Transactions of the Royal Society of London*, B 256:1–83.

———. 1982. *Mammal-like Reptiles and the Origin of the Mammals*. London: Academic Press.

Kenrick, P., and P. R. Crane. 1991. Water-conducting cells in early fossil land plants: Implications for the early evolution of tracheophytes. *Botanical Gazette* 152: 335–56.

Kenrick, P., Diane Edwards, and R. C. Dales. 1991. Novel ultrastructure in water-conducting cells of the Lower Devonian plant *Sennicaulis hippocrepiformis*. *Palaeontology* 34:751–66.

Kenrick, P., W. Remy, and P. R. Crane. 1991. The structure of water-conducting cells in the enigmatic early land plants *Stockmansella langii* Fairon-Demaret, *Huvenia kleui* Hass and Remy and *Sciadophyton* sp. Remy *et al.*, 1980. *Argumenta Palaeobotanica* 8:179–91.

Kerp, J. H. F. 1982a. Aspects of Permian paleobotany and palynology. 2. On the presence of the ovuliferous organ *Autunia milleryensis* (Renault) Krasser (Peltaspermaceae) in the Lower Permian of the Nahe area (FRG) and its relationship to *Callipteris conferta* (Sternberg) Brongniart. *Acta Botanica Neerlandica* 31:417–27.

———. 1982b. New palaeobotanical data on the "Rotliegendes" of the Nahe area (FRG). *Courier Forschungs–Institut Senckenberg* 56:7–14.

———. 1983. Aspects of Permian paleobotany and palynology. 1 *Sobernheimia jonkeri* nov. gen., nov. sp., a new fossil plant of cycadalean affinity from the Waderner Gruppe of Sobernheim. *Review of Palaeobotany and Palynology* 38: 173–83.

———. 1986. On some interesting fructifications from the Lower Permian of the Saar-Nahe area (Western Germany). *Courier Forschungs–Institut Senckenberg* 86:73–87.

Kerp, J. H. F., and J. Fichter. 1985. Die Makrofloren des saarpfälzischen Rotliegenden (?Ober–Karbon–Unter–Perm; SW Deutschland). *Mainzer Geowissenschaftliche Mitteilungen* 14:159–286.

Kerp, H., R. Poort, H. Swinkels, and R. Verwer. 1989. A conifer-dominated flora from the Rotliegend of Oberhausen (Saar-Nahe-Area). *Courier Forschungs–Institut Senckenberg* 109:137–51.

Kethley, J. B., R. A. Norton, P. M. Bonamo, and W. A. Shear. 1989. A terrestrial alicorhagiid mite (Acari: Acariformes) from the Devonian of New York. *Micropaleontology* 35:367–73.

Kevan, P. G., W. G. Chaloner, and D. B. O. Savile. 1975. Interrelationships of early terrestrial arthropods and plants. *Palaeontology* 18:391–417.

Kidston, R., and W. H. Lang. 1917. On Old Red Sandstone plants showing structure, from the Rhynie Chert bed, Aberdeenshire. Part 1. *Rhynia gwynne-vaughani*, Kidston and Lang. *Transactions of the Royal Society of Edinburgh* 51:761–84.

Kidston, R., and W. H. Lang. 1920a. On Old Red Sandstone plants showing structure, from the Rhynie Chert bed, Aberdeenshire. Part 2. Additional notes on *Rhynia gwynne-vaughani*, Kidston and Lang; with descriptions of *Rhynia major*, n. sp., and *Hornea lignieri*, n. g., n. sp. *Transactions of the Royal Society of Edinburgh* 52:603–27.

———. 1920b. On Old Red Sandstone plants showing structure, from the Rhynie Chert bed, Aberdeenshire. Part 3. *Asteroxylon mackiei*, Kidston and Lang. *Transactions of the Royal Society of Edinburgh* 52:643–80.

———. 1921a. On Old Red Sandstone plants showing structure, from the Rhynie Chert bed, Aberdeenshire. Part 4. Restorations of the vascular cryptogams, and discussions of their bearing on the general morphology of the Pteridophyta and the origin of the organisation of land-plants. *Transactions of the Royal Society of Edinburgh* 52:831–53.

———. 1921b. On Old Red Sandstone plants showing structure, from the Rhynie Chert bed, Aberdeenshire. Part 5. The Thallophyta occurring in the peat bed; the succession of the plants throughout a vertical section of the bed, and the conditions of accumulation and preservation of the deposit. *Transactions of the Royal Society of Edinburgh* 52:855–902.

King, G. M. 1981. The functional anatomy of a Permian dicynodont. *Philosophical Transactions of the Royal Society of London,* B 291:243–322.

———. 1988. *Anomodontia. Handbuch der Palaeoherpetologie 17C.* New York: Gustav Fischer Verlag. 1–174.

Kjellesvig-Waering, E. 1986. A restudy of the fossil Scorpionida of the world. *Palaeontographica Americana* 55:1–287.

Knoll, A. H. 1984. Patterns of extinction in the fossil record of vascular plants. 21–68 in M. Nitecki, ed., *Extinctions.* Chicago: University of Chicago Press.

———. 1985. Exceptional preservation of photosynthetic organisms in silicified carbonates and silicified peats. *Philosophical Transactions of the Royal Society* B311: 111–22.

———. 1986. Patterns of change in plant communities through geological time. 126–41 in J. Diamond and T. J. Case, eds., *Community Ecology.* New York: Harper and Row.

Knoll, A. H., and K. J. Niklas. 1987. Adaptation, plant evolution, and the fossil record. *Review of Palaeobotany and Palynology* 50:127–49.

Knoll, A. H., K. J. Niklas, and B. H. Tiffney. 1979. Phanerozoic land plant diversity in North America. *Science* 206:1400–1402.

Knoll, A. H., K. J. Niklas, P. G. Gensel, and B. H. Tiffney. 1984. Character diversification and patterns of evolution in early vascular plants. *Paleobiology* 10: 34–47.

Kukalova-Peck, J. 1972. Unusual structures in the Palaeozoic insect orders Megasecoptera and Palaeodictyoptera, with a description of a new family. *Psyche* 78: 306–18.

———. 1974. Wing-folding in the Paleozoic insect order Diaphanopterodea (Paleoptera), with a description of new members of the Family Elmoidae. *Psyche* 81: 315–33.

———. 1978. Origin and evolution of insect wings, and their relation to metamorphosis, as documented by the fossil record. *Journal of Morphology* 156: 53–125.

———. 1983. Origin of the insect wing and wing articulation from the arthropodan leg. *Canadian Journal of Zoology* 61:1618–69.

———. 1985. Ephemeroid wing venation based upon new gigantic Carboniferous mayflies and basic morphology, phylogeny, and metamorphosis of pterygote insects (Insecta, Ephemerida). *Canadian Journal of Zoology* 63:933–55.

Labandeira, C. C. 1990. Rethinking the diets of Carboniferous terrestrial arthropods: Evidence for a nexus of arthropod/vascular plant interactions. *Geological Society of America, Abstracts with Programs* 22(7): A265.

Labandeira, C. C., and B. S. Beall. 1990. Arthropod terrestriality. *Paleontological Society, Short Courses in Paleontology* 3 (Arthropod Paleobiology): 212–56.

Labandeira, C. C., B. S. Beall, and F. M. Hueber. 1988. Structure and inferred life habits of an Early Devonian bristletail: What does the earliest known insect tell us about the origin of insects? *Geological Society of America, Abstracts with Programs* 20 (7): A47.

Lang, W. H. 1927. Contributions to the study of the Old Red Sandstone flora of Scotland. 6. On *Zosterophyllum myretonianum*, Penh., and some other plant-remains from the Carmyllie Beds of the lower Old Red Sandstone. 7. On a specimen of *Pseudosporochnus* from the Stromness Beds. *Transactions of the Royal Society of Edinburgh* 55:443–55.

Langston, W., Jr. 1965. *Oedaleops campi* (Reptilia: Pelycosauria), a new genus and species from the Lower Permian of New Mexico, and the family Eothyrididae. *Texas Memorial Museum Bulletin* 9:1–47.

Langston, W., Jr. and R. R. Reisz. 1981. *Aerosaurus wellesi,* new species, a varanopseid mammal-like reptile (Synapsida: Pelycosauria) from the Lower Permian of New Mexico. *Journal of Vertebrate Paleontology* 1:73–96.

Laurin, M., and R. R. Reisz. 1990. *Tetraceratops* is the oldest known therapsid. *Nature* 345:249–50.

Laveine, J.–P. 1986. The size of the frond in the genus *Alethopteris* Sternberg (Pteridospermopsida, Carboniferous). *Geobios* 19:49–56.

Laveine, J.-P., Y. Lemoigne, Xingxue Xi, Xiuyuan Wu, Shanzhen Zhang, Xiuhu Zhao, Weiqing Zhu, and Jianan Zhu. 1987. Paleogeography of China in Carboniferous time in the light of paleobotanical data, in comparison with western Europe Carboniferous assemblages. *Comptes Rendues Académie des Sciences Paris,* ser. 2, 304:391–94.

Leary, R. L. 1981. Early Pennsylvanian geology and paleobotany of the Rock Island County, Illinois, area. Part 1: Geology. *Illinois State Museum, Reports of Investigations,* no. 37, 1–88.

Leary, R. L., and H. W. Pfefferkorn. 1977. An Early Pennsylvanian flora with *Megalopteris* and Noeggerathiales from west-central Illinois. *Illinois State Geological Survey, Circular* 500. 1–77.

Lebedev, O. A. 1984. The first record of a Devonian tetrapod in the USSR (in Russian). *Doklady Akademii Nauk SSSR* 278:1470–73.

Leisman, G. A. 1961. A new species of *Cardiocarpus* in Kansas coal balls. *Transactions of the Kansas Academy of Science* 64:117–22.

Lele, K. M., and J. Walton. 1961. Contributions to the knowledge of "*Zosterophyllum myretonianum*" Penhallow from the lower Old Red Sandstone of Angus. *Transactions of the Royal Society of Edinburgh* 74:469–75.

Leonardi, G. 1983. *Notopus petri* nov. gen., nov. sp.: Une empreinte d'amphibien de Devonien au Parana (Bresil). *Géobios* 16:233–39.

Lesnikowska, A. 1989. Anatomically preserved Marattiales from coal swamps of the Desmoinesian and Missourian of the midcontinent United States: Systematics, ecology, and evolution. Ph.D. diss., University of Illinois, Urbana.

———. 1990. Evidence of herbivory in tree-fern petioles from the Calhoun Coal (Upper Pennsylvanian) of Illinois. *Palaios* 5:76–80.

Lewis, G. E., and P. P. Vaughn. 1965. Early Permian vertebrates from the Cutler Formation of the Placerville area, Colorado. *United States Geological Survey Professional Paper* 503-C: 1–46.

Li X., and Yao Z. 1982. A review of recent research on the Cathaysia flora in Asia. *American Journal of Botany* 69:479–86.

Livermore, R. A., A. G. Smith, and J. C. Briden. 1985. Palaeomagnetic constraints on the distribution of continents in the Late Silurian and Early Devonian. *Philosophical Transactions of the Royal Society of London* B 309:29–56.

Lorenzo, P. 1979. Les sporophylles de *Lepidodendron dissitum* Sauver, 1848. *Geobios* 12:137–43.

Lund, R. 1978. Anatomy and relationships of the family Phlegethontiidae (Amphibia, Aïstopoda). *Annals of the Carnegie Museum* 47:53–79.

Lyons, P. C., and W. C. Darrah. 1989. Earliest conifers in North America: Upland and/or paleoclimatic indicators? *Palaios* 4:480–86.

Lyons, P. C., R. B. Finkelman, C. L. Thompson, F. W. Brown, and P. G. Hatcher. 1982. Properties, origin, and nomenclature of rodlets of the inertinite maceral group in coals of the central Appalachian Basin, USA. *International Journal of Coal Geology* 1:313–46.

Mahaffy, J. F. 1985. Profile patterns of peat and coal palynology in the Herrin (No. 6) Coal Member, Carbondale Formation, Middle Pennsylvanian of southern Illinois. *Ninth International Congress of Carboniferous Stratigraphy and Geology*, C.R. 5:25–34.

———. 1988. Vegetational history of the Springfield Coal (Middle Pennsylvanian of Illinois) and distribution patterns of a tree-fern miospore, *Thymospora pseudothiessenii*, based on miospore profiles. *International Journal of Coal Geology* 10:239–60.

Malloch, D. W., K. A. Pirozynski, and P. H. Raven. 1980. Ecological and evolutionary significance of mycorrhizal symbioses in vascular plants (a review). *Proceedings of the National Academy of Sciences* 77:2113–18.

Mamay, S. H. 1966. *Tinsleya*, a new genus of seed-bearing callipterid plants from the Permian of north-central Texas. *United States Geological Survey, Professional Paper*, 523-E: 1–15.

———. 1967. Lower Permian plants from the Arroyo Formation in Baylor County, north-central Texas. *United States Geological Survey, Professional Paper* 575-C: 120–26.

———. 1968. *Russelites*, a new genus of problematical plant from the Permian of Texas. *United States Geological Survey, Professional Paper* 593-I, 1–113.

———. 1973. *Archaeocycas* and *Phasmatocycas:* New genera of Permian cycads. *Journal of Research, United States Geological Survey* 1:687–89.

———. 1975. *Sandrewia*, n. gen., a problematical plant from the Lower Permian of Texas and Kansas. *Review of Palaeobotany and Palynology* 20:75–83.

———. 1976. Paleozoic origin of the cycads. *United States Geological Survey Professional Paper* 934, 1–48.

———. 1986. New species of Gigantopteridaceae from the Lower Permian of Texas. *Phytologia* 61:311–15.

———. 1988. *Gigantonoclea* in the Lower Permian of Texas. *Phytologia* 64:330–32.

Mamay, S. H., and W. J. Breed. 1970. Early Permian plants from the Cutler Formation in Monument Valley, Utah. *United States Geological Survey, Professional Paper* 700-B: 109–17.

Mamay, S. H., and G. Mapes. In press. Early Virgillian plant megafossils from the

Kinney Brick Company Quarry, Manzanita Mountains, New Mexico. *New Mexico Bureau of Mines and Mineral Resources, Bulletin* 138.

Mamay, S. H., and C. B. Read, 1956. Additions to the flora of the Spotted Ridge Formation in central Oregon. *United States Geological Survey Professional Paper* 274-I: 211–26.

Mamay, S. H., J. M. Miller, and D. M. Rohr. 1984. Late Leonardian plants from west Texas: The youngest Paleozoic plant megafossils in North America. *Science* 223: 279–81.

Mamay, S. H., J. M. Miller, D. M. Rohr, and W. E. Stein. 1986. *Delnortea*, a new genus of Permian plants. *Phytologia* 60:345–46.

———. 1988. Foliar morphology and anatomy of the gigantopterid plant *Delnortea abbottiae* from the Lower Permian of West Texas. *American Journal of Botany* 75:1409–33.

Mapes, G. 1987. Ovule inversion in the earliest conifers. *American Journal of Botany* 74:1205–10.

Mapes, G., and R. A. Gastaldo. 1986. Late Paleozoic non-peat accumulating floras. 115–27 in T. W. Broadhead, ed., *Land Plants, Notes for a Short Course. University of Tennessee, Department of Geological Sciences, Studies in Geology 15*.

Mapes, G., and G. W. Rothwell. 1988. Diversity among Hamilton conifers. 225–44 in G. Mapes and R. H. Mapes, eds., *Regional Geology and Paleontology of Upper Paleozoic Hamilton Quarry Area in Southeastern Kansas*. Guidebook, Twenty-second Annual Meeting, South-Central Section, Geological Society of America.

Mapes, G., G. W. Rothwell, and M. T. Haworth. 1989. Evolution of seed dormancy. *Nature* 337:645–46.

Martens, T. 1989. First evidence of terrestrial tetrapods with North American faunal elements in the red beds of Upper Rotliegendes (Lower Permian, Tambach Beds) of the Thuringian Forest (GDR): First results. *Acta Musei Reginaehradecensis* 22A: 99–104.

Matten, L. C., W. S. Lacey, and R. C. Lucas. 1980a. Studies on the cupulate seed genus *Hydrasperma* Long from Berwickshire and East Lothian in Scotland and County Kerry in Ireland. *Botanical Journal of the Linnaean Society* 81:249–73.

Matten, L. C., W. S. Lacey, B. I. May, and R. C. Lucas. 1980b. A megafossil flora from the uppermost Devonian near Ballyheigue, Co. Kerry, Ireland. *Review of Palaeobotany and Palynology* 29:241–51.

Matten, L. C., W. R. Tanner, and W. S. Lacey. 1984. Additions to the silicified Upper Devonian/Lower Carboniferous flora from Ballyheigue, Ireland. *Review of Palaeobotany and Palynology* 43:303–20.

McComas, M. A. 1988. Upper Pennsylvanian compression floras of the 7-11 Mine, Columbiana County, northeastern Ohio. *Ohio Journal of Science* 88:48–52.

McGhee, G. R. 1982. The Frasnian-Fammenian extinction event: A preliminary analysis of Appalachian marine ecosystems. *Geological Society of America, Special Paper* 190:491–500.

McGinnis, H. J. 1967. The osteology of *Phlegethontia*, a Carboniferous and Permian aïstopod amphibian. *University of California Publications in Geological Sciences* 71:1–46.

McLaren, D. J. 1982. Frasnian-Fammenian extinction. *Geological Society of America, Special Paper* 190:477–84.

————. 1983. Bolides and biostratigraphy. *Geological Society of America, Bulletin* 94:313–24.

Meyen, S. V. 1982. The Carboniferous and Permian floras of Angaraland (a synthesis). *Biological Memoirs* 7:1–110.

Mickle, J. E. 1984. *Taxonomy of specimens of the Pennsylvanian-age Marattialean fern Psaronius from Ohio and Illinois.* Illinois State Museum, Scientific Paper 19.

Millay, M. A. 1979. Studies of Paleozoic marattialeans: A monograph of the American species of *Scolecopteris*. *Palaeontographica*, Abt. B 169:1–69.

Millay, M. A., and T. N. Taylor. 1979. Paleozoic seed-fern pollen organs. *Botanical Review* 45:301–75.

————. 1980. An unusual botryopteroid sporangial aggregation from the Middle Pennsylvanian of North America. *American Journal of Botany* 67:758–73.

Milner, A. C. 1980. A review of the Nectridea (Amphibia). 377–405 in A. L. Panchen, ed., *The Terrestrial Environment and the Origin of Land Vertebrates*. New York: Academic Press.

Milner, A. R. 1980a. The tetrapod assemblage from Nýřany, Czechoslovakia. 439–96 in A. L. Panchen, ed., *The Terrestrial Environment and the Origin of Land Vertebrates*. Systematics Association Special Volume 15.

————. 1980b. The temnospondyl amphibian *Dendrerpeton* from the Upper Carboniferous of Ireland. *Palaeontology* 23:125–41.

————. 1982. Small temnospondyl amphibians from the Middle Pennsylvanian of Illinois. *Palaeontology* 25:635–64.

————. 1985. On the identity of the amphibian *Hesperoherpeton garnettense* from the Upper Pennsylvanian of Kansas. *Palaeontology* 28:767–76.

————. 1986. Dissorophoid amphibians from the Upper Carboniferous of Nýřany. 671–74 in Z. Roček, ed., *Studies in Herpetology*. Prague.

————. 1987. The Westphalian tetrapod fauna: Some aspects of geography and ecology. *Journal of the Geological Society, London* 144:395–406.

————. 1989. The relationships of the eryopoid-grade temnospondyl amphibians from the Permian of Europe. *Acta Musei Reginaehradecensis* 22A:131–37.

Milner, A. R., T. R. Smithson, A. C. Milner, M. I. Coates, and W. D. I. Rolfe. 1986. The search for early tetrapods. *Modern Geology* 10:1–28.

Miner, R. W. 1925. The pectoral limb of *Eryops* and other primitive tetrapods. *Bulletin of the American Museum of Natural History* 51:145–312.

Modesto, S. P., and R. R. Reisz. 1990. A new skeleton of *Ianthasaurus hardestii*, a primitive edaphosaur (Synapsida: Pelecosauria) from the Upper Pennsylvanian of Kansas. *Canadian Journal of Earth Sciences* 27:834–44.

Morgan, J. 1959. The morphology and anatomy of American species of the genus *Psaronius*. *Illinois Biological Monographs* 27:1–107.

Moss, J. L. 1972. The morphology and phylogenetic relationships of the Lower Permian tetrapod *Tseajaia campi* Vaughn (Amphibia: Seymouriamorpha). *University of California Publications in Geological Sciences* 98:1–72.

Moulton, J. M. 1974. A description of the vertebral column of *Eryops* based on the notes and drawings of A. S. Romer. *Breviora* 428:1–44.

Moustafa, Y. S. 1955. The skeletal structure of *Parioxys ferricolus*, Cope. *Bulletin de l'Institute d'Egypte* 36:41–76.

Murdy, W. H., and H. N. Andrews. 1957. A study of *Botryopteris globosa* Darrah. *Bulletin of the Torrey Botanical Club* 84:252–67.

Murphy, J. L. 1971. Eryopsid remains from the Conemaugh Group, Braxton County, West Virginia. *Southeastern Geology* 13:265–63.

Murry, P. A., and G. D. Johnson. 1987. Clear Fork vertebrates and environments from the Lower Permian of north-central Texas. *Texas Journal of Science* 39:253–66.

Niklas, K. J. 1982. Computer simulations of early land plant branching morphologies: Canalization of patterns during evolution? *Paleobiology* 8:196–210.

———. 1986. Large-scale changes in animal and plant terrestrial communities. 383–405 in D. M. Raup and D. Jablonski, eds., *Patterns and Processes in the History of Life*. Dahlem Conference 1986. Berlin: Springer-Verlag.

Niklas, K. J., and T. D. O'Rourke. 1982. Growth patterns of plants that maximize vertical growth and minimize internal stress. *American Journal of Botany* 69: 1367–75.

Niklas, K. J., B. H. Tiffney, and A. H. Knoll. 1980. Apparent changes in the diversity of fossil plants. *Evolutionary Biology* 12:1–89.

———. 1985. Patterns in vascular land plant diversification: An analysis at the species level. 97–128 in J. W. Valentine, ed., *Phanerozoic Diversity Patterns: Profiles in Macroevolution*. Princeton: Princeton University Press.

Olson, E. C. 1936. The dorsal axial musculature of certain primitive Permian tetrapods. *Journal of Morphology* 59:265–311.

———. 1939. The fauna of the *Lysorophus* pockets in the Clear Fork Permian, Baylor County, Texas. *Journal of Geology* 47:389–97.

———. 1941. The family Trematopsidae. *Journal of Geology* 49:149–76.

———. 1947. The family Diadectidae and its bearing on the classification of reptiles. *Fieldiana, Geology* 11:1–53.

———. 1952. The evolution of a Permian vertebrate chronofauna. *Evolution* 6: 181–96.

———. 1958. Fauna of the Vale and Choza: 14. Summary, review, and integration of the geology and the faunas. *Fieldiana, Geology* 10:397–448.

———. 1961. Food chains and the origin of the mammals. *International Colloquium on the Evolution of Lower and Unspecialized Mammals*. Koninklijke Vlaamse Academie Voor Wetenschappen, Letteren en Schone Kunsten van België, pt. 1: 97–116.

———. 1962a. Late Permian terrestrial vertebrates, USA and USSR. *Transactions of the American Philosophical Society*, n.s. 52 (2):1–224.

———. 1962b. Permian vertebrates from Oklahoma and Texas. Part 2. The osteology of *Captorhinikos chozaensis* Olson. *Oklahoma Geological Survey Circular* 59: 49–68.

———. 1965. New Permian vertebrates from the Chickasha Formation in Oklahoma. *Oklahoma Geological Survey Circular* 70:1–70.

———. 1966. Community evolution and the origin of mammals. *Ecology* 47: 291–308.

———. 1967. Early Permian vertebrates. *Oklahoma Geological Survey Circular* 74:1–111.

———. 1968. The family Caseidae. *Fieldiana, Geology* 17:225–349.

———. 1970. New and little known genera and species of vertebrates from the Lower Permian of Oklahoma. *Fieldiana, Geology* 18:357–434.

———. 1975. Permo-Carboniferous paleoecology and morphotypic series. *American Zoologist* 15:371–89.

————. 1976. The exploitation of land by early tetrapods. In A. d'A. Bellairs and C. B. Cox, eds., *Morphology and Biology of Reptiles*. Linnean Society Symposium, series no. 3:1–30.

————. 1977. Permian lake faunas: A study in community evolution. *Journal of the Palaeontological Society of India* 20:146–63.

————. 1983. Coevolution or coadaptation? Permo-Carboniferous vertebrate chronofauna. 301–38 in M. Nitecki, ed., *Coevolution*. Chicago: University of Chicago Press.

————. 1985a. A larval specimen of a trematopsid (Amphibia: Temnospondyli). *Journal of Paleontology* 59:1173–80.

————. 1985b. Nonmarine vertebrates and late Paleozoic climates. *Ninth International Congress of Carboniferous Stratigraphy and Geology*, C. R. 5:403–14.

————. 1985c. Permo-Carboniferous vertebrate communities. *Ninth International Congress of Carboniferous Stratigraphy and Geology*, C. R. 5:331–45.

————. 1991. An eryopid (Amphibia: Labyrinthodontia) from the Fort Sill fissures, Lower Permian, Oklahoma. *Journal of Vertebrate Paleontology* 11:130–32.

Olson, E. C., and J. G. Mead. 1982. The Vale Formation (Lower Permian): Its vertebrates and paleoecology. *Texas Memorial Museum Bulletin* 29:1–46.

Olson, E. C., and P. P. Vaughn. 1970. The changes of terrestrial vertebrates and climates during the Permian of North America. *Forma et Functio* 3:113–38.

Oshurkova, M. V. 1974. A facies-paleoecological approach to the study of fossilized plant remains. *Paleontological Journal* 1974:363–70.

————. 1978. Paleophytocoenocenesis as a basis of a detailed stratigraphy with special reference to the Carboniferous of the Karaganda basin. *Review of Palaeobotany and Palynology* 25:181–87.

Panchen, A. L. 1972. The interrelationships of the earliest tetrapods. 65–87 in K. A. Joysey and T. S. Kemp, eds., *Studies in Vertebrate Evolution*. Edinburgh: Oliver and Boyd.

————. 1977. On *Anthracosaurus russelli* Huxley (Amphibia: Labyrinthodontia) and the family Anthracosauridae. *Philosophical Transactions of the Royal Society of London* B 279:447–512.

————. 1981. A jaw ramus of the Coal Measure amphibian *Anthracosaurus* from Northumberland. *Palaeontology* 24:85–92.

Parrish, J. T. 1982. Upwelling and petroleum source beds, with reference to the Paleozoic. *American Association of Petroleum Geologists Bulletin* 66:750–74.

Parrish, J. M., J. T. Parrish, and A. M. Ziegler. 1986. Permian-Triassic paleogeography and paleoclimatology and implications for therapsid distributions. 109–32 in N. Hotton III, P. D. MacLean, J. Roth, and E. C. Roth, eds., *The Biology and Ecology of Mammal-like Reptiles*. Washington, D.C.: Smithsonian Institution Press.

Peabody, F. E. 1961. Annual growth zones in living and fossil vertebrates. *Journal of Morphology* 108:11–62.

Pepperburg, R. V. 1910. Preliminary notes on the Carboniferous flora of Nebraska. *Nebraska Geological Survey* (11) 3:313–30.

Peppers, R. A. 1979. Development of coal-forming floras during the early part of the Pennsylvanian in the Illinois basin. 8–14 in J. E. Palmer and R. R. Dutcher, eds., *Depositional and Structural History of the Pennsylvanian System of the Illinois Basin*, part 2. Field Trip 9, Ninth International Congress of Carboniferous Stratigraphy and Geology, Urbana, Illinois.

Peppers, R. A., and H. W. Pfefferkorn. 1970. A comparison of the floras of the Colchester (No. 2) Coal and Francis Creek Shale. 61–74 in W. H. Smith, et al., eds., *Depositional Environments in Parts of the Carbondale Formation: Western and Northern Illinois*. Illinois State Geological Survey Guidebook Series 8.

Pfefferkorn, H. W. 1980. A note on the term "upland flora." *Review of Palaeobotany and Palynology* 30:157–58.

Pfefferkorn, H. W., W. H. Gillespie, D. A. Resnick, and M. H. Scheihing. 1984. Reconstruction and architecture of medullosan pteridosperms (Pennsylvanian). *Mosasaur* 2:1–8.

Pfefferkorn, H. W., and M. C. Thomson. 1982. Changes in dominance patterns in Upper Carboniferous plant-fossil assemblages. *Geology* 10:641–44.

Phillips, T. L. 1974. Evolution of vegetative morphology in coenopterid ferns. *Annals of the Missouri Botanical Garden* 61:427–61.

———. 1979. Reproduction of heterosporous arborescent lycopods in the Mississippian-Pennsylvanian of Euramerica. *Review of Palaeobotany and Palynology* 27:239–89.

———. 1980. Stratigraphic and geographic occurrences of permineralized coal-swamp plants: Upper Carboniferous of North America. 25–92 in D. L. Dilcher and T. N. Taylor, eds., *Biostratigraphy of Fossil Plants*. Stroudsburg, Pa.: Dowden, Hutchinson and Ross.

———. 1981. Stratigraphic occurrences and vegetational patterns of Pennsylvanian pteridosperms in Euramerican coal swamps. *Review of Palaeobotany and Palynology* 32:5–26.

Phillips, T. L., and H. N. Andrews. 1968. *Biscalitheca* (Coenopteridales) from the Upper Pennsylvanian of Illinois. *Palaeontology* 11:104–15.

Phillips, T. L., and W. A. DiMichele. 1981. Paleoecology of Middle Pennsylvanian age coal swamps in southern Illinois/Herrin Coal member at Sahara Mine No. 6. 231–84 in: K. J. Niklas, ed., *Paleobotany, Paleoecology, and Evolution*, vol. 1. New York: Praeger.

———. 1992. Comparative ecology and life-history biology of arborescent lycopods in Late Carboniferous swamps of Euramerica. *Annals of the Missouri Botanical Garden*.

Phillips, T. L., and R. A. Peppers. 1984. Changing patterns of Pennsylvanian coal-swamp vegetation and implications of climatic control on coal occurrence. *International Journal of Coal Geology* 3:205–55.

Phillips, T. L., R. A. Peppers, M. J. Avcin, and P. J. Laughnan. 1974. Fossil plants and coal: Patterns of change in Pennsylvanian coal swamps of the Illinois basin. *Science* 184:1367–69.

Phillips, T. L., R. A. Peppers and W. A. DiMichele. 1985. Stratigraphic and interregional changes in Pennsylvanian coal-swamp vegetation: Environmental inferences. *International Journal of Coal Geology* 5:43–109.

Pigg, K. B., T. N. Taylor, and R. A. Stockey. 1987. Paleozoic seed ferns: *Heterangium kentuckyensis* sp. nov., from the Upper Carboniferous of North America. *American Journal of Botany* 74:1184–1204.

Pimm, S. L. 1984. The complexity and stability of ecosystems. *Nature* 307:321–26.

Pirozynski, K. A., and D. W. Malloch. 1975. The origin of land plants: A matter of mycotropism. *Biosystems* 6:153–64.

Ponomarenko, A. G. 1969. Historical development of Archostemata beetles (in Russian). *Trudy Palaeontologicheskogo Instituta Akademiya Nauk SSSR* 125:1–240.

Pratt, L. M., T. L. Phillips, and J. M. Dennison. 1978. Evidence of non-vascular land plants from the early Silurian (Llandoverian) of Virginia, USA. *Review of Palaeobotany and Palynology* 25:121–49.

Pruvost, P. 1919. Le faune continentale du terrain houiller du Nord de la France. *Mém. Expl. Cart. Géol. Det. France, Paris.*

Pryor, J. S. 1988. Sampling methods for quantitative analysis of coal-ball plants. *Palaeogeography, Palaeoclimatology, Palaeoecology* 63:313–26.

Ramanujam, C. G. K., G. W. Rothwell, and W. N. Stewart. 1974. Probable attachment of the *Dolerotheca* companulum to a *Myeloxylon-Alethopteris* type frond. *American Journal of Botany* 61:1057–66.

Rasnitsyn, A. P. 1980. Origin and evolution of Hymenoptera (in Russian). *Trudy Paleontologicheskogo Instituta, Akademiya Nauk SSSR* 174:1–190.

Raymond, A. 1985. Floral diversity, phytogeography, and climatic amelioration during the Early Carboniferous (Dinantian). *Paleobiology* 11:293–309.

———. 1987. Interpreting ancient swamp communities: Can we see the forest in the peat? *Review of Palaeobotany and Palynology* 52:217–31.

———. 1988. Paleoecology of a coal-ball deposit from the Middle Pennsylvanian of Iowa dominated by cordaitalean gymnosperms. *Review of Palaeobotany and Palynology* 53:233–50.

Raymond, A., W. C. Parker, and S. F. Barrett. 1985a. Early Devonian phytogeography. 129–67 in B. H. Tiffney, ed., *Geological Factors and the Evolution of Plants.* New Haven: Yale University Press.

Raymond, A., W. C. Parker, and J. T. Parrish. 1985b. Phytogeography and paleoclimate of the Early Carboniferous. 169–222 in B. H. Tiffney, ed., *Geological Factors and the Evolution of Plants.* New Haven: Yale University Press.

Raymond, A., and T. L. Phillips. 1983. Evidence for an Upper Carboniferous mangrove community. 19–30 in H. J. Teas, ed., *Tasks for Vegetation Science.* The Hague: Dr. W. Junk.

Rayner, R. J. 1983. New observations on *Sawdonia ornata* from Scotland. *Transactions of the Royal Society of Edinburgh, Earth Sciences* 74:79–93.

———. 1984. New finds of *Drepanophycus spinaeformis* Goppert from the Lower Devonian of Scotland. *Transactions of the Royal Society of Edinburgh, Earth Sciences* 75:353–63.

Read, C. B. 1934. A flora of Pottsville age from the Mosquito Range, Colorado. *United States Geological Survey Professional Paper* 185-D: 74–96.

———. 1947. Pennsylvanian floral zones and floral provinces. *Journal of Geology* 55:271–79.

Read, C. B., and S. H. Mamay. 1964. Upper Paleozoic floral zones and floral provinces of the United States. *United States Geological Survey Professional Paper* 434-K: 1–35.

Read, C. B., and C. W. Merriam. 1940. A Pennsylvanian flora from central Oregon. *American Journal of Science* 238:107–11.

Read, W. F. 1943. Environmental significance of a small deposit in the Texas Permian. *Journal of Geology* 51:473–87.

Reisz, R. R. 1972. Pelycosaurian reptiles from the Middle Pennsylvanian of North

America. *Bulletin of the Museum of Comparative Zoology, Harvard University* 144:27–62.

———. 1975. Pennsylvanian pelycosaurs from Linton, Ohio, and Nýřany, Czechoslovakia. *Journal of Paleontology* 49:522–27.

———. 1980. A protorothyridid captorhinomorph reptile from the Lower Permian of Oklahoma. *Royal Ontario Museum Life Sciences Contributions* 121:1–16.

———. 1981. A diapsid reptile from the Pennsylvanian of Kansas. *University of Kansas Museum of Natural History Special Publication* 7:1–74.

———. 1990. Geology and paleontology of the Garnett quarry. *Kansas Geological Survey, Open File Report* 90-24:43–48.

Reisz, R. R., and D. Baird. 1983. Captorhinomorph "stem" reptiles from the Pennsylvanian coal-swamp deposits of Linton, Ohio. *Annals of Carnegie Museum* 52:393–411.

Reisz, R. R., and D. S Berman. 1986. *Ianthasaurus hardestii* n. sp., a primitive edaphosaur (Reptilia: Pelycosauria) from the Upper Pennsylvanian Rock Lake Shale near Garnett, Kansas. *Canadian Journal of Earth Sciences* 23:77–91.

Reisz, R. R., D. S Berman, and D. Scott. 1984. The anatomy and relationships of the Lower Permian reptile *Araeoscelis*. *Journal of Vertebrate Paleontology* 4:57–67.

Reisz, R. R., M. J. Heaton, and B. R. Pynn. 1982. Vertebrate fauna of Late Pennsylvanian Rock Lake Shale near Garnett, Kansas: Pelycosauria. *Journal of Paleontology* 56:741–50.

Remy, W. 1975. The floral changes at the Carboniferous-Permian boundary in Europe and North America. 305–52 in J. A. Barlow, ed., *The Age of the Dunkard*. West Virginia Geological and Economic Survey.

———. 1982. Lower Devonian gametophytes: Relation to the phylogeny of land plants. *Science* 215:1625–27.

Remy, W., and H. Hass. 1991a. Ergänzende Beobachtungen an *Lyonophyton rhyniensis*. *Argumenta Palaeobotanica* 8:1–27.

———. 1991b. *Kidstonophyton discoides* nov. gen., nov. spec., ein Gametophyt aus dem Chert von Rhynie (Unterdevon, Schottland). *Argumenta Palaeobotanica* 8:29–45.

———. 1991c. *Langiophyton mackiei* nov. gen., nov. spec., ein Gametophyt mit Archegoniophoren aus dem Chert von Rhynie (Unterdevon, Schottland). *Argumenta Palaeobotanica* 8:69–117.

Remy, W., and R. Remy. 1980. Devonian gametophytes with anatomically preserved gametangia. *Science* 208:295–96.

———. 1980b. *Lyonophyton rhyniensis* nov. gen. et nov. sp., ein Gametophyt aus dem Chert von Rhynie (Unterdevon, Schottland). *Argumenta Palaeobotanica* 6:37–72.

Remy, W., R. Remy, H. Hass, S. Schultka, and F. Franzmeyer. 1980a. *Sciadophyton* Steinmann—ein Gametophyt aus dem Siegen. *Argumenta Palaeobotanica* 6:73–94.

Remy, W., S. Schultka, and H. Hass. 1991. *Calyculiphyton blanai* nov. gen., nov. spec., ein Gametophyt aus dem Ems. *Argumenta Palaeobotanica* 8:119–45.

Remy, W., S. Schultka, H. Hass, and F. Franzmeyer. 1980b. *Sciadophyton*-Bestände im Siegen des Rheinischen Schiefergebirges als Beleg für festländische Bedingungen. *Argumenta Palaeobotanica* 6:95–114.

Retallack, G. J. 1981. Fossil soils: Indicators of ancient terrestrial environments.

55–102 in K. J. Niklas, ed., *Paleobotany, Paleoecology, and Evolution*, vol. 1. New York: Praeger.

———. 1985a. Fossil soils as grounds for interpreting the advent of large plants and animals on land. *Philosophical Transactions of the Royal Society of London*, B 309: 108–42.

———. 1985b. Descriptions, chemical, and petrographic data pertinent to the advent of large plants and animals on land. Open file report. Department of Geology, University of Oregon, Eugene, Oregon.

———. 1986. The fossil record of soils. 1–57 in V. P. Wright, ed., *Paleosols: Their Recognition and Interpretation*. Oxford: Blackwell Scientific.

Retallack, G. J., and D. L. Dilcher. 1988. Reconstructions of selected seed ferns. *Annals of the Missouri Botanical Garden* 75: 1010–57.

Retallack, G. J., and C. R. Feakes. 1987. Trace fossil evidence for Late Ordovician animals on land. *Science* 235: 61–63.

Rex, G. M. 1986. The preservation and palaeoecology of the Lower Carboniferous silicified plant deposits at Esnost, near Autun, France. *Géobios* 19: 773–800.

Rex, G. M., and A. C. Scott. 1987. The sedimentology, paleoecology, and preservation of the Lower Carboniferous plant deposits at Pettycur, Fife, Scotland. *Geological Magazine* 124: 43–66.

Richardson, E. S., and R. G. Johnson. 1971. The Mazon Creek faunas. *Proceedings of the North American Palaeontological Convention* 1: 1222–35.

Ricqlès, A. de, and J. R. Bolt. 1983. Jaw growth and tooth replacement in *Captorhinus aguti* (Reptilia: Captorhinomorpha): A morphological and histological analysis. *Journal of Vertebrate Paleontology* 3: 7–24.

Riek, E. F. 1976. An entomobryid collembolan (Hexapoda: Collembola) from the Lower Permian of South Africa. *Palaeontologia Africana* 19: 141–43.

Rieppel, O. 1980. The edopoid amphibian *Cochleosaurus* from the Middle Pennsylvanian of Nova Scotia. *Palaeontology* 23: 143–49.

Rolfe, W. D. I. 1980. Early invertebrate terrestrial faunas. 117–57. in A. L. Panchen, ed., *The Terrestrial Environment and the Origin of Land Vertebrates*. New York: Academic Press.

———. 1985a. Aspects of the Carboniferous terrestrial arthropod community. *Ninth International Congress of Carboniferous Stratigraphy and Geology*, C.R. 5: 303–16.

Rolfe, W. D. I. 1985b. Early terrestrial arthropods: A fragmentary record. *Philosophical Transactions of the Royal Society* B 309: 207–18.

Rolfe, W. D. I., G. P. Durant, A. E. Fallick, A. J. Hall, D. J. Large, A. C. Scott, T. R. Smithson, and G. M. Walkden. 1990. An early terrestrial biota preserved by Visean vulcanicity in Scotland. In M. G. Lockley and A. Rice, eds., *Volcanism and Fossil Biotas*. Geological Society of America, Special Paper 244: 13–24.

Rolfe, W. D. I., and J. K. Ingham. 1967. Limb structure, affinity, and diet of the Carboniferous "centipede" *Arthropleura*. *Scottish Journal of Geology* 3: 118–24.

Romer, A. S. 1922. The locomotor apparatus of certain primitive and mammal-like reptiles. *Bulletin of the American Museum of Natural History* 46: 517–606.

———. 1927. Notes on the Permo-Carboniferous reptile *Dimetrodon*. *Journal of Geology* 35: 673–89.

———. 1928. A skeletal model of the primitive reptile *Seymouria*, and the phylogenetic position of that type. *Journal of Geology* 36:248–60.

———. 1944. The Permian cotylosaur *Diadectes tenuitectus*. *American Journal of Science* 242:139–44.

———. 1945. The Late Carboniferous vertebrate fauna of Kounova (Bohemia) compared with that of the Texas redbeds. *American Journal of Science* 243:417–42.

———. 1946. The primitive reptile *Limnoscelis* restudied. *American Journal of Science* 244:148–88.

———. 1948. Relative growth in pelycosaurian reptiles. 45–55 in A. L. Du Toit, ed., *Robert Broom Commemorative Volume*. Special Publication of the Royal Society of South Africa.

———. 1952. Late Pennsylvanian and Early Permian vertebrates of the Pittsburgh–West Virginia Region. *Annals of the Carnegie Museum* 33:47–112.

———. 1957. The appendicular skeleton of the Permian embolomerous amphibian *Archeria*. *Contributions, Museum of Paleontology, University of Michigan* 13:103–59.

———. 1969. The cranial anatomy of the Permian amphibian *Pantylus*. *Breviora* 314:1–37.

———. 1973. Permian reptiles. 159–67 in A. Hallam, ed., *Atlas of Palaeobiogeography*. Amsterdam: Elsevier Scientific.

Romer, A. S., and L. I. Price. 1940. Review of the Pelycosauria. *Geological Society of America Special Papers* 28:1–538.

Roth, E. A. 1955. The anatomy and modes of preservation of the genus *Cardiocarpus spinatus* Graham. *University of Kansas Science Bulletin* 37:151–74.

Rothwell, G. W. 1972. Evidence of pollen tubes in Paleozoic pteridosperms. *Science* 175:772–74.

———. 1977. Evidence of a pollination-drop mechanism in Paleozoic pteridosperms. *Science* 198:1251–52.

———. 1981. The Callistophytales (Pteridospermopsida): Reproductively sophisticated gymnosperms. *Review of Palaeobotany and Palynology* 32:103–21.

———. 1989. *Botryopteris forensis*, an epiphyte on *Psaronius*. *American Journal of Botany* 76 (6): 206–7.

Rothwell, G. W., and G. Mapes. 1988. Vegetation of a Paleozoic conifer community. *Kansas Geological Survey Guidebook Series* 6:213–23.

Rothwell, G. W., and S. E. Scheckler. 1988. Biology of ancestral gymnosperms. 85–134 in C. B. Beck, ed., *Origin and Evolution of Gymnosperms*. New York: Columbia University Press.

Rothwell, G. W., and A. C. Scott. 1985. Ecology of the Lower Carboniferous plant remains from Oxroad Bay, East Lothian, Scotland. *American Journal of Botany* 72 (6): 899.

Rothwell, G. W., and S. Warner. 1984. *Cordaixylon dumusum* n. sp. (Cordaitales) 1. Vegetative structures. *Botanical Gazette* 145:275–91.

Rowe, N. P., and J. Galtier. 1989. A Lower Carboniferous plant assemblage from La Serre (Montagne Noire, France). Part 1. *Review of Palaeobotany and Palynology* 61:239–71.

Rowley, D. B., A. Raymond, J. T. Parrish, A. L. Lottes, C. R. Scotese, and A. M. Ziegler. 1985. Carboniferous paleogeographic, phytogeographic, and paleoclimatic reconstructions. *International Journal of Coal Geology* 5:7–42.

Russell-Sigogneau, D. 1989. *Theriodontia I. Handbuch der Palaeoherpetologie 17B/1.* New York: Gustav Fischer Verlag.

Rust, B. R., M. R. Gibling, and A. S. Legun. 1984. Coal deposition in an anastomosing-fluvial system: The Pennsylvanian Cumberland Group south of Joggins, Nova Scotia, Canada. *Special Publication of the International Association of Sedimentologists* 7:105–20.

Sander, P. M. 1987. Taphonomy of the Lower Permian Geraldine bonebed in Archer County, Texas. *Palaeogeography, Palaeoclimatology, Palaeoecology* 61:221–36.

Sawin, H. J. 1941. The cranial anatomy of *Eryops megacephalus. Bulletin of the Museum of Comparative Zoology at Harvard College* 88:405–63.

Scheckler, S. E. 1978. Ontogeny of progymnosperms. 2. Shoots of Upper Devonian Archaeopteridales. *Canadian Journal of Botany* 56:3136–70.

———. 1985. Origins of the coal swamp biome: Evidence from the southern Appalachians. *Geological Society of America, Abstracts with Programs* 17(2):134.

———. 1986a. Floras of the Devonian-Mississippian transition. *University of Tennessee, Department of Geological Sciences, Studies in Geology* 15:81–96.

———. 1986b. Geology, floristics, and paleoecology of Late Devonian coal swamps from Appalachian Laurentia (USA). *Annales de la Société géologique de Belgique* 109:209–22.

———. 1986c. Old Red continent facies in the Late Devonian and Early Carboniferous of Appalachian North America. *Annales de la Société géologique de Belgique* 109:223–36.

Scheckler, S. E., and H. E. Beeler. 1984. Early Carboniferous coal swamp floras from eastern USA (Virginia). *Second International Organization of Palaeobotany Conference, Edmonton, Alberta, Abstracts,* 36.

Scheckler, S. E., and S. A. Hill. 1989. Fossil plants of the high Arctic. *Virginia Academy of Sciences. Abstracts.*

Scheihing, M. H., and H. W. Pfefferkorn. 1984. The taphonomy of land plants in the Orinoco Delta: A model for the incorporation of plant parts in clastic sediments of Late Carboniferous age of Euramerica. *Review of Palaeobotany and Palynology* 41:205–80.

Schneider, J. 1978. Zur Taxonomie und Biostratigraphie der Blattodea (Insecta) aus dem Oberkarbon und Perm der DDR. *Freiberger Forschungshefte* (C) 340:1–152.

———. 1980. Zur Taxonomie der jungpaläozoischen Neorthroblattinidae (Insecta, Blattodea). *Freiberger Forschungshefte* (C) 348:31–40.

———. 1982. Entwurf einer biostratigraphischen Zonengliederung mittels der Spiloblattinidae (Blattodea, Insecta) für das kontinentale euramerische Permokarbon. *Freiberger Forschungshefte* (C) 345:27–47.

Schweitzer, H.-J. 1980. *Die Gattungen* Taeniocrada White *und* Sciadophyton *Steinmann im Unterdevon des Rheinlandes.* Bonner Palaeobotanische Mitteilungen, no. 5.

Schweitzer, H.-J. 1981. *Der Generationswechsel rheinischer Psilophyten.* Bonner Palaeobotanische Mitteilungen, no. 8.

———. 1983. Die Unterdevonflora des Rheinlandes. *Palaeontographica,* Abt. B. 189:1–138.

Scotese, C. R., R. Van der Voo, and S. F. Barrett. 1985. Devonian base maps. *Philosophical Transactions of the Royal Society of London* B 309:29–56.

Scott, A. C. 1977. A review of the ecology of Upper Carboniferous plant assemblages, with new data from Strathclyde. *Palaeontology* 20:447–73.

———. 1978. Sedimentological and ecological control of Westphalian B plant assemblages from West Yorkshire. *Proceedings of the Yorkshire Geological Society* 41:461–508.

———. 1979. The ecology of some Coal Measures floras from northern Britain. *Proceedings of the Geologists Association* 90:97–116.

———. 1980. The ecology of some Upper Paleozoic floras. 87–115 in A. L. Panchen, ed., *The Terrestrial Environment and the Origin of Land Vertebrates.* Systematics Association Special Volume 15. London: Academic Press.

———. 1984. Studies on the sedimentology, palaeontology, and palaeoecology of the middle Coal Measures (Westphalian B, Upper Carboniferous) at Swillington, Yorkshire. Part 1. Introduction. *Transactions of the Leeds Geological Association* 10:1–16.

———. 1988. Volcanoes, fires, and the Lower Carboniferous vegetation of Scotland. *Open University Geological Society Journal* 8.2:27–31.

———. 1990. Preservation, evolution, and extinction of plants in Lower Carboniferous volcanic sequences in Scotland. 25–38 in M. G. Lockley and A. Rice, eds., *Volcanism and Fossil Biotas.* Geological Society of America, Special Paper 244.

Scott, A. C., and W. G. Chaloner. 1983. The earliest fossil conifer from the Westphalian B of Yorkshire. *Proceedings of the Royal Society* B 220: 163–82.

Scott, A. C., and J. Galtier. 1985. Distribution and ecology of early ferns. *Proceedings of the Royal Society of Edinburgh* B 86: 141–49.

Scott, A. C., W. G. Chaloner, and S. Paterson. 1985. Evidence of pteridophyte-arthropod interactions in the fossil record. *Proceedings of the Royal Society of Edinburgh* B 86: 461–508.

Scott, A. C., J. Galtier, and G. Clayton. 1984a. Distribution of anatomically preserved floras in the Lower Carboniferous of Western Europe. *Transactions of the Royal Society of Edinburgh, Earth Sciences* 75:311–40.

———. 1985b. A new late Tournaisian (Lower Carboniferous) flora from the Kilpatrick Hills, Scotland. *Review of Palaeobotany and Palynology* 44:81–99.

Scott, A. C., B. Meyer-Berthaud, J. Galtier, G. M. Rex, S. A. Brindley, and G. Clayton. 1986. Studies on a new Lower Carboniferous flora from Kingswood near Pettycur, Scotland. 1. Preliminary report. *Review of Palaeobotany and Palynology* 48:161–80.

Scott, A. C., and G. M. Rex. 1987. The accumulation and preservation of Dinantian plants from Scotland and its borders, 329–44 in J. Miller, A. E. Adams, and V. P. Wright, ed., *European Dinantian Environments.* Geological Journal Special Issue no. 12.

Scott, A. C., and T. N. Taylor. 1983. Plant/animal interactions during the Upper Carboniferous. *Botanical Review* 49:259–307.

Selden, P. A. 1986. Ricinuleids: Living fossils? *Actas 10 Congreso Internacional de Aracnología, Jaca, Spain* 1:425.

Sellards, E. H. 1908. Fossil plants of the Upper Paleozoic of Kansas. *Kansas Geological Survey Report* 9:434–67.

Sharov, A. G. 1968. Phylogeny of orthopteroid insects. *Trudy Palaeontologicheskogo*

Instituta Akademiya Nauk SSSR 118:1–218. English translation from Russian, 1971, Israel Program for Scientific Translation.

————. 1973. Morphological features and way of life of the Palaeodictyoptera (in Russian). *Doklady 24 Chtenii Pamiati N.A. Cholodkovkoga* 1:49–63.

Shear, W. A. 1986. A fossil fauna of early terrestrial arthropods from the Givetian (upper Middle Devonian) of Gilboa, New York, USA. *Actas 10 Congreso Internacional Aracnología, Jaca, Spain* 1:387–92.

Shear, W. A., P. M. Bonamo, J. D. Grierson, W. D. I. Rolfe, E. L. Smith, and R. A. Norton. 1984. Early land animals in North America: Evidence from Devonian-age arthropods from Gilboa, New York. *Science* 224:492–94.

Shear, W. A., and J. Kukalova-Peck. 1990. The ecology of Paleozoic terrestrial arthropods: The fossil evidence. *Canadian Journal of Zoology* 68:1807–34.

Shear, W. A., J. M. Palmer, J. A. Coddington, and P. M. Bonamo. 1989. A Devonian spineret: Early evidence of spiders and silk use. *Science* 246:479–81.

Shear, W. A., P. A. Selden, W. D. I. Rolfe, P. M. Bonamo, and J. D. Grierson. 1987. New terrestrial arachnids from the Devonian of Gilboa, New York. *American Museum Novitates* 2901:1–74.

Sherwood-Pike, M. A., and I. Gray. 1985. Silurian fungal remains: Probable records of the class Ascomycetes. *Lethaia* 18:1–20.

Smith, A. V. H., and M. A. Butterworth. 1967. Miospores in the Coal Seams of the Carboniferous of Great Britain. *Special Paper in Palaeontology*, no. 1:1–324.

Smith, R. M. H. 1987. Helical burrow casts of therapsid origin from the Beaufort Group (Permian) of South Africa. *Palaeogeography, Palaeoclimatology, Palaeoecology* 60:155–170.

Smithson, T. R. 1985a. Scottish Carboniferous amphibian localities. *Scottish Journal of Geology* 21:123–42.

————. 1985b. The morphology and relationships of the Carboniferous amphibian *Eoherpeton watsoni* Panchen. *Zoological Journal of the Linnean Society* 85:317–410.

————. 1989. The earliest known reptile. *Nature* 342:676–78.

Solem, A., and E. L. Yochelson. 1979. North American Paleozoic land snails with a summary of other Paleozoic non-marine snails. *United States Geological Survey, Professional Paper* 1072:1–42.

Stanley, S. M. 1988. Paleozoic mass extinctions: Shared patterns suggest global cooling as a common cause. *American Journal of Science* 288:334–52.

Stebbins, G. L. 1952. Aridity as a stimulus to plant evolution. *American Naturalist* 86:33–44.

Stewart, W. N., and T. Delevoryas. 1956. The medullosan pteridosperms. *Botanical Review* 22:45–80.

Stidd, B. M. 1980. The current status of medullosan seed ferns. *Review of Palaeobotany and Palynology* 32:63–101.

Stidd, B. M., and T. L. Phillips. 1973. The vegetative anatomy of *Schopfiastrum decussatum* from the Middle Pennsylvanian of the Illinois basin. *American Journal of Botany* 60:463–74.

Størmer, L. 1970. Arthropods from the Lower Devonian (Lower Emsian) of Alken an der Mosel, Germany. Part 1. *Senckenbergiana Lethaea* 51:335–69.

———. 1976. Arthropods from the Lower Devonian (Lower Emsian) of Alken an der Mosel, Germany. Part 5. *Senkenbergiana Lethaea* 57:87–183.

Stovall, J. W. 1950. A new cotylosaur from north central Oklahoma. *American Journal of Science* 248:46–54.

Stovall, J. W., L. I. Price, and A. S. Romer. 1966. The postcranial skeleton of the giant Permian pelycosaur *Cotylorhynchus romeri*. *Harvard University, Bulletin of the Museum of Comparative Zoology* 135:1–30.

Stschegolev, A. K. 1975. Die Entwicklung der Pflanzenbedeckung im Süden des Europäischen Teils der UdSSR, vom Ende des Mittelkarbons bis zum Perm. Umfang und Gliederung des oberen Karbons (Stefan). *C.R. Seventh International Congress Carboniferous Stratigraphy and Geology (Krefeld)* 4:275–80.

Sues, H.-D. 1986. Locomotion and body form in early therapsids (Dinocephalia, Gorgonopsia, and Therocephalia). 61–70 in N. Hotton et al., eds., *The Ecology and Biology of Mammal-like Reptiles*. Washington, D.C.: Smithsonian Institution Press.

Sues, H.-D., and J. A. Boy. 1988. A procynosuchid cynodont from central Europe. *Nature* 331:523–24.

Sumida, S. S. 1989a. The appendicular skeleton of the Early Permian genus *Labidosaurus* (Reptilia, Captorhinomorpha, Captorhinidae) and the hind limb musculature of captorhinid reptiles. *Journal of Vertebrate Paleontology* 9:295–313.

———. 1989b. New information on the pectoral girdle and vertebral column in *Lupeosaurus* (Reptilia, Pelycosauria). *Canadian Journal of Earth Sciences* 26: 1343–49.

———. 1990. Vertebral morphology, alternation of neural spine height, and structure in Permo-Carboniferous tetrapods, and a reappraisal of primitive modes of terrestrial locomotion. *University of California Publications in Zoology* 122:1–129.

Tanner, W. R. 1983. A fossil flora from the Beartooth Butte Formation of northern Wyoming. Ph.D. diss., Southern Illinois University, Carbondale.

Taylor, T. N. 1965. Paleozoic seed studies: A monograph of the genus *Pachytesta*. *Palaeontographica*, Abt. B. 117:1–46.

———. 1982. The origin of land plants: A paleobotanical perspective. *Taxon* 31: 155–77.

———. 1990. Fungal associations in the terrestrial paleoecosystem. *Trends in Ecology and Evolution* 5:21–25.

Taylor, T. N., and A. C. Scott. 1983. Interactions of plants and animals during the Carboniferous. *BioScience* 33:488–93.

Taylor, T. N., and W. N. Stewart. 1964. The Paleozoic seed *Mitrosperum* in American coal balls. *Palaeontographica*, Abt. B 115:51–58.

Thayer, D. W. 1985. New Pennsylvanian lepospondyl amphibians from the Swisshelm Mountains, Arizona. *Journal of Paleontology* 59:684–700.

Tidwell, W. D. 1967. Flora of the Manning Canyon Shale. Part 1: A lowermost Pennsylvanian flora from the Manning Canyon Shale, Utah, and its stratigraphic significance. *Brigham Young University Geology Studies* 14:3–66.

———. 1988. A new Upper Pennsylvanian or Lower Permian flora from southeastern Utah. *Brigham Young University Geology Studies* 35:33–56.

Tidwell, W. D., D. A. Medlyn, and A. D. Simper. 1974. Flora of the Manning Canyon Shale. Part 2: Lepidodendrales. *Brigham Young University Geology Studies* 21: 119–46.

Tiffney, B. H., and K. J. Niklas. 1985. Clonal growth in land plants: A paleobotanical perspective. 35–66 in J. B. C. Jackson, L. W. Buss, and R. E. Cook, eds., *Population Biology and Evolution of Clonal Organisms*. New Haven: Yale University Press.

Timms, J. D., and T. C. Chambers. 1984. Rhyniophytina and Trimerophytina from the early land floras of Victoria, Australia. *Palaeontology* 27:265–79.

Trivett, M. L., and G. W. Rothwell. 1985. Morphology, systematics, and paleoecology and Paleozoic fossil plants: *Mesoxylon priapi*, sp. nov. (Cordaitales). *Systematic Botany* 10:205–23.

———. 1988. Modelling the growth architecture of fossil plants: A Paleozoic filicalean fern. *Evolutionary Trends in Plants* 2:25–29.

Valentine, J. W. 1980. Determinants of diversity in higher taxonomic categories. *Paleobiology* 6:444–50.

Valentine, J. W., and T. D. Walker. 1987. Extinctions in a model taxonomic hierarchy. *Paleobiology* 13:193 207.

Vaughn, P. P. 1955. The Permian reptile *Araeoscelis* restudied. *Bulletin of the Museum of Comparative Zoology at Harvard College* 113:303–467.

———. 1958. On the geologic range of the labyrinthodont amphibian *Eryops*. *Journal of Paleontology* 32:918–22.

———. 1964. Vertebrates from the Organ Rock Shale of the Cutler Group, Permian of Monument Valley and vicinity, Utah and Arizona. *Journal of Paleontology* 38:567–83.

———. 1969. Upper Pennsylvanian vertebrates from the Sangre de Cristo Formation of central Colorado. *Los Angeles County Natural History Museum, Contributions to Science* 164:1–28.

———. 1970. Lower Permian vertebrates of the Four Corners and the Midcontinent as indices of climatic differences. *Proceedings of the North American Paleontological Convention* 1969 D: 388–408.

———. 1971. A *Platyhystrix*-like amphibian with fused vertebrae, from the Upper Pennsylvanian of Ohio. *Journal of Paleontology* 45:464–69.

———. 1972. More vertebrates, including a new microsaur, from the Upper Pennsylvanian of central Colorado. *Los Angeles County Natural History Museum, Contributions to Science* 223:1–30.

Veevers, J. J., and C. McA. Powell. 1987. Late Paleozoic glacial episodes in Gondwanaland reflected in transgressive-regressive depositional sequences in Euramerica. *Geological Society of America Bulletin* 98:475–87.

Vermeij, G. J. 1987. *Evolution and Escalation*. Princeton: Princeton University Press.

Visscher, H., J. H. F. Kerp, and J. A. Clement-Westerhof. 1986. Aspects of Permian palaeobotany and palynology. 6. Towards a flexible system of naming Palaeozoic conifers. *Acta Botanica Neederlandica* 35:87–99.

Visser, J. N. J. 1987. The palaeogeography of part of southwestern Gondwana during the Permo-Carboniferous glaciation. *Palaeogeography, Palaeoclimatology, Palaeoecology* 61:205–19.

————. 1989. The Permo-Carboniferous Dwyka Formation of southern Africa: Deposition by a predominantly subpolar marine ice sheet. *Palaeogeography, Palaeoclimatology, Palaeoecology* 70:377–91.

Warren, A. A., R. Jupp, and B. R. Bolton. 1986. Earliest tetrapod trackway. *Alcheringa* 10:183–86.

Warren, J. W., and N. A. Wakefield. 1972. Trackways of tetrapod vertebrates from the Upper Devonian of Victoria, Australia. *Nature* 238:469–70.

Watson, D. M. S. 1954. On *Bolosaurus* and the origin and classification of reptiles. *Bulletin of the Museum of Comparative Zoology at Harvard College* 111:295–449.

Wellstead, C. F. 1982. A Lower Carboniferous aïstopod amphibian from Scotland. *Palaeontology* 25:193–208.

Werneburg, R. 1989. Some notes to systematics, phylogeny, and biostratigraphy of labyrinthodont amphibians from the Upper Carboniferous and Lower Permian in central Europe. *Acta Musei Reginaehradecensis* (A) 22:117–29.

White, D. 1912. The characters of the fossil plant *Gigantopteris* Schenk and its occurrence in North America. *United States National Museum Proceedings* 41:493–516.

————. 1929. *Flora of the Hermit Shale, Grand Canyon, Arizona.* Carnegie Institute of Washington 405.

————. 1933. Some features of the Early Permian flora of America. *Sixteenth International Geological Congress* 1:679–89.

White, T. E. 1939. Osteology of *Seymouria baylorensis* Broili. *Bulletin of the Museum of Comparative Zoology at Harvard College* 85:323–410.

Willard, D. A. In press. Vegetational patterns in the Springfield Coal (Middle Pennsylvanian, Illinois Basin): Comparison of miospore and coal-ball records. *Geological Society of America Special Paper.*

Williston, S. W. 1911. *American Permian Vertebrates.* Chicago: University of Chicago Press.

Winston, R. B. 1983. A late Pennsylvanian upland flora in Kansas: Systematics and environmental implications. *Review of Palaeobotany and Palynology* 40:5–31.

————. 1986. Characteristic features and compaction of plant tissues traced from permineralized peat to coal in Pennsylvanian coals (Desmoinesian) from the Illinois Basin. *International Journal of Coal Geology* 6:21–41.

————. 1988. Paleoecology of Middle Pennsylvanian–age peat-swamp plants in Herrin Coal, Kentucky, USA. *International Journal of Coal Geology* 10:203–38.

————. 1989. Identification of plant megafossils in Pennsylvanian-age coal. *Review of Palaeobotany and Palynology* 57:265–76.

————. 1990. Implications of paleobotany of Pennsylvanian-age coal of the central Appalachian basin for climate and coal-bed development. *Geological Society of America Bulletin* 102:1720–26.

Winston, R. B., and R. W. Stanton. 1989. Plants, Coal, and Climate in the Pennsylvanian of the central Appalachians. 118–26 in C. B. Cecil et al., eds. *Carboniferous Geology of the Eastern United States.* International Geological Congress, Field Trip Guidebook, T143.

Wnuk, C. 1985. The ontogeny and paleoecology of *Lepidodendron rimosum* and *Lepidodendron bretonense* trees from the Middle Pennsylvanian of the Bernice Basin, Sullivan County, Pennsylvania. *Palaeontographica,* Abt. B. 195:153–81.

————. 1986. Preliminary observations on the paleoecological significance and taphonomic history of underclay floras. 89–98 in W. E. Cox and W. F. Snyder, chairpersons, New River Symposium, Sponsored by New River Gorge National River (National Park Service), Wytheville Community College and West Virginia Department of Culture and History.

———. 1989. The ontogeny and paleoecology of the Middle Pennsylvanian arborescent lycopod *Bothrodendron punctatum*, Bothrodendraceae (Western Middle Anthracite Field, Shamokin Quadrangle, Pennsylvania). *American Journal of Botany* 76:966–80.

Wnuk, C., and H. W. Pfefferkorn. 1984. The life habits and paleoecology of Middle Pennsylvanian medullosan pteridosperms based on an *in situ* assemblage from the Bernice Basin (Sullivan County, Pennsylvania, USA). *Review of Palaeobotany and Palynology* 41:329–51.

———. 1987. A Pennsylvanian-age terrestrial storm deposit: Using plant fossils to characterize the history and process of sediment accumulation. *Journal of Sedimentary Petrology* 57:212–21.

Wood, S. P., A. L. Panchen, and T. R. Smithson. 1985. A terrestrial fauna from the Scottish Lower Carboniferous. *Nature* 314:355–56.

Woodrow, D. L., F. W. Fletcher, and W. F. Ahrnsbrak. 1973. Paleogeography and paleoclimate at the deposition sites of the Devonian Catskill and Old Red facies. *Geological Society of America Bulletin* 84:3051–64.

Wootton, R. J. 1976. The fossil record and insect flight. 235–54 in R. C. Rainey, ed., *Insect Flight*. New York: John Wiley and Sons.

———. 1981. Paleozoic insects. *Annual Review of Entomology* 26:319–44.

Wright, V. P. 1985. The precursor environment for vascular plant colonization. *Philosophical Transactions of the Royal Society of London* B 309: 143–45.

Zajíc, J., and S. Stamberg. 1985. Summary of Permocarboniferous freshwater fauna of the limnic basins of Bohemia and Moravia. *Acta Musei Reginaehradecensis* (A) 20:61–82.

Zhang Shanzhen, J.-P. Laveine, Y. Lemoigne, and Ding Hui. 1987. Fossil plants from the Penchi Formation (Carboniferous) in Taiyuan area, Shanxi Province, North China. *Revue de Paléobiologie* 6:5–17.

Zhao Xiu-hu, and Wu Xiu-yuan, 1979. Carboniferous macrofloras of southern China. *Nanjing Institute of Geology and Paleontology, Academia Sinica, Nanjing, China* 1–8.

Ziegler, A. M. 1990. Phytogeographic patterns and continental configurations during the Permian Period. 363–79 in W. S. McKerrow and C. R. Scotese, eds., *Palaeozoic Palaeogeography and Biogeography*. Geological Society of London Memoir 12.

Ziegler, A. M., R. K. Bambach, J. T. Parrish, S. F. Barrett, E. H. Gierlowski, W. C. Parker, A. Raymond, and J. J. Sepkoski. 1981. Paleozoic biogeography and climatology. 231–66 in K. J. Niklas, ed., *Paleobotany, Paleoecology, and Evolution*, vol. 2. New York: Praeger.

Zwan, C. J. van der. 1981. Palynology, phytogeography, and climate of the Lower Carboniferous. *Palaeogeography, Palaeoclimatology, and Palaeoecology* 33: 279–310.

S I X   Mesozoic and Early Cenozoic
            Terrestrial Ecosystems

Scott L. Wing and Hans-Dieter Sues, RAPPORTEURS

IN COLLABORATION WITH BRUCE H. TIFFNEY, RICHARD K. STUCKY, DAVID B.
    WEISHAMPEL, ROBERT A. SPICER, DAVID JABLONSKI, CATHERINE E. BADGLEY,
    MARK V. H. WILSON, AND WARREN L. KOVACH

1 INTRODUCTION

As discussed in chapter 5, the basic trophic structure of modern terrestrial ecosystems was established no later than the Late Permian, when vertebrate herbivores joined the terrestrial food pyramid for the first time. Although the fundamental template for subsequent terrestrial communities was created during the first 150-to-200 My of life on land, the 200-My interval encompassed by the Mesozoic and early Cenozoic records major shifts in vegetational structure, in the functional attributes of herbivorous animals, and in the degree and nature of inferred interactions between plants and herbivores, as well as the two greatest mass-extinction episodes in the history of terrestrial ecosystems. We have included the Paleocene and Eocene with the Mesozoic because this facilitates comparison of terrestrial ecosystems before and after the Cretaceous/Tertiary (K/T) extinctions.

Mesozoic and early Cenozoic terrestrial ecosystems are temporally intermediate between the first integrated ecosystems of the Late Permian and middle to late Tertiary ecosystems that have close living analogues. Taxonomic modernization during this interval was "progressive" (though hardly uniformly so), but our paleoecological inferences suggest this was not true for the structure of terrestrial vegetation or the dynamics of plant-animal interactions. For example, the dynamics of some late Eocene ecosystems may be more analogous to those of the later Mesozoic than of the Paleocene and early Eocene ecosystems that immediately preceded them. Similarly, in some aspects of trophic relationships and vertebrate-plant interactions, the Paleocene may more closely resemble the late Paleozoic than the Late Cretaceous. These major fluctuations in the nature of Mesozoic and early Cenozoic ecosystems contrast with the more directional progression of the Paleozoic and may give us greater insight into the response of terrestrial ecosystems to major perturbations.

Increasing complexity and interdependence of ecological systems during the Paleozoic seems to have been related to evolutionary innovations among terrestrial animals and plants (e.g., development of tree habit in plants, herbivory in animals). During the Mesozoic and early Cenozoic, there were periods of stability and change in type of herbivory and architecture of vegetation that appear to be related to climatic shifts, major extinctions, and perhaps cybernetic properties of ecosystems, though evolutionary innovations continued to play an important role in expanding the array of ecological interactions between terrestrial organisms. The contrast between the Paleozoic and Mesozoic-Paleogene can be seen as the difference between observing the behavior of a system during its assembly and observing its response to perturbation once it is established.

Data on the paleoecology and spatiotemporal distribution of fossil plants and herbivorous animals can be culled from various reference works (vertebrates in general: Carroll 1988; dinosaurs: Weishampel et al. 1990; insects: Hennig 1981; plants: Taylor 1981; Stewart 1983; Meyen 1987), but they are rarely integrated with one another. We think this review of the combined animal and plant data from an evolutionary perspective has generated new questions. It will be clear from the temporal, geographic, and environmental distribution of the fossil localities discussed here that many inferences about past ecological interactions would benefit from more evidence. Perhaps the single most important defect of the fossil record is that insects and other terrestrial invertebrates are poorly known, or are described only in systematic terms, making it difficult to evaluate their possible ecological interactions with other land organisms. In present-day ecosystems, herbivory by insects and other invertebrates (including acarids, tardigrades, myriapods, nematodes and gastropods) is approximately of the same order of magnitude as that by vertebrates, although the latter generally have a far greater impact on the structure of vegetation (Crawley 1983). Because modern types of arthropod herbivory probably were established by the mid-Pennsylvanian (Labandeira 1990; Scott 1991), the selective impact of arthropod herbivory on plants may have changed relatively little since the early Mesozoic. Thus, our scant knowledge of the fossil record of terrestrial invertebrates may not distort too seriously our understanding of changes in herbivory during the Mesozoic and early Cenozoic. In the modern world, terrestrial arthropods also play a major role in plant reproduction, and though there have been recent advances in interpreting past pollination interactions (e.g., Crepet and Friis 1987), these are based more on plant fossils than on arthropod fossils, leaving open the possibility that interpretations will change substantially when better information is available on the pollinators themselves.

This report is divided into three main parts. Sections 2–5 summarize the geographic, temporal, and environmental distribution of important fossil assemblages of terrestrial plants and herbivorous tetrapods and insects, and provide a chronological summary of first-order inferences about changing vegetation and faunas during the Mesozoic and early Cenozoic. In syntheses following the reviews of the fossil record for each of the Mesozoic periods and for the Paleocene-Eocene, we highlight the outstanding paleoecological characteristics of the respective floras and faunas of the Triassic, Jurassic, Cretaceous, and Paleogene, and we make some inferences concerning the ecological interactions between contemporaneous animals and plants. We also relate ecological inferences to prominent evolutionary events among animals and plants, to extinctions and diversifications, and to paleoclimatic changes. We place special emphasis on times and places where ecological and evolutionary changes are strongly concordant or discordant and on offering explanations for these patterns. We also make predictions about what will be found in the fossil record if our explanations are correct.

In §6 we discuss how various kinds of ecological interactions appear to have changed through time. In §7 we use examples to take up the connection between environmental conditions and inferred community structure, the implications of change in community structure and dynamics for the evolution of individual lineages (and vice versa), and the way in which large-scale perturbations might have affected ecological structure and evolutionary opportunity.

2 TRIASSIC BIOTAS

The history of Mesozoic terrestrial ecosystems is marked by a series of large-scale perturbations. The basis for the boundary between the Paleozoic and Mesozoic is the largest marine extinction of the Phanerozoic. However, in contrast to the estimates of a 91% to 97% loss of species in shallow-water marine faunas (Raup 1986), estimates of the diversity decline in terrestrial organisms range from about 20% for species of plants (Niklas et al. 1980) to a questionable 50% for genera of tetrapods (Benton 1985, 1987; Padian and Clemens 1985). Disappearance of most Permian pollen types exactly at the Permian-Triassic boundary in Israel and reduced diversity in Early Triassic assemblages are thought to reflect a preservational or depositional hiatus rather than a catastrophic extinction (Eshet 1990). The relatively smaller decrease in diversity among terrestrial organisms, particularly plants, should not detract attention from the very high rates of turnover at the generic and familial level; for example, similarities between Permian and Triassic floras are mostly restricted to family and higher taxonomic levels. Also, unlike some of the extinctions in the shallow-water marine fauna, floristic turnover during the Permian-Triassic interval appears to have been a prolonged process. Worldwide it took place over a period of at least 25 my, although in any one place it could have been more rapid (Frederiksen 1972; Knoll 1984; Eshet 1990). On the basis of our present understanding of the tempo and paleoecological characteristics (at least among plants) of the Permian/Triassic extinctions, it seems most probable that they reflect large-scale climatic and paleogeographic alterations rather than a sudden catastrophic mortality (Knoll 1984; Traverse 1988). This is consistent with recent understanding of the Permian/Triassic extinctions in the marine realm (Erwin 1990).

Correlation of early Mesozoic continental deposits generally is difficult (see Olsen and Sues 1986), and chronostratigraphic resolution is poor. This is a consequence of several factors, including the occurrence of important fossil assemblages in relatively condensed stratigraphic sections and especially poor correlation with the marine standard sequences (Olsen and Sues 1986). Consequently it is difficult to determine whether some differences in Triassic terrestrial biotas represent change through time or geographic variation controlled by climate or other regional factors.

The maximum development of Pangaea occurred at approximately the Middle/Late Triassic boundary (Veevers 1989; comparable to Early Jurassic paleogeography shown in fig. 6.1), and throughout the period this landmass moved slowly northward. In spite of the contiguity of the continental masses forming Pangaea, Triassic faunas and floras were markedly provincial, probably reflecting monsoonal climatic zonation and extreme seasonality caused by continentality and the symmetrical placement of Pangaea about the equator (Parrish et al. 1982; Crowley et al. 1989; Kutzbach and Gallimore 1989). Triassic Pangaean biotas are commonly divided into northern (Laurasian) and southern (Gondwanan) realms, with several cases of documented overlap, especially in India where a Gondwanan flora (Dobruskina 1987) occurs together with characteristically Laurasian tetrapods (Olsen and Sues 1986). During the Middle to Late Triassic, Laurasian floras were divided into Siberian and Eurasian provinces. However, endemism in both floras (Meyen 1987) and vertebrate faunas (Olsen and Sues 1986) significantly decreased during the Triassic. High-latitude climates were relatively warm, an inference based in part on the common occurrence of large amphibians in Early Triassic assemblages from Australia, Antarctica, eastern Greenland, and Spitzbergen. Coal-rich sequences have been recorded from both high northern and high southern paleolatitudes; these reflect climates that were wetter than those of lower paleolatitudes.

The biotic character of the Triassic is probably linked to dry or highly seasonal climates associated with the coalescence of Pangaea. Such features include: (1) the continued increase in diversity of gymnospermous seed plants, particularly of xeromorphic groups of scale-leaved conifers and thick-cuticled seed ferns and cycadophytes; (2) the increase in diversity and abundance of diapsid (presumed uric-acid-exreting, more water-efficient), rather than synapsid (presumed urea-excreting, less water-efficient), amniotes (Robinson 1971); and (3) decrease in diversity of many groups of plants and herbivorous tetrapods, especially during the time interval from the late Carnian to Hettangian (Olsen and Sues 1986). The existence, timing, and causes of plant and vertebrate extinctions at the Carnian/Norian boundary have been debated, but the most recent work links these events to a late Carnian wet period followed by a return to dry climates in the early Norian (Simms and Ruffell 1989).

Like its beginning, the end of the Triassic was marked by a substantial extinction among terrestrial vertebrates. Unlike the Permian/Triassic extinctions, however, the Triassic/Jurassic extinctions occurred over a relatively short period of time, and during the inital breakup of Pangaea. Typical Early Jurassic assemblages have been estimated to occur as little as 100 ky to 200 ky after the Triassic/Jurassic boundary (Olsen et al. 1987), although the extinctions may have occurred over a period of up to 2 my. Olsen et al. (1987) suggest that the terminal Triassic extinctions were associated with the bolide impact that created the large Manicouagan crater (> 70 km in diameter) in Quebec (Olsen et al. 1987). More recent observations of impact ejecta associated with bivalve extinctions near the Triassic/Jurassic boundary in marine deposits in Europe (Newton and McRoberts 1990) and of an increase in fern spore abundance in earliest Jurassic palynofloras of North America (Fowell 1990) offer some support for the theory that the extinctions were impact related

2.1 Triassic Vegetation of Laurasia

Laurasian Triassic floras can be characterized generally as a mixture of primitive conifers (Voltziaceae and Lebachiaceae) along with various cycadaleans, bennettitaleans, "pteridosperms," ferns (including some of tree habit), and sphenopsids (Dobruskina 1987). Ginkgoaleans (e.g., *Sphenobaiera*) were more abundant and diverse at northerly paleolatitudes. Most of the conifers and ginkgoaleans are thought to have been medium-sized to large trees that perhaps formed a diffuse canopy. Mesozoic conifer seeds were generally small and winged, suggesting that they were not animal dispersed and that they required well-lit environments for germination and early growth (Tiffney 1986a). Conifers were probably the "big trees" in many Triassic landscapes. This idea is based on the observation that almost all of the large silicified trunks of this age are attributable to conifers. (Note, however, that *Araucarioxylon* is a form genus for rather generalized and primitive dense wood that may belong to a wide variety of gymnosperms.) Although the diversity of plant habits among Triassic conifers may have been considerably greater than it is in the present-day angiosperm-dominated world, there is little fossil evidence for dramatically different forms (e.g., herbs, lianas, floating aquatics). Ash (1987a) has reported a possible cordaitalean of small stature from the Late Triassic of the southwestern United States, but it is unclear whether the fossils represent a mature shrubby or herbaceous plant or the sapling of a tree in which the wood has not been preserved.

Many living Araucariaceae form diffuse canopies. Some extant species of *Araucaria* and *Agathis* have leaves with sharp apexes that persist on the trunk

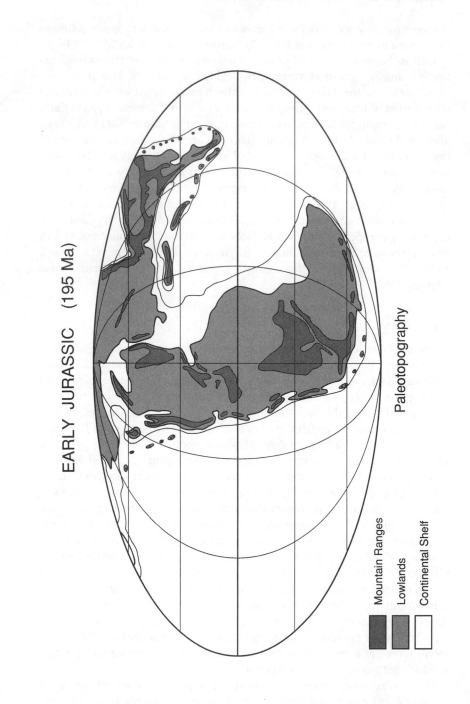

EARLY JURASSIC (195 Ma)

Paleotopography

Mountain Ranges

Lowlands

Continental Shelf

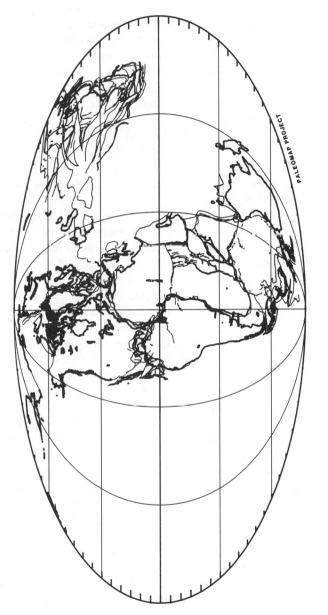

Plate Positions

Fig. 6.1. Generalized continental positions and orography in Early Jurassic, similar to those of the Late Triassic. (Courtesy of C. R. Scotese and the PaleoMap Project, University of Texas, Arlington)

and main branches, gradually becoming surrounded by secondary tissue. These "armored" trunks and branches may be (or may have been) a defense against vertebrate herbivores, but it is not known if this is a retained primitive trait that also would have characterized Mesozoic araucarians.

Primitive ginkgoaleans (e.g., *Sphenobaiera* and *Glossophyllum*) were probably small-seeded, short, underbrush trees in conifer forests (Krassilov 1975a). During the Mesozoic the lineage increased in size, a trend accompanied by the appearance of larger seeds, longer petioles, less dissected leaves, and deciduousness (Krassilov 1975b). The sole extant species, *Ginkgo biloba,* is notable for having seeds consisting of a fleshy, odoriferous outer layer surrounding an inner stony layer. At maturity, the seeds are shed en masse from the tree. It has been suggested that these features are relict adaptations for dispersal by terrestrial reptiles (saurochory *sensu* Van der Pijl 1982), although the persistence of *Ginkgo* throughout the Tertiary argues against complete dependence on dinosaurian dispersers.

Most living cycads are relatively short, stout, and unbranched. Exceptions include species of *Macrozamia* that reach 20 m in height and branch several times. Living cycads characteristically have very slow growth rates but display a wide range of habits spanning epiphytes, mangroves, and fire-tolerant shrubs (Norstog 1987). The structure of cycad seeds and the manner of their presentation both match the suite of traits postulated to be associated with reptile dispersal (van der Pijl 1982). Cycad foliage is toxic to mammals (Chamberlain 1935).

The range of habits and growth forms among Mesozoic cycads is likely to have been greater than what is seen even in living species (Spicer 1987). Some Mesozoic cycads had short trunks, but others were more slender and branched; the existence of a relatively slender form in the Late Triassic (*Leptocycas*) raises the possibility that this is the primitive condition for the whole group (Delevoryas and Hope 1976). The lax cones of *Beania* and *Androstrobus* indicate that wind pollination was much better developed in some Mesozoic cycads than in any extant species.

The widespread continental climates of the Late Permian and early Mesozoic may have selected against pteridophytes with water-dependent sexual reproduction. The Paleozoic-Mesozoic transition was an interval of turnover among pteridophyte groups. The Marattiaceae and Osmundaceae carried over from the Paleozoic, but the early Mesozoic was a time of diversification of many modern "lower" fern groups, such as Schizaeaceae, Osmundaceae, Dipteridaceae, and Gleicheniaceae.

Mesozoic marattiacean ferns appear to have been more similar to the extant genera in the family than to Paleozoic forms such as *Psaronius*. Living species are tropical, understory trees with short, unbranched stems up to 60 cm in diameter and thick, pinnately compound leaves up to 4 m long. All extant species are specialized for growth on wet, shaded forest floors. The extant families Osmundaceae, Dipteridaceae, and perhaps Polypodiaceae are also known from Mesozoic deposits. Although many of the living species in these families occupy wet, shaded spots under forest canopies, there is no direct evidence for the habits or habitats of the Mesozoic species.

Ferns in the families Matoniaceae and Gleicheniaceae may have been major colonists of open, dry, or even nutrient-poor substrates during much of the Mesozoic, perhaps forming vegetation analogous to savannas or grasslands now dominated by herbaceous angiosperms (Coe et al. 1987). Matoniaceae were rhizomatously spreading, relatively thick-cuticled ferns less than 2 m high. Living Gleicheniaceae generally have similar growth habits and are colonists of disturbed tropical habitats. Although the Gleicheniaceae are virtually absent in the Triassic, the family has been recognized in the late Paleozoic, and *Gleichenia*-like plants became common in the Jurassic and Cretaceous (Taylor 1981).

Both sphenopsids and lycopsids decreased in diversity, and both became increasingly restricted to herbaceous forms during the Triassic. Sphenopsid fossils are particularly abundant and are often preserved in situ, probably because they tended to grow on freshly deposited substrates that are common in active depositional settings. All Mesozoic sphenopsids were based on the same basic body plan as present-day *Equisetum,* having unbranched central axes bearing whorls of leaves. They reproduced by spores, although rhizomatous growth also characterizes the group. Extant species of *Equisetum* contain numerous large, siliceous phytoliths in their tissues, which may serve as an herbivore deterrent, but there is no direct evidence for these structures in the fossils. During the early Mesozoic, axes of genera such as *Neocalamites* were 10–30 cm thick and may have attained heights of 10 m. The genus *Schizoneura* (Triassic-Early Jurassic) had an axis 2 cm wide and up to 2 m tall. Smaller equisetaleans are also abundant in Mesozoic floras. The anatomy and distribution of Mesozoic sphenopsids is consistent with primary colonization of open or disturbed damp habitats, but their rhizomatous growth and moderate size may have allowed them to form dense thickets that could have inhibited colonization by later successional forms.

Early Triassic coastal vegetation in the northern part of Laurasia and Australia was dominated by stands of the lycopsid *Pleuromeia*. *Pleuromeia* was a genus of unbranched, herbaceous plants 1–2 m tall and about 10 cm in diameter. *Pleuromeia* stems were covered with narrow leaves about 10 cm long and were terminated by an apical cone. These plants were heterosporous and dioecious. They occur in extensive monospecific assemblages together with freshwater-to-marine invertebrates, suggesting they formed dense growths along coasts of lakes, estuaries, and oceans (Retallack 1975). The *Pleuromeia* associations represent the last vegetation type in which lycopsids were the major structural element, although species of *Pleuromeia*-like plants survived later into the Triassic.

During the latter half of the Triassic, east-west provinciality decreased in the equatorial region of Laurasia, and cycadaleans, bennettitaleans, and more advanced conifers (including Pinaceae and early Cheirolepidiaceae) became more abundant in many areas. Palynofloras from eastern North America show major reductions in diversity in the Carnian and again at the Triassic/Jurassic boundary (Cornet and Olsen 1985; Olsen and Sues 1986; Boulter et al. 1988). Late Triassic floras of Eurasia typically contain a diverse mixture of peltaspermaceous seed ferns, bennettitaleans, cycadaleans, conifers, ginkgoaleans, and a number of ferns, including species in the Marattiaceae, Osmundales, Dipteridaceae, and Matoniaceae (Vakhrameev et al. 1978). Peltaspermaceae were probably small-to-moderate-sized plants with small seeds and bipinnate fronds about 0.5 m long. Foliage of the peltasperm *Lepidopteris ottonis* had a thick cuticle with stomata on both sides; this seed fern occurs in near-shore marine sediments, suggesting that it was a mangrove.

Specific Laurasian Plant Assemblages

The Late Triassic flora of the Chinle Formation in the area of the Petrified Forest National Park in Arizona is late Carnian to early Norian in age (Olsen and Sues 1986; Ash et al. 1986; Litwin 1986) and has been reconstructed as representing three vegetation types (Gottesfeld 1972). Proximal floodplains were occupied by low-statured vegetation dominated by the bennettitalean *Otozamites* and the ferns *Phlebopteris, Clathropteris,* and *Todites. Schilderia,* a buttressed tree possibly related to the living Gnetales, also occupied proximal floodplain areas. Large *Neocalamites* (30 cm stem diameter and at least 6 m tall) grew along the channels in disturbed sites. Slightly more distal floodplains supported forests of *Araucarioxylon* trees 40–60 m high. Close spacing of in situ stumps and the rarity of branching (even on long sections of trunks) led Gottesfeld (1972) to suggest that the *Araucarioxylon* forests were dense. Palynomorphs provide evidence of an understory of ferns and at least one bennettitalean. The relatively unbranched nature of *Araucarioxylon* might imply self-pruning in a low-light subcanopy habitat (Gottesfeld 1972). However, sparse canopies are common in some living swamp conifers, apparently as a consequence of stressed conditions associated with waterlogged soils (Brown 1981). The existence of an "upland" flora (i.e., vegetation existing on surfaces slightly higher than the floodplain) consisting of cordaites (*Dadoxylon*), ginkgoaleans, and caytonialean seed ferns is inferred from palynofloras and rare specimens of fossil wood (Gottesfeld 1972). Elements of the Petrified Forest National Park flora not discussed by Gottesfeld include another probable araucarian, *Woodworthia* (generally smaller than *Araucarioxylon* and with persistent short shoots on the trunk), a cycad (*Lyssoxylon*), a variety of ferns (including two tree ferns), and several herbaceous lycopsids (Ash 1972c).

Late Triassic plant assemblages approximately coeval with those of the Petrified Forest are known from two other depositional settings in the southwestern United States, lake shore and oxidized floodplain deposits (Ash 1978, 1987b). The Ciniza lake-bed assemblages probably were deposited in a shallow body of water of limited areal coverage and represent vegetation growing in the immediate vicinity of the lake margin (Ash 1978). The flora is of low diversity (8–14 species) but includes four types of possibly araucarian conifer foliage (*Pagiophyllum*), a bennettitalean (*Nilssoniopteris*), and the enigmatic gymnosperm *Dinophyton* (Ash 1978). The vegetation on the lake margin is thought to have been dominated by *Equisetites, Phlebopteris,* and *Nilssoniopteris* (the tallest plant in this community). Somewhat dryer sites farther away from the lake margin apparently supported conifer forest with an understory of the ferns *Todites* and *Clathropteris* and a subcanopy level including bennettitaleans and caytoniaceous seed ferns (Ash 1978). The relative rarity of ferns in these assemblages may reflect the tendency of ferns not to abscise leaves, thus releasing fewer leaves for transport into lacustrine sediments.

The "red bed" flora described by Ash (1987b) is possibly Norian in age and thus somewhat younger than that of the Ciniza Lake Beds or the Petrified Forest, but compositional differences appear to be related more to depositional environment than to time. The red-bed flora is extremely depauperate; only seven genera are known from six localities: *Neocalamites, Pagiophyllum, Araucarioxylon, Pelourdea, Woodworthia, Schilderia,* and *Sanmiguelia.* Another curiosity of this flora is the in situ burial of nearly whole plants at several sites (Ash 1987b). Ash interpreted some of these assemblages as representing low-diversity, low-statured vegetation growing on river levees. Alternatively, in a location with frequent depositional events, small in situ plants could represent saplings of species that attained larger size in more stable settings.

The Late Triassic floras of the American Southwest provide some of the best available examples of early Mesozoic vegetation. Generally these floras are of moderate to low diversity and contain relatively few species that are likely to have been large trees, though some conifers clearly formed tall forests. Matoniaceous, osmundaceous, and other ferns were evidently the most abundant and diverse element of the understory, particularly in wet soils. Bennettitales were probably important, both as low-statured plants of open settings and as understory elements in conifer forest. Most calamites were probably colonizers of disturbed sites.

Another Late Triassic flora, probably of late Norian age, was described from the Scoresby Sound area of eastern Greenland (Harris 1937 and references therein). The fossiliferous deposits were formed in abandoned fluvial channels. Although this specific environment has not been identified as producing megafossil assemblages in the Chinle Formation, the Scoresby Sound assemblages and most of the Chinle floral occurrences presumably are derived from wet floodplain vegetation, and thus differences in composition are not likely to reflect purely taphonomic effects. The most striking difference between the Scoresby Sound and American assemblages is in simple numbers of species (200 vs. 60), but unfortunately it is hard to determine the degree to which this reflects differences in sampling and taxonomic treatment. The Scoresby Sound flora was collected from over 70 "major" plant beds (Harris 1937) and presumably even more individual localities distributed through a significant stratigraphic interval, while the Chinle flora appears to have been collected from between 10 and 20 major sites (Ash 1972). Thus, the much greater species richness of the Scoresby Sound flora may result in part from better sampling of heterogeneous floodplain vegetation, and from combining floras across a longer time interval. However, the presence of coal beds in the Scoresby Sound sequence and of many species with mesomorphic cuticles (even in groups, such as the conifers, that commonly display some degree of xeromorphy) suggests that the Late Triassic floras from eastern Greenland grew under a wetter climate than that characteristic of the southwestern United States during the Late Triassic. This climatic difference has been predicted on the basis of the proximity of East Greenland to the Arctic Basin and its paleo-latitudinal position at about 45 to 50 degrees north (Parrish et al. 1982). Therefore the greater number of species in the Scoresby Sound flora could reflect a genuinely more diverse, mesic vegetation.

Regardless of the relative diversity of the Chinle and Scoresby Sound floras, it is clear that the Greenland assemblages reflect a rich and diverse forest vegetation. Harris (1937) mentions large pieces of poorly preserved fossil wood that demonstrate the presence of trees, but this also is documented by the compression flora. Tiffney (1986a) measured and tabulated the seeds described by Harris and reported a mean seed size of 188 mm^3, with a maximum of 1,600 mm^3 and a minimum of 0.64 mm^3 (based on 30 form species). The larger seeds in this flora generally were borne by the cycads and ginkgoaleans, with bennettitaleans and seed ferns producing medium-to-small-sized seeds. As a whole, the Scoresby Sound flora has somewhat smaller seed sizes than the summed Triassic floras (Tiffney 1986a).

All of the major Triassic plant groups occur in the Scoresby Sound flora, including lycopsids, sphenopsids, ferns, ginkgoaleans, bennettitaleans, cycadaleans, pteridosperms, and conifers. Diversity is also very high in groups likely to have had an arborescent habit. Based on foliar material alone, Harris (1937) reported 28 conifer species, 20 cycads, 19 bennettitaleans, 10 ginkgoaleans, 8 pteridosperms, and 4 czekanowskialeans. Thus, even assuming that all of the ferns and sphenopsids can be classified accurately as herbaceous, about 90 species of woody plants with shrub-to-tree habit grew on the Scoresby Sound floodplains. The Scoresby stratigraphic sequence represents a considerable interval of time, and therefore not all of these species necessarily coexisted. Nevertheless the number of species identified in the Scoresby flora is more similar to the number found in the angiosperm-dominated floras of the

Late Cretaceous and Tertiary than in most other Mesozoic floras. This implies that during the Late Triassic complex and diverse forest vegetation may have flourished under warm, wet climates, even at latitudes that dictated substantial seasonal variation in day length and a moderately low angle of solar radiation.

2.2 Triassic Vegetation of Gondwana

Toward the end of the Permian in the Southern Hemisphere, the *Glossopteris* flora declined. Early Triassic vegetation in the same areas came to be dominated by the corystospermaceous seed fern *Dicroidium*, although a transitional flora characterized by a high abundance of the seed fern *Thinnfeldia* is known from a few places (Retallack 1977; Anderson and Anderson 1983a, b). *Dicroidium* played a leading role in a variety of vegetation types, from heath to broad-leaved forest to dry woodland, and is sometimes found in virtually monospecific stands (Retallack and Dilcher 1988). Voltziacean and primitive podocarpaceous conifers also were part of Gondwanan Triassic floras, and they became more important during the Middle to Late Triassic, along with other peltaspermaceous seed ferns and some Laurasian groups of cycadaleans and ginkgos. At or near the end of the Triassic, the *Dicroidium* flora declined and was replaced by a more cosmopolitan conifer-bennettitalean flora.

Specific Gondwanan Plant Assemblages

Retallack (1977) reported on several Late Permian through Triassic assemblages from eastern Australia and identified a series of vegetational types based on floral composition, depositional environment, and paleosol characteristics. In much of Gondwana, Early Triassic coastal vegetation consisted of monodominant stands of small, arborescent lycopsids such as *Pleuromeia* and *Gregicaulis,* but, by the Middle to Late Triassic, these same environments were dominated by the seed fern *Pachypteris* (Retallack 1975, 1977; Anderson and Anderson 1983b). In the Early Triassic (Scythian and early Anisian) of eastern Australia, delta-top and lagoon-margin woody vegetation comprised sphenopsids (*Neocalamites*), ferns (*Cladophlebis*), seed ferns such as *Dicroidium, Lepidopteris,* and *Pachypteris,* and the cycadophyte *Taeniopteris. Neocalamites* was probably most important in early succession. *Taeniopteris lentriculiformis* is inferred to have been a slender-branched coastal scrub and was an important part of coastal vegetation through the early and middle Anisian. Assemblages dominated by the primitive conifer *Voltziopsis* also are known from the Early Triassic of this area and may represent a different kind of lowland forest vegetation than those assemblages characterized by *Dicroidium.* Vegetation of drier areas is reconstructed as a woodland dominated by *Dicroidium* and *Xylopteris* associated with *Sphenobaiera* and *Czekanowskia.* The most taxonomically diverse and structurally complex vegetation reported by Retallack (1977) from eastern Gondwana was late Anisian to early Ladinian broadleaf forest composed of sphenopsids, ferns, seed ferns, cycadaleans, ginkgoaleans, and conifers. This forest included *Neocalamites, Phyllotheca, Asterotheca, Cladophlebis, Coniopteris, Dictyophyllum, Dicroidium, Lepidopteris, Pachypteris, Johnstonia, Tetraptilon, Sphenobaiera, Ginkgoites, Phoenicopsis, Rissikia, Taeniopteris, Pterophyllum,* and *Pseudoctenis. Phoenicopsis* sometimes occurred as a monodominant in levee settings. The assemblages described by Retallack (1977) suggest a high degree of vegetational and landscape heterogeneity in east Gondwana during the first part of the Triassic.

Early to Middle Triassic assemblages are also known from South Africa (Anderson and Anderson 1983a, b). The seed ferns *Dicroidium* and *Lepidopteris,* along with sphenopsids and lycopsids, are reported as dominant in a wet or seasonally wet inland basin environment. Interpreted vegetational associations include forest or woodland dominated by *Dicroidium* and *Lepidopteris* that grew on levees or other elevated areas, monospecific sphenopsid stands in wet areas, and monospecific lycopsid stands fringing bodies of water (Anderson and Anderson 1983b). Middle Triassic (Anisian-Ladinian?) floras from the Molteno Formation are known from more than 70 sites and include some 115 species; individual sites produce from 1 to more than 60 species, with a mean of about 10 (Anderson and Anderson 1983b). The Molteno sequence was deposited in an inland basin at a paleolatitude of about 55–60 degrees S under a seasonal but moist climate. In distance from the paleoequator, overall climate, and in some of the depositional settings preserved, the Molteno floral assemblages show similarities to the late Norian Scoresby Sound flora of Greenland. Overall the Molteno flora is dominated by *Dicroidium,* the conifer *Heidiphyllum,* the ginkgophyte *Sphenobaiera,* and *Neocalamites.* Anderson and Anderson (1983b) recognized 25 associations of fossil plants that they interpreted to represent four types of vegetation: (1) monospecific *Neocalamites* thickets in wet areas; (2) monodominant or low-diversity assemblages of *Heidiphyllum* in slightly less wet areas, sometimes found directly above *Neocalamites* assemblages; (3) low-diversity shrubby vegetation of *Sphenobaiera;* and (4) moderately diverse (>20 species) forest or woodland on more elevated ground, dominated by *Dicroidium.* The most diverse assemblages are dominated by species of *Dicroidium* (50%–60% of specimens) but also typically include *Neocalamites,* various ferns such as *Todites, Cladophlebis,* and *Dictyophyllum,* the ginkgophyte *Sphenobaiera,* the seed ferns *Lepidopteris, Glossopteris,* and *Yabeiella,* the bennettitalean *Pseudoctenis,* the cycadophyte *Taeniopteris,* and the conifers *Rissikia* and *Heidiphyllum.* The predominance of 1 or 2 species of *Dicroidium* at diverse localities, with many rare forms, may imply that the relative abundance of the common species was enhanced by deciduousness and/or large size, both of which have been suggested for *Dicroidium* on morphological grounds (discussed under the name *Umkomasia granulata* by Retallack and Dilcher [1988]). Although the diversity of the

Molteno floras is not spread across all major taxonomic groups to the degree that it is in the Scoresby Sound flora, the South African assemblages are nevertheless quite diverse for local floodplain floras. The relatively high diversity again seems to suggest that under moist, though seasonal, climates, Triassic floras may have attained greater diversity and structural complexity than they did under monsoonally dry or arid climates of the low-latitude regions.

2.3   Triassic Faunas

At least five orders of Paleozoic insects with probable plant-sucking habits did not survive into the Triassic. Continuing from the Permian were the Orthoptera (at least some of which were herbivorous), Hemiptera, Thysanoptera (including possible pollen feeders), and Coleoptera (which did not yet include definitely herbivorous forms). Chewing stick insects (Phasmida) and xyeloid hymenopterans (which have herbivorous larvae) made their first appearance in the fossil record.

Triassic terrestrial vertebrate faunas were a mixture of groups that persisted from the late Paleozoic and the stem groups or earliest known representatives of most extant major taxa. The establishment of herbivory in several lineages of amniote tetrapods during the late Paleozoic was followed by a great radiation of herbivorous synapsids in the Permian. Although this group suffered a major decrease in diversity toward the end of the Permian, the two major groups of Late Permian herbivores, Dicynodontia and Procolophonoidea, were still represented in Early Triassic faunas.

Early Triassic herbivorous tetrapods in general seem to have been smaller than those of the Late Permian, at least in Gondwana, where the fossil record is most complete (Benton 1983). These forms (e.g., those from the *Lystrosaurus*-Zone fauna) show a limited range of body sizes (0.5 to < 2.0 m body length). The dominance and diversity of large herbivores increased during the Middle and Late Triassic, especially with the appearance of prosauropod dinosaurs in the Carnian. Early Triassic faunas retained the dicynodonts, the dominant Late Permian group of small-to-large quadrupedal herbivores. The dicynodont *Lystrosaurus* was the characteristic herbivore of many Early Triassic tetrapod assemblages, especially in the Gondwanan realm. Dicynodonts had robustly built, presumably akinetic skulls. The typically edentulous upper and lower jaws were covered by horny beaks (rhamphothecae), between which plant fodder was cut by retraction of the mandible (Crompton and Hotton 1967; King 1981). Some Middle and Late Triassic dicynodonts, such as *Dinodontosaurus,* probably reached body weights in excess of 1,000 kg ("megaherbivores" *sensu* Owen-Smith 1987) and were very common in Gondwanan faunas (Cox 1965). Foraging heights ranged between 1 and 2 meters above the ground.

The Middle Triassic also witnessed a radiation of gomphodont cynodonts

(Diademodontidae, Trirachodontidae, and the probably paraphyletic Traversodontidae). They formed a fairly diverse group with a considerable range in body size and were especially diverse in the Gondwanan realm. A few Carnian traversodont cynodonts, such as *Exaeretodon* from Argentina and Brazil and *Scalenodontoides* from southern Africa, reached a length of 3 m. They had large, massive skulls and large molariform teeth that met in complex occlusion and show extensive wear. They employed bilateral mastication and a palinal power stroke (Crompton 1972; Crompton and Attridge 1986). These features, combined with evidence for muscular cheeks, imply that these cynodonts were capable of oral processing of very tough vegetation.

Rhynchosaurs were another important group of late Middle to Late Triassic medium-sized (1–2 m body length) herbivores that employed a high degree of oral food processing. These archosauromorph diapsids are characterized by large tooth plates and powerful development of the jaw adductor muscles (Sill 1971; Benton 1983). The lower teeth cut into a groove in the maxillary dentition. Plant material was probably positioned across the maxillary groove and cut when the lower teeth were brought diagonally into occlusal contact from a posterior to anterior position (precision-shear style of mastication; Benton 1983, 1984). Rhynchosaurs were particularly diverse in the Gondwanan realm, where they were the most abundant herbivorous forms in many tetrapod assemblages (Bonaparte 1982; Benton 1983). Their quadrupedal posture suggests foraging ranges of up to a meter above the ground; Benton (1983) noted that the structure of the limb skeleton suggests rhynchosaurs might have dug up food items such as rhizomes. (Some Mesozoic and Cenozoic equisetaleans had thick rhizomes [Brown 1962], and these may have been one food source for rhynchosaurs and other Mesozoic "rooting" herbivores.)

Another important group in Late Triassic faunas were the Stagonolepididae, which were heavily armored, possibly herbivorous archosaurian reptiles up to 3 or 4 meters long. They had simple, leaf-shaped teeth and blunt snouts that may have been covered in part by a horny beak, or rhamphotheca (Walker 1961). Jaw motion was orthal, with teeth passing adjacent to one another to slice food items.

The Procolophonidae were a group of small primitive amniotes that persisted as a common element from the Late Permian to the Late Triassic. Their robust, presumably akinetic skulls typically had heterodont dentitions with transversely expanded posterior teeth that show evidence for some tooth-to-tooth occlusion. The jaw adductor masculature was prominently developed, and the mandibular symphysis was robust. Mastication consisted of an orthal puncture-crushing chewing stroke (Weishampel and Norman 1989).

The Trilophosauridae are only known with certainty from the Late Triassic of North America (Gregory 1945). The relatively broad snout probably supported a horny beak. The posterior teeth were transversely expanded and rather high-crowned, and the pattern of tooth wear suggests orthal jaw motion. The volume of the adductor musculature was large relative to the occlusal surface (Weishampel and Norman 1989). Foraging ranges were restricted to within 1 meter above the ground.

The appearance of prosauropod dinosaurs in the Carnian marked the first radiation of high-browsing terrestrial herbivores (Bakker 1978; Galton 1986). Prosauropods were facultatively bipedal forms up to 10 m long and >1000 kg in weight that could have fed at heights of up to 4 m above the ground. They had large but lightly constructed skulls and little tooth wear, suggesting that their teeth were used merely for gathering food items (Galton 1985, 1986). Mechanical breakdown of the fodder was accomplished in a gastric mill through the action of gastroliths (Raath 1974). The genus *Plateosaurus* is especially abundant in late Norian assemblages from Germany (Galton 1986), and melanorosaurid prosauropods form a major element in Norian faunas from Argentina (Bonaparte 1982) and southern Africa (Galton 1986).

The Late Triassic radiation of ornithischian dinosaurs produced a number of low-browsing (within 1 meter above the ground) herbivorous forms. The Fabrosauridae, the most primitive known ornithischians, were small (1–3 m) bipedal dinosaurs. Best documented by the Early Jurassic *Lesothosaurus* from southern Africa, these animals had lightly built skulls with beaks (Thulborn 1970a). The development of nearly vertical wear facets on the maxillary and dentary teeth (Thulborn 1970b; Weishampel 1984b; Weishampel and Norman 1989) indicates some degree of oral food processing (orthal slicing). As was the case in all ornithischians, the gut capacity of fabrosaurs presumably was substantially increased relative to overall body size by the retroversion of the pubes. Large gut capacity may have been important in allowing fermentation of plant material. Foraging probably took place within a meter above the ground.

The late Carnian and Norian produced a major turnover, not only in the groups of herbivorous tetrapods present but also in the nature of vertebrate herbivory. Early and Middle Triassic herbivorous tetrapod faunas were composed largely of forms that foraged within a meter above the ground. Some of these animals had dental and jaw forms consistent with sophisticated oral processing of plant material, such as highly organized puncture crushing (rhynchosaurs and trilophosaurs) and incorporation of a palinal power stroke into the masticatory cycle (dicynodonts and gomphodont cynodonts) (see Weishampel and Norman 1989). Although the timing, tempo, and cause of a possible Late Triassic faunal turnover and the Triassic/Jurassic mass extinction among tetrapods are still controversial (Benton 1985, 1986; Olsen and Sues 1986; Olsen et al. 1987), by the end of the Triassic the functional attributes of the dominant herbivores were markedly different from those of earlier forms (Crompton and Attridge 1986; Galton 1986). Norian herbivore faunas were characterized by species with much larger body size and higher browsing range, and by the predominance of prosauropod dinosaurs that used a gastric mill to break down plant material.

2.4 Triassic Ecological and Evolutionary Trends

The Permian/Triassic extinctions severely reduced the diversity of terrestrial vertebrates, and floral turnover appears to have been rapid during this interval. However, the extinctions do not seem to have affected particular habitat or trophic groups differentially, as has been documented for the K/T vertebrate extinctions, in which small and aquatic vertebrates suffered much lower rates of extinction (e.g., Hutchison and Archibald 1986). Similar herbivorous adaptations and some of the same clades were present among Late Permian and Early Triassic tetrapods, and there was no dramatic change in body size distribution. At least among Gondwanan plants, a major taxonomic change occurred from the *Glossopteris* flora to the *Dicroidium* flora, but the vegetational consequences of this are not fully understood. No short-term increase in early successional plants of the type observed in the earliest Jurassic and earliest Paleocene has been noted in the earliest Triassic. The absence of a strong ecological pattern in the Permian/Triassic extinctions of plants and continental vertebrates suggests that the extinction episode did not result in a major restructuring of terrestrial ecosystems. The similar biotic patterns at the end of the Paleozoic and the beginning of the Mesozoic may reflect the existence of a similar range of terrestrial environments, which in turn reflect a continuity of climatic conditions induced by the continued presence of Pangaea. The similarity of Late Permian and Early Triassic terrestrial ecosystems also may be related to a gradual or stepwise, rather than catastrophic, pattern of extinction that would have allowed guild structure to remain intact during taxonomic turnover. This would have conserved the existing structure and dynamics of communities, thereby restricting the opportunities for the establishment of new ways of making a living. Regardless of the importance of Permian/Triassic extinctions for shallow-water marine faunas, they do not appear to have been a major landmark in the history of terrestrial ecosystems. This is implicit in the recognition of a Paleophytic/Mesophytic boundary within the Permian period (Frederiksen 1972; Traverse 1988).

Both faunas and floras underwent considerable high-level taxonomic turnover during the Triassic, particularly beginning in the Carnian. Floristic changes include the initial rise of the *Dicroidium* flora in the Early Triassic of Gondwana, an increase in conifer dominance and decrease in floral diversity in equatorial regions during the Middle to Late Triassic, and the demise of the *Dicroidium* flora in Gondwana during the Late Triassic. Among vertebrate herbivores, there was the diversification of low-statured, herbivorous, urea-excreting, nonmammalian synapsids with complex masticatory adaptations during the Early and Middle Triassic, and their almost complete replacement by high-browsing, uric-acid-excreting dinosaurian herbivores with gastric mills in the Late Triassic. Some of these faunal and floral trends were probably related to the development of monsoonal or seasonally dry climates in the low-latitude regions of the world (Parrish et al. 1982; Parrish and Peterson 1988; Crowley et al. 1989), although it has been argued that a "pluvial" phase in the mid-to-late Carnian, followed by a return to dry conditions in the early Norian, was responsible for faunal and floral turnover, as well as extinctions in marine faunas (Simms and Ruffell 1989).

The faunal and floral changes that occurred during the Triassic apparently involved major modifications of trophic relationships or ecological roles in some cases, but replacement on an ecological theme in others. Throughout the Triassic, many of the herbivore species were quadrupedal forms that foraged within a meter of ground level and had body weights of between 10 and 1,000 kg, although smaller herbivores such as procolophonids and possibly sphenodontids (< 10 kg) were also numerous (Fraser and Sues, unpublished data). Today large terrestrial herbivores tend to be most diverse and abundant in relatively open vegetation (e.g., Eisenberg 1981, chap. 20), which is typical of regions that do not receive heavy, year-round rainfall. Smaller, generally arboreal, frugivore/herbivores are more important in dense forest vegetation typical of wet climates (Eisenberg 1981, chap. 20; Emmons et al. 1983). Body-size distribution among herbivores in many Triassic faunas is more consistent with open, low-statured vegetation than with closed forests, as are many paleoclimatic indicators. In some areas (e.g., parts of the Chinle Formation of Arizona), there is reasonably good paleobotanical evidence for denser forest vegetation. However, these forests are believed to have coexisted with nearby more open vegetation (Gottesfeld 1972), which may have formed the habitat of larger herbivores (J. M. Parrish 1989).

The weight of the evidence from functional morphology and body size favors the idea that Early and Middle Triassic herbivore faunas were dominated by generalist browsers that would have relied on plant productivity within about 1 meter of the ground. This implies that herbaceous pteridophytes, low-statured cycadaleans or seed ferns, and seedlings of all types received the brunt of browsing activity. Plants that dropped fleshy, perhaps strong-smelling, seeds on the ground in large quantities could also have provided an important food source for larger herbivores. This combination of features has been termed the saurochorous dispersal syndrome in extant plants (Van der Pijl 1982). There is little direct evidence for saurochory in Triassic plants, but animal dispersal is known in modern conifers and, to a lesser extent, in ginkgoaleans and cycads.

The increase in browsing height from 1 to 4 meters that occurred with the advent of prosauropod dinosaurs in the Late Triassic may well have been a trophic change of great ecological and evolutionary significance. Increased browsing height would have exposed plants to vertebrate herbivory over a longer period of their life cycle, thus directly affecting competitive interactions between plants and the amount of energy invested in defense mechanisms, as well as aspects of life history, such as the optimal age for reproduction and seed size. The ability to tap plant resources at higher levels also may have been a factor in the radiation of herbivorous dinosaurs.

During the Triassic, six major methods of food processing occurred in vertebrate herbivores: orthal pulping, orthal slicing, orthal puncture crushing, propalinal grinding, transverse grinding, and gut processing (Weishampel and Norman 1989). The Late Triassic records reduction in diversity of clades using propalinal grinding, orthal pulping, and orthal puncture crushing, with an increase in the number of gut-processing species (fig. 6.2; Weishampel and Norman 1989). Much has been made of this general shift from oral to gastric-mill processing of food in early Mesozoic herbivores (Galton 1986; Crompton and Attridge 1986), but though the change in method of processing plant fodder was structurally striking, its effect on vegetation is difficult to assess. All large herbivores would be expected to have a substantial impact on vegetation regardless of how the plant matter was processed and digested, and both synapsid dentitions and prosauropod gastric mills should have been capable of dealing with tough plant material. However, if the shift from oral to gastric processing was also related to changes in body size and metabolic rate, then

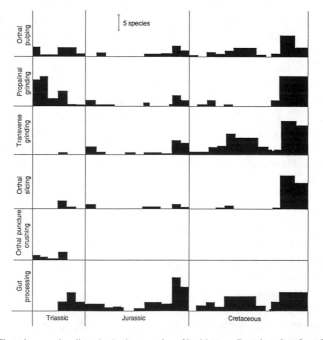

Fig. 6.2. Changing species diversity in 6 categories of herbivores. Based on data from Weishampel and Norman 1989.

the processing method may have indirectly affected the rate, amount, and type of vegetation consumed. Oral and gastric processing methods might also be expected to affect seed viability differently, although there is no direct evidence that large herbivores were involved in seed dispersal at this time.

Although the dominant interaction between herbivorous tetrapods and plants in the Triassic probably was generalized browsing, the existence of smaller herbivores raises the possibility of specialized vertebrate seed dispersal as well. There exists as yet no solid evidence for arboreal herbivores at any time during the Triassic. The only possible herbivores small enough to have been facultatively arboreal were the haramiyid synapsids, which first appear in the Norian (Lillegraven et al. 1979), and some sphenodontian lepidosaurs. Thus, during the entire Early and Middle Triassic, woody plants more than a few meters in height and with enough structural strength to resist felling would have been essentially free from vertebrate herbivory, although, of course, phytophagous insects would have been undeterred by height. Herbivorous arthropods were well established as arboreal primary consumers during this same interval. Although gliding, and therefore presumably arboreal, insectivorous tetrapods are known from the Late Triassic, the evolution of canopy-living vertebrate herbivores may not have occurred until the Late Cretaceous or Early Tertiary. Arthropods, insectivorous tetrapods, and herbivorous tetrapods apparently came to live in the canopy in the same sequence in which they assumed importance in terrestrial ecosystems; arthropods and carnivorous vertebrates exploited resources in previously unoccupied portions of ecosystems in advance of vertebrate herbivores.

The causes for the turnover in faunal composition during the Triassic (Bonaparte 1982) have been a subject of considerable debate, although there is little evidence for the various scenarios proposed to date. Replacement traditionally has been attributed to competition between synapsids and later-appearing archosaurian reptiles such as dinosaurs, with the newcomers succeeding as a result of superior locomotory abilities or physiological characteristics that were favored under the drier climates and more open vegetation of the later Triassic (see Charig 1984). Benton (e.g., 1986) has suggested that competition did not play a major role in faunal change. Instead, he implicated climatic and vegetational shifts in the extinction of nonmammalian synapsids and rhynchosaurs, which left empty adaptive space into which archosaurs could diversify. These explanations may not be mutually incompatible, because they both invoke an underlying cause of climatic and vegetational change (Zawiskie 1986). The success of prosauropods may have resulted from successful competition for low-level browse coupled with their unique access to high-level plant resources, which were unavailable to lower-statured animals. Very large body size and high mobility might also have been advantageous in permitting seasonal migration and greater ability to withstand short-term environmental fluctuations.

As mentioned above, the Triassic ended with a major extinction among terrestrial vertebrates. The relative suddenness of this event (200 k.y.?) makes it difficult to explain in terms of the long-term paleoclimatic changes that are associated with other Triassic biotic turnovers. The impact hypothesis presented by Olsen et al. (1987) is consistent with the tempo of the extinctions, and the size of the Manicouagan crater in Quebec is approximately as large as that predicted for a bolide with a diameter of 10 km (Silver and Schultz 1982)—the same size as is estimated for the end-Cretaceous bolide. The Triassic/Jurassic extinction was fairly severe in terms of percentage of tetrapod families lost (Olsen and Sues 1986). Shocked quartz crystals at the Triassic/Jurassic boundary (Newton and McRoberts 1990) and greatly increased fern spore abundance in the earliest Jurassic (Fowell 1990) also suggest parallels with the K/T event, but the ecological pattern of the two vertebrate extinctions is not similar.

One of the most significant differences between the two extinctions is in the survival of herbivorous vertebrates. Two clades of megaherbivores (Anchisauridae and Melanorosauridae) and a number of small-to-medium-sized herbivores, such as some sphenodontians, primitive ornithischians, and haramiyid synapsids, survived the Triassic/Jurassic boundary (Olsen and Sues 1986). This is in sharp contrast to the K/T extinctions, which wiped out all medium-to-large vertebrate herbivores. The immediate dependence of large herbivorous tetrapods on plant productivity would be expected to make them particularly vulnerable to even short interruptions in primary productivity, compared with smaller omnivores or carnivore/scavengers. Therefore the different survival rates of large herbivores across the Triassic/Jurassic and K/T boundaries suggest two possibilities: either our understanding of prosauropod diets is faulty (i.e., they were not obligate herbivores; Cooper 1981), or the nature and magnitude of ecological disruption at the Triassic/Jurassic boundary differed substantially from that at the K/T boundary. Ecological differences between the two extinctions, both of which are linked to bolide impacts of similar size, suggest that other environmental factors must play an important role in mediating the biological effects of such bolide impacts.

3 JURASSIC BIOTAS

Although the rifting of Pangaea commenced in the Late Triassic and continued slowly throughout the Early Jurassic, dispersal of the major continental areas and separation by ocean basins probably did not occur until the Middle or early Late Jurassic (Veevers 1989; fig. 6.3). In the absence of impassable barriers to dispersal, Jurassic faunas and floras included numerous cosmopolitan elements. Generally speaking, Gondwanan and Laurasian biotas were less distinctive during the Jurassic than they had been in the Triassic.

Global climates continued to be warm during the Jurassic, as indicated by high-latitude floras from Siberia (Vakhrameev 1970; Vakhrameev et al. 1978; Hallam 1984), evidence from the marine realm, and absence of glacial deposits at high latitudes (e.g., Frakes 1986). Frakes (1986) suggested that subtropical climates may have extended as far northward as 60 degrees latitude. The distribution of climatically sensitive sediments such as coals and evaporites indicates that aridity and seasonal aridity were widespread during the Early and Middle Jurassic in low-to-middle latitudes, particularly in western Pangaea. Major Lower to Middle Jurassic coal beds in higher paleolatitudes of both the Southern and Northern Hemisphere (e.g., Ordos Basin of northern China: E. A. Johnson et al. 1989) suggests these areas had wet climates (Parrish et al. 1982; Hallam 1984). Increasing aridity in southern Laurasia during the Middle and Late Jurassic has been attributed to a breakdown of the monsoonal circulation that typified the Late Triassic and part of the Early Jurassic (Parrish and Doyle 1984). Strong latitudinal variation in climate, particularly rainfall, continued throughout most of the Jurassic and was probably responsible for minor north-south floral provincialism (Doyle and Parrish 1984), although it does not seem to have barred continuing interchange between the Northern and Southern Hemispheres, especially among vertebrates.

3.1 Jurassic Vegetation

Jurassic vegetation was composed of a mixture of various woody gymnospermous groups and largely herbaceous pteriodophytes, a pattern similar to that of the Triassic. Conifers continued to be the most diverse large trees; species of the extant families Araucariaceae, Cephalotaxaceae, Pinaceae, Podocarpaceae, Taxaceae, and Taxodiaceae, together with members of the extinct family Cheirolepidiaceae, numerically dominated many Jurassic assemblages. *Ginkgo*-like plants also continued to be important in many floras, particularly in northern mid-to-high latitudes. During the Jurassic, the bennettitaleans became the most important group of small trees and shrubs, while seed ferns and marattiacean ferns apparently declined in abundance and diversity. Dicksoniaceous ferns and caytoniaceous seed ferns also may have been relatively successful groups of small tree stature (Wesley 1973). Among herbaceous groups, lycopsids continued to be of relatively minor importance, and *Neocalamites* was largely replaced by the smaller *Equisetum*. Osmundaceous, matoniaceous, and dipteridaceous ferns were probably the dominant herbs in many floras (Wesley 1973).

During the Early Jurassic on the southern continents, Gondwanan floras came to resemble more closely those of the northern continents. Typically equatorial groups like Bennettitales and Cheirolepidiaceae, along with dipteridaceous and matoniaceous ferns, became more important components of the Gondwanan flora. Major differences between Jurassic floras of Gondwana

LATE JURASSIC (152 Ma)

Paleotopography

Mountain Ranges

Lowlands

Continental Shelf

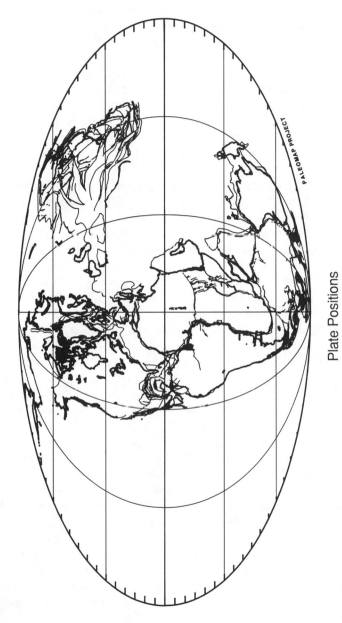

Plate Positions

Fig. 6.3. Generalized continental positions and orography in the Late Jurassic. (Courtesy of C. R. Scotese and the PaleoMap Project, University of Texas, Arlington)

and Laurasia included the relative paucity of Ginkgoales and Czekanowskiales and the greater importance of podocarpaceous conifers in the southern
regions, although podocarps are known from Laurasia as well.

Two groups of plants appear to have dominated low-latitude vegetation during the Jurassic: the conifer family Cheirolepidiaceae and the cycadophytic
Bennettitales. The Cheirolepidiaceae first appeared in the Late Triassic but
became a major element in palynofloras (*Classopollis* [*Corollina*]) and megafloras (e.g., some species of *Brachyphyllum* and *Cupressinocladus*) in the
Jurassic and Early Cretaceous (Barnard 1973). These conifers apparently
ranged in stature from tall forest trees (Alvin 1982, 1983) to small bushes
(Vakhrameev 1970). Direct evidence on the stature of *Cupressinocladus* from
the Purbeck Beds (Late Jurassic) of southern England shows that this particular genus was moderately tall, sparsely branched (Francis 1983), and in some
cases had multiple, thick trunks arising from a single rootstock (Spicer, unpublished data). Most cheirolepidiaceous conifers had small, short, pointed
leaves similar to those of many living conifers, but in some genera, such as
Frenelopsis and *Pseudofrenelopsis,* the main photosynthetic organs were unbranched, cylindrical, succulent-looking shoots covered with thick cuticles
showing sunken or otherwise protected stomata (Alvin 1983). Large accumulations of this "foliage" on some bedding planes, coupled with shoot morphology, wood anatomy, and other ecological considerations (Alvin et al.
1981), raise the possibility that some species were deciduous. The strongly
xeromorphic attributes of many cheirolepidiaceous conifers and their worldwide abundance at low paleolatitude (<40 degrees) sites with sedimentary indicators of aridity suggest they were dominants in dry-climate vegetation
(Vakhrameev 1970; Upchurch and Doyle 1981; Alvin et al. 1978; Francis
1983). Indeed the relative abundance of *Classopollis* in palynofloras has been
used as an index of aridity (e.g., Vakhrameev 1970). The abundance of
Cheirolepidiaceae in certain marginal marine rocks probably indicates that
some species were halophytic (Watson 1977; Daghlian and Person 1977).
However, the group was diverse both in number of species and in morphology
and no doubt occupied a wide array of situations in many kinds of vegetation
(Batten 1974; Alvin et al. 1978; Upchurch and Doyle 1981).

Although many bennettitaleans were pachycaulous (thick-stemmed) relative to most extant angiosperms, they exhibited a range of growth forms.
Short, stout forms, typified by the Cretaceous *Cycadeoidea,* had globose,
0.5-to-3-m-tall trunks that contained large amounts of soft tissue but were
covered with tough, persistent leaf bases and bore large, pinnately compound,
leathery leaves near the apex. Intermediate types such as *Zamites gigas* and
some *Williamsonia* species were several meters tall with sparsely branched
trunks about 30 cm in diameter that bore pinnately compound foliage near the
top. Species of *Wielandiella* and *Williamsoniella* were small, more highly
branched plants with stems a few centimeters in diameter, bearing a small

number of thick-textured, simple or pinnately-lobed (pinnatified) leaves (clustered at the nodes in *Wielandiella*). This range of growth habits implies that the group played a variety of vegetational roles. However, no large tree-sized bennettitaleans are known, and other aspects of their morphology and inferred biology are less variable.

Most bennettitalean foliage shows xeromorphic features such as small, reflexed pinnae, sunken stomata, and papillae or hairs on the leaf surfaces. A number of bennettitalean reproductive structures are known, and though the Late Triassic *Vardekloeftia* did not have specializations for insect pollination (Pedersen et al. 1989), many later species are thought to have been either self-pollinated or insect-pollinated (Crepet 1974; Crepet and Friis 1987). Bennettitales are thus another gymnospermous group that, along with various seed ferns and cycads, was insect-pollinated. Most bennettitaleans had small seeds with thin, hairlike projections; pollination, fertilization, and embryo development probably followed one another in rapid succession (Doyle 1978). Bennettitalean receptacles were fleshy, bore closely packed seeds, and disintegrated at maturity (Harris 1969). Bennettitalean receptacle cuticle has been recovered from Jurassic coprolites (Hill 1976), suggesting predation and possible dispersal by vertebrates (endozoochory), a mode of dispersal consistent with the fleshy receptacle of bennettitaleans (Harris 1973). Abiotic dispersal of bennettitaleans also has been suggested, partly on the basis of the small size of the seeds (Tiffney 1986a).

Retallack and Dilcher (1981) postulated that bennettitaleans were early successional plants. Crane (1987) summarized the probable biology of bennettitaleans and concurred that they were shrubby plants that occupied open habitats. The "armored" appearance of the stem in the short and medium forms of Bennettitales may have protected them from fire damage, and their seeds may have been released by fire (Harris 1973). At least some members of the group probably were deciduous (Delavoryas 1968; Krassilov 1975b). Krassilov (1981) maintained that many of these plants were ecologically analogous to extant shrubs; certainly the small size of their seeds means there would have been little food stored for the embryo, and thus the requirement of high light conditions for successful seedling establishment. Some taxa have been suggested as bog shrubs (*Nilssoniopteris;* Krassilov 1975b) or even mangroves (some species of *Ptilophyllum;* Krassilov 1975b).

Other groups of plants that achieved importance in Jurassic vegetation include the caytoniaceous seed ferns and the dicksoniaceous and cyatheaceous ferns. Like most of the other Mesozoic seed fern groups, Caytoniaceae have been reconstructed as shrub-to-small-tree-sized plants. *Caytonia* had palmately compound leaves ("*Sagenopteris*"). Its seeds were carried in an inrolled, fleshy receptacle, and both seeds and receptacles have been found in coprolites (Harris 1964). However, the structure and abundance of dispersed seeds in some deposits suggest dispersal through abiotic means rather than as a result of consumption by animals. This is another group for which the importance of animal dispersal is unresolved.

Living Dicksoniaceae and Cyatheaceae are tall tree ferns that bear very large, dissected, pinnately compound leaves and usually occupy moist-to-wet tropical forests. Most species are part of the upper understory or lower canopy vegetation, although some Cyatheaceae may reach heights of 20 m and become part of the canopy in forests of moderate stature (LaPasha and Miller 1984; Tryon and Tryon 1982). Fossil members of these two taxa are known from stem petrifactions and leaf compression/impressions that generally resemble the extant forms. Fossil dicksoniaceous tree ferns have been described as important understory elements in conifer forest, but also as participants in coastal marsh vegetation (Krassilov 1975b; LaPasha and Miller 1984).

Specific Plant Assemblages

Probably the best-known lower-latitude Jurassic flora is that from the Middle Jurassic of Yorkshire, England (e.g., Harris 1961, 1964, 1969, 1979, 1983; Harris, Millington, and Miller 1974; Spicer and Hill 1979). Over 80 genera of plants have been described from this flora, but because the flora comprises many separate assemblages from a variety of depositional environments and distributed through a considerable thickness of section, it is difficult to interpret this number in terms of the true species richness of the vegetation at any particular moment in time. A Yorkshire locality reported by Hill et al. (1985) has yielded 31 organ-species that probably represent 23 biological species. The common elements of the local floral assemblage are two species of equisetaleans, the bennettitalean *Williamsonia, Czekanowskia,* and the conifer *Elatides.* These are associated with rarer specimens of ferns, lycopsids, cycads, ginkgoaleans, and conifers. The Hasty Bank locality exhibits several lithologies, each with somewhat different floras; but it generally is dominated by *Equisetum, Nilssonia* (Cycadales), *Pachypteris* (Corystospermaceae), *Ptilophyllum* (Bennettitales), *Sphenobaiera* (Ginkgoales), and *Brachyphyllum* (Cheirolepidiaceae) (Spicer and Hill 1979). Because of the xeromorphic features of its leaves and its tendency to occur in monodominant assemblages along with brackish-to-marine fauna, the species of *Pachypteris* is thought to have been a mangrove (Harris 1983).

The seeds of the Yorkshire flora were measured and tabulated by Tiffney (1986a), who reported a mean size of 304 mm^3, maximum of 3050 mm^3, and minimum of 1.0 mm^3 (based on a sample of 19). Compared to the summed Jurassic data, the Yorkshire flora has relatively small seeds. The larger seeds in the flora were produced by cycads and ginkgoaleans.

The impression of Middle Jurassic equatorial vegetation derived from Yorkshire is that the major trees were cheirolepidiaceous conifers (producers of *Classopollis*) and rarer ginkgoaleans (such as *Baiera*), and that bennettitaleans, seed ferns, czekanowskialeans, and cycads formed the undergrowth in conifer forests or grew in open, low-stature vegetation with ferns and sphenopsids.

Several Late Jurassic assemblages from France have been reviewed by Barale (1981). The total diversity of these floras is about 60 species, with the most diverse localities producing 30–40 forms (approximately 25–35 species). As in Yorkshire, most of these Kimmeridgian French assemblages are dominated by cheirolepidiaceous conifers and bennettitaleans. However, at two localities these groups account for less than 40% of the specimens, and here both ferns (*Stachypteris, Sphenopteris*) and pteridosperms (*Pachypteris, Raphidopteris*) are more abundant. It is difficult to assess whether the association of ferns and seed ferns records an original vegetation type, but the carbonate mud sediment from which the plant material was derived is consistent with deposition in a lagoonal setting (Barale 1981). If the depositional environment has been inferred correctly, the fossil plant assemblages probably experienced relatively little transport and may represent a low-statured coastal vegetation.

On the basis of the "fossil forests" of the Purbeck Formation, where in situ bennettitalean stems occur scattered among the dominant cheirolepidiaceous conifers (Francis 1983), it appears that the typical low-latitude Middle Jurassic vegetation continued unabated into the latest part of the period. A Late Jurassic assemblage from northwestern China also contains silicified trees in growth position, in this case showing the presence of a mature forest (trees up to 2.5 m in diameter) composed exclusively of conifers (*Araucarioxylon*) with seasonal growth rings in their wood (McKnight et al. 1990).

Although the Morrison Formation of the western United States is more noted for its fossil vertebrates, there are scattered reports of plant compression fossils and pollen and spores from the northern portion of the outcrop area in Montana (R. W. Brown 1956; J. T. Brown 1975; Miller 1987; Pocock 1962; Dodson et al. 1980), and of silicified wood from Utah (Arnold 1962; Tidwell and Rushforth 1970; Medlyn and Tidwell 1975). The compression floras are dominated by probable bennettitaleans such as *Zamites* and *Nilssonia* but also contain seed ferns (*Sagenopteris*), ferns (e.g., *Cladophlebis* and *Coniopteris*), conifers (e.g., *Podozamites, Pityophyllum, Pagiophyllum*), and ginkgos. Cycadophyte remains are more abundant than those of conifers at all localities, and Miller (1987) interpreted Morrison vegetation as relatively open with scattered conifers and an understory of cycadophytes, tree ferns, and ferns. Similar vegetation is thought to have extended northward from Montana into southern Canada (Miller 1987).

Krassilov (1975b) divided the Jurassic flora from Kamenka in the Donets Basin, Russia, into a series of environmental-vegetational units. Mangrove and coastal marshes were dominated by monospecific thickets of large *Equisetites*, along with small *Coniopteris* (Dicksoniaceae) and the bennettitalean

Ptilophyllum, the last being a putative mangrove. Pioneer arborescent vegetation consisted of *Podozamites* and *Czekanowskia,* with the scrambling fern *Klukia* (Schizaeaceae). Wet "bog" forests were largely composed of *Nilssoniopteris* (Bennettitales) and *Pityophyllum,* with *Elatides* (conifer) being more important near channel margins; the lower story here consisted of *Todites* (Osmundaceae), *Caytonia, Ptilophyllum,* and *Dictyophyllum* (fern). Finally the higher-ground forest was dominated by *Ginkgoites* and *Phoenicopsis.* Whether these separate assemblages represent distinct vegetation types or seral stages of vegetation in the same area is not yet determined (Krassilov 1987). Krassilov (1975b) reconstructed the Jurassic genus *Phoenicopsis* as a large tree. Ginkgoaleans having *Ginkgo*-like leaves were almost certainly deciduous, and many had moderate-to-large, fleshy seeds that could well have been animal dispersed (Tiffney 1986a).

More northerly floras typically had a higher component of conifers other than Cheirolepidiaceae. As in the Donets Basin flora, ginkgoaleans and Czekanowskiales were also important groups in high-latitude Northern Hemisphere vegetation (Smiley 1969; Spicer 1987). Generally, ferns, sphenophytes and cycadaleans formed the understory in these forests, although there was a poleward decrease in cycad diversity. Podocarpaceous conifers apparently were an important component of higher-latitude Southern Hemisphere floras, as perhaps were the Pentoxylales, a gymnosperm group of uncertain taxonomic affinity.

A latest Jurassic or earliest Cretaceous flora from Hope Bay on the Antarctic peninsula provides one of the best examples of high southern latitude floras (65-to-70 degrees S) during the middle of the Mesozoic (Gee 1989). The assemblage includes a total of 43 species that probably grew under a warm-temperate climate. Ferns are the most diverse element of the flora (15 spp. in the genera *Coniopteris, Dictyophyllum,* and *Todites,* among others), with bennettitaleans (9 spp.), conifers (8 spp.), and seed ferns (7 spp.) making up most of the remaining species. Cycads were a minor element of the flora.

3.2   Jurassic Faunas

Numerous groups of herbivorous insects were present in the Jurassic, including sawflies, leafhoppers (Cicadelloidea and Fulgoroidea), plant hoppers (Psylloidea), shield bugs (Pentatomoidea), plant bugs (Cimicoidea), and herbivorous representatives of the Orthoptera. In addition, most of the extant groups of Coleoptera appeared in the Jurassic, as did the earliest undoubted thrips (Thysanoptera).

Early Jurassic herbivorous tetrapod assemblages were dominated by prosauropods, cynodont synapsids, and a variety of mostly small ornithischian dinosaurs. The latter included, in addition to the Fabrosauridae, the Heterodontosauridae and the armored Scelidosauridae. The Heterodontosauridae were bipedal forms that foraged 1 meter above the ground. The skull was sturdily built and had a well-developed dentition with robust, closely spaced cheek teeth with distinct wear facets along the length of the tooth row (Weishampel 1984a). These dinosaurs would have been capable of processing relatively tough plant material (Weishampel 1984a; Galton 1986). The Scelidosauridae were quadrupedal forms characterized by robust, leaf-shaped teeth with high-angle double wear facets (Weishampel and Norman 1989). Their foraging range was probably within 1 meter above the ground.

The Tritylodontidae were small to medium-sized, extremely mammal-like cynodont synapsids, which persisted well into the Middle Jurassic. They had multicuspid cheek teeth that met in complex occlusion and an extensive palinal power stroke, well suited for shredding fibrous plant material (Crompton 1972; Sues 1986). These quadrupedal forms foraged within a meter above ground level. The structure of the shoulder girdle and forelimb suggests capability for digging, and wear on the enlarged incisors is consistent with digging for underground plant parts (Sues 1984).

By the Middle Jurassic, dinosaur-dominated herbivore faunas were well developed. Such assemblages are most thoroughly documented from the Middle to Late Jurassic of China (Dong et al. 1983) and from the Late Jurassic of western North America (Coombs 1975) and Tanzania (Dodson et al. 1980). The largest species in these faunas were the sauropods, which presumably could have browsed at heights up to 10 m in a quadrupedal posture (and possibly even higher if they could have assumed a tripodal stance; Bakker 1978; Alexander 1985). Some sauropods weighed more than 50 tons and were the largest terrestrial herbivores of all time (Coombs 1975). Sauropods are first known from the Early Jurassic, by which time they already had attained large body size (Jain et al. 1975). Sauropod dentitions consist of peglike or spatulate teeth that show few signs of tooth wear, implying that oral processing of plant material was not extensive and that, as in prosauropods, the teeth were used in cropping. Food was broken down mechanically in a gastric mill, and the capacious abdominal cavity suggests the presence of a voluminous gut (Farlow 1987).

The second most abundant element in these herbivore assemblages were the armored Stegosauria. These heavily built quadrupeds were up to 7 m long, weighed up to 5 t, and had relatively narrow, elongate snouts with simple spatulate teeth. The teeth occasionally show obliquely inclined wear facets that suggest some form of orthal puncture crushing (Weishampel and Norman 1989). Stegosaurs would have fed mostly within a meter of the ground, but Bakker (1978) has argued that they were capable of a tripodal stance to forage at heights of 2–3 meters above ground level.

An important radiation of mostly smaller, facultatively (if not obligately) bipedal herbivores was represented by ornithopod dinosaurs, including *Camp-*

tosaurus (length up to 7 m and weight up to 500 kg), Dryosauridae (3–4 m long), and Hypsilophodontidae (1–3 m long). These forms show well-developed capabilities for oral food processing involving a transverse power stroke (Weishampel 1984a; Norman and Weishampel 1985; Weishampel and Norman 1989). Advanced ornithopods, much like the Heterodontosauridae, had closely spaced cheek teeth that, with continuing wear, formed long and relatively wide occlusal surfaces. Furthermore, movement along skull sutures permitted inclusion of a transverse power stroke (Weishampel 1984a; Norman and Weishampel 1985). These ornithopods foraged mainly within 1–2 m (perhaps up to 4 m) above the ground and were very abundant in many communities.

Even smaller Jurassic herbivores and/or omnivores included sphenodontids, which were relatively ubiquitous, and by Mid to Late Jurassic times, lizards and docodont and multituberculate mammals (Lillegraven et al. 1979). Multituberculates developed distinctive dentitions with capabilities for orthal slicing and palinal grinding (Krause 1982). All these forms would have foraged close to the ground, but at least some of them may have been facultatively arboreal, thus dramatically increasing their foraging ranges.

3.3   Jurassic Ecological and Evolutionary Trends

The ecological dominants of Jurassic terrestrial herbivore faunas were the very large, high-browsing sauropods. Simply by virtue of size, these dinosaurs must have had considerable caloric requirements even if they had lower metabolic rates than large herbivorous mammals (Coe et al. 1987). Analogy to extant ecosystems with "megaherbivores" suggests that contemporary vegetation should have produced large amounts of leafy material and should have had a high capacity for regrowth. A further prediction is that evolution in a variety of plant lineages (particularly those with slower growth, delayed maturation, and/or large seeds) would have led to the development of mechanical or chemical mechanisms to deter herbivory (Bakker 1978; Janzen and Martin 1982; Janzen 1986; Coe et al. 1987; Owen-Smith 1987; Wing and Tiffney 1987).

Judging from the structure of fossil leaves and, for some taxa, the habits of living relatives, the major groups of Jurassic trees (araucariacean and cheirolepidiacean conifers, cycads, bennettitaleans, ginkgoaleans) probably did not produce large quantities of foliage, nor is there evidence that they had high growth rates that would have allowed regrowth following feeding events. The ubiquity of mechanical and chemical defenses cannot be judged from known fossil data, but the common occurrence of spiny foliage among living araucarians and cycads and their tendency to retain dead foliage around the trunk raise the possibility that these were important defenses for their Jurassic relatives. Many living cycads also have toxic foliage (Chamberlain 1935).

Furthermore, the thick texture of many Jurassic leaves may have been an herbivore deterrent as well as an adaptation to dry climates. On the whole, it seems more probable that Jurassic trees had mechanical or chemical defences against herbivores than that they were capable of rapid growth and high productivity. Paradoxically, in light of the supposed high-browsing adaptations of sauropods, the Jurassic plant groups most likely to have produced abundant vertebrate fodder were low-statured (e.g., gleicheniaceous, schizaeaceous, and matoniaceous ferns, sphenopsids, possibly certain seed ferns, and many cycadophytes).

The contrast between high diversity of large herbivores and probable low productivity of terrestrial vegetation is illustrated well by the Morrison Formation biota. In Montana, the presence of coals suggests a fairly wet climate during Morrison time, and open, fern- and seed-fern-dominated vegetation (Miller 1987) may have been productive enough to support dinosaur herbivore faunas. However, most vertebrate remains come from more southerly strata of the Morrison Formation, where paleosols and lithological features have been interpreted as indicating a seasonally dry climate in southern Wyoming (Dodson et al. 1980). Recent work on the sedimentological characteristics of the upper part of the formation has raised the possibility that the climate during Morrison time may have been extremely arid at times, especially in Utah and Colorado (Turner and Fishman 1991). Although high diversity and biomass of large herbivores is seen today in seasonally dry regions like the Serengeti (Sinclair and Norton-Griffiths 1979), it is based largely on the productivity of herbaceous angiosperms (especially grasses). It is not clear what groups of Mesozoic plants would have been able to maintain high levels of productivity in a seasonally arid or arid climate.

The conflict between animal and plant evidence suggests two (not necessarily mutually exclusive) possibilities. First, the degree of productivity of Mesozoic plants has been underestimated because of inadequate knowledge of their biologies. This is not unlikely, given the difficulty of estimating the amount of edible material produced by extinct plants, especially for lineages that are entirely extinct, such as Cheirolepidiaceae. Second, the browsing pressure on Mesozoic vegetation has been overestimated, because either per individual consumption or total herbivore density has been overestimated. Estimates of per capita consumption depend strongly on the contentious matter of dinosaurian metabolic rates (see Olson and Thomas 1980), and furthermore, physiological parameters may have varied among the various taxa of herbivorous dinosaurs. There is no accepted method for estimating standing biomass of herbivores directly from fossil assemblages, and some authors have assumed that fossil communities had the same total rates of plant consumption as do present-day systems. This assumption, coupled with a questionably low figure for dinosaur metabolic rate, led Coe et al. (1987) to suggest that the standing herbivore biomass in the ecosystem represented by the Upper Jurassic Morrison Formation was 20 times the equivalent figure for Amboseli Park, Kenya. It seems equally plausible that this Jurassic ecosystem had low primary productivity and low population density of large herbivores.

4 CRETACEOUS BIOTAS

4.1 Early Cretaceous Vegetation

Earliest Cretaceous (Neocomian) floras generally lack angiosperms and bear many similarities to the preceding Late Jurassic assemblages. By Barremian to Aptian times, floras showed a substantial difference between equatorial regions and higher latitudes to the north and south. Brenner (1976) recognized four main floral provinces based on pollen assemblages: northern Laurasia, southern Laurasia, northern Gondwana, and southern Gondwana. (Figure 6.4 shows reconstructed paleogeography for the early Late Cretaceous, which is broadly similar to that for the Early Cretaceous.) Lower-latitude palynofloral assemblages from the Barremian-Aptian show that angiosperms had appeared and were undergoing significant diversification, although pollen similar to that of the extant xerophytic gymnosperm *Ephedra* was also abundant in the northern Gondwana area (Brenner 1976; Doyle et al. 1982; Crane and Lidgard 1989). Climate in the southern Laurasian and northern Gondwanan provinces is inferred to have been seasonally dry on the basis of palynological and sedimentary indicators (Brenner 1976; Ziegler et al. 1987). Fern and conifer palynomorphs were of greater importance in the northern Laurasian and southern Gondwanan regions, which are thought to have had cooler and wetter climates (Brenner 1976; Ziegler et al. 1987).

Although angiosperms were a significant component in low-latitude palynofloras of the Early Cretaceous, they were slow to dominate in middle-to-high latitudes (Brenner 1976; Crabtree 1987; Drinnan and Crane 1990). At mid latitudes, angiosperms became a significant part of disturbed riparian plant communities during Aptian and Albian times but did not develop understory tree/shrub habits until the Cenomanian (Doyle and Hickey 1976; Hickey and Doyle 1977). Angiosperms were even less important in high-latitude regions. For instance, an Early Cretaceous (Aptian) flora from Koonwarra, Australia, containing one of the oldest angiosperm megafossils (Taylor and Hickey 1990), is thought to represent a forest dominated by ginkgoaleans and podocarpalean conifers with an understory of ferns, pentoxylaleans, and sphenophytes (Drinnan and Chambers 1986). The presence of only a single type of angiosperm pollen in the Koonwarra palynoflora of 62 taxa (Dettmann 1986) and the herbaceous character of the single known angiosperm megafossil imply that flowering plants formed a minor component of this particular vegetation. The low diversity of angiosperms is typical of Early Cretaceous Australian floras; there were only 6 species of angiosperms known from Neocomian-through-Albian collections from southern Victoria at 70-to-80 degrees S paleolatitude (Douglas and Williams 1982). The Australian pattern is also seen in small Neocomian-Albian floras from South Africa, which lack angiosperm megafossils and have only a few types of angiosperm pollen (Anderson and Anderson 1983a, 1985; Drinnan and Crane 1990). Just a few species of angiosperms occur in a conifer-cycadophyte assemblage from volcaniclastic Aptian rocks in southern Argentina (Romero and Archangelsky 1986), although angiosperm pollen is known from Barremian assemblages in Argentina (Archangelsky 1980). Angiosperms were also rare or absent in middle northern latitudes during the Aptian (e.g., LaPasha and Miller 1984) and were essentially absent from Arctic floras until near the end of the Albian (Spicer 1990). The 20–30 my lag between the first appearance of flowering plants in equatorial regions and their rise to importance in mid- to high-latitude vegetation has been attributed to slow evolution of cold tolerance in angiosperms (e.g., Brenner 1976).

During the Early Cretaceous, wetland areas still supported abundant equisetaleans and small lycopsids such as *Isoetites* (Batten 1974; Kovach and Dilcher 1985). Osmundaceous ferns also were common in wet areas, as are the living species, although at least one Cretaceous species in this family is thought to have been an epiphyte (Harris 1961).

Ferns of the families Schizaeaceae and Gleicheniaceae were rhizomatously spreading plants less than 2 m high that were probably major colonizers of open, perhaps even dry and/or low-nutrient, substrates during much of the Mesozoic. Ferns of these families may have formed vegetation analogous to savannas or grasslands now dominated by herbaceous angiosperms (Coe et al. 1987). Fusinized leaves of these families are abundant at some Lower Cretaceous localities (Alvin 1974; Harris 1981), and their spores are common elements in palynofloras (e.g., Batten 1974). Furthermore, most living species of Gleicheniaceae and Schizaeaceae are thicket-forming or scrambling types that occupy open, sometimes fire-disturbed, habitats in tropical-to-subtropical areas (Tryon and Tryon 1982). One particularly enigmatic element of Early Cretaceous vegetation was *Tempskya*, which had a large pseudo-trunk, composed of many small, intertwining stems, that was up to 40 cm thick but only 4.5 m tall (Read and Brown 1936). The ecological role of *Tempskya* is difficult to infer, because it so little resembles any other living or fossil form.

Conifers, cycads, cycadeoids, and other gymnosperms such as Cheirolepidiaceae and Czekanowskiales continued from the Jurassic into the Cretaceous without major changes.

Specific Plant Assemblages

Good examples of Early Cretaceous vegetation without angiosperms come from western North America (LaPasha and Miller 1984), eastern Siberia (Krassilov 1973, 1975a), and from the Wealden of western Europe (e.g., Daber

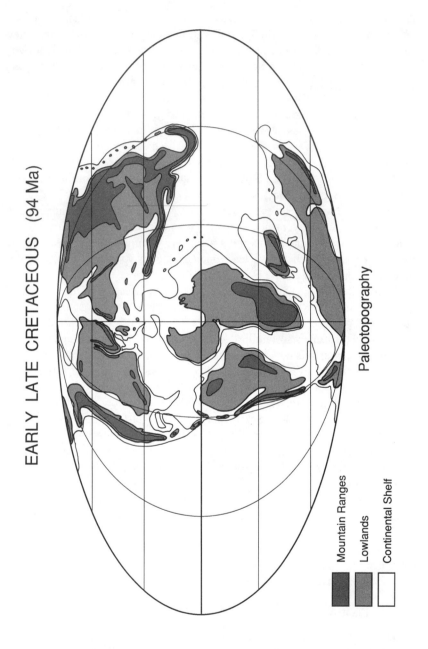

EARLY LATE CRETACEOUS (94 Ma)

Paleotopography

Mountain Ranges

Lowlands

Continental Shelf

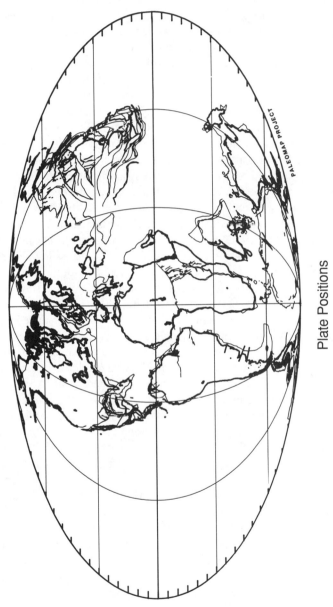

Plate Positions

Fig. 6.4. Generalized continental positions and orography in the Late Cretaceous. (Courtesy of C. R. Scotese and the PaleoMap Project, University of Texas, Arlington)

1968; Batten 1974; Oldham 1976; Riegel et al. 1986; Pelzer 1987). The Kootenai flora from the upper Aptian of Montana contains 25 species of vascular plants, the most common of which are ferns (*Coniopteris, Acrostichopteris, Sphenopteris*, and *Cladophlebis*), caytoniaceous pteridosperms (*Sagenopteris*), and conifers (*Athrotaxites, Elatides*, and *Elatocladus*). LaPasha and Miller (1984) interpreted the vegetation to have been a conifer- (*Athrotaxites*) dominated swamp forest with a high understory of *Coniopteris. Sagenopteris williamsii* also is thought to have been a swamp plant. Several species of ferns, two species of *Sagenopteris, Ginkgo*, and the conifers *Elatides* and *Elatocladus* are thought to have occupied slightly better drained areas. Remains of the bennettitalean *Zamites* are rare and appear strongly allochthonous.

Early Cretaceous floras from the Vladivostok region of the far east of Russia described by Krassilov (1973) are floristically somewhat similar to those from the Kootenai Formation. The coal-swamp assemblage consists of conifers (*Athrotaxites* and *Cephalotaxus*), ferns (*Cladophlebis*), and bennettitaleans (*Nilssoniopteris*). Stream borders were occupied by *Zamiophyllum, Subzamites, Nilssonia*, and *Podocarpus*. The floodplain supported a diverse vegetation including many species of ferns, conifers, and cycadaleans, but the dominant forms were the cycads *Ctenis* and *Cycadites*. Krassilov also noted that the majority of conifers occurring in the cycadophyte-dominated assemblages were probably relatively small trees, suggesting that interfluves may have held a low-stature, chaparral type vegetation.

Wealden pollen assemblages from southern England show a reduced importance of cheirolepidiaceous pollen compared to Jurassic floras, but these conifers probably still formed an important part of coastal vegetation (Batten 1974; Watson 1977). Batten (1974) envisioned the vegetation as a savannalike open forest with coniferous trees and a fern/lycopsid herb layer. Alvin (1983) reconstructed one of the dominant conifers of the Wealden, *Pseudofrenelopsis parceramosa*, as a moderate-to-large tree with whorled branching and xeromorphic shoots that might have been shed during times of water stress. However, he found no evidence that *P. parceramosa* was halophytic (salt-tolerant). In a study of dispersed cuticles from Wealden deltaic deposits in southern England, Oldham (1976) found that conifer remains were most abundant, followed by bennettitaleans, cycads, ginkgoaleans, and seed ferns. Among the conifers, Cheirolepidiaceae were the most ubiquitous and abundant, Cupressaceae were sporadically abundant, Taxodiaceae were rare, and only pollen of Araucariaceae and Pinaceae was found. Oldham suggested that the vegetation of the Wealden might have been loosely comparable to that of southern Florida today, with extensive low-lying flooded marshes interspersed with slightly higher ground occupied by more diverse forest. In some floodplain areas with oxidized soils, *Weichselia, Phlebopteris*, and *Gleichenites* appear to have formed fern heaths or savannas that were basically devoid of arborescent vegetation, probably as a result of frequent burning (Harris 1981). Similar *Weichselia*-dominated assemblages are also known from the Wealden of northern Germany (Daber 1968), and the Early Cretaceous of Venezuela (P. Crane, pers. comm., 1990).

Wealden sediments in Germany preserve a number of nearly autochthonous vegetational associations (Pelzer 1987; Pelzer and Wilde 1987). Pollen floras indicate that peat substrates were dominated by gymnosperms, and megafloras from these environments consist mainly of leaves of the conifer *Abietites*. Natural levees in the lower delta plain supported plant communities including *Ruffordia*, *Zamites*, and *Sphenolepis*. Wet back-swamp settings, indicated by carbonaceous shale deposition, preserve a flora containing *Matonidium*, *Nilssonia*, *Sagenopteris*, *Zamites*, *Ginkgoites*, and *Dictyophyllum* (Riegel et al. 1986; Pelzer and Wilde 1987). On the basis of sedimentological and paleobotanical evidence, Pelzer and Wilde (1987) inferred a shift from semiarid conditions in the Late Jurassic of northwest Europe to wetter, although still seasonally dry, climate in the Early Cretaceous.

4.2   Late Cretaceous Vegetation

During the early Late Cretaceous, angiosperm diversity increased explosively, especially in middle-to-high paleolatitudes, and by the end of the Late Cretaceous 50% to 80% of the species in typical local fossil assemblages were flowering plants (Crabtree 1987; Lidgard and Crane 1990). This proportional gain in diversity for flowering plants was largely at the expense of cycadophytes and ferns (broadly speaking), while the proportion of conifer species in local assemblages was relatively unchanged (Lidgard and Crane 1988, 1990). The increase in within-assemblage angiosperm diversity during the Late Cretaceous is less noticeable in palynofloras than in megafloras (Lidgard and Crane 1990).

Although angiosperms became much more diverse, abundant, and widespread in the Late Cretaceous, there is little evidence that, prior to the Campanian and Maastrichtian, they achieved the range of habits and ecological roles they occupy today. Occurrences of Cretaceous angiosperm wood are rare in comparison to those of flowers, leaves, and pollen from the same time, and in comparison with coniferous wood (Wing and Tiffney 1987). Cretaceous angiosperm wood is generalized and shows little anatomical variation (E. Wheeler, pers. comm., 1991). These observations are consistent with the idea that mid-Cretaceous angiosperms tended to be small trees. Angiosperm seeds in the Late Cretaceous were very small, which also is consistent with derivation from relatively "weedy" plants that dominated mostly disturbed, early successional settings (Tiffney 1984). Angiosperms dominated early successional forests and were locally abundant in brackish habitats (Retallack and

Dilcher 1981, 1985; Upchurch and Dilcher 1990); they also may have occupied the understory of relatively open forests. (See Crabtree 1987 and Crane 1987 for opposing views on the importance of angiosperms in late successional vegetation.) Conifers, including podocarpaceous, araucariaceous, and taxodiaceous forms, apparently were the dominant large trees. Wolfe and Upchurch (1987a) reviewed a large suite of floras from the Late Cretaceous of North America and concluded that the small leaf size of most angiosperms, coupled with the rarity of drip tips and of liana leaf types, implied that multistratal rain forest of the modern type was not present. Rather, leaf physiognomy suggests an open woodland vegetation, reflecting a relatively dry climate (Wolfe and Upchurch 1987a, b).

The pollination biology of Late Cretaceous angiosperms was distinctly less modern than in the early Tertiary, with most flowers being actinomorphic (radially symmetrical) and probably pollinated by either wind or a broad range of insects (Friis and Crepet 1987; Crepet and Friis 1987). However, advanced eusocial bees are known from the Campanian (Michener and Grimaldi 1988), and the morphology of Cenomanian lauraceous flowers is consistent with bee pollination (Drinnan et al. 1990).

The Cheirolepidiaceae underwent a drastic decline in diversity at about the Cenomanian/Turonian boundary, especially at low latitudes (P. Crane, pers. comm., 1990), and went extinct later in the Cretaceous. During the same interval, Pinaceae became more important (Krassilov 1978b, 1981). Cycadophytes were still present, though rare in most assemblages, and ferns and sphenophytes continued to be common herbaceous elements. Based on the abundance of their megaspores, *Isoetes*-like lycopsids were important elements in wetland or floodplain floras (Kovach and Dilcher 1985; Kovach 1988).

High-latitude Late Cretaceous floras from Alaska indicate a somewhat different kind of vegetation. In these assemblages, angiosperms were less abundant than at midlatitudes, and the forests mostly consisted of taxodiaceous and podozamitean conifers, along with a number of ginkgoaleans and cycadaleans and the usual fern-sphenophyte ground cover. Most or all of the species are thought to have been deciduous. Conifer diversity was initially fairly low in higher latitudes of the Northern Hemisphere but increased sharply during the mid-Cretaceous. The wood structure of some species indicates very rapid growth in the season of light, and coals reflect a high rate of productivity (Spicer and Parrish 1986; Parrish and Spicer 1988b). Similar observations apply to conifer-dominated assemblages from the Antarctic peninsula (Francis 1986).

By Maastrichtian times, climatic deterioration had caused a dramatic drop in diversity in these high-latitude assemblages and also in the stature and productivity of the vegetation (Spicer and Parrish 1986). Fossil wood and oxygen isotope data derived from marine mollusk shells of the Antarctic peninsula also record declining temperatures from the Santonian-Campanian interval to the Maastrichtian (Francis 1989; Pirrie and Marshall 1990). Similar climatic trends for both northern and southern high latitudes imply the Late Cretaceous cooling was a global event.

Specific Plant Assemblages

Although Late Cretaceous floras are numerous (e.g., Crabtree 1987; Mc-Clammer and Crabtree 1989), few have been studied from a vegetational or paleoecological perspective. One exception is the flora of the Blackhawk Formation (probably Campanian) in central Utah (Parker 1976). Autochthonous plant fossils from the Blackhawk Formation occur in two major sedimentary environments: swamps and "bottomlands." The swamp vegetation was dominated by a taxodiaceous conifer, *Sequoia cuneata*, and a deciduous dicot, *Rhamnites eminens*, but 8 other conifers (including species of *Brachyphyllum*, *Moriconia*, and *Protophyllocladus*) and 8 other species of dicots were also moderately abundant in the swamp assemblages and were probably canopy or subcanopy trees. The palm *Geonomites*, the cycadophyte *Nageiopsis*, and several ferns were abundant locally and probably were understory or forest floor plants. In contrast to the swamps, the assemblages from slightly drier "bottomland" areas are dominated in numbers and species by dicots. The bottomland vegetation was probably dominated by species of *Cercidiphyllum*, *Platanus*, and *Dryophyllum*, although 10 other dicot species occur in moderate abundance. The 2 conifers found in the bottomland assemblages are rare elements. It is interesting to note that even as late as the Campanian, conifers were still vegetational dominants in swamp settings, and that 7 or 8 species co-occurred in the same local area; this is a far-higher local species richness than is seen in Paleogene or contemporary conifer swamps.

The Santonian-age Mgachi locality on Sakhalin island, Russia, has produced an assemblage typical of the Late Cretaceous "North Pacific refugium" (Vakhrameev 1987), in which archaic forms were especially common, particularly *Sequoia* (Taxodiaceae), followed by *Araucarites* (Araucariaceae), *Protophyllocladus* (Podocarpaceae?), *Araliaephyllum* (platanoid angiosperm), and *Cupressinocladus* (conifer), along with 3 taxa of angiosperms, 3 ferns, and 1 taxon each of cycad, ginkgoalean, and caytonialean that together make up about 22% of the specimens studied (Krassilov 1975a). These relative abundances probably indicate a conifer forest with ferns, cycads, and seed ferns either in the understory or in nearby riparian habitats.

A sequence of floras from the North Slope of Alaska span the Late Cretaceous and provide evidence for cooling climate and less-diverse vegetation during this interval of time (Parrish and Spicer 1988a, b; Spicer and Parrish 1990a, b; Frederiksen et al. 1988). Latest Albian or Cenomanian assemblages from the Nanushuk Group have produced 67 species of angiosperms, 18 ferns, perhaps as many as 12 species of conifers, 4 ginkgoaleans, and 2 cyadophytes. Angiosperm fossils are most abundant in fluvial channel and lake margin settings, although they also occur in ponded-water environments (Spicer and Parrish 1990b). By the Coniacian, the ginkgoaleans were reduced to 1 species and the cycadophytes were absent. The physiognomy of these Coniacian assemblages has been compared to extant low montane mixed coniferous forest growing under mean annual temperatures of less than 13 degrees C. By the Campanian and Maastrichtian, the diversity and productivity of high Arctic vegetation had both been decreased. Only 2 species of conifers are known as megafossils, fossil trunks are generally smaller in diameter, and a sizable proportion of the angiosperm diversity may have been in herbaceous species (Spicer and Parrish 1990b, Frederiksen et al. 1988). Equisetaleans probably were the dominant component of the herb layer in Campanian-Maastrichtian vegetation of the North Slope (Brouwers et al. 1987).

4.3   Cretaceous Faunas

Although some evidence exists for the presence of Lepidoptera (or at least very closely related taxa) in the Jurassic, the oldest unambiguous records (referable to the Micropterigidae) date from the Early Cretaceous (Whalley 1986). Adults of extant micropterigids subsist on pollen, whereas their caterpillars feed on mosses, liverworts, and detrital organic matter. Evidence for angiosperm leaves mined by lepidopterans also is recorded from the Late Cretaceous (Dilcher in Whalley 1986; Gall and Tiffney 1983). Aphids (Homoptera: Aphidoidea), short-horned grasshoppers (Orthoptera: Acridoidea), and gall wasps (Hymenoptera: Cynipidae) all first appear in the Cretaceous. The oldest-known termites (Isoptera) and ants (Formicoidea) are Late Cretaceous in age (Hennig 1981; Wilson 1987). Of great significance to the ecology and evolution of flowering plants is the presence of advanced eusocial bees in the Late Cretaceous (Michener and Grimaldi 1988).

By the Early Cretaceous, tetrapod faunas appear to show pronounced differences between Northern and Southern Hemisphere assemblages (Bonaparte 1987, 1990), although some interchange is still evident. In the Northern Hemisphere, the main change among herbivorous tetrapods was a decrease in diversity and abundance of high-browsing sauropods (Weishampel and Norman 1989). Stegosaurs were also very much reduced in importance in the Early Cretaceous. Another group of large armored quadrupedal ornithischians, the Ankylosauria, slowly increased in diversity and abundance throughout the Early Cretaceous (Maryanska 1977; Coombs 1978). These dinosaurs reached a length of up to 6 m and a weight of up to 2 t. They had massive skulls with fusion of most cranial bones. The teeth are small, spatulate, and very simple. Although no clear pattern of tooth wear is apparent, many teeth show obliquely inclined wear facets, suggesting some sort of puncture crushing (Russell 1940; Weishampel and Norman 1989). Ankylosaurs had been present as a small part of some Jurassic faunas and did not diversify until the Cretaceous.

The Ornithopoda underwent a rapid radiation during the Cretaceous. Some Early Cretaceous ornithopods such as *Iguanodon* and *Tenontosaurus* reached larger body size (length up to 10 m and weight up to 2 t) and thus presumably foraged at higher levels above the ground than did their Jurassic predecessors. They continued the trend toward elaboration of dental occlusion for oral processing of food (Norman and Weishampel 1985; Weishampel and Norman 1989).

The first Ceratopsia appeared in the Cretaceous. They included the bipedal Psittacosauridae from the Early Cretaceous (Osborn 1923) and the quadrupedal Protoceratopsidae from the Late Cretaceous of East Asia and western North America (Russell 1970; Maryanska and Osmolska 1974). The closely packed, interlocking teeth of psittacosaurs had wear facets oriented approximately 60 degrees to the horizontal, and it would appear that the chewing stroke involved both palinal and orthal movements of the mandible (Weishampel and Norman 1989). In the Protoceratopsidae, the wear facets are virtually vertically oriented, suggesting strictly orthal, slicing jaw motion, much as in the Ceratopsidae (Ostrom 1966; Weishampel and Norman 1989).

Small herbivores and/or omnivores of this time period included multituberculates and broad-toothed sphenodontians, all of which presumably foraged within the first meter above the ground. At least some of the multituberculates were arboreal (Krause and Jenkins 1983) and thus had substantially increased vertical foraging ranges. The Early Cretaceous sphenodontian lepidosaur *Toxolophosaurus* has transversely broadened and closely packed teeth, and wear facets suggest a back-to-front chewing stroke (Throckmorton et al. 1981).

The main difference between Northern and Southern Hemisphere tetrapod faunas is that in the Southern Hemisphere sauropods apparently continued to be the dominant element among the herbivores, with hadrosaurid ornithopods being quite rare (Bonaparte 1987). Recently recovered mammalian assemblages from the Upper Cretaceous of Patagonia are notable for the complete lack of eutherian and metatherian taxa and the presence of many endemic taxa, including the hypsodont *Gondwanatherium* (Bonaparte 1986, 1990).

Late Cretaceous assemblages of large herbivores in the Northern Hemisphere were dominated, in the numbers of both specimens and species, by hadrosaurid ornithopods (length up to 12 m and weight up to 4 t) and ceratopsid ceratopsians (length up to 10 m and weight up to 9 t). The former group had a circumboreal distribution, whereas ceratopsids were apparently restricted to North America. Both hadrosaurids and ceratopsids had elaborate dentitions with closely packed, interlocking teeth. Unlike the condition in hadrosaurids, the nearly vertical orientation of the wear facets in ceratopsids (and protoceratopsids) suggests that the power stroke was restricted to an orthal slicing movement (Ostrom 1966; Weishampel and Norman 1989). The browsing range for the quadrupedal ceratopsids was probably within one or two meters above the ground. Taphonomic evidence suggests that both ceratopsids (Currie and Dodson 1984) and hadrosaurids (J. R. Horner, pers. comm., 1989) may have foraged as herds.

Pachycephalosaurs were another group of small-to-medium-sized bipedal ornithischians that first appeared in the Early Cretaceous (Maryanska and Osmolska 1974; Sues and Galton 1987). They have relatively simple spatulate teeth with some wear, suggesting some form of puncture crushing, and a rather extensive gut capacity. They probably foraged within one to three meters above ground level, and all had relatively simple teeth that would have sufficed for cropping but must have relied on other means of breaking down plant fodder, such as gut fermentation (Farlow 1987). The major difference between Campanian-Maastrichtian and earlier tetrapod assemblages is the abundance and diversity of small mammalian herbivores, mostly multituberculates. These mouse-to-woodchuck-sized animals had dentitions well suited to processing discrete food items such as fruits and seeds (Krause 1982). By Maastrichtian times, the taxonomic diversity of small mammalian herbivores, including eutherians, equaled or exceeded that of dinosaurian herbivores in some Northern Hemisphere faunas. Gondwanan herbivore faunas were dominated by the large sauropods of the family Titanosauridae, which included a number of armored forms. Only a few hadrosaurs are known from the Late Cretaceous of South America (Bonaparte 1987), and the mammalian faunas are highly endemic (Bonaparte 1986, 1990).

4.4 Cretaceous Ecological and Evolutionary Trends

The Early Cretaceous appearance of flowering plants marks the change from the Mesophytic to the Cenophytic. Although this appearance, and the subsequent radiation of angiosperms during the mid-Cretaceous, are undoubtedly among the most significant evolutionary events in the history of life on land, the ecological consequences are less striking. It has long been held that the initial radiation of flowering plants (Barremian-Albian) took place largely in disturbed or ephemeral habitats (Hickey and Doyle 1977). The impressive taxonomic diversification of angiosperms during the mid-Cretaceous and their tendency to make up the great majority of species in local floras of Cenomanian and later age (Crabtree 1987; Lidgard and Crane 1988, 1990) have led to the tacit belief that angiosperms dominated terrestrial vegetation by the early Late Cretaceous.

However, high diversity within a clade does not necessarily indicate it has an important role in structuring vegetation (e.g., Orchidaceae or Asteraceae in extant vegetation), and few studies have quantified the abundance of Cretaceous angiosperm remains in addition to their diversity. Because leaf number has a fairly high correlation with stem basal area of a species (Burnham et al. 1992), the relative biomass of angiosperms vs. other plants in the Cretaceous could be assessed roughly by comparison of relative abundances of leaf fossils. Such data have been published for the Late Cretaceous Blackhawk Formation flora, in which 4 of the 9 taxa represented by more than 100 specimens (out of a total 7,400 specimens) are conifers or cycadophytes (Parker 1976). These numbers imply that in this local area, gymnosperms were codominant with flowering plants, even in the last half of the Late Cretaceous.

Other lines of evidence, including angiosperm diaspore size distributions (Tiffney 1984), angiosperm leaf physiognomy (Wolfe and Upchurch 1987a, b), angiosperm trunk size (Wing and Tiffney 1987), and angiosperm palynomorph diversity (Crabtree 1987), are also consistent with the suggestion that flowering plants were not major canopy elements until at least the Campanian or Maastrichtian. In spite of this probable lag in becoming important canopy trees, there is little doubt that, from the Albian onward, angiosperms were extremely abundant in disturbed habitats such as riparian corridors or recently burned areas. The evolutionary radiation of angiosperms might thus have had an important effect on the rate of colonization and productivity of disturbed sites.

The increase in the relative abundance and diversity of angiosperms in disturbed areas more or less coincides with an increase in the relative abundance and diversity of two groups of large herbivores: the hadrosaurid ornithopods and ceratopsid ceratopsians (Bakker 1978; Wing and Tiffney 1987; Weishampel and Norman 1989). Both groups of dinosaurs were adapted to extensive oral processing of plant food and may have foraged in herds. The coincidence of the two radiations may reflect diffuse coevolution, in which increased abundance and productivity of disturbed-site vegetation led to the success of these two herbivore groups, and in which herbivore disturbance in turn favored the success of relatively "weedy" plants, such as angiosperms (Bakker 1978). Given this scenario, the pattern of co-occurrence of herbivore communities dominated by hadrosaurids and ceratopsids and floras dominated by weedy angiosperms should persist at finer geographic and temporal scales. A possible test case is represented by the Late Cretaceous biotas of South America, where titanosaurid sauropods, rather than low-browsing hadrosaurs and ceratopsians, were the dominant herbivores. If the coevolutionary scenario is valid, then South American Late Cretaceous vegetation should have been significantly different from coeval North American vegetation. Specifically, areas of disturbed angiosperm-dominated vegetation should have been less common, either because South American angiosperms were less weedy in their habits or because they were a less important element in the vegetation.

Limited evidence from the southern continents does indicate that angiosperms were very slow to invade high southern latitudes, perhaps not becoming diverse there until the Cenomanian or later (e.g., Drinnan and Crane 1990).

5   PALEOGENE BIOTAS

5.1   *The K/T Boundary and Earliest Paleocene Vegetation*

Earliest Paleocene vegetation is known best from palynological assemblages. In North America these assemblages record a low-diversity flora with few angiosperms that contrasts sharply with latest Cretaceous, angiosperm-rich pollen floras. In several parts of western North America, palynofloras from the first few centimeters above the boundary consist of just a few kinds of fern spores, forming the "fern spike" (Orth et al. 1981; Nichols et al. 1986). In sections displaying a clear fern spike, angiosperm pollen does not predominate again until 10–20 cm above the K/T boundary (Tschudy and Tschudy 1986). These palynofloral changes in western North America are thought to indicate the nearly instantaneous destruction of diverse, angiosperm-dominated latest Cretaceous vegetation, its replacement by low-diversity, early successional vegetation dominated by a few species of ferns, then the return to angiosperm-dominated vegetation with a rather different composition (Orth et al. 1981; Tschudy et al. 1984; Tschudy and Tschudy 1986). However, a fern spike is not present at several apparently continuous K/T sections in southern Canada (e.g., Lerbekmo and St. Louis 1986; Lerbekmo et al. 1987).

Earliest Paleocene megafloras have been described from New Mexico, Colorado, and North Dakota. The megafloras from New Mexico and Colorado show a small increase in diversity and angiosperm dominance in the decimeters above the boundary, but the angiosperms are inferred to have been early successional taxa (Wolfe and Upchurch 1986, 1987b). Megafloras in this area continued to be of low diversity for at least the first million years of the Paleocene. This led Wolfe and Upchurch (1986, 1987a) to propose the process of "quasi-succession," a sequence of changes analogous to ecological succession but requiring several orders of magnitude more time because of the continental scale of the initiating disturbance. In contrast to the lengthy recovery phase in the southern Rocky Mountains, earliest Paleocene assemblages from western North Dakota show a more rapid increase in diversity, even though extinction levels at the K/T boundary were very high (K. R. Johnson et al. 1989; K. R. Johnson 1991; Johnson and Hickey 1991). In all probability the interregional differences in the expression of the fern spike and characteristics of the earliest Paleocene megafloras reflect differences in ecological recovery following the K/T extinctions that were induced by varying local sedimentary environments, local climatic differences, and/or taphonomic effects. These kinds of effects have been documented in Alberta, Canada, where intraregional differences in Maastrichtian to early Paleocene climate exerted a strong influence on palynofloral composition and sedimentary characteristics (Jerzykiewicz and Sweet 1988).

Although earliest Paleocene pollen assemblages from near-shore marine environments in Japan show an increase in fern spores, as well as pine pollen, relative to latest Cretaceous palynofloras from the same area (Saito et al. 1986), the fern spike is not as pronounced as that seen in many sections in the Rocky Mountains. Furthermore, the increase in pine pollen may reflect taphonomic enrichment of this buoyant pollen type in marine sediments rather than a true increase in pine (Spicer 1989). The limited data from Japan are consistent with a less severe destruction of vegetation at the K/T boundary and more rapid recovery in the Paleocene. Southern Hemisphere palynofloras show even less dramatic patterns of change at the K/T boundary. Palynofloras from terrestrial sections in New Zealand show only moderate levels of turnover across the K/T boundary (Raine 1988), and palynofloras from near-shore marine sediments on Seymour Island off Antarctica show almost no extinction through the latest Cretaceous–earliest Paleocene interval (Askin 1988). These southern high-latitude early Paleocene assemblages represent coniferous forest vegetation (mostly Podocarpaceae) with ferns and a variety of angiosperms, including some *Nothofagus* (southern beech). Generally, southern high-latitude, Paleocene palynofloras are less diverse than those of the Late Cretaceous, but this probably reflects long-term cooling rather than destruction of vegetation at the K/T boundary (Askin 1988).

Whatever the local variations in the pattern of recovery in western North America were, it seems clear that K/T devastation of terrestrial biotas was less severe in other parts of the world. This is consistent with geological and mineralogical data indicating the bolide(s) struck the earth near North America, perhaps at the northern end of the Yucatan Peninsula, and/or in Iowa (Bohor et al. 1987; Hildebrand and Boynton 1990; Hildebrand et al. 1990; Izett 1990). There is now little disagreement that the K/T bolide impact(s) caused high levels of local-to-regional extinction and ecological disruption in western North America, but the longer-term effects of the event on terrestrial vegetation are less well understood. Evidence for a sharp but long-lasting transition to wetter climates over most of North America in the early Paleocene has been seen in many studies (e.g., Hickey 1984; Fastovsky and McSweeney 1987, Retallack et al. 1987; Wolfe and Upchurch 1987b; Lehman 1990), and Wolfe and Upchurch (1987b) have argued the shift may be a direct consequence of the K/T bolide impact. Wolfe (1987) maintained that preferential extinction of evergreen species at the K/T boundary (because they were less able to survive a short period of cold temperatures generated by the impact cloud) left Northern Hemisphere floras permanently enriched in deciduous lineages. Hickey (1980) argued that the high proportion of deciduous species in many Paleocene floras was a consequence of colder climate. Confirmation of Wolfe's hypothesis in particular awaits better independent documentation of paleoclimatic trends, because continental paleoclimate inferences for the Paleocene depend largely on physiognomic features of the vegetation, which Wolfe (1987) maintained were influenced by the K/T extinction.

5.2 Paleocene-Eocene Vegetation

Paleocene megafloral assemblages from high northern latitudes demonstrate the existence of forests dominated by deciduous conifers of the bald cypress family (Taxodiaceae), with an accessory component of ginkgos and deciduous dicots that probably grew in riparian habitats (e.g., Hickey et al. 1983). Eocene forests in the Canadian Arctic are preserved as standing trunk fields with in situ litter (peat) beneath them (Basinger et al. 1988; Francis 1991). Eocene swamp forests also were dominated by taxodiaceous conifers (*Glyptostrobus* and *Metasequoia*), but presumed evergreen species of *Pinus* (pine), *Picea* (spruce), and *Abies* (fir), as well as the deciduous *Larix* (larch), by then had become important in these settings. The swamp conifers had trunk diameters of 1.5 m or more, growth rings comparable to those of extant temperate-zone conifers, and inferred heights of up to 50 m (Basinger et al. 1988). The large overall size and thick growth rings of the trees imply a productive ecosystem. The Eocene high Arctic assemblages also contain *Ginkgo* and a number of dicot species in the Betulaceae (birch family), Juglandaceae (walnuts), Platanaceae (sycamores), and Fagaceae (oaks). The dicots probably grew in riparian areas (Basinger et al. 1988).

Throughout the Northern Hemisphere, early-to-middle Paleocene mid-latitude floras are known for their relatively low taxonomic diversity and homogeneity, for example, late Paleocene assemblages from Mongolia, China, the western United States, and Scotland have many similar forms (Brown 1962; Mukulbekov 1988; Guo 1985; Boulter and Kvacek 1989; Crane et al. 1990). As in contemporaneous Arctic assemblages, the dominants in wet or swampy areas were the taxodiaceous conifers *Metasequoia* and *Glyptostrobus* (e.g., Hickey 1980; Fastovsky and McSweeney 1991), although better-drained substrates supported a limited variety of most deciduous dicots (e.g., Hickey 1977, 1980). These floras may represent a rough equivalent to extant deciduous, hardwood forests, although the existence of some large herbaceous forms (e.g., *Zingiberopsis* in the ginger family) implies they were not subject to winter temperatures far below freezing. In southern North America, fossil leaf assemblages from swamp settings tend to be dominated more by dicots than by conifers (Wolfe and Upchurch 1987b), and silicified Paleocene trunks from fluvial swamp sediments in southern Texas are dicots (Wheeler 1991); but in more northerly areas, wood in such settings generally is coniferous.

By latest Paleocene times, evergreen broad-leaved angiosperms became more common in many Northern Hemisphere floras (e.g., Hickey 1980; Wolfe and Upchurch 1987a). The earliest well-documented Tertiary fruit, seed, and wood assemblages are late Paleocene to early Eocene in age and indicate that there existed a wide variety of angiosperms with a "modern" range of seed sizes (Tiffney 1984; but see Knobloch and Mai 1986). These forms were broadly seral, and it is highly probable that the species with larger diaspores were biotically dispersed and capable of establishing and maintaining populations under a closed canopy forest.

Paleocene assemblages from 46 degrees S latitude in Argentina are thought to represent a suite of vegetational types including mangrove, swamp forest, tropical rain forest, montane rain forest, and savanna-sclerophyllous forest (Romero 1986). Few Paleocene floras from Australia have been described in paleoclimatic or paleoecological terms. However, a series of late Paleocene floras from southeastern New South Wales, at a paleolatitude of about 57 degrees S, includes both palynomorphs and fossil wood (Taylor et al. 1990). Podocarpaceous and araucarian conifers are the most abundant components of these palynofloras, although angiosperms are more diverse and include *Nothofagus* and members of the Proteaceae, among others. Taylor et al. (1990) suggest these palynofloras were derived from cool-temperate rain forests that had a coniferous canopy, and this is supported by the abundance of conifer wood relative to angiosperm wood. As in North American Late Cretaceous and Paleocene floras, conifers were particularly dominant in swamps and on peat substrates, although the Australian conifers were podocarps rather than bald cypresses.

The acme of tropicality and of angiosperm-dominated closed forest was probably during the early Eocene, when subtropical vegetation may have extended to 60 degrees N latitude, and full tropical, multistratal rain forests occurred as far north as 30 degrees N latitude (Wolfe 1985). (Fig. 6.5 shows reconstructed middle Eocene paleogeography.) In northern Europe, the fruit and seed assemblages from the Eocene London Clay record a diverse flora that was taxonomically, and perhaps vegetationally, most similar to those now occurring in tropical Southeast Asia. Vegetational types identified from the London Clay flora include mangrove swamp with abundant *Nypa* (Palmae) and associated *Ceriops* (Rhizophoraceae), as well as highly diverse, dense paratropical forest and probable gallery forest with diverse lianas of the families Menispermaceae and Icacinaceae (Collinson and Hooker 1987). However, the London Clay flora also retained members of typical Paleocene Holarctic lineages that are now temperate in their distribution, such as *Cornus* and Betulaceae (Collinson 1983). Animal and plant fossils characteristic of the mangrove biota also occur together in middle Eocene sediments of southern Texas (Westgate and Gee 1990).

Eocene megafloras from China have been summarized recently by Guo (1985). Eocene assemblages tend to be very diverse and to contain a high pro-

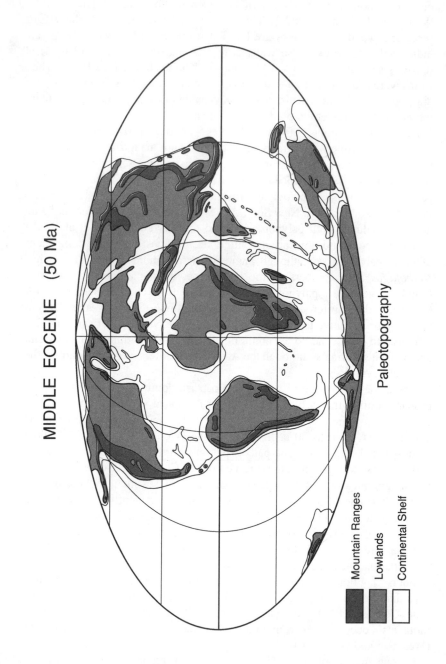

MIDDLE EOCENE (50 Ma)

Paleotopography

Mountain Ranges

Lowlands

Continental Shelf

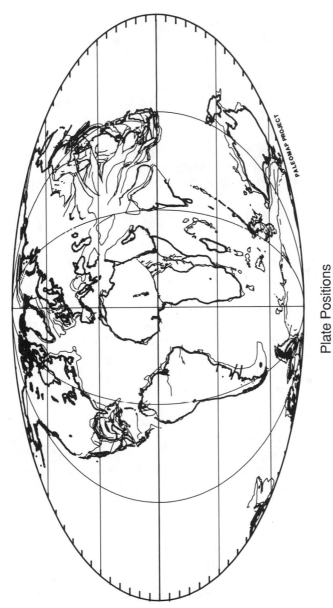

Plate Positions

Fig. 6.5. Generalized continental positions and orography in the middle Eocene. (Courtesy of C. R. Scotese and the PaleoMap Project, University of Texas, Arlington)

PALEOMAP PROJECT

portion of broad-leaved evergreen forms. Generally the leaf physiognomy of Eocene floras suggests somewhat drier climate than during the Paleocene, although a variety of vegetational types have been recognized in different parts of China. Humid subtropical forests consisting of a mixture of deciduous and evergreen dicots, and deciduous conifers existed in the north, and these shared many taxa with coeval floras of North America: the ferns *Lygodium* and *Salvinia,* the conifers *Glyptostrobus* and *Metasequoia,* and a number of angiosperms, including *Cercidiphyllum, Cinnamomum, Alnus,* and palms. Dry subtropical forests characterized central China. These assemblages are somewhat less diverse than those from farther north, and leaf texture tends to be more coriaceous. Humid tropical forests occupied the southern part of the country. Over two-thirds of the species in these assemblages are broad-leaved evergreen forms, and there is particularly high diversity in the Lauraceae (laurel family). Conifers apparently were very rare in the humid tropical forests (Guo 1985).

During the Eocene, Australia, which was then between 30 and 60 degrees S, supported diverse, angiosperm-dominated, subtropical-to-tropical rainforest (Christophel and Greenwood 1989; Hill 1982). Megafossil assemblages from several middle Eocene localities in Anglesea, southeastern Australia, contain species with nearest living relatives now restricted to tropical rain forest in northern Queensland, including *Podocarpus,* the cycad *Bowenia,* and dicots in the Casuarinaceae and Proteaceae. Leaf physiognomic characteristics of these floras are also consistent with tropical or warm subtropical forests (Christophel and Greenwood 1989). Interlocality differences in floral composition at Anglesea suggest these forest were heterogeneous in species composition (Christophel et al. 1987). Eocene megafloras from Tasmania suggest somewhat cooler, though still subtropical, climatic conditions. Lowland rain forests in Tasmania included Araucariaceae, Casuarinaceae, and *Nypa* (Hill and Bigwood 1987), and vegetation from somewhat higher elevations included *Nothofagus* (Hill 1984).

Although Eocene floras from Argentina indicate subtropical to fully tropical forest, some taxa suggest seasonal dryness (Romero 1986). Beginning in the middle Eocene and continuing through to the Oligocene, decreases in the percentage of species with entire-margined leaves and in species with large leaves imply cooler and/or more seasonal climates in southern South America (Romero 1986). Africa appears to have suffered a restriction of wet tropical forests beginning in the middle Tertiary, though perhaps this did not affect much of the continent until the Miocene (Axelrod and Raven 1978). Late Eocene floras from northern Egypt suggest wet or seasonally wet tropical forest (Wing, unpublished data from Kasr El Sagha Fm.).

Drying and warming climate in the eastern Rocky Mountain region of North America during the middle and later Eocene resulted in the loss of conifers even in wetter habitats at lower elevations (Wing 1987), although pinaceous conifers became important in montane forests in the Rocky Mountains and farther west (Wolfe and Wehr 1987; Axelrod 1966). The increasing seasonal dryness suggests vegetation that was probably more open, more deciduous, and of lower stature (MacGinitie 1953, 1969; Leopold and MacGinitie 1972; Wing 1987), although it is not known whether herbaceous angiosperms were an important part of this vegetation. The geographic extent of the drying trend is as yet unknown, but it is clearly established that mesic tropical-to-subtropical forest persisted along the Pacific and Gulf coasts of North America throughout the Eocene (MacGinitie 1941; Wolfe 1978; Frederiksen 1988). A global decline in temperatures and increase in seasonality about 33 Ma resulted in a shift to more deciduous broad-leaved types of vegetation in coastal western North America (Wolfe 1978), and floral evidence from Europe also suggests a major drop in temperature at or near the end of the Eocene (Collinson et al. 1981; Cavelier et al. 1981; Collinson and Hooker 1987; Medus and Pairis 1990). The suddenness and synchroneity of the vegetational shift in Europe has not been precisely determined, although it appears to have been a fairly gradual process (Collinson 1991).

Mean seed size in later Tertiary temperate-latitude floras shows a slight decline, which perhaps reflects the increasing abundance and diversity of herbaceous angiosperms (Tiffney 1984). Although clear evidence for the development of grassland vegetation is not present before the mid-late Miocene, megafossils of grasses that are modern in appearance are known from the latest Eocene Florissant flora (MacGinitie 1953), and extant tribes apparently were established by the early Eocene (Crepet and Feldman 1989).

5.3 Paleogene Faunas

The fossil record suggests a continuous increase in the number of insect families throughout the Paleocene. The appearance of modern moths and butterflies (Monotrysia and Ditrysia) in the middle Eocene provided important new pollination agents for the angiosperms. Formicoid ants underwent a considerable radiation during the latest Cretaceous or earliest Paleocene (Wilson 1987).

The beginning of the Paleocene witnessed the rise of mammals. Mammalian diversity in western North America quickly rose from 20 to 45 genera within 250,000 years of the K/T boundary, and 2 million years into the Paleocene, it reached 70 genera (Archibald 1983). Placental herbivores were initially rather small and fairly similar to each other in overall appearance. Many had specialized anterior dentitions (large and/or procumbent incisors and canines and enlarged premolars), which implies they may have selected and processed small food items such as fruits, seeds, or small prey. Larger forms in these early Tertiary faunas, generally heavy-bodied, short-limbed Pantodonta, lacked conspicuous dental specializations; this suggests that they utilized a wide variety of vegetational resources. Specialized browsers were not a significant component of the fauna (Stucky 1990).

The transition from the Paleocene to the Eocene in North America is characterized by a significant turnover in mammalian faunas due to immigrations of new taxa and declines in the diversity through extinction of archaic forms (Rose 1981, 1984; Stucky 1990). Many herbivorous groups with extant representatives made their first appearance in the fossil record across the Paleocene/Eocene boundary, including rodents at the end of the Paleocene. The early Eocene represented a period of major biotic interchange between North America and Europe, resulting in homogenization of mammalian faunas in the Northern Hemisphere (McKenna 1983). Both groups of modern ungulates, the even-toed Artiodactyla and the odd-toed Perissodactyla, appeared in the Northern Hemisphere at the beginning of the Eocene. A major radiation of these herbivorous mammals commenced in the early Eocene and was accompanied by a trend toward larger body size (Rose 1984; Stucky 1989). Artiodactyls were initially less numerous and diverse than the perissodactyls but underwent an extensive radiation later in the Eocene (Cifelli 1981). Development of cutting ridges (lophodonty) appeared in the early Eocene among perissodactyls but development of crescentic shearing crests (selenodonty) did not become common among artiodactyls until the late middle Eocene. Mammalian assemblages from the late Paleocene and Eocene of Patagonia are dominated by a variety of endemic ungulates, especially notoungulates (McKenna 1980), which are characterized by the pattern of lophs on their molar teeth and reach their maximum diversity in the Oligocene. In North America the highest alpha-diversity levels for mammals were reached during the Paleocene and in the early-middle Eocene interval; assemblages from the later Eocene and Oligocene have fewer species and are more likely to be dominated by single species, perhaps indicating the clumped distribution of animal populations that is today characteristic of open vegetation (Stucky 1990). The transition from the late Eocene to the Oligocene (Chadronian and Orellan Provincial Ages) is marked by major changes in the composition of mammalian faunas and continuing increases in cursoriality and lophodonty in many lineages (particularly those of Asian origin: Webb 1977). The Chadronian-Orellan faunal turnover in North America appears to correlate with the Eocene/Oligocene boundary as recognized in the marine realm and with the sharp increase in temperature seasonality seen in terrestrial floras (Swisher and Prothero 1990). However, the Chadronian-Orellan event does not represent a major extinction, but rather a continuation of the trend established in the late Eocene (Stucky 1989, 1990).

5.4   Paleogene Ecological and Evolutionary Trends

Starting in the middle-to-late Paleocene, fruit and seed assemblages document a great range of sizes of angiosperm diaspores; some of the larger ones were almost certainly biotically dispersed (Tiffney 1984). Paleocene and Eocene diversification in the Juglandaceae (Tiffney 1986b; Manchester 1987), and possibly in the Fagaceae and Podocarpaceae, produced more species with diaspores suited to dispersal by small mammals. Beginning in the early-to-middle Paleocene and continuing into the early Eocene, the guild of small, omnivorous/herbivorous, arboreal/scansorial mammals began to show a high diversity of species with specialized anterior dentitions, generally featuring the hypertrophy of premolars or incisors, and crushing molar teeth (Periptychidae, Plesiadapiformes, Paromomyidae, Microsyopidae, Carpolestidae, Rodentia, and Multituberculata).

The concomitant increase in the number of large, probably biotically dispersed angiosperms and the number of species of fruit- and seed-eating mammals could represent the initial coevolutionary radiation of what is now an extremely common and important kind of relationship, especially in tropical vegetation (Wing and Tiffney 1987). If this does indeed represent a coevolutionary spiral, then the relationship should be apparent when examined at finer temporal and geographic scales. Present knowledge of Paleocene fruit and seed floras is based on a small number of floras that are not well correlated with mammalian faunas. Late Paleocene and early Eocene biotas from southern England (Collinson and Hooker 1987), however, are consistent with this scenario, and similar data could be assembled for the western United States. If the coevolutionary explanation is correct, diaspore size and structure should show changes as early as the Puercan/Torrejonian boundary.

Another possible example of coordinated change in faunas and floras is seen in the late Eocene of the Rocky Mountain region of North America. Paleobotanical evidence suggests a shift toward an increasingly open savanna-like vegetation with plants of lower stature (MacGinitie 1969; Leopold and MacGinitie 1972; Wing 1987). There was a concurrent, initially quite subtle trend among mammalian herbivores toward increased body size, lophodonty, and cursoriality, and a decline of arboreal forms (Webb 1977; Stucky 1990). This may represent another case of diffuse coevolution, in which the development of more open vegetation (presumably the result of increased seasonality in mean humidity and annual temperature; Wolfe 1978) created opportunities for many lineages to develop features of open-country bulk feeders. Large herbivores, in turn, may have played a significant role in maintaining suitable conditions for weedy plants, particularly herbaceous angiosperms (Wing and Tiffney 1987). The generality of this pattern must be examined on a finer temporal and geographic scale. Floras on the Pacific and Gulf coasts of North America lagged behind those of the eastern Rocky Mountain region in this change, retaining closed semitropical vegetation. Therefore the faunas in this area should prove to be more "archaic" in their ecological characteristics. In fact, mammalian assemblages from Texas and California do appear to retain higher proportions of arboreal species later than those in the northern midcontinent (Stucky 1989, 1990).

It has been argued that Paleogene mammalian faunas in South America developed adaptations to open settings earlier than those in North America (Stebbins 1981), but more recent work on the chronology and composition of mammalian assemblages in southern South America suggests that, as in North America, open-country faunas were primarily a latest Eocene and later phenomenon (Marshall and Cifelli 1990). In Europe a different set of mammalian lineages also exhibited evolutionary trends congruent to those observed in western North America at roughly the same period of time. A prediction and test of the North American data is that the corresponding European floras should reflect the same types of vegetational change; some evidence suggests that this is indeed the case (Cavelier et al. 1981; Legendre 1987).

6   MESOZOIC AND EARLY CENOZOIC ECOLOGICAL CHANGES

Much of the preceding text has been an attempt to describe the structure of terrestrial communities typical of certain intervals in the past and to infer interactions such as herbivory, dispersal, pollination, and competition that occurred in those long-extinct communities. What follows is a summary of how we think some of these interactions have changed over the whole span of the Mesozoic and early Cenozoic.

6.1   *Changes in Herbivory*

The basic mechanisms of arthropod herbivory (chewing, leaf mining, sucking, etc.) seem to have been established during the late Paleozoic (see chap. 5). At present, limited data prevent us from seeing patterns of change or stasis in arthropod damage is known from the fossil record, including leaf mines (e.g., Crane and Jarzembowski 1980), galls (Larew 1986), and chewed leaves (e.g., Grande 1984), which should permit changes in arthropod feeding types to be compared with changes in observed damage through time.

Through the Mesozoic and early Cenozoic, there were substantial changes in the body-size distribution of herbivorous tetrapods, the height above ground at which they fed ("browseline" of Coe et al. 1987), and the characteristics of their food-processing structures (e.g., Weishampel and Norman 1989). Mesozoic herbivore faunas characteristically seem to have had a high proportion of their diversity in larger body-size categories, but species richness overall was much less than in Cenozoic assemblages. This pattern began to change in the Late Cretaceous, and during the Paleocene and Eocene small herbivores were much more diverse than large ones. Not until the latter half of the late Eocene did very large herbivores again become a diverse faunal element, but body size did not approach that of the dinosaurian megaherbivores of the Mesozoic. Changes in feeding height roughly followed those in body size, although feeding above 1 or 2 meters did not develop until the Late Triassic. The browseline was pushed to perhaps as much as 10 or 12 meters above the ground during the Jurassic. During the Late Cretaceous, the diversity and abundance of high-browsing herbivores diminished in the Northern Hemisphere, while tetrapods feeding between 1 and 3 meters above the ground became more abundant and diverse. From the K/T boundary until the mid-Cenozoic there were no high-browsing herbivores. Terrestrial browsers feeding at or above 2 meters did not reappear until the Oligocene.

These changes in herbivore body-size distribution and in masticatory and locomotor characteristics are consistent with an overall pattern of: (1) browsing herbivory through much of the Mesozoic; (2) a shift to more frugivory-granivory in the Paleocene-Eocene, following the extinction of large browsers at the K/T boundary; and (3) a partial return to browsing in the mid Tertiary. Herbivores of larger body size tend to respond to vegetation as a finer-grained resource than do small herbivores (Peters 1983; Crawley 1983) and thus would be expected to create a different kind of selective regime for plants. In particular, large herbivores eat a broader range of plant tissues, including those with low unit energy, and are more likely to cause significant damage to the plant in a single feeding episode (e.g., Croze 1974). These feeding habits may favor a variety of plant traits including mechanical defenses such as thorns (Janzen 1986), strong compensatory growth (Stebbins 1981; McNaughton 1984), and secondary compounds. Most small vertebrate herbivores are selective folivores (e.g., McNab 1978), frugivores-granivores (Fleming et al. 1987), nectivores, gummivores, or omnivores and are therefore more likely to view plant resources in a coarse-grained manner and to be particularly interested in tissues with higher unit energy. Herbivory by small vertebrates might be expected to elicit a broadly similar set of plant defenses, but probably with a greater emphasis on protection of seeds and less emphasis on compensatory growth.

6.2 Changes in Dispersal

The wide taxonomic distribution of animal dispersal among seed plants, as well as direct and indirect evidence from the fossil record, suggests that fruits and seeds have been a significant food resource for tetrapods since at least the Late Permian. Vertebrate dispersal, seed morphology consistent with vertebrate dispersal, or seed remains in vertebrate coprolites are known among living and/or fossil members of Ginkgoales, Cycadales, Caytoniaceae, Bennettitales, and Gnetales (Hill 1976; Van der Pijl 1982; Weishampel 1984b; Tiffney 1986a; Crane 1987; Norstog 1987). Some Mesozoic conifers were probably animal dispersed (Tiffney 1986a), as is true of a number of distantly related species of extant conifers (e.g., species of *Pinus, Juniperus,* and

Podocarpus). Indeed, animal (fish?) dispersal may have been characteristic of some pteridosperms and cordaites as early as the Carboniferous (Tiffney 1986a). Facultative animal dispersal is primitive for many branches of the seed-plant clade above the level of Lyginopteridales (cycads, conifers, ginkgos, many angiosperm lineages, and various Mesozoic "seed fern" groups); this implies that the interaction is easily developed in one form or another.

Although some form of frugivory and dispersal may have been common in terrestrial ecosystems from the Late Permian onward, evidence from the fossil record strongly indicates major differences in the type of frugivory and animal dispersal characteristic of the Mesozoic versus the Cenozoic. Direct evidence for animal consumption of seeds is extremely rare in the fossil record, and even then such evidence only proves that the seeds were eaten by an animal, not that they were dispersed by one. Therefore changes in dispersal must be inferred from morphological correlates in plants or their possible dispersers. For plants the correlates include diaspore size (large diaspores generally being biotically dispersed; see Primack 1987 and references therein), evidence for fleshy tissue around the seed (endozoochory), hooks or barbs (epizoochory), and wings or hair tufts (anemochory). For vertebrates the correlates include body size (see preceding discussion on herbivory) and dental specializations. On the basis of these traits, Tiffney (1986a) argued that dispersal during much of the Mesozoic was largely by abiotic means or as a result of generalized endozoochory carried out by large herbivores as a by-product of foliage browsing. Tiffney also suggested that the kind of dispersal typical of living birds and mammals in tropical forests, in which dispersers consume a fleshy fruit enclosing a few large seeds or a number of small seeds, was not common until the early Cenozoic. Such dispersal interactions are of major importance in modern tropical forests; 50–90% of shrubs and trees rely on vertebrate dispersers, and up to 80% of mammalian and avian biomass can consist of frugivores (Fleming et al. 1987).

6.3   Changes in Pollination

As with animal dispersal, insect pollination is taxonomically widespread in extant seed plants, most certainly was present in several Mesozoic groups, and may have been present as early as the Carboniferous (Dilcher 1979; chap. 5). Of the Mesozoic groups, some, perhaps most, Bennettitales were insect pollinated (Crepet 1972; Crepet and Friis 1987). Gnetales, which were diverse and abundant in low-latitude floras in the Early Cretaceous (Doyle et al. 1982; Crane and Upchurch 1987; Lidgard and Crane 1990), probably included many insect-pollinated species, judging from the widespread occurrence of insect pollination in extant species. Similarly, living cycads are known to have elaborate insect pollination systems (Norstog 1987), and it seems likely that Mesozoic cycads were insect-pollinated. Ginkgos and conifers are the only major groups of seed plants common in the Mesozoic that generally lack features consistent with insect pollination. The apparent absence of insect pollination in ginkgoaleans may reflect the present low diversity of the clade and incomplete knowledge of the extinct species. In contrast, there exists no evidence for insect pollination in the moderately diverse living conifers or in fossils of this group.

Given that insect pollination is ancient and taxonomically widespread, it seems probable that it was a common interaction in preangiosperm Mesozoic vegetation. The insect groups implicated in preangiosperm pollination were beetles, sawflies, flies, or other ancient groups (Crepet and Friis 1987). In the context of the pollination biology of extant angiosperms, these insects, especially beetles, are usually considered to be "primitive" pollinators (e.g., Crepet 1984). However, some beetles may participate in rather sophisticated pollination interactions of the kind typical of more "advanced" pollinators (e.g., Young 1988). The presence of advanced pollination interactions in older insect groups at least raises the possibility that complex and sophisticated pollination could have arisen prior to the Late Cretaceous (even though there is little direct evidence), but it would not necessarily have conformed to the modes of advanced pollination seen in living angiosperms.

6.4 Changes in Plant Competition

Mesozoic vegetation has been reconstructed as low-statured and open, perhaps because of seasonal rainfall and fires. A diverse fauna of large herbivores represents an additional element that could have increased the frequency and scale of disturbance, analogous to the role of elephants in some present-day African forests (Jones 1955). Clearly, habitat openings had to be more extensive during the Jurassic and Cretaceous than in the Paleogene, given the large body size of dinosaurian herbivores. If disturbance events in the Mesozoic tended to be more frequent and of larger spatial scale than in many contemporary forested areas, direct plant-plant competition could have been reduced and mediated by selective removal of individuals less resistant to large vertebrate herbivory or damage by fire.

It is not clear what effects such a disturbance regime would have had on the species diversity of late Mesozoic vegetation. It has been proposed that over ecological time, selective herbivory can maintain the richness of a plant community by preventing competitively superior species from excluding those that are slower-growing but less appetizing (the "intermediate disturbance hypothesis" of Grime [1973] and Connell [1978]). We do not know to what degree dinosaurian herbivory was selective. It may be inappropriate, furthermore, to apply an ecological-time theory to geological-time observations. For instance, in modern dry-country grasslands with an evolutionarily long history of grazing, competition seems to be little reduced by current grazing, because most plant species have similar growth habits and response to damage. In contrast, in "subhumid" grasslands (where a wider variety of life-forms is present), competition is reduced and diversity is increased at intermediate levels of herbivory (Milchunas et al. 1988). These observations suggest that the effects of herbivory on plant competition and diversity may depend as much on the evolutionary history of the particular system as on the actual amount or kind of herbivory.

The loss of large herbivorous tetrapods at the K/T boundary should have substantially affected disturbance regimes of many kinds of vegetation. Increased rainfall (Retallack et al. 1987; Wolfe and Upchurch 1987b; Lehman 1990) also would have reduced disturbance created by fire, although rainfall can increase the rate of formation of small tree-fall gaps (Brokaw 1985). These rapid decreases in the frequency and scale of disturbance could well have increased the importance of direct competitive interactions between plants. Removal of vertebrate herbivores has been demonstrated to have strong short-term effects on vegetational structure (McNaughton 1979; Thornton 1971; Hatton and Smart 1984), and strong long-term effects also have been inferred (Owen-Smith 1987). Although the effects of insect herbivory and tree falls might have remained constant or even increased following the K/T extinctions, it seems likely that in many environments there would have been an initial decrease in diversity as species with greater competitive abilities excluded those that invested more energy in defending against nonexistent large vertebrate herbivores or rare fire. Over "evolutionary" time, the lowered disturbance regime and presumed increase in competition might have favored a wider range of competitive and "stress-tolerant" strategies (*sensu* Grime 1977), leading to larger seeds, greater shade tolerance, and more "quantitative" chemical defense (Denslow 1987; Coley et al. 1985).

7   DISCUSSION AND CONCLUSION

In this final section, we use examples to elucidate the connection between environmental conditions and inferred community structure, the implications of change in community structure and dynamics for the evolution of individual lineages (and vice versa), and the way in which large-scale perturbations might affect ecological structure and evolutionary opportunities.

7.1   *Precipitation, Diversity Gradients, and Vegetational Structure*

Richness and standing biomass of living terrestrial vegetation is strongly influenced by four kinds of environmental variables: precipitation, temperature, light, and nutrients. Nutrient levels often vary on quite small spatial scales and are difficult to estimate for fossil floras (although low available nutrients are characteristic of many peat-forming environments), but estimates of the other factors are frequently available. In the present day, seasonal fluctuations in light and temperature increase strongly with higher latitude, making it difficult to separate their effects. However, both are probably major factors in strongly depressing alpha species richness at higher (particularly polar) latitudes.

Recently much emphasis has been placed on the role of temperature fluctuation as a causal factor in extinctions and other long-term biotic changes in the marine realm (Stanley 1984). The record of terrestrial life during the Mesozoic, however, strongly documents the importance of rainfall. This is partly because the Mesozoic and early Cenozoic were times of relatively low equator-to-pole temperature gradients, but it also reflects the importance of rainfall in determining the productivity and structure of terrestrial ecosystems.

Rainfall is high and relatively aseasonal in many equatorial regions as well as in some midlatitude areas that have moderate seasonality of light and temperature. The most diverse vegetation is found in tropical forests, where rainfall, light, and temperature are high with little seasonal fluctuation. The vegetation also has high standing biomass. Vegetation growing under high, aseasonal rainfall but seasonal light and temperature (i.e., temperate rain forest) can achieve exceptionally high biomass but has relatively lower diversity (Franklin 1988). In tropical latitudes where light and temperature are high with low seasonal fluctuations, areas with low or highly seasonal rainfall support vegetation of reduced standing biomass and reduced richness (deciduous forest to savanna to grassland).

The Mesozoic to early Cenozoic interval differed strongly from the present in that temperatures were warmer, especially at the poles (e.g., Parrish 1987). Furthermore, the existence of Pangaea, with its attendant monsoonal circulation, and the absence of polar fronts that would have confined the Intertropical Convergence Zone to equatorial regions created a global climatic regime where the low-to-mid-latitude regions of the earth experienced persistently to seasonally arid conditions for essentially the whole Mesozoic and early Cenozoic (Robinson 1971; Frakes 1979; Parrish et al. 1982; Ziegler et al. 1987; Parrish and Peterson 1988). (Monsoonal circulation was reduced in the later Jurassic in Laurasia and in the Early Cretaceous in Gondwana [Parrish et al. 1986; Parrish and Peterson 1988].) Coal deposition, and by inference wetter climates, returned to many midlatitude areas in the Late Cretaceous and Paleocene, but evidence for such climates in the equatorial zone is not widespread until the mid-Tertiary (Ziegler et al. 1987).

Thus, in contrast to the modern situation in which warm temperatures and high levels of precipitation and light are found generally in the equatorial regions, the equatorial region would have been drier during most of the Mesozoic–early Cenozoic. Rainfall would have been higher (and presumably evapo-transpiration somewhat lower) at midlatitudes, but even in an ice-free world, temperatures would have been lower, and light seasonality would have had the same gradient as today (Axelrod 1984). Thus, productivity of terrestrial vegetation was probably lower near the equator than it was at midlatitudes (Ziegler et al. 1983). Given the interaction of rainfall, temperature, and light in the Mesozoic world, both richness and standing biomass probably reached a maximum in midlatitudes, with the stature and diversity of equatorial vegetation reduced by low or highly seasonal rainfall, and that of high-latitude (polar) vegetation reduced by light seasonality and cool temperatures. Limited evidence from Late Triassic floras supports a high midlatitude diversity hump. The most species-rich vegetation of the Mesozoic might be expected to be less diverse than present tropical rain forest, because the tropical areas of the earth's surface, which had the highest levels of temperature and light, were receiving less rainfall than today, and the midlatitude areas with high rainfall had lower temperatures and seasonal illumination.

7.2   Long-Term Effects of Precipitation on Plant-Herbivore Dynamics

The nature of plant-animal interactions and the trophic characteristics of herbivore faunas are influenced strongly by climate through the intermediary of vegetational structure and productivity. Where mean annual temperature and/ or precipitation are low or strongly seasonal, individual plants tend to be small in stature and vegetation has low standing biomass (there is a more than 500-fold difference in above-ground standing biomass between some tundras and forests in North America; see Franklin 1988 and Bliss 1988). Although low or highly seasonal temperature and precipitation have an enormous effect on standing biomass levels, they have less effect on net productivity of vegetation, particularly net productivity of leaves. For example, the standing biomass of dense, *Prosopis*-dominated, Sonoran desert scrub is one-tenth to one-hundredth that of most eastern deciduous forests, but its annual production of foliage is less by only about one-third to one-fifth (MacMahon 1988; Greller 1988). Jordan (1971) compared a number of vegetation types worldwide and noted that the maximum efficiency of energy storage in short-term structures (i.e., leaves) was achieved in systems where the major primary producers were annual plants, even though the highest overall rates of productivity were in tropical forests. Much of this difference between overall productivity and productivity of leaves is accounted for by the greater proportion of energy devoted to wood production in forest vegetation.

A number of the vegetational variables influenced by climate are significant to herbivores: the productivity of leafy tissue, the proportion of wood to leaves, the height above ground of the canopy, the total standing biomass, and the size of spatial openings in the vegetation. Since wood is not digestible to vertebrate herbivores, these aspects of vegetation in turn have a particularly large influence on the characteristics of vertebrate herbivore faunas. For example, in the present day, areas with a high biomass and diversity of large vertebrate herbivores also have highly seasonal climate (Coe et al. 1976; Coe 1983). This is presumably because vegetation under this kind of climate is usually low in stature, dominated by herbaceous species, and produces large quantities of leafy tissue, thus providing a large resource pool for herbivores restricted to feeding within a few meters of the ground. Large grazing or browsing herbivores may have a reciprocal effect on vegetation (McNaughton 1984), favoring the success of plant species with a "ruderal" (Grime 1977) to "large gap" (Denslow 1987) strategy, i.e., those with fast growth rates, many small seeds, and high reproductive capacity. Furthermore, Oksanen (1988) argued that herbivores have their most profound influence on vegetational structure in ecosystems with moderate-to-low total productivity (e.g., tundra, grasslands), because under these conditions predator populations may be insufficient to regulate herbivore numbers effectively.

In contrast, high, aseasonal rainfall favors a large standing biomass of plants, much of which is woody support tissue that does not provide a direct resource for large herbivores. In dense forest vegetation, it is small, arboreal, frugivorous and insectivorous vertebrates rather than large terrestrial ones that are diverse and abundant. This is consistent with an ecosystem in which primary productivity is available largely in the canopy or after being cycled through populations of arthropod detritivores and herbivores.

As we have seen, many of the observed or inferred characteristics of low-to-mid-latitude Triassic through Early Cretaceous terrestrial organisms and ecosystems are consistent with dry or seasonally dry climates: plants with xeromorphic features of foliage and cuticles, increasing dominance of seed plants over pteridophytes, little evidence for closed-canopy vegetation, importance of fusainized plant material in many fossil deposits, declining importance of amphibians, diversification of tetrapod groups excreting uric acid rather than urea, and evolution of large body size in many groups of herbivorous tetrapods. The general similarity of Triassic (perhaps even Late Permian) through Early Cretaceous terrestrial communities in these features suggests that rainfall may have been a controlling factor in terrestrial community structure during the Mesozoic. Dry or seasonally dry conditions favored savanna-like rather than rain-forest-like interactions between plants and herbivores.

The low diversity of large herbivores and the importance of closed vegetation types during the early Paleocene stand in stark contrast with the situation in the Mesozoic. From a paleoecological perspective, this appears to be the most dramatic structural shift to take place in terrestrial ecosystems from their full establishment in the mid-Permian to the present day. Throughout the Paleocene and into the Eocene, the characteristics of most terrestrial biotas (at least in the well-studied midlatitude regions) are consistent with a dense forest vegetation, as are the inferred interactions of herbivores and plants. Precipitation appears to have been high in the mid-to-high latitudes, but the equatorial area is thought to have remained drier until the mid-Tertiary (Ziegler et al. 1987). It is unfortunate that Paleogene floras and faunas of equatorial regions are poorly known because these would provide an important test of the effect of rainfall on vegetational structure and plant-herbivore interactions.

7.3   Coevolutionary Limits on Vegetational Richness

The development of animal pollination and dispersal of plants has played a major role in theories developed to explain changes in vegetational richness through time. These ideas have been formulated explicitly by Niklas et al. (1980) and Knoll (1986). Knoll used number of species per described flora as a proxy for alpha species richness and found that *mean* richness had increased substantially through time, but with long plateau periods prior to the development of seeds in the Late Devonian and the major diversification of flowering plants in the Late Cretaceous. Knoll (1986) hypothesized that these plateaus were broken by the evolution of the seed habit and of advanced insect pollination, respectively. Crepet (1984) also highlighted the importance of insect pollination in the radiation of the angiosperms.

Number of species per described flora may be a poor proxy for alpha diversity for a number of reasons, including the large effect of sampling intensity on number of species (Burnham et al. 1990). Even assuming that the described pattern of change in *mean* alpha richness is correct, Knoll's causal explanation is called into question by the data presented on *maximum* species number per flora. If the rarity of the seed habit in Paleozoic floras and less-elaborate insect pollination in Mesozoic floras limited alpha richness levels, then it is difficult to explain why the most diverse floras from the Late Carboniferous and Late Triassic are nearly as diverse as the most diverse Late Cretaceous floras. Biotic limitations should affect maximum diversity as much as or more than mean diversity, so the very great richness of even a few Late Carboniferous and Late Triassic floras is not consistent with a biotically determined ceiling on alpha richness. Explanations of diversity change that rely solely on biotic interactions and innovations ignore the potential role of climate. Given that seasonally dry climates prevailed for most of the Mesozoic in the low latitudes, and given the strong influence of moisture on alpha diversity (e.g., Gentry 1988), it is possible that the Mesozoic plateau in mean diversity is related more to climatic than biotic limitations (although clearly these are not independent factors). Furthermore, the problems involved in estimating alpha diversity of past vegetation and the unknown strength of the climatic effect make it premature to use lower levels of alpha diversity to argue that Mesozoic plant-animal interactions were less intricate or less likely to enhance diversity than those of the present.

7.4   Comparative Paleoecology of Mass Extinctions

Seven significant Mesozoic–early Cenozoic extinctions were identified by Raup and Sepkoski (1986) in an analysis of marine animal families: Permian/Triassic (actually Dzhulfian), Triassic/Jurassic, Early Jurassic ("Pliensbachian"), Jurassic/Cretaceous, mid-Cretaceous (Cenomanian/Turonian), Cretaceous/Tertiary (Maastrichtian), and terminal Eocene. The "Pliensbachian" extinction is probably regionally confined and actually a Toarcian marine event (A. Hallam, pers. comm., 1986). Likewise, significant faunal changes in the marine realm at the Cenomanian/Turonian (Raup and Sepkoski 1986) and at the end of the Eocene (Stucky 1990) do not appear to be paralleled by comparable changes in terrestrial biotas. The remaining periods do represent times of increased extinction and turnover on land, although with the exception of the K/T boundary, correlation with the marine extinctions is rather imprecise. Further refinement in the data, especially at the substage level, will certainly permit more accurate documentation of the patterns of extinction. Traditionally mass extinctions have been compared on the basis of their magnitude and duration, with less consideration of their selectivity in an ecological sense (but see Jablonski 1986 and Raup 1986). Below we summarize briefly the ecological signatures of the Permian/Triassic, Triassic/Jurassic, Cretaceous/Tertiary, and terminal Eocene extinctions.

Particular emphasis is placed on the fate of large herbivorous tetrapods during these periods of extinction. Such animals should be particularly sensitive even to short-term cessation of primary productivity, such as is predicted by impact models of mass extinction, because large herbivores are especially dependent on a continuous supply of low-unit-energy plant food (e.g., Demment and Van Soest 1985), and because physiologically and mechanically they are poorly equipped to shift to alternative food sources (e.g., carrion, insects, detritus). Periods of climatic deterioration (i.e., more seasonally dry or cool climate) may actually increase habitat and resources for large herbivores by favoring ruderal plants, so (in contrast to an impact-generated halt in primary productivity) they should not necessarily lead to preferential extinction of large herbivores.

For terrestrial plants, the Permian/Triassic extinction might be better described as the Permian-Triassic transition—a long period of turnover (Late Carboniferous and much of the Permian) during which there was extensive replacement of higher-level taxa. The demonstrated diachroneity of the transition in a time frame of millions of years and its relation to large-scale continental movements probably indicate a climatic origin (Knoll 1984; Ziegler 1989). Although the *Dicroidium* flora replaced the *Glossopteris* flora in Gondwana, much of the shift toward gymnosperm dominance had already taken place by the Permian/Triassic boundary. The vegetational consequences of taxonomic change in terrestrial floras at the Permian/Triassic boundary are hard to reconstruct but do not seem to have been profound. At present there is no evidence for a pulsed plant extinction event in the Late Permian or at the Permian/Triassic boundary. It has been claimed that tetrapod extinctions were more severe (Benton 1985; Padian and Clemens 1985), but ecological characteristics such as size distribution and inferred trophic adaptations of herbivores were similar across the Permian/Triassic boundary. The general similarity of

Late Permian and Early Triassic terrestrial biotas may reflect similar global climatic regimes during those time intervals.

The Triassic/Jurassic vertebrate extinctions were severe and were apparently confined to a relatively short (< 1 My) interval (Olsen and Sues 1986; Benton 1986; Olsen et al. 1987). Observation of a fern spike at the Triassic/Jurassic boundary throughout the Newark Supergroup in eastern North America has led to the hypothesis of a mass-kill event and interruption of terrestrial productivity analogous to that hypothesized for the K/T boundary (Fowell 1990). Furthermore, palynofloral turnover at the Triassic/Jurassic boundary is sharp and substantial, and the dramatic increase in the importance of Cheirolepidiaceae in the Early Jurassic indicates vegetational differences between the two periods. To some extent, smaller herbivorous tetrapods may have had higher survival rates than large ones. However, at least two groups of prosauropod "megaherbivores" are known from both the Late Triassic and the Early Jurassic. Given the sensitivity of large-bodied herbivores to disturbance in primary productivity, the survival of these prosauropods and many other tetrapods may place limits on the severity of the effects of the bolide impact hypothesized by Olsen et al. (1987). As with the Permian/Triassic boundary, global climate does not seem to have undergone a major transition during this interval, although monsoonal conditions may have reached a peak in the Late Triassic (Parrish and Peterson 1988).

The K/T extinction is the only event of the Mesozoic and early Cenozoic that has an unmistakable ecological component: total extinction of large terrestrial vertebrates and radical changes in body-size distribution among herbivorous vertebrates, in herbivore trophic adaptations, and in angiosperm diaspore size, lasting for many millions of years following the extinction event. Furthermore, it is almost certain that devastation of vegetation in western North America and the subsequent appearance of early successional vegetation directly above the K/T boundary were causally related to the impact of an extraterrestrial object. The notion of a bolide impact, first advanced by Alvarez and others, is increasingly supported by geochemical anomalies and features such as shocked quartz in boundary strata. Important questions remain, however, about its long-term consequences and the relative importance of the direct physical consequences of the impact as opposed to climatic changes (independent of, induced by, or enhanced by the bolide) and ecological reorganizations resulting from extinction and evolutionary re-radiation of terrestrial organisms.

The unique combination of rapid, major extinction in many groups of terrestrial organisms, especially dinosaurs, and the appearance of structurally very different terrestrial ecosystems following the extinctions imply that the K/T extinction had a unique cause or combination of causes. Climatic change may have been involved in the long-term changes. Increased coal deposition in the mid-to-high latitudes during the Late Cretaceous suggests more (or more evenly distributed) annual rainfall (Spicer et al., 1990), and coal deposition in inland areas increased again in the Paleocene. The additional rainfall may have been one cause for the spread of denser, closed-canopy forests, the greater abundance and diversity of small herbivores, and the decreased diversity of large herbivores.

Many important issues remain unresolved: How long did it take terrestrial ecosystems to re-equilibrate? What was the role of changed climatic conditions in creating new types of ecosystems? Which effects of the bolide impact were most likely to have reverberated for periods of millions of years? The concept of quasi-succession (Wolfe and Upchurch 1986, 1987b) draws a direct analogy between ordinary ecological succession and processes that took place over millions of years following the K/T bolide impact. Wolfe (1987) also argued that the dominantly deciduous vegetation of the northern Rocky Mountains during the Paleocene resulted from higher survival of deciduous species at the K/T boundary. These ideas emphasize the direct effects of the bolide impact. An alternative (though not mutually exclusive) idea is that K/T extinctions forced the development of new ecological relationships between plants and animals, resulting in long-term evolutionary and ecological changes (e.g., Wing and Tiffney 1987).

A layer with anomalously high iridium concentrations and levels containing microtektites demonstrate an unusual input of extraterrestrial matter to the earth near the Eocene/Oligocene boundary (Ganapathy 1982; Alvarez et al. 1982). It is now clear, however, that there were several separate impact events during which microtektites were deposited, and that all of these events occurred around 35–36 Ma, thus predating the Eocene/Oligocene boundary (Miller et al. 1991), as well as the temperature decline observed in terrestrial floras (Wolfe 1978). Furthermore, the ecological and temporal pattern of terrestrial vertebrate extinctions in North America is not consistent with a bolide-induced extinction. Much of the terrestrial extinction attributed to the Eocene/Oligocene boundary appears to have taken place over an extended interval, rather than at the boundary itself (Prothero 1985; Stucky 1990), and diversity of large browsing and grazing herbivores increased during this time (Stucky 1990). Extensive penecontemporaneous faunal interchange also has been reported in Europe, where changes in the distribution of vertebrate body sizes associated with the extinctions have been interpreted as evidence for a shift to more open, drier habitats (Legendre 1987). Floras from the interval between 30 and 35 Ma indicate a great increase in deciduous, temperate species on the Pacific coast of North America, which is believed to reflect a substantial decrease in temperature and increase in temperature seasonality at about 33 Ma (Wolfe 1978). This temperature change in the terrestrial realm is approximately synchronous with oxygen isotope and marine faunal evidence for decreasing ocean temperatures, which probably resulted from the establishment of the Antarctic ice cap (e.g., Kennett 1977). Although major changes in world climate and the structure of terrestrial ecosystems seem to have been concentrated during a five-million-year period from about 36 to 33 Ma, there is no good evidence for catastrophic change at or near the Eocene/Oligocene boundary.

7.5 Ecological Context and Evolutionary Opportunity

It is widely accepted that a major perturbation can "clear" ecological space and permit the explosive adaptive radiation of formerly unimportant groups through a mechanism analogous to ecological release (e.g., Van Valen 1978; Erwin et al. 1987). Following the origin of the largest-scale evolutionary innovations, similar events may have occurred, for instance, the "Cambrian explosion" of marine metazoan life has been attributed to a breakthrough into previously unexploited modes of life (Erwin et al. 1987). Perhaps similar events occurred following the first invasions of land. As shown in chapter 5, a number of evolutionary-ecological breakthroughs continued to occur in terrestrial biotas through much of the Paleozoic. However, the increasingly crowded world of the Mesozoic and Cenozoic may have been rather different. To what degree have evolutionary radiations followed upon evolutionary innovations that broke the existing ecological order? To what extent has external disruption of the ecosystem been necessary before evolutionary innovations could be "captured" by a subsequent radiation? How different is diversification within an unperturbed ecosystem from diversification in a decimated one? To what degree is morphological innovation the cause, and to what degree is it the result, of changes in ecological interaction?

Answering such questions will require a high level of temporal resolution, an understanding of the branching sequence of the evolutionary radiation, and knowledge of its ecological context. In the absence of these we can cite what may be end members of a spectrum. For example, the Early-to-mid-Cretaceous diversification of angiosperms appears to have taken place in a relatively intact ecosystem. The paleoecological indicators used by us suggest that, in general, Early Cretaceous terrestrial ecosystems were similar in many ways to those of the Jurassic and Triassic. In contrast, the Paleocene diversification of mammals was an event unprecedented in the previous 150 My of mammalian history, and it occurred immediately following a major disruption of vertebrate communities at the end of the Cretaceous. Both of these radiations rapidly produced a large number of species (angiosperms: Lidgard and Crane 1988, 1990; mammals: Van Valen 1978; Archibald 1983). Even though both groups are now major components of most terrestrial communities, the mammals appear to have become ecologically diverse and important far more rapidly (1–5 My?) than the angiosperms (10 My?). The rapid evolutionary radiation of mammals into different ecological roles may have been a consequence of the relatively unoccupied adaptive landscape in the earliest Paleocene.

REFERENCES

Alexander, R. McN. 1985. Mechanics of posture and gait in some large dinosaurs. *Zoological Journal of the Linnean Society* 83:1–25.

Alvarez, W., F. Asharo, H. V. Michel, and L. W. Alvarez. 1982. Iridium anomaly approximately synchronous with terminal Eocene extinctions. *Science* 216:886–88.

Alvin, K. L. 1974. Leaf anatomy of *Weichselia* based on fusainized material. *Palaeontology* 17:587–98.

———. 1982. Cheirolepidiaceae: Biology, structure, and paleoecology. *Review of Palaeobotany and Palynology* 37:71–98.

———. 1983. Reconstruction of a Lower Cretaceous conifer. *Botanical Journal of the Linnean Society* 86:169–76.

Alvin, K. L., C. J. Fraser, and R. A. Spicer. 1981. Anatomy and palaeoecology of *Pseudofrenelopsis* and associated conifers in the English Wealden. *Palaeontology* 24:759–78.

Alvin, K. L., R. A. Spicer, and J. Watson. 1978. A *Classopolis*-containing male cone associated with *Pseudofrenelopsis*. *Palaeontology* 21:847–56.

Anderson, J. M., and H. M. Anderson. 1983a. *Palaeoflora of Southern Africa*, vol. 1, part 1, *Introduction*. Rotterdam: A. A. Balkema.

———. 1983b. *Palaeoflora of Southern Africa: Molteno Formation (Triassic)*. Vol. 1, part 2, *Dicroidium*. Rotterdam: A. A. Balkema.

———. 1985. *Palaeoflora of Southern Africa: Prodromus of South African Megafloras—Devonian to Lower Cretaceous*. Rotterdam: A. A. Balkema.

Archangelsky, S. 1980. Palynology of the Lower Cretaceous in Argentina. *Proceedings of the Fourth International Palynological Conference, Lucknow* 2:425–28.

Archibald, J. D. 1983. Structure of the K-T mammal radiation in North America: Speculations on turnover rates and trophic structure. *Acta Palaeontologica Polonica* 28:7–17.

Archibald, J. D., R. F. Butler, E. H. Lindsay, W. A. Clemens, and L. Dingus. 1982. Upper Cretaceous–Paleocene biostratigraphy and magnetostratigraphy, Hell Creek and Tullock Formations, northeastern Montana. *Geology* 10:153–59.

Archibald, J. D., and W. A. Clemens. 1982. Late Cretaceous extinctions. *American Scientist* 70:377–85.

Arnold, C. A. 1962. A *Rhexoxylon*-like stem from the Morrison Formation of Utah. *American Journal of Botany* 49:883–86.

Ash, S. R. 1972a. Late Triassic plants from the Chinle Formation in northeastern Arizona. *Palaeontology* 15:598–618.

———. 1972b. *Marcouia*, gen. nov., a problematical plant from the late Triassic of the southwestern USA. *Palaeontology* 15:423–29.

———. 1972c. Plant megafossils of the Chinle Formation. *Bulletin of the Museum of Northern Arizona* 47:59–73.

———. 1978. Plant megafossils. 23–42 in S. R. Ash, ed., Geology, paleontology, and paleoecology of a Late Triassic lake, western New Mexico. *Brigham Young University Geology Studies* 25.

———. 1982. Occurrence of the controversial plant fossil *Sanmiguelia* cf. *S. lewisi* Brown in the Upper Triassic of Utah. *Journal of Paleontology* 56:751–54.

———. 1987a. Growth habit and systematics of the Upper Triassic plant *Pelourdea poleonsis*, southwestern USA. *Review of Palaeobotany and Palynology* 51:37–49.

————. 1987b. The Upper Triassic red bed flora of the Colorado Plateau, western United States. *Journal of the Arizona-Nevada Academy of Science* 22:95–105.

Ash, S. R., R. Litwin, and R. Long. 1986 Biostratigraphic correlation of the Chinle Fm. (Late Triassic) of the Colorado Plateau: A progress report. *Geological Society of America, Abstracts with Program* 18:338.

Askin, R. A. 1988. The palynological record across the Cretaceous/Tertiary transition on Seymour Island, Antarctica. 155–62 in R. M. Feldman and M. O. Woodburne, eds., Geology and Paleontology of Seymour Island, Antarctic Peninsula. *Geological Society of America Memoir* 169.

Axelrod, D. I. 1966. The Eocene Copper Basin flora of northeastern Nevada. *University of California Publications in Geological Sciences* 59:1–125.

————. 1984. An interpretation of Cretaceous and Tertiary biota in polar regions. *Palaeogeography, Palaeoclimatology, Palaeoecology* 45:105–147.

Axelrod, D. I., and P. Raven. 1978. Late Cretaceous and Tertiary vegetation history of Africa. 77–130 in M. J. A. Werger and A. C. Van Bruggen, eds., *Biogeography and Ecology of Southern Africa.* The Hague: Dr. W. Junk.

Bakker, R. T. 1978. Dinosaur feeding behaviour and the origin of flowering plants. *Nature* 274:661–63.

Barale, G. 1981. *La Paléoflore Jurassique du Jura Français: Etude systématique, aspects stratigraphiques et paléoecologiques.* Documents de Laboratoire de Géologie, Lyon 81:1–467.

Barnard, P. W. 1973. Mesozoic floras. 175–87 in N. F. Hughes, ed., *Organisms and Continents through Time.* Special Papers in Palaeontology 12.

Basinger, J. F., and D. L. Dilcher. 1984. Ancient bisexual flowers. *Science* 224: 511–13.

Basinger, J. F., E. E. McIver, and B. A. Lepage. 1988. The fossil forests of Axel Heiberg Island. *Muskox* 36:50–55.

Batten, D. J. 1974. Wealden palaeoecology from the distribution of plant fossils. *Proceedings of the Geologists' Association* 85:433–58.

Benton, M. J. 1983. The Triassic reptile *Hyperodapedon* from Elgin: Functional morphology and relationships. *Philosophical Transactions of the Royal Society of London,* B 302:605–720.

————. 1984. Tooth form, growth, and function in Triassic rhynchosaurs. *Palaeontology* 27:737–76.

————. 1985. Patterns in the diversification of Mesozoic non-marine tetrapods and problems in historical diversity analysis. *Special Papers in Palaeontology* 33: 185–202.

————. 1986. The Late Triassic tetrapod extinction events. 303–20 in K. Padian, ed., *The Beginning of the Age of Dinosaurs.* New York: Cambridge University Press.

————. 1987. Mass extinctions among families of non-marine tetrapods: The data. *Mémoires de la Société Géologique de France* n.s. 150:21–32.

Bliss, L. C. 1988. Arctic tundra and polar desert biome. 1–32 in M. G. Barbour and W. D. Billings, eds., *North American Terrestrial Vegetation.* New York: Cambridge University Press.

Bohor, B. F., P. J. Modreski, and E. E. Foord. 1987. Shocked quartz in the Cretaceous-Tertiary boundary clays: Evidence for a global distribution. *Science* 236: 705–9.

Bonaparte, J. F. 1982. Faunal replacement in the Triassic of South America. *Journal of Vertebrate Paleontology* 28:362–71.

———. 1986. A new and unusual Late Cretaceous mammal from Patagonia. *Journal of Vertebrate Paleontology* 6:264–70.

———. 1987. History of the terrestrial Cretaceous vertebrates of Gondwana. *Actas IV Congreso Argentino de Paleontologia y Bioestratigrafia, Mendoza,* vol. 2:63–95.

———. 1990. New Late Cretaceous Mammals from the Los Alamitos Formation, northern Patagonia. *National Geographic Research* 6:63–93.

Boulter, M. C., and Z. Kvaček. 1989. The Paleocene flora of the Isle of Mull. *Special Papers in Palaeontology* 42:1–149.

Boulter, M. C., R. A. Spicer, and B. A. Thomas. 1988. Patterns of plant extinction from some palaeobotanical evidence. 1–36 in G. P. Larwood, ed., *Extinction and Survival in the Fossil Record.* Systematics Association Special Volume 34.

Brenner, G. J. 1976. Middle Cretaceous floral provinces and early migrations of angiosperms. 23–47 in C. B. Beck, ed., *Origin and Early Evolution of Angiosperms.* New York: Columbia University Press.

Brokaw, N. V. L. 1985. Treefalls, regrowth, and community structure in tropical forests. 53–69 in S. T. A. Pickett and P. S. White, eds., *The Ecology of Natural Disturbance and Patch Dynamics.* New York: Academic Press.

Brouwers, E. M., W. A. Clemens, R. A. Spicer, T. A. Ager, L. D. Carter, and W. V. Sliter. 1987. Dinosaurs on the North Slope, Alaska: High latitude, latest Cretaceous environments. *Science* 237:1608–10.

Brown, J. T. 1975. Upper Jurassic and Lower Cretaceous ginkgophytes from Montana. *Journal of Paleontology* 47:724–30.

Brown, R. W. 1956. Fossil plants and the Jurassic-Cretaceous boundary in Montana and Alberta. *American Association of Petroleum Geologists Bulletin* 30:238–48.

———. 1962. Paleocene flora of the Rocky Mountains and Great Plains. *U.S. Geological Survey Professional Paper* 375:1–119.

Brown, S. 1981. A comparison of the structure, primary productivity, and transpiration of cypress ecosystems in Florida. *Ecological Monographs* 51:403–27.

Burnham, R. J., and R. A. Spicer. 1986. Forest litter preserved by volcanic activity at El Chichon, Mexico: A potentially accurate record of the preeruption vegetation. *Palaios* 1:158–61.

Burnham, R. J., S. L. Wing, and G. G. Parker. 1990. Plant diversity and the fossil record: How reliable are the estimates? *International Congress of Systematic and Evolutionary Biology 4, Abstracts.*

———. 1992. Reflection of temperate forest composition and structure in the litter: Implications for the fossil record. *Paleobiology* 18:34–53.

Carroll, R. L. 1988. *Vertebrate Palaeontology and Evolution.* New York: H. W. Freeman.

Cavelier, C., J.-J. Chateauneuf, C. Pomerol, D. Rabussier, M. Renard, and C. Vergnaud-Grazzini. 1981. The geological events at the Eocene/Oligocene boundary. *Palaeogeography, Palaeoclimatology, Palaeoecology* 36:223–48.

Chamberlain, C. J. 1935. *Gymnosperms: Structure and Evolution.* Chicago: University of Chicago Press.

Charig, A. J. 1984. Competition between therapsids and archosaurs during the Triassic period: A review and synthesis of current theories. *Symposia of the Zoological Society of London* 52:597–628.

Christophel, D. C., and Greenwood, D. R. 1989. Changes in climate and vegetation in Australia during the Tertiary. *Review of Palaeobotany and Palynology* 58:95–109.

Christophel, D. C., W. K. Harris, and A. K. Syber. 1987. The Eocene flora of the Anglesea locality, Victoria. *Alcheringa* 11:303–23.

Cifelli, R. L. 1981. Patterns of evolution among the Artiodactyla and Perissodactyla (Mammalia). *Evolution* 35:433–40.

Clemens, W. A. 1983. Mammalian evolution during the Cretaceous-Tertiary transition: Evidence for gradual, non-catastrophic patterns of biotic change. *Acta Palaeontologica Polonica* 28:55–61.

Coe, M. 1983. Large herbivores and food quality. 345–68 in J. A. Lee, S. McNeil, and I. H. Rorison, eds., *Nitrogen as an Ecological Factor*. British Ecological Society Symposium 22.

Coe, M. J., D. H. Cumming, and J. Phillipson. 1976. Biomass and production of large African herbivores in relation to rainfall and primary production. *Oecologia* 22: 341–54.

Coe, M. J., D. L. Dilcher, J. O. Farlow, D. M. Jarzen, and D. A. Russell. 1987. Dinosaurs and land plants. 225–58 in E. M. Friis, W. G. Chaloner, and P. R. Crane, eds., *The Origins of Angiosperms and Their Biological Consequences*. New York: Cambridge University Press.

Coley, P. D., J. P. Bryant, and F. S. Chapin III. 1985. Resource availability and plant antiherbivore defense. *Science* 230:895–99.

Collinson, M. E. 1983. *Fossil Plants of the London Clay*. Palaeontological Association Field Guides to Fossils 1.

———. 1990. Plant evolution and ecology during the early Cainozoic diversification. 1–98 in J. A. Callow, ed., *Advances in Botanical Research* 17. London: Academic Press.

———. 1991. Vegetational and floristic changes around the Eocene/Oligocene boundary in western and central Europe. In D. R. Prothero and W. A. Berggren, eds., *Eocene-Oligocene Climatic and Biotic Evolution*. Princeton: Princeton University Press.

Collinson, M. E., K. Fowler, and M. C. Boulter. 1981. Floristic changes indicate a cooling climate in the Eocene of southern England. *Nature* 291:315–17.

Collinson, M. E., and J. J. Hooker. 1987. Vegetational and mammalian faunal changes in the Early Tertiary of southern England. 259–304 in E. M. Friis, W. G. Chaloner, and P. R. Crane, eds., *The Origins of Angiosperms and Their Biological Consequences*. New York: Cambridge University Press.

Connell, J. H. 1978. Diversity in tropical rainforests and coral reefs. *Science* 199: 1302–10.

Coombs, W. P., Jr. 1975. Sauropod habits and habitats. *Palaeogeography, Palaeoclimatology, Palaeoecology* 17:1–33.

———. 1978. The families of the ornithischian dinosaur order Ankylosauria. *Palaeontology* 21:143–70.

Cooper, M. R. 1981. The prosauropod dinosaur *Massospondylus carinatus* Owen from Zimbabwe: Its biology, mode of life, and phylogenetic significance. *Occasional Papers of the National Museum and Monuments of Zimbabwe, Series B, Natural Science* 6:689–840.

Cornet, B., and P. E. Olsen. 1985. A summary of the biostratigraphy of the Newark Supergroup of eastern North America, with comments on early Mesozoic provinciality. 67–81 in R. Weber, ed., *Symposio sobre Flores del Triasico Tardio, su Fitografia y Paleoecologia, Memoria.* Proceedings, Third Latin-American Congress on Paleontology (1984). Universidad Nacional Autonoma de Mexico.

Cox, C. B. 1965. New Triassic dicynodonts from South America, their origins and relationships. *Philosophical Transactions of the Royal Society of London,* B 248: 457–516.

Crabtree, D. R. 1987. Angiosperms of the northern Rocky Mountains: Albian to Campanian (Cretaceous) megafossil floras. *Annals of the Missouri Botanical Garden* 74:707–47.

Crane, P. R. 1987. Vegetational consequences of the angiosperm diversification. 107–44 in E. M. Friis, W. G. Chaloner, and P. R. Crane, eds., *The Origins of Angiosperms and Their Biological Consequences.* New York: Cambridge University Press.

Crane, P. R., E. M. Friis, and K. R. Pedersen. 1986. Lower Cretaceous angiosperm flowers: Fossil evidence on early radiation of dicotyledons. *Science* 232:852–54.

Crane, P. R., and E. A. Jarzembowski. 1980. Insect leaf mines from the Paleocene of southern England. *Journal of Natural History* 14:629–36.

Crane, P. R., and S. Lidgard. 1989. Angiosperm diversification and paleolatitudinal gradients in Cretaceous floristic diversity. *Science* 246:675–78.

Crane, P. R., S. R. Manchester, and D. L. Dilcher. 1990. A preliminary survey of fossil leaves and well-preserved reproductive structures from the Sentinel Butte Formation (Paleocene) near Almont, North Dakota. *Fieldiana,* n.s., 20:1–63.

Crane, P. R., and G. R. Upchurch. 1987. *Drewria potomascensis* gen. et sp. nov., an Early Cretaceous member of Gnetales from the Potomac Group of Virginia. *American Journal of Botany* 74:1722–36.

Crawley, M. J. 1983. *Herbivory: The Dynamics of Plant-Animal Interactions.* Berkeley: University of California Press.

Crepet, W. L. 1972. Investigations of North American cycadeoids: Pollination mechanisms in *Cycadeoidea. American Journal of Botany* 59:1048–56.

———. 1974. Investigations of North American cycadeoids: The reproductive biology of *Cycadeoidea. Palaeontographica,* B 148: 144–69.

———. 1979. Some aspects of the pollination biology of middle Eocene angiosperms. *Review of Palaeobotany and Palynology* 27:213–38.

———. 1981. The status of certain families of the Amentiferae during the middle Eocene and some hypotheses regarding the evolution of wind pollination in dicotyledonous angiosperms. 103–28 in K. J. Niklas, ed., *Paleobotany, Paleoecology, and Evolution,* vol. 1. Ithaca: Praeger.

———. 1984. Advanced (constant) insect pollination mechanisms: Pattern of evolution and implications vis-à-vis angiosperm diversity. *Annals of the Missouri Botanical Garden* 71:607–30.

Crepet, W. L., and G. D. Feldman. 1989. Paleocene/Eocene grasses from the southeastern USA. Abstract. *American Journal of Botany* 76:161.

Crepet, W. L., and E. M. Friis. 1987. The evolution of insect pollination in angiosperms. 181–201 in E. M. Friis, W. G. Chaloner, and P. R. Crane, eds., *The Origins of Angiosperms and Their Biological Consequences.* New York: Cambridge University Press.

Crepet, W. L., and D. W. Taylor. 1985. The diversification of the Leguminosae: First fossil evidence of the Mimosoideae and Papilionoideae. *Science* 228:1087–89.

Crompton, A. W. 1972. Postcanine occlusion in cynodonts and tritylodontids. *Bulletin of the British Museum of Natural History (Geology)* 21:27–71.

Crompton, A. W., and J. Attridge. 1986. Masticatory apparatus of the larger herbivores during Late Triassic and Early Jurassic times. 223–36 in K. Padian, ed., *The Beginning of the Age of Dinosaurs.* New York: Cambridge University Press.

Crompton, A. W., and N. Hotton III. 1967. Functional morphology of the masticatory apparatus of two dicynodonts (Reptilia: Therapsida). *Postilla* 109:1–51.

Crowley, T. J., W. T. Hyde, D. A. Short. 1989. Seasonal cycle variations on the supercontinent Pangaea. *Geology* 17:457–60.

Croze, H. 1974. The Seronera bull problem. 2. The trees. *East African Wildlife Journal* 12:29–47.

Currie, P., and P. Dodson. 1984. Mass death of a herd of ceratopsian dinosaurs. 61–66 in W.-E. Reif and F. Westphal, eds., *Second Symposium on Mesozoic Terrestrial Ecosystems, Short Papers.* Tübingen: Attempto Verlag.

Daber, R. 1968. A *Weichselia-Stiehleria*-Matoniaceae community within the Quedlinburg Estuary of Lower Cretaceous age. *Botanical Journal of the Linnean Society* 61:75–85.

Daghlian, C. P. 1978. Coryphoid palms from the lower and middle Eocene of southeastern North America. *Palaeontographica,* B 166:44–82.

Daghlian, C. P., and R. Person. 1977. Cuticular analysis of *Frenelopsis* from the Lower Cretaceous of Texas. *American Journal of Botany* 64:564–72.

———. 1981. A review of the fossil record of monocotyledons. *Botanical Review* 47:517–55.

Delevoryas, T. 1968. Some aspects of cycadeoid evolution. *Botanical Journal of the Linnean Society* 61:137–46.

———. 1969. Glossopterid leaves from the middle Jurassic of Oaxaca, Mexico. *Science* 165:895–96.

Delevoryas, T., and R. C. Hope. 1976. More evidence for a slender growth habit in Mesozoic cycadophytes. *Review of Palaeobotany and Palynology* 21:93–100.

———. 1981. More evidence for conifer diversity in the Upper Triassic of North Carolina. *American Journal of Botany* 68:1003–7.

Demment, M. W., and P. J. Von Soest. 1985. A nutritional explanation for body-size patterns of ruminant and nonruminant herbivores. *American Naturalist* 125:641–72.

Denslow, J. S. 1987. Tropical rainforest gaps and tree species diversity. *Annual Review of Ecology and Systematics* 18:431–51.

Dettmann, M. E. 1986. Early Cretaceous palynoflora of subsurface strata correlative with the Koonwarra Fossil Bed, Victoria. 79–110 in P. A. Jell and J. Roberts, eds., *Plants and invertebrates from the Lower Cretaceous Koonwarra fossil bed, South Gippsland, Victoria. Memoirs of the Association of Australasian Palaeontologists* 3.

Dilcher, D. L. 1979. Early angiosperm reproduction: An introductory report. *Review of Palaeobotany and Palynology* 27:291–328.

Dilcher, D. L., and W. L. Kovach. 1986. Early angiosperm reproduction: *Caloda delevoryana* gen. et sp. nov., a new fructification from the Dakota Formation (Cenomanian) of Kansas. *American Journal of Botany* 73:1230–37.

Dobruskina, I. A. 1987. Phytogeography of Eurasia during the Early Triassic. *Palaeogeography, Palaeoclimatology, Palaeoecology* 58:75–86.

Dodson, P., A. K. Behrensmeyer, R. T. Bakker, and J. S. McIntosh. 1980. Taphonomy and paleoecology of the dinosaur beds of the Jurassic Morrison Formation. *Paleobiology* 6:208–32.

Dong, Z., S. Zhou, and Y. Zhang. 1983. The dinosaurian remains from Sichuan Basin, China (in Chinese with English summary). *Palaeontologia Sinica,* n.s., C, 23:1–145.

Douglas, J. G., and G. E. Williams. 1982. Southern polar forests: The Early Cretaceous floras of Victoria and their palaeoclimatic significance. *Palaeogeography, Palaeoclimatology, Palaeoecology* 39:171–85.

Doyle, J. A. 1978. Origin of angiosperms. *Annual Review of Ecology and Systematics* 9:365–92.

Doyle, J. A., and M. J. Donoghue. 1986. The origin of angiosperms: A cladistic approach. 17–49 in E. M. Friis, W. G. Chaloner, and P. R. Crane, eds., *The Origins of Angiosperms and Their Biological Consequences.* New York: Cambridge University Press.

Doyle, J. A., and L. J. Hickey. 1976. Pollen and leaves from the mid-Cretaceous Potomac Group and their bearing on early angiosperm evolution. 139–206 in C. B. Beck, ed., *Origin and Early Evolution of Angiosperms.* New York: Columbia University Press.

Doyle, J. A., and J. T. Parrish. 1984. Jurassic–Early Cretaceous plant distributions and paleoclimatic models. *International Organization of Paleobotany Conference 3, Abstracts.*

Doyle, J. A., S. Jardine, and A. Doerenkamp. 1982. *Afropollis,* a new genus of early angiosperm pollen, with notes on the Cretaceous palynostratigraphy and paleoenvironments of northern Gondwana. *Bulletin Centres Récherches et Exploration— Production Elf Aquitaine* 6: 39–117.

Drinnan, A. N., and T. C. Chambers. 1986. Flora of the Lower Cretaceous Koonwarra Fossil Bed (Korumburra Group), South Gippsland, Victoria. 1–77 in P. A. Jell and J. Roberts, eds., Plants and invertebrates from the Lower Cretaceous Koonwarra fossil bed, South Gippsland, Victoria. *Memoirs of the Association of Australasian Palaeontologists* 3.

Drinnan, A. N., and P. R. Crane. 1990. Cretaceous paleobotany and its bearing on the biogeography of austral angiosperms. 192–219 in T. N. Taylor and E. L. Taylor, eds., *Antarctic Paleobiology: Its Role in the Reconstruction of* Gondwana. New York: Springer-Verlag.

Drinnan, A. N., P. R. Crane, E. M. Friis, and K. R. Pedersen. 1990. Lauraceous flowers from the Potomac Group (mid-Cretaceous) of eastern North America. *Botanical Gazette* 151:370–84.

Eisenberg, J. F. 1981. *The Mammalian Radiations: An Analysis of Trends in Evolution, Adaptation, and Behavior.* Chicago: University of Chicago Press.

Emmons, L. H., A. Gautier-Hion, and G. Dubost. 1983. Community structure of the frugivorous-folivorous forest mammals of Gabon. *Journal of Zoology, London* 199:209–22.

Erwin, D. H. 1990. The end-Permian mass extinction. *Annual Reviews of Ecology and Systematics* 21:69–91.

Erwin, D. H., J. W. Valentine, and J. J. Sepkoski, Jr. 1987. A comparative study of diversification events: The early Paleozoic versus the Mesozoic. *Evolution* 41: 1177–86.

Eshet, Y. 1990. Paleozoic-Mesozoic palynology of Israel. 1. Palynological aspects of the Permo-Triassic succession in the subsurface of Israel. *Geological Survey of Israel Bulletin* 81:1–57.

Farlow, J. O. 1987. Speculations about the diet and digestive physiology of herbivorous dinosaurs. *Paleobiology* 13:60–72.

Fastovsky, D. E., and K. McSweeney. 1987. Paleosols spanning the Cretaceous-Paleogene transition, eastern Montana and western North Dakota. *Palaios* 2: 282–95.

———. 1991. Paleocene paleosols of the petrified forests of Theodore Roosevelt National Park, North Dakota: A natural experiment in compound pedogenesis. *Palaios* 6:67–80.

Fleming, T. H., R. Breitwisch, and G. H. Whitesides. 1987. Patterns of tropical frugivore diversity. *Annual Review of Ecology and Systematics* 18:91–109.

Fowell, S. J. 1990. Palynological evidence for a Triassic-Jurassic boundary event, Newark Supergroup. *Geological Society of America, Abstracts with Programs:* 355.

Frakes, L. A. 1979. *Climates through Geologic Time.* Amsterdam: Elsevier.

———. 1986. Mesozoic-Cenozoic climatic history and causes of the glaciation. 33–48 in K. J. Hsü, ed., *Mesozoic and Cenozoic Oceans.* Geodynamics Series, American Geophysical Union and Geological Society of America, vol. 15.

Francis, J. E. 1983. The dominant conifer of the Jurassic Purbeck Formation. *Palaeontology* 26:277–94.

———. 1986. Growth rings in Cretaceous and Tertiary wood from Antarctica and their palaeoclimatic implications. *Palaeontology* 29:665–84.

———. 1989. Palaeoclimatic significance of Cretaceous-early Tertiary fossil forests of the Antarctic Peninsula. In M. R. A. Thomson, J. A. Crame, and J. W. Thomson, eds., *Geological Evolution of Antarctica.* Cambridge: Cambridge University Press.

———. 1991. Arctic Eden. *Natural History* 57–63.

Franklin, J. F. 1988. Pacific northwest forests. 104–30 in M. G. Barbour and W. D. Billings, eds., *North American Terrestrial Vegetation.* New York: Cambridge University Press.

Frederiksen, N. O. 1972. The rise of the mesophytic flora. *Geoscience and Man* 4:17–28.

———. 1980. Mid-Tertiary climate of southwestern United States: The sporomorph evidence. *Journal of Paleontology* 54:728–39.

———. 1988. Sporomorph biostratigraphy, floral changes, and paleoclimatology: Eocene and earliest Oligocene of the eastern Gulf Coast. *U.S. Geological Survey Professional Paper* 1448.

Frederiksen, N. O., T. A. Ager, and L. E. Edwards. 1988. Palynology of Maastrichtian and Paleocene rocks, lower Colville region, North Slope of Alaska. *Canadian Journal of Earth Sciences* 25:512–27.

Friis, E. M., P. R. Crane, and K. R. Pedersen. 1986. Floral evidence for Cretaceous chloranthoid angiosperms. *Nature* 320:163–64.

Friis, E. M., and W. L. Crepet. 1987. Time of appearance of floral features. 145–79 in E. M. Friis, W. G. Chaloner, and P. R. Crane, eds., *The Origins of Angiosperms and Their Biological Consequences.* New York: Cambridge University Press.

Gall, L. F., and B. H. Tiffney. 1983. A fossil noctuid moth egg from the late Cretaceous of eastern North America. *Science* 219:507–9.

Galton, P. M. 1985. The diet of prosauropod dinosaurs from the Late Triassic and Early Jurassic. *Lethaia* 18:105–23.

————. 1986. Herbivorous adaptations of Late Triassic and Early Jurassic dinosaurs. 203–21 in K. Padian, ed., *The Beginning of the Age of Dinosaurs*. New York: Cambridge University Press.

Ganapathy, R. 1982. Evidence for a major meteorite impact on the earth 34 million years ago: Implications for Eocene extinctions. *Science* 216:885–86.

Gee, C. T. 1989. Revision of the Late Jurassic/Early Cretaceous flora from Hope Bay, Antarctica. *Palaeontographica*, B 213:149–214.

Gentry, A. L. 1988. Changes in plant community diversity and floristic composition on environmental and geographical gradients. *Annals of the Missouri Botanical Garden* 75:1–34.

Gilmore, J. S., J. D. Knight, C. J. Orth, C. L. Pillmore, and R. H. Tschudy. 1984. Trace element patterns at a non-marine Cretaceous-Tertiary boundary. *Nature* 307:224–28.

Gottesfeld, A. S. 1972. Paleoecology of the lower part of the Chinle Formation in the Petrified Forest. *Bulletin of the Museum of Northern Arizona* 47:59–73.

Grande, L. 1984. Paleontology of the Green River Formation, with a review of the fish fauna. 2d ed. *Geological Survey of Wyoming Bulletin* 63:1–333.

Gregory, J. T. 1945. Osteology and relationships of *Trilophosaurus*. University of Texas Publications no. 4401:273–359.

Greller, A. M. 1988. Deciduous forest. 288–316 in M. G. Barbour and W. D. Billings, eds., *North American Terrestrial Vegetation*. New York: Cambridge University Press.

Grime, J. P. 1973. Control of species density in herbaceous vegetation. *Journal of Environmental Management* 1:151–67.

————. 1977. Evidence for the existence of three primary strategies in plants and its relevance to ecological and evolutionary theory. *American Naturalist* 111:1169–94.

Guo Shuangxing. 1985. Preliminary interpretation of Tertiary climate by using megafossil floras in China. *Palaeontologia Cathayana* 2:169–75.

Hallam, A. 1984. Continental humid and arid zones during the Jurassic and Cretaceous. *Palaeogeography, Palaeoclimatology, Palaeoecology* 47:195–223.

Harris, T. M. 1937. The Fossil Flora of Scoresby Sound, East Greenland, Part 5. *Meddelser om Gronland* 112:1–114.

————. 1961. *The Yorkshire Jurassic Flora. 1. Thallophyta-Pteridophyta*. London: British Museum (Natural History).

————. 1964. *The Yorkshire Jurassic Flora. 2. Caytoniales, Cycadales and Pteridosperms*. London: British Museum (Natural History).

————. 1969. *The Yorkshire Jurassic Flora. 3. Bennettitales*. London: British Museum (Natural History).

————. 1973. *The Strange Bennettitales*. Nineteenth Sir Albert Charles Seward Memorial Lecture, Birbal Sahni Institute of Paleobotany, 1970.

————. 1979. *The Yorkshire Jurassic Flora. 4. Coniferales*. London: British Museum (Natural History).

————. 1981. Burnt ferns from the English Wealden. *Proceedings of the Geologists' Association* 92:47–58.

————. 1983. The stem of *Pachypteris papillosa* (Thomas and Bose) Harris. *Botanical Journal of the Linnean Society* 86:149–59.

Harris, T. M., W. Millington, and J. Miller. 1974. *The Yorkshire Jurassic Flora. 5.*

Ginkgoales and Czekanowskiales. London: British Museum (Natural History).

Harrison, J. L. 1962. The distribution of feeding habits among animals in a tropical rain forest. *Journal of Animal Ecology* 31:53–63.

Hatton, J. C., and N. O. E. Smart. 1984. The effect of long-term exclusion of large herbivores on soil nutrient status in Murchison Falls National Park, Uganda. *African Journal of Ecology* 22:23–30.

Hennig, W. 1981. *Insect Phylogeny.* Chichester: John Wiley and Sons.

Hickey, L. J. 1977. Stratigraphy and paleobotany of the Golden Valley Formation (early Tertiary) of western North Dakota. *Geological Society of America Memoir* 150:1–181.

———. 1980. Paleocene stratigraphy and flora of the Clark's Fork Basin. 33–49 in P. D. Gingerich, ed., Early Cenozoic paleontology and stratigraphy of the Bighorn Basin, Wyoming. *University of Michigan Papers on Paleontology* 24.

———. 1984. Changes in the angiosperm flora across the Cretaceous-Tertiary boundary. 279–313 in W. A. Berggren and J. A. Van Couvering, eds., *Catastrophes and Earth History.* Princeton: Princeton University Press.

Hickey, L. J., and J. A. Doyle. 1977. Early Cretaceous fossil evidence for angiosperm evolution. *Botanical Review* 43:3–104.

Hickey, L. J., R. M. West, M. R. Dawson, D. K. Choi. 1983. Arctic terrestrial biota: Paleomagnetic evidence of age disparity with mid-northern latitudes during the Late Cretaceous and Early Tertiary. *Science* 221:1153–56.

Hildebrand, A. R., and W. V. Boynton. 1990. Proximal Cretaceous-Tertiary boundary impact deposits in the Caribbean. *Science* 248:843–47.

Hildebrand, A. R., D. A. Kring, W. V. Boynton, G. T. Penfield, and M. Pilkington. 1990. Cretaceous/Tertiary boundary impact site(s) between the Americas. *Geological Society of America, Abstracts with Programs:* 280.

Hill, C. R. 1976. Coprolites of *Ptilophyllum* cuticles from the Middle Jurassic of North Yorkshire. *Bulletin of the British Museum (Natural History), Geology* 27: 289–94.

Hill, C. R., D. T. Moore, J. T. Greensmith, and R. Williams. 1985. Palaeobotany and petrology of a Middle Jurassic ironstone bed at Wrack Hills, North Yorkshire. *Proceedings of the Yorkshire Geological Society* 45:277–92.

Hill, R. S. 1982. The Eocene megafossil flora of Nerriga, New South Wales, Australia. *Palaeontographica,* B 181:44–77.

Hill, R. S. 1984. Tertiary *Nothofagus* macrofossils from Cethana, Tasmania. *Alcheringa* 8:81–86.

Hill, R. S., and A. J. Bigwood. 1987. Tertiary gymnosperms from Tasmania: Araucariaceae. *Alcheringa* 10:325–36.

Hutchison, J. H., and J. D. Archibald. 1986. Diversity of turtles across the Cretaceous/Tertiary boundary in northeastern Montana. *Palaeogeography, Palaeoclimatology, Palaeoecology* 55:1–22.

Izett, G. A. 1990. Mineralogic data indicate the K-T boundary impact occurred on continental crust probably near Manson, Iowa. *Geological Society of America Abstracts with Programs,* 280.

Jablonski, D. 1986. Background and mass extinctions: The alternation of macroevolutionary regimes. *Science* 231:129–33.

Jain, S. L., T. S. Kutty, and T. Roy-Chowdhury. 1975. The sauropod dinosaur from the Lower Jurassic Kota Formation of India. *Proceedings of the Royal Society of London,* B 188:221–28.

Janzen, D. H. 1980. When is it coevolution? *Evolution* 34:611–12.

———. 1986. Chihuahuan desert nopaleras: Defaunated big mammal vegetation. *Annual Review of Ecology and Systematics* 17:595–636.

Janzen, D. H., and P. S. Martin. 1982. Neotropical anachronisms: The fruit the gomphotheres ate. *Science* 215:19–27.

Jerzykiewicz, T., and A. R. Sweet. 1988. Sedimentological and palynological evidence of regional climate changes in the Campanian to Paleocene sediments of the Rocky Mountain Foothills, Canada. *Sedimentary Geology* 59:29–76.

Johnson, E. A., S. Liu, and Y. Zhang. 1989. Depositional and tectonic controls on the coal-bearing Lower to Middle Jurassic Yan'an Formation, southern Ordos Basin, China. *Geology* 17:1123–26.

Johnson, K. R. In press. Leaf-fossil evidence for extensive floral extinction at the Cretaceous-Tertiary boundary, North Dakota, USA. *Cretaceous Research.*

Johnson, K. R., and L. J. Hickey. 1991. Megafloral change across the Cretaceous/ Tertiary boundary in the northern Great Plains and Rocky Mountains, USA. 433–44 in V. L. Sharpton and P. D. Ward, eds., Global Catastrophes in Earth History: An Interdisciplinary Conference on Impacts, Volcanism, and Mass Mortality. *Geological Society of America Special Paper* 247.

Johnson, K. R., D. J. Nichols, M. Attrep, Jr., and C. J. Orth. 1989. High-resolution leaf-fossil record spanning the Cretaceous/Tertiary boundary. *Nature* 340:708–11.

Jones, E. W. 1955. Ecological studies on the rain forest of southern Nigeria, 4. The plateau forest of the Okomu Forest Reserve. *Journal of Ecology* 43:564–94.

Jordan, C. F. 1971. Productivity of a tropical forest and its relation to a world pattern of energy storage. *Journal of Ecology* 59:127–42.

Kennett, J. P. 1977. Cenozoic evolution of Antarctic glaciation, the Circum-Antarctic Ocean, and their impact on global paleoceanography. *Journal of Geophysical Research* 82:3843–59.

King, G. M. 1981. The functional anatomy of a Permian dicynodont. *Philosophical Transactions of the Royal Society of London,* B 291:243–322.

Knobloch, E., and D. H. Mai. 1986. *Monographie der Früchte und Samen in der Kreide von Mitteleuropa.* Rozpravy, Ustredniho Ustavu Geologickeho, Svazek 47: 1–219.

Knoll, A. H. 1984. Patterns of extinction in the fossil record of vascular plants. 21–68 in M. H. Nitecki, ed., *Extinctions.* Chicago: University of Chicago Press.

———. 1986. Patterns of change in plant communities through time. 126–44 in J. Diamond and T. J. Case, eds., *Community Ecology.* New York: Harper and Row.

Knoll, A. H., and K. J. Niklas. 1987. Adaptation, plant evolution, and the fossil record. *Review of Palaeobotany and Palynology* 50:127–49.

Kovach, W. I. 1988. Quantitative palaeoecology of megaspores and other dispersed plant remains from the Cenomanian of Kansas, USA. *Cretaceous Research* 9: 265–83.

Kovach, W. I., and D. L. Dilcher. 1985. Morphology, ultrastructure, and paleoecology of *Paxillitriletes vittatus* sp. nov. from the mid-Cretaceous (Cenomanian) of Kansas. *Palynology* 9:85–94.

Krassilov, V. A. 1973. Climatic changes in eastern Asia as indicated by fossil floras.

1. Early Cretaceous. *Palaeogeography, Palaeoclimatology, Palaeoecology* 13: 261–73.

―――. 1975a. Climatic changes in eastern Asia as indicated by fossil floras. 2. Late Cretaceous and Danian. *Palaeogeography, Palaeoclimatology, Palaeoecology* 17:157–72.

―――. 1975b. *Paleoecology of Terrestrial Plants*. New York: John Wiley and Sons.

―――. 1978a. Araucariaceae as indicators of climate and paleolatitudes. *Review of Palaeobotany and Palynology* 26:113–24.

―――. 1978b. Late Cretaceous gymnosperms from Sakhalin and the terminal Cretaceous event. *Palaeontology* 21:893–905.

―――. 1981. Changes of Mesozoic vegetation and the extinction of dinosaurs. *Palaeogeography, Palaeoclimatology, Palaeoecology* 34:207–24.

―――. 1987. Palaeobotany of the Mesophyticum: State of the art. *Review of Palaeobotany and Palynology* 50:231–54.

Krause, D. W. 1982. Jaw movement, dental function, and diet in the Paleocene multituberculate *Ptilodus*. *Paleobiology* 8:265–81.

Krause, D. W., and F. A. Jenkins, Jr. 1983. The postcranial skeleton of North American multituberculates. *Bulletin of the Museum of Comparative Zoology, Harvard University* 150:199–246.

Kutzbach, J. E., and R. G. Gallimore. 1989. Pangaean climate: Megamonsoons of the megacontinent. *Journal of Geophysical Research* 94:3341–57.

Labandeira, C. C. 1990. Rethinking the diets of Carboniferous terrestrial arthropods: Evidence for a nexus of arthropod/vascular plant interactions. *Geological Society of America, Abstracts with Programs:* 265.

LaPasha, C. A., and C. N. Miller, Jr. 1984. Flora of the Early Cretaceous Kootenai Formation in Montana: Paleoecology. *Palaeontographica*, B 194:109–30.

―――. 1985. Flora of the Early Cretaceous Kootenai Formation in Montana: Bryophytes and tracheophytes excluding conifers. *Palaeontographica*, B 196:111–45.

Larew, H. G. 1986. The fossil gall record: A brief summary. *Proceedings of the Entomological Society of Washington* 88:385–88.

Legendre, S. 1987. Analysis of mammalian communities from the late Eocene and Oligocene of southern France. *Palaeovertebrata* 16:191–212.

Lehman, T. M. 1990. Paleosols and the Cretaceous/Tertiary transition in the Big Bend region of Texas. *Geology* 18:362–64.

Leopold, E. B., and H. D. MacGinitie. 1972. Development and affinities of Tertiary floras in the Rocky Mountains. 147–200 in A. Graham, ed., *Floristics and Paleofloristics of Asia and Eastern North America*. Amsterdam: Elsevier.

Lerbekmo, J. F., and R. M. St. Louis. 1986. The terminal Cretaceous iridium anomaly in the Red Deer Valley, Alberta, Canada. *Canadian Journal of Earth Sciences* 23:120–24.

Lerbekmo, J. F., A. R. Sweet, and R. M. St. Louis. 1987. The relationship between the iridium anomaly and palynological floral events at three Cretaceous-Tertiary boundary localities in western Canada. *Geological Society of America Bulletin* 99:325–30.

Lidgard, S., and P. R. Crane. 1988. Quantitative analyses of the early angiosperm radiation. *Nature* 331:344–46.

―――. 1990. Angiosperm diversification and Cretaceous floristic trends: A comparison of palynofloras and leaf macrofloras. *Paleobiology* 16:77–93.

Lillegraven, J. A., Z. Kielan-Jaworowska, and W. A. Clemens, eds. 1979. *Mesozoic Mammals: The First Two-Thirds of Mammalian History.* Berkeley: University of California Press.

Litwin, R. J. 1986. The palynostratigraphy and age of the Chinle and Moenave formations, southwestern USA. Ph.D. diss., Pennsylvania State University.

MacGinitie, H. D. 1941. *A Middle Eocene Flora from the Central Sierra Nevada.* Carnegie Institution of Washington Publication 534:1–178.

———. 1953. *Fossil Plants of the Florissant Beds, Colorado.* Carnegie Institution of Washington Publication 599:1–198.

———. 1969. The Eocene Green River Flora of Northwestern Colorado and Northeastern Utah. *University of California Publications in Geological Sciences* 83:1–202.

———. 1974. An Early Middle Eocene Flora from the Yellowstone-Absaroka Volcanic Province, Northwestern Wind River Basin, Wyoming. *University of California Publications in Geological Sciences* 108:1–103.

MacMahon, J. A. 1988. Warm deserts. 232–64 in M. G. Barbour and W. D. Billings, eds., *North American Terrestrial Vegetation.* New York: Cambridge University Press.

Makulbekov, N. M. 1988. Paleogene flora of the south Mongolia (in Russian). *Transactions of the Joint Soviet-Mongolian Paleontological Expedition* 35.

Manchester, S. R. 1981. Fossil plants of the Eocene Clarno nut beds. *Oregon Geology* 43:75–81.

———. 1986. Vegetative and reproductive morphology of an extinct plane tree (Platanaceae) from the Eocene of western North America. *Botanical Gazette* 147:200–226.

———. 1987. The fossil history of the Juglandaceae. *Monographs in Systematic Botany of the Missouri Botanical Garden* 21:1–137.

Manchester, S. R., and P. R. Crane. 1983. Attached leaves, inflorescences, and fruits of *Fagopsis,* an extinct genus of fagaceous affinity from the Oligocene Florissant flora of Colorado, USA. *American Journal of Botany* 70:1147–64.

Marshall, L. G., and R. L. Cifelli. 1990. Analysis of changing diversity patterns in Cenozoic land mammal age faunas, South America. *Palaeovertebrata* 19:169–210.

Maryanska, T. 1977. Ankylosauridae (Dinosauria) from Mongolia. *Palaeontologia Polonica* 37:85–151.

Maryanska, T., and H. Osmolska. 1974. Pachycephalosauria, a new suborder of ornithischian dinosaurs. *Palaeontologia Polonica* 30:45–102.

McClammer, J. U., and D. R. Crabtree. 1989. Post-Barremian (Early Cretaceous) to Paleocene paleobotanical collections in the western interior of North America. *Review of Palaeobotany and Palynology* 57:221–32.

McKenna, M. C. 1980. Early history and biogeography of South America's extinct land mammals. 43–77 in R. L. Ciochan and A. B. Chiarelli, eds., *Evolutionary Biology of the New World Monkeys and Continental Drift.* New York: Plenum Press.

———. 1983. Holocene landmass rearrangement, cosmic events, and Cenozoic terrestrial organisms. *Annals of the Missouri Botanical Garden* 70:459–89.

McKnight, C. L., S. A. Graham, A. R. Carroll, Q. Gan, D. L. Dilcher, M. Zhao, and Y. H. Liang. 1990. Fluvial sedimentology of an Upper Jurassic petrified forest assemblage, Shishu Formation, Junggar Basin, Xinjiang, China. *Palaeogeography, Palaeoclimatology, Palaeoecology* 79:1–9.

McNab, B. K. 1978. Energetics of arboreal folivores: Physiological problems and ecological consequences of feeding on a ubiquitous food supply. 53–62 in G. G.
Montgomery, ed., *The Ecology of Arboreal Folivores,* Washington, D.C.: Smithsonian Institution Press.

McNaughton, S. J. 1979. Grassland-herbivore dynamics. 46–81 in A. R. E. Sinclair
and M. Norton-Griffiths, eds., *Serengeti: Dynamics of an Ecosystem.* Chicago:
University of Chicago Press.

———. 1984. Grazing lawns: Animals in herds, plant form, and coevolution. *American Naturalist* 124:863–86.

Medlyn, D. A., and W. D. Tidwell. 1975. Conifer wood from the Upper Jurassic of
Utah Part 1: *Xenoxylon morrisonense* sp. nov. *American Journal of Botany* 62:
203–8.

Medus, J., and J. L. Pairis. 1990. Reworked pollen assemblages and the Eocene-
Oligocene boundary in the Paleogene of the western external French Alps. *Palaeogeography, Palaeoclimatology, Palaeoecology* 81:59–78.

Meyen, S. V. 1987. *Fundamentals of Palaeobotany.* London: Chapman and Hall.

Michener, C. D., and D. A. Grimaldi. 1988. The oldest fossil bee: Apoid history, evolutionary stasis, and antiquity of social behavior. *Proceedings of the National Academy of Sciences USA* 85:6424–26.

Milchunas, D. G., O. E. Sala, and W. K. Lauenroth. 1988. A generalized model of
the effects of grazing by large herbivores on grassland community structure. *American Naturalist* 132:87–106.

Miller, C. N., Jr. 1987. Land plants of the northern Rocky Mountains before the
appearance of flowering plants. *Annals of the Missouri Botanical Garden* 74:
692–706.

Miller, C. N., Jr., and C. A. LaPasha. 1984. Flora of the Early Cretaceous Kootenai
Formation in Montana: Conifers. *Palaeontographica,* B 193:1–17.

Miller, K. G., W. A. Berggren, J. Zhang, and A. A. Palmer-Jordan. 1991. Biostratigraphy and isotope stratigraphy of upper Eocene microtektites at site 612: How
many impacts? *Palaios* 6:17–38.

Mitter, C., B. Farrell, and B. Wiegmann. 1988. The phylogenetic study of adaptive
zones: Has phytophagy promoted insect diversification? *American Naturalist* 132:
107–28.

Newton, C. R., and C. A. McRoberts. 1990. Bivalve extinction at the Triassic-
Jurassic boundary: Association with abrupt sedimentological changes and presence
of shocked minerals. *Geological Society of America, Abstracts with Programs*
1990: A356.

Nichols, D. J., D. M. Jarzen, C. J. Orth, and P. Q. Oliver. 1986. Palynological and
iridium anomalies at Cretaceous-Tertiary boundary, south-central Saskatchewan.
Science 231:714–17.

Niklas, K. J. 1978. Coupled evolutionary rates and the fossil record. *Brittonia* 30:
373–94.

Niklas, K. J., B. H. Tiffney, and A. H. Knoll. 1980. Apparent changes in the diversity
of fossil plants. *Evolutionary Biology* 12:1–89.

———. 1983. Patterns in vascular land plant diversification. *Nature* 303:614–16.

Norman, D. B., and D. B. Weishampel. 1985. Ornithopod feeding mechanisms: Their
bearing on the evolution of herbivory. *American Naturalist* 126:151–64.

Norstog, K. 1987. Cycads and the origin of insect pollination. *American Scientist* 75:270–79.

Oksanen, L. 1988. Ecosystem organization: Mutualism and cybernetics or plain Darwinian struggle for existence? *American Naturalist* 131:424–44.

Oldham, T. C. B. 1976. Flora of the Wealden plant debris beds of England. *Palaeontology* 19:437–502.

Olsen, P. E., N. H. Shubin, and M. H. Anders. 1987. New Early Jurassic tetrapod assemblages constrain Jurassic-Triassic tetrapod extinction event. *Science* 237: 1025–29.

Olsen, P. E., and H.-D. Sues. 1986. Correlation of continental Late Triassic and Early Jurassic sediments, and patterns of the Triassic-Jurassic tetrapod transition. 321–351 in K. Padian, ed., *The Beginning of the Age of Dinosaurs*. New York: Cambridge University Press.

Olson, E. C., and R. D. K. Thomas, eds. 1980. *A Cold Look at the Hot-Blooded Dinosaurs*. AAAS Symposium Volume 28. Boulder: Westview Press.

Orth, C. J., J. S. Gilmore, J. D. Knight, C. L. Pillmore, R. H. Tschudy, and J. E. Fassett. 1981. An iridium abundance anomaly at the palynological Cretaceous-Tertiary boundary in northern New Mexico. *Science* 214:1341–42.

Osborn, H. F. 1923. Two Lower Cretaceous dinosaurs of Mongolia. *American Museum Novitates* 95:1–10.

Ostrom, J. H. 1966. Functional morphology and evolution of the ceratopsian dinosaurs. *Evolution* 20:290–308.

Owen-Smith, N. 1987. Pleistocene extinctions: The pivotal role of megaherbivores. *Paleobiology* 13:351–62.

Padian, K., and W. A. Clemens. 1985. Terrestrial vertebrate diversity: Episodes and insights. 41–96 in J. W. Valentine, ed., *Phanerozoic Diversity Patterns: Profiles in Macroevolution*. Princeton: Princeton University Press.

Parker, L. R. 1976. The paleoecology of the fluvial coal-forming swamps and associated floodplain environments in the Blackhawk Formation (Upper Cretaceous) of central Utah. 99–116 in A. T. Cross and E. B. Maxfield, eds., Aspects of coal geology, northwest Colorado Plateau: Some geologic aspects of coal accumulation, alteration, and mining in western North America. *Brigham Young University Geology Studies* 22.

Parrish, J. M. 1989. Vertebrate paleoecology of the Chinle Formation (Late Triassic) of the southwestern United States. *Palaeogeography, Palaeoclimatology, Palaeoecology* 72:227–47.

Parrish, J. M., J. T. Parrish, and A. M. Ziegler. 1986. Permian-Triassic paleogeography and paleoclimatology and implications for therapsid distribution. 109–31 in N. Hotton III, P. D. MacLean, J. J. Roth, and E. C. Roth, eds., *The Ecology and Biology of Mammal-like Reptiles*. Washington, D.C.: Smithsonian Institution Press.

Parrish, J. T. 1987. Global palaeogeography and palaeoclimate of the Late Cretaceous and Early Tertiary. 51–73 in E. M. Friis, W. G. Chaloner, and P. R. Crane, eds., *The Origins of Angiosperms and Their Biological Consequences*. New York: Cambridge University Press.

Parrish, J. T., and J. A. Doyle. 1984. Predicted evolution of global climate in Late Jurassic-Cretaceous time. *International Organization of Paleobotany Conference 3, Abstracts*.

Parrish, J. T., and F. Peterson. 1988. Wind directions predicted from global circulation models and wind directions determined from eolian sandstones of the western United States: A comparison. *Sedimentary Geology* 56: 261–82.

Parrish, J. T., and R. A. Spicer. 1988a. Late Cretaceous terrestrial vegetation: A near-polar temperature curve. *Geology* 16:22–25.

————. 1988b. Middle Cretaceous wood from the Nanushuk Group, central North Slope, Alaska. *Palaeontology* 31:19–34.

Parrish, J. T., A. M. Ziegler, and C. R. Scotese. 1982. Rainfall patterns and the distribution of coals and evaporites in the Mesozoic and Cenozoic. *Palaeogeography, Palaeoclimatology, Palaeoecology* 40:67–101.

Pederson, R. K., P. R. Crane, and E. M. Friis. 1989. The morphology and phylogenetic significance of *Vardekloeftia* Harris (Bennettitales). *Review of Palaeobotany and Palynology* 60:7–24.

Pelzer, G. 1987. Transition from fluvial to littoral environments in the Wealden facies (lowermost Cretaceous) of northwest Germany. 177–81 in P. J. Currie and E. H. Koster, eds., *Fourth Symposium on Mesozoic Terrestrial Ecosystems, Short Papers.* Occasional Paper of the Tyrrell Museum of Palaeontology 3.

Pelzer, G., and V. Wilde. 1987. Klimatische Tendenzen wahrend der Ablagerung der Wealden-Fazies in Nordwesteuropa. *Geologisches Jahrbuch* A96:239–63.

Peters, R. H. 1983. *The Ecological Implications of Body Size.* Cambridge: Cambridge University Press.

Peterson, G. L., and P. L. Abbott. 1979. Mid-Eocene climatic change, southwestern California and northwestern Baja California. *Palaeogeography, Palaeoclimatology, Palaeoecology* 26:73–87.

Pirrie, D., and J. D. Marshall. 1990. High-latitude Late Cretaceous paleotemperatures: New data from James Ross Island, Antarctica. *Geology* 18:31–34.

Pocock, S. A. J. 1962. Microfloral analysis and age determination of strata at the Jurassic-Cretaceous boundary in the western Canadian plains. *Palaeontographica,* B 111:1–95.

Primack, R. B. 1987. Relationships among flowers, fruits, and seeds. *Annual Review of Ecology and Systematics* 18:409–30.

Prothero, D. R. 1985. North American mammalian diversity and Eocene-Oligocene extinctions. *Paleobiology* 11:389–405.

Raath, M. A. 1974. Fossil vertebrate studies in Rhodesia: Further evidence of gastroliths in prosauropod dinosaurs. *Arnoldia (Zimbabwe)* 7:1–7.

Raine, J. I. 1988. The Cretaceous/Cainozoic boundary in New Zealand terrestrial sequences. *Abstracts of the Seventh International Palynological Congress, Brisbane,* 137.

Raup, D. M. 1986. Biological extinction in earth history. *Science* 231:1528–33.

Raup, D. M., and J. J. Sepkoski. 1986. Periodic extinctions of families and genera. *Science* 231:833–36.

Read, C. B., and R. W. Brown. 1936. American Cretaceous ferns of the genus *Tempskya. U.S. Geological Survey Professional Paper* 186-F:105–31.

Retallack, G. J. 1975. The life and times of a Triassic lycopod. *Alcheringa* 1:3–29.

————. 1977. Reconstructing Triassic vegetation of eastern Australia: A new approach to the biostratigraphy of Gondwanaland. *Alcheringa* 1:247–77.

Retallack, G. J., and D. L. Dilcher. 1981. A coastal hypothesis for the dispersal and rise to dominance of flowering plants. 27–77 in K. J. Niklas, ed., *Paleobotany, Paleoecology, and Evolution,* vol. 2. Ithaca: Praeger.

——. 1985. Cretaceous angiosperm invasion of North America. *Cretaceous Research* 7:227–52.

——. 1988. Reconstructions of selected seed ferns. *Annals of the Missouri Botanical Gardens* 75:1010–57.

Retallack, G. J., G. D. Leahy, and M. D. Spoon. 1987. Evidence from paleosols for ecosystem changes across the Cretaceous/Tertiary boundary in eastern Montana. *Geology* 15:1090–93.

Riegel, W., et al. 1986. Erste Ergebnisse einer paläobotanischen Grabung in der fluviatilen Wealden-Fazies des Osterwaldes bei Hannover. *Courier Forschungs-Institut Senckenberg* 86:137–70.

Robinson, P. L. 1971. A problem of faunal replacement on Permo-Triassic continents. *Palaeontology* 14:131–53.

Romero, E. J. 1986. Paleogene phytogeography and climatology of South America. *Annals of the Missouri Botanical Garden* 73:449–61.

Romero, E. J., and S. Archangelsky. 1986. Early Cretaceous angiosperm leaves from southern South America. *Science* 234:1580–82.

Rose, K. D. 1981. Composition and species diversity in Paleocene and Eocene mammal assemblages: An empirical study. *Journal of Vertebrate Paleontology* 1: 367–88.

——. 1984. Evolution and radiation of mammals in the Eocene, and the diversification of modern orders. 110–27 in P. D. Gingerich and C. E. Badgley, eds., *Mammals: Notes for a Short Course.* University of Tennessee, Department of Geology, Studies in Geology, vol. 8.

Russell, D. A. 1970. A skeletal reconstruction of *Leptoceratops gracilis* from the upper Edmonton Formation. *Canadian Journal of Earth Sciences* 7: 181–84.

Russell, L. S. 1940. *Edmontonia rugosidens* (Gilmore), an armoured dinosaur from the Belly River Series of Alberta. *University of Toronto Geological Studies Series* 43:1–28.

Saito, T., T. Yamanoi, and K. Kaiho. 1986. Devastation of the terrestrial flora at the end of the Cretaceous in the Boreal Far East. *Nature* 323:253–56.

Scott, A. C. 1991. Evidence for plant-arthropod interactions in the fossil record. *Geology Today* 7:58–61.

Sill, W. D. 1971. Functional morphology of the rhynchosaur skull. *Forma et Functio* 4:303–18.

Silver, L. T., and P. H. Schultz, eds. 1982. Geological implications of impacts of large asteroids and comets on Earth. *Geological Society of America Special Paper* 190:1–528.

Simms, M. J., and A. H. Ruffell. 1989. Synchroneity of climatic change and extinctions in the Late Triassic. *Geology* 17:265–68.

Sinclair, A. R. E., and M. Norton-Griffiths. 1979. *Serengeti: Dynamics of an Ecosystem.* Chicago: University of Chicago Press.

Smiley, C. J. 1969. Floral zones and correlations of Cretaceous Kukpowruk and Corwin Formations, northwestern Alaska. *Bulletin of the American Assocation of Petroleum Geologists* 53:482–502.

Spicer, R. A. 1987. The significance of the Cretaceous flora of northern Alaska for the reconstruction of the climate of the Cretaceous. *Geologisches Jahrbuch* A96: 265–91.

————. 1989. Plants at the Cretaceous-Tertiary boundary. *Philosophical Transactions of the Royal Society of London,* B 325:291–305.

————. 1990. Reconstructing high-latitude Cretaceous vegetation and climate: Arctic and Antarctic compared. 27–36 in T. N. Taylor and E. L. Taylor, eds., *Antarctic Paleobiology, Its Role in the Reconstruction of Gondwana.* New York: Springer-Verlag.

Spicer, R. A., and C. R. Hill. 1979. Principal component and correspondence analyses of quantitative data from a Jurassic plant bed. *Review of Paleobotany and Palynology* 28:273–99.

Spicer, R. A., and J. T. Parrish. 1986. Paleobotanical evidence for cool north polar climates in middle Cretaceous (Albian-Cenomanian) time. *Geology* 14:703–6.

————. 1990a. Late Cretaceous–early Tertiary palaeoclimates of northern high latitudes: A quantitative view. *Journal of the Geological Society of London* 147: 329–41.

————. 1990b. Latest Cretaceous woods of the central North Slope, Alaska. *Palaeontology* 33:225–42.

Spicer, R. A., J. T. Parrish, and P. R. Grant. In press. Evolution of vegetation and coal-forming environments in the Late Cretaceous of the North Slope of Alaska: A model for polar coal deposition at times of global warmth. In P. J. McCabe and J. T. Parrish, eds., *Controls on the Distribution and Quality of Cretaceous Coals.* Geological Society of America Special Paper.

Stanley, S. M. 1984. Temperature and biotic crises in the marine realm. *Geology* 12:205–8.

Stebbins, G. L. 1981. Coevolution of grasses and herbivores. *Annals of the Missouri Botanical Garden* 68:75–86.

Stewart, W. N. 1983. *Palaeobotany and the Evolution of Plants.* Cambridge: Cambridge University Press.

Stucky, R. K. 1989. The anatomy of continental diversity: Mammalian faunal dynamics of the North American Eocene. In N. Stenseth, ed., *Coevolution in Ecosystems: The Red Queen Hypothesis.* Cambridge: Cambridge University Press.

————. 1990. Evolution of land mammal diversity in North America during the Cenozoic. 375–432 in H. H. Genoways, ed., *Current Mammalogy* 2. New York: Plenum Press.

Sues, H.-D. 1984. Inferences concerning locomotion and feeding in the Tritylodontidae (Synapsida). 231–36 in W.-E. Reif and F. Westphal, eds., *Second Symposium on Mesozoic Terrestrial Ecosystems, Short Papers.* Tübingen: Attempto Verlag.

————. 1986. The skull and dentition of two tritylodontid synapsids from the Lower Jurassic of western North America. *Bulletin of the Museum of Comparative Zoology, Harvard University* 151:217–68.

Sues, H.-D. and P. M. Galton. 1987. Anatomy and classification of the North American Pachycephalosauria (Dinosauria: Ornithischia). *Palaeontographica,* A 198: 1–40.

Swisher, C. C., III, and D. R. Prothero. 1990. Single-crystal ^{40}Ar/^{39}Ar dating of the Eocene-Oligocene transition in North America. *Science* 249:760–62.

Taylor, D. W., and W. L. Crepet. 1987. Fossil floral evidence of Malpighiaceae and an early plant-pollinator relationship. *American Journal of Botany* 74:274–86.

Taylor, D. W., and L. J. Hickey. 1990. An Aptian plant with attached leaves and flowers: Implications for angiosperm origin. *Science* 247:702–3.

Taylor, G., E. M. Truswell, K. G. McQueen, and M. C. Brown. 1990. Early Tertiary palaeogeography, landform evolution, and palaeoclimates of the southern Monaro, N.S.W., Australia. *Palaeogeography, Palaeoclimatology, Palaeoecology* 78:109–34.

Taylor, T. N. 1981. *Paleobotany: An Introduction to Fossil Plant Biology.* New York: McGraw-Hill.

Thomas, B. A., and R. A. Spicer. 1987. *The Evolution and Palaeobiology of Land Plants.* London and Sydney: Croom Helm.

Thornton, D. D. 1971. The effect of complete removal of hippopotamus on grassland in the Queen Elizabeth National Park, Uganda. *East African Wildlife Journal* 9: 47–55.

Throckmorton, G. S., J. A. Hopson, and P. Parks. 1981. A redescription of *Toxolophosaurus cloudi* Olson, a Lower Cretaceous herbivorous sphenodontid reptile. *Journal of Paleontology* 55:586–97.

Thulborn, R. A. 1970a. The skull of *Fabrosaurus australis,* a Triassic ornithischian dinosaur. *Palaeontology* 13:414–32.

———. 1970b. Tooth wear and jaw action in the Triassic ornithischian dinosaur *Fabrosaurus. Journal of Zoology, London* 164:165–79.

Tidwell, W. D., S. R. Ash, and L. R. Parker. 1981. Cretaceous and Tertiary floras of the San Juan Basin. 307–32 in S. G. Lucas, K. Rigby, Jr., and B. Kues, eds., *Advances in San Juan Basin Paleontology.* Albuquerque: University of New Mexico Press.

Tidwell, W. D., and S. R. Rushforth. 1970. *Osmundacaulis wadei,* a new osmundaceous species from the Morrison Formation (Jurassic) of Utah. *Bulletin of the Torrey Botanical Club* 97:137–44.

Tiffney, B. H. 1984. Seed size, dispersal syndromes, and the rise of the angiosperms: Evidence and hypothesis. *Annals of the Missouri Botanical Garden* 71:551–76.

———. 1985. The Eocene North Atlantic land bridge: Its importance in Tertiary and modern phytogeography of the northern hemisphere. *Journal of the Arnold Arboretum* 66:243–73.

———. 1986a. Evolution of seed dispersal syndromes according to the fossil record. 273–305 in D. R. Murray, ed., *Seed Dispersal.* North Ryde, N.S.W.: Academic Press Australia.

———. 1986b. Fruit and seed dispersal and the evolution of the Hamamelidae. *Annals of the Missouri Botanical Garden* 73:394–416.

Traverse, A. 1988. Plant evolution dances to a different beat: Plant and animal evolutionary mechanisms compared. *Historical Biology* 1:277–301.

Tryon, R. M., and A. F. Tryon. 1982. *Ferns and Allied Plants, with Special Reference to Tropical America.* New York: Springer-Verlag.

Tschudy, R. H., C. L. Pillmore, C. J. Orth, J. S. Gilmore, and J. D. Knight. 1984. Disruption of the terrestrial plant ecosystem at the Cretaceous-Tertiary boundary, western Interior. *Science* 225:1030–32.

Tschudy, R. H., and B. D. Tschudy. 1986. Extinction and survival of plant life following the Cretaceous-Tertiary boundary event, Western Interior, North America. *Geology* 14:667–70.

Turner, C. E., and N. S. Fishman. 1991. Jurassic Lake T'oo'dichi': A large, saline lake, Morrison Formation, eastern Colorado Plateau. *Geological Society of America Bulletin* 103:538–58.

Upchurch, G. R., Jr., and D. L. Dilcher. 1990. Cenomanian angiosperm leaf mega-fossils, Dakota Formation, Rose Creek locality, Jefferson County, southeastern Nebraska. *U.S. Geological Survey Bulletin* 1915:1–55.

Upchurch, G. R., Jr., and J. A. Doyle. 1981. Paleoecology of the conifers *Frenelopsis* and *Pseudofrenelopsis* (Cheirolepidiaceae) from the Cretaceous Potomac Group of Maryland and Virginia. 167–202 in R. Romans, ed., *Geobotany*, vol. 2.

Upchurch, G. R., Jr., and J. A. Wolfe. 1986. Mid-Cretaceous to Early Tertiary vegetation and climate: Evidence from fossil leaves and woods. 75–105 in E. M. Friis, W. G. Chaloner, and P. R. Crane, eds., *The Origins of Angiosperms and their Biological Consequences*. Cambridge: Cambridge University Press.

Vakhrameev, V. A. 1970. Range and palaeoecology of Mesozoic conifers, the Cheirolepidiaceae (in Russian). *Palaeontologiskhii Zhurnal* 1970:19–34.

———. 1987. Climates and the distribution of some gymnosperms in Asia during the Jurassic and Cretaceous. *Review of Palaeobotany and Palynology* 51:205–12.

Vakhrameev, V. A., I. A. Dobruskina, S. V. Meyen and E. D. Zaklinskaja. 1978. *Palaeozoische und mesozoische Floren Eurasiens und die Phytogeographie dieser Zeit*. Jena: VEB Gustav Fischer Verlag.

Van der Pijl, L. 1982. Principals of Dispersal in Higher Plants. 3d ed. Berlin: Springer-Verlag.

Van Valen, L. 1978. The beginning of the age of mammals. *Evolutionary Theory* 4:45–80.

Veevers, J. J. 1989. Middle/Late Triassic (230 + 5 Ma) singularity in the stratigraphic and magmatic history of the Pangean heat anomaly. *Geology* 17:784–87.

Walker, A. D. 1961. Triassic reptiles from the Elgin area: *Stagonolepis, Dasygnathus*, and their allies. *Philosophical Transactions of the Royal Society of London*, B, 244:103–204.

Watson, J. 1977. Some Lower Cretaceous conifers of the Cheirolepidiaceae from the USA and England. *Palaeontology* 20:715–49.

———. 1983. Two Wealden species of *Equisetum* found *in situ*. *Acta Palaeontologica Polonica* 28:265–69.

Webb, S. D. 1977. A history of savanna vertebrates in the New World, Part 1: North America. *Annual Review of Ecology and Systematics* 8:355–80.

Weishampel, D. B. 1984a. Evolution of jaw mechanisms in ornithopod dinosaurs. *Advances in Anatomy, Embryology, and Cell Biology* 87:1–110.

Weishampel, D. B. 1984b. Interactions between Mesozoic plants and vertebrates: Fructifications and seed predation. *Neues Jahrbuch für Geologie und Paläontologie, Abhandlungen* 167:224–50.

Weishampel, D. B., P. Dodson, and H. Osmolska. 1990. *The Dinosauria*. Berkeley: University of California Press.

Weishampel, D. B., and D. B. Norman. 1989. Vertebrate herbivory in the Mesozoic: Jaws, plants, and evolutionary metrics. 87–100 in J. O. Farlow, ed., *Paleobiology of the Dinosaurs*. Geological Society of America Special Paper 238.

Wesley, A. 1973. Jurassic plants. 329–38 in A. Hallam, ed., *Atlas of Palaeobiogeography*. Amsterdam: Elsevier.

Westgate, J. W., and C. T. Gee. 1990. Paleoecology of a middle Eocene mangrove biota (vertebrates, plants, and invertebrates) from southwest Texas. *Palaeogeography, Palaeoclimatology, Palaeoecology* 78:163–77.

Whalley, P. 1986. A review of current fossil evidence of Lepidoptera in the Mesozoic. *Biological Journal of the Linnean Society* 28:253–71.

Wheeler, E. A. 1991. Paleocene dicotyledonous trees from Big Bend National Park, Texas: Variability in wood types common in the Late Cretaceous and early Tertiary, and ecological inferences. *American Journal of Botany* 78:658–71.

Wilson, E. O. 1987. The earliest known ants: An analysis of the Cretaceous species and an inference concerning their social organization. *Paleobiology* 13:44–53.

Wing, S. L. 1980. Fossil floras and plant-bearing beds of the central Bighorn Basin. 119–25 in P. D. Gingerich, ed., Early Cenozoic Paleontology and Stratigraphy of the Bighorn Basin, Wyoming. *University of Michigan Papers on Paleontology* 24.

———. 1987. Eocene and Oligocene floras and vegetation of the Rocky Mountains. *Annals of the Missouri Botanical Garden* 74:748–84.

Wing, S. L., and T. M. Bown. 1985. Fine-scale reconstruction of late Paleocene–early Eocene paleogeography in the Bighorn basin of northern Wyoming. 93–105 in R. M. Flores and S. S. Kaplan, eds. *Cenozoic Paleogeography of the West-Central United States*. Rocky Mountain Section, Society of Economic Paleontologists and Mineralogists, Boulder.

Wing, S. L., and B. H. Tiffney. 1987. The reciprocal interaction of angiosperm evolution and tetrapod herbivory. *Review of Palaeobotany and Palynology* 50:179–210.

Wolfe, J. A. 1968. Paleogene biostratigraphy of nonmarine rocks in King County, Washington. *U.S. Geological Survey Professional Paper* 571:1–33.

———. 1975. Some aspects of plant geography of the Northern Hemisphere during the late Cretaceous and Tertiary. *Annals of the Missouri Botanical Garden* 62:264–79.

———. 1977. Paleogene floras from the Gulf of Alaska region. *U.S. Geological Survey Professional Paper* 997:1–108.

———. 1978. A paleobotanical interpretation of Tertiary climates in the Northern Hemisphere. *American Scientist* 66:694–703.

———. 1985. Distribution of major vegetational types during the Tertiary. *Geophysical Monographs* 32:357–75.

———. 1987. Late Cretaceous–Cenozoic history of deciduousness and the terminal Cretaceous event. *Paleobiology* 13:215–26.

Wolfe, J. A., and G. R. Upchurch, Jr. 1986. Vegetation, climatic, and floral changes at the Cretaceous-Tertiary boundary. *Nature* 324:148–52.

———. 1987a. Leaf assemblages across the Cretaceous-Tertiary boundary in the Raton Basin, New Mexico and Colorado. *Proceedings of the National Academy of Sciences USA* 84:5096–5100.

———. 1987b. North American nonmarine climates and vegetation during the Late Cretaceous. *Palaeogeography, Palaeoclimatology, Palaeoecology* 61:33–77.

Wolfe, J. A., and W. Wehr. 1987. Middle Eocene dicotyledonous plants from Republic, northeastern Washington. *U.S. Geological Survey Bulletin* 1597:1–25.

Young, H. J. 1988. Neighborhood size in a beetle-pollinated tropical aroid: Effects of low density and asynchronous flowering. *Oecologia* 76:461–66.

Zawiskie, J. M. 1986. Terrestrial vertebrate faunal succession during the Triassic.

353–62 in K. Padian, ed., *The Beginning of the Age of Dinosaurs*. New York: Cambridge University Press.

Ziegler, A. M. 1989. Phytogeographic patterns and continental configurations during the Permian Period. 363–79 in W. S. McKerrow, and C. R. Scotese, eds., *Paleozoic Palaeogeography and Biogeography*. Geological Society Memoir 12.

Ziegler, A. M., A. L. Raymond, T. C. Gierlowski, M. A. Horrell, D. B. Rowley, and A. L. Lottes. 1987. Coal, climate, and terrestrial productivity: The present and early Cretaceous compared. 25–29 in A. C. Scott, ed., *Coal and Coal-Bearing Strata: Recent Advances*. Geological Society Special Publication 32.

Ziegler, A. M., C. R. Scotese, and S. F. Barrett. 1983. Mesozoic and Cenozoic paleogeographic maps. 240–52 in P. Brosche and J. Sunderland, eds., *Tidal Friction and the Earth's Rotation*, vol. 2. Berlin: Springer-Verlag.

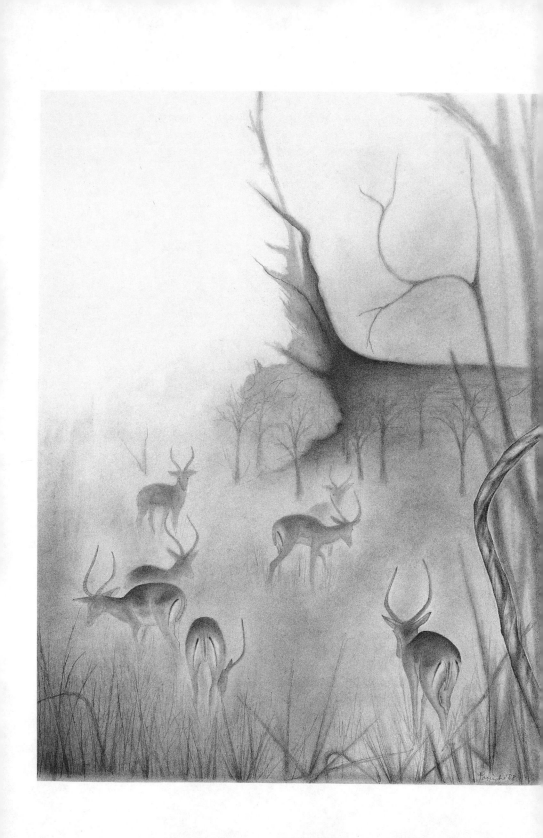

S E V E N   Late Cenozoic Terrestrial Ecosystems

Richard Potts and Anna K. Behrensmeyer, RAPPORTEURS

IN COLLABORATION WITH RALPH E. TAGGART, W. GEOFFREY SPAULDING, JUDITH
A. HARRIS, BLAIRE VAN VALKENBURGH, LARRY D. MARTIN, JOHN D.
DAMUTH, AND ROBERT FOLEY

1 INTRODUCTION

The purpose of this chapter is (1) to characterize the paleoecological trends of
the last 34 million years, which we will refer to simply as the late Cenozoic;
(2) to summarize the paleoecological histories of the major continental masses
in order to show the varied expression of the trends; and (3) to consider the
contributions of the Cenozoic perspective to the study of terrestrial paleoecology.

In § 1 we introduce certain generalizations and assumptions that typify paleoecological studies of the Oligocene through Quaternary time interval, or
the past 34 million years. In § 2 we identify the key ecological trends of this
interval, recount the prominent changes in climates, floras, and faunas, and
trace the origins, radiations, and extinctions of major groups of plants and
animals. In § 3 is a chronicle of the paleontological records of five continents:
North America, South America, Africa, Eurasia, and Australia. The large
body of information provided in § 3 illustrates the continental and regional
variations on the main ecological themes introduced previously. In § 4 we se-
lectively review the processes of ecological change that can be inferred on the
basis of paleontological evidence from all continental areas, and in § 5 we examine the general contributions of the late Cenozoic perspective.

As in other time intervals, late Cenozoic floras and faunas experienced distinct pulses of species originations and extinctions. Ancient communities
were invaded by exotic taxa and developed their own endemic signatures by
local evolution. The ecosystems of today, when examined from certain structural perspectives and levels of analysis, are not necessarily like those in existence 34 million years ago. Models and explanations of paleoecological
change during this recent period tend to have the same flavor as accounts of
ecological shifts in the Mesozoic and Paleozoic, often relying upon correlations among floras, faunas, and climates to fuel ideas as to how and why ecological change occurred. The temporal and geographic density of late
Cenozoic paleontological data, however, affords more refined tests of correla-

MIDDLE MIOCENE (14 Ma)

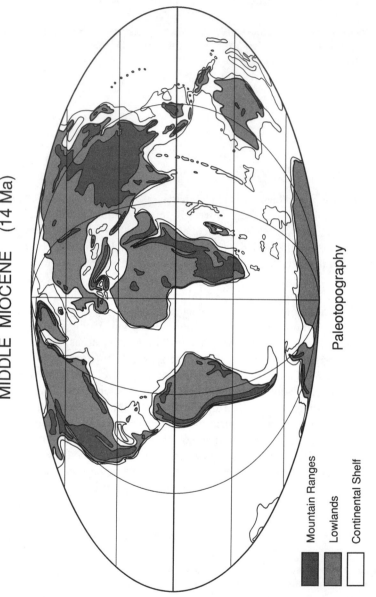

Paleotopography

Mountain Ranges

Lowlands

Continental Shelf

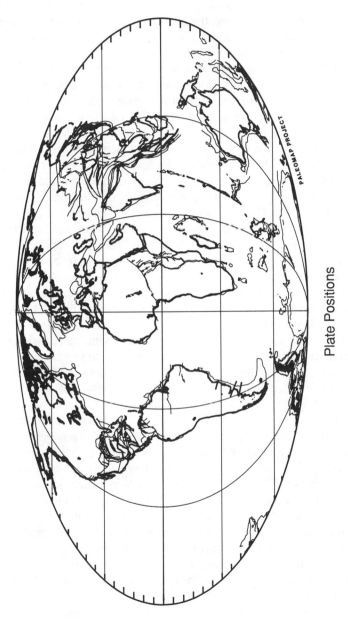

Plate Positions

Fig. 7.1. Global paleogeographic reconstruction for the middle Miocene (circa 13 Ma). (Courtesy of C. R. Scotese and the PaleoMap Project, University of Texas, Arlington).

tions and more precise assessment of the contemporaneity of shifts in paleoecological variables than is possible for most earlier times. Since ideas
about ecological history in a geological perspective are based on analysis of
correlations and contemporaneity, the finer time-space resolution available in
late Cenozoic studies offers a unique perspective on the grand scale of ecological history considered in this volume.

Although there may be differences in opinion regarding how or whether the
study of the late Cenozoic biota differs from that of earlier time periods, two
factors influence any description of paleoecological history during this time
interval. The first is proximity to the present. The Oligocene through Quaternary interval ends with the present, when living organisms and their associations with one another and with certain physical parameters are directly
observable. Just as in the Mesozoic and Paleozoic, ecological or functional
characterizations of Cenozoic species (see chaps. 3 and 4) should rely upon
principles of functional morphology and contextual information. Yet because
a larger number of late Cenozoic species have close modern relatives, reference is commonly made to modern analogues in the ecological characterization of paleofaunas and floras. A few studies, for example, have attempted to
identify changes in Miocene to early Pleistocene environments on the basis of
ecological diversity spectra, i.e., the distributions of ecomorphic attributes
(e.g., body size, type of locomotion) of animals that comprise paleontological
assemblages (Andrews et al. 1979; Andrews and Nesbit-Evans 1979; Van
Couvering 1980; Artemiou 1983). Not only has it been tempting in such studies
to categorize ecologically the paleontological taxa according to their closest
modern relatives, but the habitat types and diversity spectra serving as models
for the Miocene have also been limited to those existing today. In careful and
innovative studies of ecological change such as these, wholesale comparisons
between ancient and modern biotas have assumed that the associations observable in the present constitute a reasonable range of models for ancient ecological communities. Although such studies have sought differences as well as
similarities between ancient and modern diversity spectra, interpretation still
relies heavily on uniformitarian indicators of specific habitats. Functional
(ecomorphic) analysis should offer a more reliable approach than taxonomic
affinity, especially since the latter can seriously misdirect interpretation of
habitat structure and other environmental variables (e.g., Solunias et al. 1988;
Solunias and Dawson-Saunders 1988). The point is that most descriptions of
late Cenozoic ecological history have been rife with implicit and explicit comparisons of specific paleotaxa and even entire habitats to supposed modern
counterparts. This is much less true, for better or for worse, in studies of earlier time intervals.

The second factor that influences any account of late Cenozoic history is
that this interval is very brief compared with the time periods examined in
chapters 5 and 6, yet it involves a fossil record that is relatively dense temporally and geographically. Characterizations of faunas and floras during the

Paleozoic, for instance, typically encompass much larger time spans and may be unable to consider the fine spatial scales that are available late in the Cenozoic. For much of this interval, variations in the pattern and pace of biotic change often can be documented both between and within continents. Because of such variations, we have chosen to summarize the ecological history of the past 34 my continent by continent. The detail contained in these histories exemplifies the large-scale themes of ecological change on each continent. Even so, this compilation only touches upon the evidence from the thousands of fossil localities that contribute to our understanding of late Cenozoic ecological history.

2 SYNOPSIS OF LATE CENOZOIC ECOLOGICAL HISTORY

2.1 Overview of Environmental Change

As noted in chapter 6, the early Paleogene was marked by a very low diversity of large herbivores and the predominance of closed vegetation (at least in mid-latitude zones), and this was in remarkable contrast to the late Mesozoic. This shift in animal and plant community composition and structure indeed may represent the most dramatic change in terrestrial ecosystems since the mid-Permian (chap. 5). In this light, the dominant ecological theme of the late Cenozoic has been the redevelopment and diversification of ecosystems that include large herbivores and more open, lower biomass vegetation. Forests have remained one of the most significant vegetational types on earth (29% of the total land surface today; see Atjay et al. 1979). Moreover, the late Cenozoic does not exhibit a simple, directional history toward the expansion of grasslands and other open terrains at the expense of closed vegetation. Shifts in biomes since the Oligocene have been temporally complex and spatially mosaic. Nonetheless, the most evident outcome worldwide during the late Cenozoic has been the diminishing of forests and the concomitant expansion of open woodlands, grasslands, and deserts.

Variations in the amplitude of glacial cycles, in the development of polar ice, and in other factors also resulted in a climatic history during this same period that was neither simple nor directional. Nevertheless the proliferation of open vegetation since the Miocene has been associated with overall global cooling, which interacted with localized factors (e.g., orogeny, rifting) that enhanced seasonality in rainfall and/or temperature. The establishment of steep thermal gradients from the poles to the tropics has been one of the primary climatic shifts since the Eocene (e.g., Laporte and Zihlman 1983; Street 1981; Traverse 1988, 289). Overall characterization of the late Cenozoic as cooler and drier uses the early Eocene tropical expansion as a point of reference. We can say on the basis of oceanic paleotemperature curves (Shackleton and Kennett 1975a, 1975b; Kennett 1985; Miller et al. 1987) that the warmest periods since the Eocene occurred in the early Miocene, when global temperatures reached only the minimum level estimated for most of the early Cenozoic (65–34 Ma). After 16 Ma, sporadic buildup of the East Antarctic ice sheet and intensification of oceanic and atmospheric circulation appear to have steepened worldwide temperature gradients and created cooler, drier climates overall, though these would have oscillated with warmer global temperatures (Woodruff et al. 1981; Van Zinderen Bakker and Mercer 1986, Barry et al. 1990).

A global cooling trend and evidence of drying are recognized in some paleontological records even prior to the early Oligocene. The Madro-Tertiary and Cordilleran floras of North America suggest relative aridity during the Eocene (Axelrod 1958; Axelrod and Raven 1985; see also Wing 1987). Marine microfaunas in Italy and from cores in the Indian, Atlantic, and Pacific Oceans indicate a cooling trend from the middle Eocene to the Eocene/Oligocene boundary, accompanied by lowered productivity and decreased diversity in calcareous planktonic and benthic faunas. Although no catastrophic extinctions occurred, the marine faunas indicate that climatic deterioration intensified worldwide from the late Eocene to early Oligocene (Nocchi et al. 1988; Corliss et al. 1984). The earliest documented extensive buildup of glacial ice in Antarctica during the Cenozoic, by at least 38 Ma, also may be related to an Eocene-Oligocene temperature decrease. A worldwide marine regression event during the mid-Oligocene has been proposed on the basis of canyon cutting on continental shelves and a basinward shift of strandlines (Haq et al. 1987; Vail and Haq 1988), which would correspond with a major climatic event at that time. Revised chronology for the Eocene-Oligocene boundary (34 Ma compared with previous placement at ca. 36.5 Ma) suggests, however, that extinctions of terrestrial animals were dispersed over the period 40–30 Ma, and that climatic events traditionally associated with the Oligocene may now be placed within the Eocene or at the Eocene-Oligocene boundary (Swisher and Prothero 1990; contra Prothero 1985; Prothero and Berggren in press).

Global cooling again accelerated during the late Neogene with numerous reversals superimposed upon the overall trend, on all continental masses (Berggren and Van Couvering 1974; Frakes 1979; Kennett 1978; LeMasurier 1972a, 1972b; Robin 1988; Denton et al. 1971; Shackleton 1987). According to Shackleton and Opdyke (1977), continental environments prior to 3.2 Ma were relatively stable climatically for long periods compared with the continually fluctuating climates (on scale of 10^4–10^5 yrs) after this time. Diatoms and fossil wood in lower Pliocene rocks of the Transantarctic Mountains, however, imply instability in the late Neogene glacial history of Antarctica and episodes of widespread climatic warming during this period (Webb et al. 1984; Pickard et al. 1988; Webb 1990). A dramatic climatic event of global significance has been identified at 2.4 Ma—"A truly glacial interval, with an ice volume similar to maxima during the middle Pleistocene" (Shackleton et al. 1984). Periodic development of ice sheets over Northern Hemisphere continents began by 2.4 Ma, which coincides with an abrupt increase in global ice as indicated by planktonic $\delta^{18}O$ values (Shackleton et al. 1984; Prentice and Denton 1988; Webb 1990).

Major temperature variations during the past 0.9 Ma, probably regulated by Milankovitch cycles (e.g., Hays et al. 1976; Imbrie et al. 1984), resulted in displacements of Quaternary snow lines to a maximum of 1000 m below present-day snow lines (equilibrium line altitudes) between 50°N and 40°S, drops in global sea level of 100–150 m, and decreases in global climates of 6° C during glacial maxima (Jouzel et al. 1987; Robin 1988). While orbital variations may have been responsible for cyclical climatic shifts over a much lengthier history, it is the large amplitude of climatic oscillations over the past 1 million years that is especially striking. Long-term records of global and regional climates have been sought from deep sea cores, terrestrial pollen records, loess sequences, glacial and lacustrine sedimentary sequences, sea level changes, and nonmarine oxygen isotope records (e.g., Kukla 1989). In general, the correlations among these records (especially between terrestrial and marine sequences) are poorly known, though concordance between terrestrial (pollen) and marine (foraminifera) estimates of temperature has been demonstrated (Van Campo et al. 1990). Further success has been claimed in linking glacial-interglacial, arid-moist, and deposition-erosion cycles; this linkage has served to place faunas and floras in their large-scale climatic context (e.g., Kukla 1977, 1987; Van Zinderen Bakker 1976; Sarnthein 1978; Heusser and Shackleton 1979; Smith 1984; Roberts 1984; Street-Perrott and Harrison 1984).

While studies of late Cenozoic temperature-glacial variations and global cooling traditionally have focused on the high latitudes of the Northern Hemisphere, patterns of climatic change and their effect on biotas appear to have been equally complex in tropical latitudes. Roberts (1984), among others, hypothesizes that during phases of glacial aridity the rain forests in central Africa and the Amazon shrank to isolated refugia, whereas those in Sumatra remained intact (see Flenley 1979; Maloney 1980); disruption of Southeast Asian rain forests also occurred repeatedly, however, as a result of high interglacial sea levels. In contrast with previous ideas about "pluvials," many low-latitude regions experienced aridity when cold (glacial) periods occurred in higher latitudes, and moist conditions (evidenced by high lake levels) appear to have coincided with interglacials (Street-Perrott and Harrison 1984). This correlation between tropical- and temperate-latitude climates is accepted at present for the Quaternary of Africa (Street and Grove 1979; Van Zinderen Bakker 1976). Australia also may follow this pattern (e.g., Dodson 1977); tropical pollen studies in South America are too limited to determine a correlation (Livingstone and Van der Hammen 1978; Van Zinderen Bakker 1978). Geomorphological evidence further suggests that tropical fluvial systems responded in a highly variable manner to climatic fluctuations during the Quaternary (Street 1981). The overall conclusion that can be drawn about the

Global Change in the Late Cenozoic

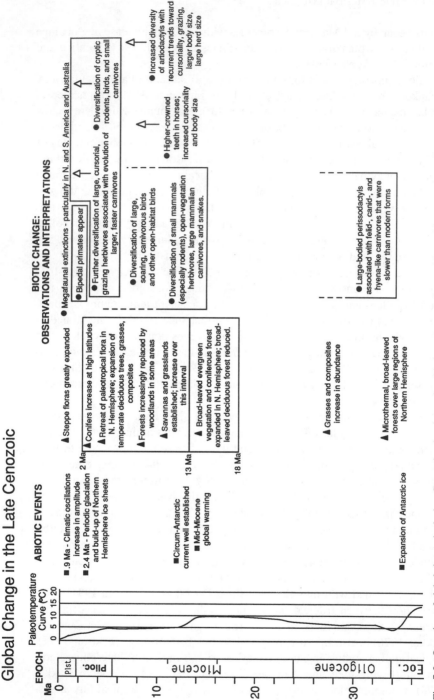

Fig. 7.2. Overview of global change in the late Cenozoic, showing major abiotic events and observations and interpretations of biotic change from the late Eocene to the Recent. Paleotemperature curve is based on composite records from the Atlantic DSDP sites (Miller et al. 1987). Square symbols indicate abiotic events, triangles refer to floras, and dots refer to faunas in figures 7.1–7.6

tropics during the Quaternary is that climates were quite unstable, and this undoubtedly affected the distribution and probably the turnover of tropical biotas.

A parallel relationship between temperature and moisture has been proposed for Europe: glacial cold corresponded with aridity, interglacial warmth with climatic wetness (e.g., Van der Hammen et al. 1971). The cold-dry association appears to be reversed, however, in a number of areas in North America (e.g., central and southwestern United States), where high lake stands were correlated with glacials (Wells 1979; Benson 1978; Spaulding et al. 1983; Spaulding and Graumlich 1986). Although large-scale global climatic trends can be identified (fig. 7.2), complex regional variations and temporal fluctuations are quite characteristic of the climatic record of the latter part of the Cenozoic.

2.2   Ecological Trends of the Late Cenozoic

Numerous taxa of terrestrial animals and plants have originated, radiated, and become extinct since the Eocene. Despite widespread changes in the taxonomic composition of biotas, only a small number of major ecological trends occurred. These trends were represented as parallel developments on different continents and in different taxonomic groups (fig. 7.2; See appendix tables A-7.1–A-7.4 for plant and animal nomenclature).

An important theme in the history of plant communities during the late Cenozoic was the spread of grasses and the alteration of vegetation structure involving the development of large, continuous tracts of grassland. Associated responses in the evolution of animals during this period are largely noted among the mammals, and include: (1) increased development of hypsodonty and dietary specialization related to grazing; (2) development of cursoriality and other locomotor characteristics related to movement in open habitats, including increase in foraging radius; (3) a trend toward large body size among herbivores, which was related to the increased need to process low-quality grasses and browse in large amounts; (4) diversification of small-bodied herbivores (especially rodents and birds), including fossorial forms, based upon survival by cryptic behavior and need for shelter in open terrain; and (5) diversification of predators into large forms with locomotor and social specializations for capturing large, cursorial herbivores, and small forms that specialized on smaller herbivores.

These major ecomorphic characterizations of the late Cenozoic represent enhancements of adaptive variations that were present in the Eocene. Although clear evidence for the development of grasslands does not appear prior to the mid–late Miocene, modern tribes of grasses are known from megafossils from the late Paleocene–early Eocene (Crepet and Feldman 1988). As noted in chapter 6, a shift toward open vegetation and an early radiation of artiodactyls and perissodactyls occurred by the late Eocene. Increases in herbivore body size probably related to "bulk feeding" in open environments occurred by the Oligocene, and certain groups of fossorial specialists indicating more open terrain diversified in the early Cenozoic (e.g., the edentates). Thus, what might appear to be new functional developments in terrestrial paleoecological history of the late Cenozoic are largely amplifications upon ecological themes already in existence by the Oligocene.

Late Cenozoic Plants

FOREST, WOODLAND, AND GRASSLAND   The late Cenozoic is often characterized as the time of grassland expansion and forest contraction, and there is a tendency to think of ancient vegetation as one or the other, open or closed. Though woodlands are recognized as intermediate ecologically and physiognomically, their role in the forest-grassland transition is not well understood.

For modern East Africa, woodland has been defined as having a canopy cover of more than 20%, wooded grassland as having between 2% and 20%, and grassland as having under 2% (Pratt et al. 1966). The term "savanna" is often used to encompass all of these, to the point where tree cover becomes continuous (Bourliere 1983). The range of tree cover between 20% and 100% defines a broad spectrum of possible habitats within which both grasses and trees could be significant resources for different kinds of herbivores. Although these intermediate ranges of tree cover may have existed throughout the late Cenozoic, modern analogues with such mixtures of trees and grass appear to be rare. This is worth keeping in mind during the following discussions of patterns of ecological change in plants and animals over the past 34 My.

MODERNIZATION OF PLANTS   Plant megafossils contribute importantly to the interpretation of Oligocene and Miocene botanical history, but efforts to reconstruct floral settings of the late Cenozoic have been dominated by palynological studies. According to a classic study of fossil pollen (Muller 1981), 16% of 141 extant families of angiosperms had first pollen appearances in the Oligocene, 5% in the early Miocene, 2% in the middle Miocene, 13% in the late Miocene, 5% in the Pliocene, and 0% in the Pleistocene. All families recognized in the fossil pollen record at 20 Ma are living today, and by 10 Ma all genera in the fossil record represent extant forms (Traverse 1988). Muller (1981) describes the period from the Oligocene to the Pliocene as the last phase in the development of the modern pollen record. This phase is characterized by an initial decline in the rate of taxonomic diversification, followed by an increased rate in the late Miocene and Pliocene (though comparisons with modern floras may result in the recognition of more taxa during this later time period). The main feature of this phase is the diversification of Compositae, which began in the Oligocene, and the marked diversification of the order Myrtales and of orders characterized by herbaceous forms.

FOREST VEGETATION   The end of the Eocene was marked by extensive vegetational change, inferred from analyses of fossil leaf assemblages. Microthermal (table A-7.1) broad-leaved deciduous forest extended southward into regions formerly covered with evergreen species (Wolfe 1985, 1987) and for the first time occupied large regions of the Northern Hemisphere. Taxonomic diversity declined in middle- and high-latitude forests. After this initial Eocene-Oligocene deterioration, diversity increased during the Oligocene. Broad-leaved evergreen vegetation became restricted to 35° latitude around the equator (compared with 70° in the early Eocene), and megathermal, multistratal vegetation was confined to 15° latitude (compared with 60–65° during the early Eocene and 20–25° today, in the Northern Hemisphere) around the equator (Wolfe 1980, 1985; Wing 1987). Oligocene pollen floras include forms that persisted from the Eocene (e.g., nyssoid pollen, many triporates, and numerous forms representing extant families but preserved far from the modern ranges of those families). The pollen record also shows the expansion of temperate deciduous forest. Grasses (Gramineae) and composites (Compositae) begin to occur in some abundance but were not dominant floral elements (Traverse 1988).

Wet-to-monsoonal tropical forest is evidenced in the early Oligocene flora from the Fayum, Egypt. Beyond this locality, however, low-latitude vegetation during the mid-Cenozoic is very poorly known. It is probable that Fagaceae, Moraceae, Leguminosae, and Dipterocarpaceae were common in these floras, and the mangrovelike palm *Nypa,* which grows today in brackish water and which originated at least by the early Eocene, was present in Southeast Asia during the Oligocene (Flenley 1979). Based on the pantropical occurrence of certain genera, the few data that exist suggest that low-latitude floras during the early and mid-Cenozoic were all quite similar. Miocene megafossil floras from tropical Africa, Asia, and South America suggest that rain forest was present throughout most of the low-latitude zones, though this inference is undoubtedly an oversimplification (Flenley 1979). More open habitats probably occurred in regions disturbed by recurring volcanism (Bishop 1968; Pickford 1983).

During the early Miocene, the high latitudes of Eurasia were occupied by mixed northern hardwood forest, characterized by large leaves comparable in size to those of the polar deciduous forests of the Eocene, and by leaves larger than those of the midlatitude deciduous trees during the Miocene (Wolfe 1985). During the early–middle Miocene warming trend (18–13 Ma), broad-leaved evergreen vegetation and coniferous forest both expanded slightly at the expense of broad-leaved deciduous forest. After 13 Ma, a decline in temperature from mid-Miocene through Pliocene is associated with a retreat of paleotropical flora in the Northern Hemisphere and expansion of temperate deciduous trees, grasses, composites, and other herbaceous dicots, and of conifers at high altitudes and latitudes. The areal extent of microthermal broad-leaved deciduous forest was gradually reduced during the Neogene and replaced (e.g., in Siberia and Alaska) by coniferous forest, taiga, and birch forest. Forests in southwestern North America and the Mediterranean area were replaced by woodlands during the Miocene (Wolfe 1985).

Angiosperm floras exhibited two important responses to cooling and increased seasonality in the midlatitudes. First, trees and shrubs with deciduous habits became dominant. Second, herbaceous, probably annual, monocots and dicots evolved and diversified; this is thought to correspond to the expansion of grasslands and semiarid to arid conditions (Tiffney 1984). Gramineae and Compositae are known to have woody, putatively more primitive representatives confined to the tropics and subtropics, whereas herbaceous members appear in abundance only in the late Neogene. Chenopods, which diversified primarily during the Pliocene and Quaternary, consist of herbs and shrubs. Pollen of Cyperaceae, a group which has only herbaceous members, also is mainly Pliocene-Quaternary in age (Traverse 1988; Wolfe 1985). A decline in closed forests relative to open vegetation apparently was associated with a decrease in mean diaspore size during the second half of the Tertiary (Tiffney 1984).

EXPANSION OF LOW BIOMASS VEGETATION   The development of all low-standing biomass vegetation types, including savanna, steppe, tundra, and desert, may have been Miocene or younger. Savannas are inferred in Venezuela, southwest North America, eastern Africa, southern Asia, and central Australia during the mid-Miocene; all of these cases occurred in low-to-mid-latitude areas that are now dry. Grasslands and steppes in most midlatitude zones, however, appeared later, as suggested by the continued presence of broad-leaved deciduous forests during the mid-Miocene in Asia and little evidence of extensive grasslands prior to the late Miocene in much of North America (fig. 7.2). The rising Himalayas produced steppe vegetation in central Asia, while steppe developed in western North America during the Pliocene (Wolfe 1985). It appears, however, that massive change to such vegetation in midlatitude North America and Eurasia did not begin until the mid–late Pliocene, possibly coinciding with the arrival of continental ice sheets (Leopold and Wright 1985; Traverse 1982, 1988). Some paleobotanists believe that the North American midcontinent prairie was not well developed until the late Pleistocene (Axelrod 1985). Isotopic studies indicate that grasslands on the scale of the modern Serengeti were not present in East Africa until the middle Pleistocene (Cerling and Hay 1986; Cerling et al. 1988). In northern latitudes tundra is first recorded ca. 2–3 Ma (Wolfe 1985). On the basis of the absence of pollen of Cactaceae in the Tertiary, Wolfe (1985) proposes that true desert vegetation did not occur before the Quaternary, though evidence from northern Africa may indicate otherwise (see § 3.3).

According to Wolfe (1978, 1985), the spread of low-biomass vegetation during the Neogene is linked to two parallel trends in climate: a consistent decline in mean annual temperature in high latitudes and an increase in mean temperature in low latitudes. Consequently, latitudinal temperature gradients were enhanced and subtropical high-pressure systems were intensified, causing summer drought along some continental margins. The spread of steppes in midlatitudes, however, was associated with increased elevation of particular mountain ranges. Pliocene pollen assemblages of North America and Eurasia are characterized by an increase in conifers, though angiosperms are dominant in most of the floras (except in high-latitude zones). This suggests widespread cooler climates but not aridity (Traverse 1988).

During the Quaternary, plants in mid-to-high latitudes migrated with the spread and retreat of glacial ice. The effect on species is thought to have been greater in Europe than in North America, because the east-west trending mountains of Europe are supposed to have limited the southerly migrations of plants, while the north-south oriented ranges of North America may have permitted migration and thus involved lower levels of extinction (Thomas and Spicer 1987). In tropical latitudes, the restriction of some plants to refugia during arid periods may have caused extinction of many tropical species; however, Traverse (1988) points out that no massive changes in floras at the generic level have occurred at low latitudes since pre-Messinian Miocene. This emphasizes the fact that the last part of the Cenozoic has not been a period of unusual floral turnover, in spite of its highly unstable climatic conditions and large shifts in the spatial distribution of vegetation.

Late Cenozoic Animals

NONMAMMALIAN FAUNA

Insects. Essentially all groups of modern insects had their origin by the early Cenozoic, and approximately 75% of the genera known from the mid-Tertiary still exist today (Carpenter and Burnham 1985). Insects, especially ants and termites (McBrearty 1990), comprise a large part of the animal biomass in open vegetation habitats that have developed during the late Cenozoic. The important contribution by insects to these habitats evidently entailed diversification during this interval. The poor fossil record of arthropods remains a vexing problem, with the result that virtually all pre-Quaternary paleoecological studies exclude them from consideration. The existing fossil remains deserve more concerted study, however, and it is likely that additional specimens could be obtained from specific taphonomic contexts (see chap. 2). The usefulness of beetles in interpreting Quaternary climates is discussed by Coope (1967) and Birks and Birks (1980).

Amphibians and Reptiles. Since the early Tertiary, no increase or decrease in the diversity of amphibians or reptiles has been evident on the scale seen in the Paleozoic and Mesozoic, though a slight rise in amphibian diversity apparently has occurred since the Miocene (Carroll 1977). The fundamental functional-ecological attributes of amphibians evolved well before the late Cenozoic, and only 6 of 26 modern families of amphibians with a fossil record had their first appearance in the late Cenozoic.

The freshwater pelomedusid turtles had already appeared by the early Cretaceous, though the advanced chelids (known today by seven genera in South

America, Australia, and New Guinea) are not known prior to the Miocene. The largest-known turtle, the pelomedusid *Stupendemys* from Venezuela, lived during the Pliocene (Wood 1976; Carroll 1988).

The modern functional attributes of lizards had originated by the Cenozoic; certain groups (e.g., varanoids) were represented by very large-bodied forms in the last part of the Cenozoic (e.g., *Megalania* from the Pleistocene of Australia). The basic adaptive characteristics of modern crocodiles evolved by the mid-Mesozoic, and modern crocodylian taxa first occurred in the Cretaceous. The modern crocodiles (suborder Eusuchia) were far more diverse during the early Cenozoic than today, and their decline is attributed to climatic deterioration since the Paleogene (Carroll 1988). Snakes, on the other hand, were a rapidly evolving group of reptiles during the late Cenozoic, and modern families have had major radiations since the early Miocene. Diverse feeding mechanisms were developed that facilitated consumption of warm-blooded prey, swallowed whole using highly mobile upper and lower jaws and palate. The recent evolutionary success of snakes thus appears to be tied to the late Cenozoic diversification of small mammals (Carroll 1988).

Birds.   A number of orders of birds either originated or diversified during the late Cenozoic (see review by Olson 1985; Carroll 1988). Groups that had their origin prior to the Oligocene include the pigeons (Columbiformes), cuckoos (Cuculiformes), fowls (Galliformes), rollers and kingfishers (Coraciiformes), and long-legged wading birds (Charadriiformes). Examples of birds that had their first appearance or radiation during the late Cenozoic are nocturnal forms that possessed gaping mouths for catching insects (Caprimulgiformes); species that possessed specializations for tree-trunk foraging and for grasping twigs (e.g., woodpeckers, parrots); modern forms of swifts; and modern genera of ducks and swans (Olson 1985; Carroll 1988). Passerine birds (Passeriformes) are the dominant avian elements of continental communities today and are first known ca. 30 Ma (Olson 1985). They were not important components of the avian fauna in the Northern Hemisphere until the Miocene, but at that time they rapidly replaced the previously dominant arboreal, perching birds (the Coraciiformes). An overall increase in bird diversity starting in the early Miocene and peaking in the middle Miocene reflects the radiation of passerines and land birds that exploited alpine and xeric habitats (Unwin 1988).

Diurnal predators—falcons, eagles, and hawks—are known in their modern form only since the Miocene, though the earliest members of these raptorial bird taxa date back to the Oligocene and late Eocene in Europe. Owls (Stigiformes), although known in the Paleocene, were more common in the late Tertiary; modern genera appeared in the Miocene, and the largest owl occurred during the Pleistocene. The evolution of enormous vulturelike birds in the New World (Vulturidae and Teratornithidae), which took place from the late Miocene through the Quaternary, involved the development of specialized adaptations for soaring and for consuming very large prey. This development appears to be correlated with the spread of open habitats in North and South America (Carroll 1988; Olson 1985). The evolution of the modern ostrich genus *Struthio* is also associated with open vegetation habitats (where it feeds on plants, insects, and small vertebrates). *Struthio* is known as early as the middle Miocene (Shipman 1986) and ranged throughout Eurasia and Africa during the Pliocene and Pleistocene.

MAMMALIAN FAUNA   Although the number of species of birds and reptiles exceeds that of mammals today and probably has throughout the Cenozoic, the structural and ecological diversification of mammals during the Cenozoic parallels that of the dinosaurs during the Mesozoic (Padian and Clemens 1985). By the early Eocene, most of the living orders of placental mammals were already well represented.

Rodents.   Rodents diversified during the Paleogene, and numerous modern families appeared in the late Eocene and Oligocene. The most diverse group of rodents living today, the muroids, includes two families; the cricetids (voles and hamsters) appeared in the late Eocene (China) and early Oligocene (North America), and the murids (rats and mice) are first known from the middle Miocene (southern Asia).

Primates.   The oldest primates of modern appearance diversified during the Eocene; these were the lemurlike adapids and the tarsierlike omomyids, which were represented extensively throughout North America and Europe. A decrease in diversity of these primates was associated with climatic deterioration toward the end of the Eocene, and only 2 Eocene genera out of 50 or so occurred in the early Oligocene. Primates first appeared in Africa at this time and were present in South America by the late Oligocene. These primates and late Eocene–early Oligocene forms in Southeast Asia represent primitive members of the New World and Old World higher primates. In the Old World, radiations of apelike primates occurred in the early and middle Miocene, and monkeys underwent a major radiation beginning in the mid-Miocene. The largest-bodied forms of Old World apes, monkeys, and prosimians appeared in the Pleistocene to Recent (e.g., the hominoids *Gorilla* and *Gigantopithecus;* the middle Pleistocene *Theropithecus* monkey; the lemur *Megaladapis*). The presence of these forms appears to be related to coarse, high-bulk plant diets. Several different species of bipedal hominid evolved during the late Miocene through early Pleistocene in East and South Africa. Although several lineages within each taxonomic group of Old World primates developed savanna-living forms, none of the South American primates appear to have followed this pattern (Szalay and Delson 1979; Rose and Fleagle 1981; Fleagle 1988).

Perissodactyls.   Perissodactyls, represented today by horses, rhinoceroses, and tapirs, had their main radiation (5 superfamilies and 14 families) during the Eocene (Carroll 1988). In contrast with artiodactyls, they experienced an overall reduction in diversity during the late Cenozoic, although the pattern of decline was not steady. Certain groups—e.g., rhinos, chalicotheres, and horses—experienced periods of diversification from the late Oligocene through the Miocene (Janis 1989). Anatomical modifications in rhinos and horses indicate an increase in cursoriality compared with more primitive ancestors.

Although very large ungulates are known from the end of the Cenozoic, especially the Pleistocene, large-bodied perissodactyls are also known from the Oligocene, including the brontotheres of Asia and North America, the gigantic rhinoceratid *Paraceratherium* (*Baluchitherium*), and the rhino-sized *Diceratherium*. Dental wear suggests that certain of these forms depended upon fruit, and their demise by the end of the Oligocene may mean that fruit resources decreased in abundance (Janis 1989).

Among the most unusual perissodactyls were the large, clawed chalicotheres, including *Chalicotherium* from the Miocene of Europe, which possessed limb proportions and pelvic modifications enabling a semiorthograde stance like that of a gorilla or sloth. This perissodactyl may have browsed from tall trees, using its forelimbs to bring vegetation to its mouth, and/or dug up roots and tubers using claws on its rear limbs (Coombs 1983). Analysis of postcranial bones indicates that although no such large, standing, clawed herbivores live today, certain mammals of the Northern Hemisphere, South America, and Australia converged on this ecomorphic specialization (Janis and Damuth 1991).

The adaptive modifications of equids paralleled the spread of open habitats, with specializations in the limb for running and in the teeth and jaw musculature for transverse chewing motions effective in processing coarse, grassy vegetation. Several clades derived from the middle Miocene genus *Merychippus* exhibited anatomical changes that characterize this overall trend in equid evolution (Janis 1989). Teeth became high crowned, apparently to resist wear from silica particles in grass and from grit that increased in forage as ground-level feeders occupied drier, more open habitats (Janis 1988). In addition, the jaws and face became deeper to accommodate the long roots of the cheek teeth; the high-crowned teeth incorporated cementum; a loph pattern similar to that of modern horses appeared; and the muzzle became elongated and extended the reach of enlarged incisors (Radinsky 1984). Much of the body weight seems to have been supported on the central of three toes. All of these specializations of equids derived from *Merychippus* point to relatively rapid cursoriality and grazing that evolved in concert with an expansion of open vegetation. Miocene diversification of equids resulted in up to 10 contemporaneous species in some faunas in North America (MacFadden 1988) and 8 or more in Eurasia (Bernor et al. 1990). Specializations for grazing and cursoriality were exemplified by species of *Hipparion,* which spread from North America into Eurasia and Africa, and in the equines with the eventual emergence of *Equus* during the Pliocene. *Equus* also dispersed and diversified in both the New and Old World.

Overall diversity patterns in perissodactyls therefore appear to follow climatically induced changes in vegetation. Medium-sized, somewhat cursorial, browsing rhinos diversified during the early Miocene. With increasing aridity and decreasing tree cover during the mid-Miocene, larger-bodied rhinos and grazing equids came to dominate the perissodactyl fauna. Whereas rhinos remained abundant throughout the Miocene, equids have been the dominant perissodactyl worldwide since the late Miocene (Janis 1989).

Artiodactyls.   Artiodactyls were well represented in the Eocene, when they first became evident in the fossil record. Of the 10 modern families of artiodactyls, however, 3 appeared first in the middle-to-late Oligocene, and 7 in the Miocene. The evolution of selenodont dentition (crescent-shaped cusps) in the Eocene coincides with the extensive radiation of the ruminants during this period. Selenodonty has generally been associated with the processing of coarse plant foods, which can be accomplished more effectively with crescent-shaped cusps than with the bunodont cusps of more primitive, more omnivorous artiodactyls. During the early and late Oligocene, the selenodont artiodactyls radiated (oreodonts in North America, anthracotheres in the Old World, traguloids across the Northern Hemisphere) (Janis 1989) and may have formed large herds in the open vegetation areas (Carroll 1988). Camels are known from the late Eocene and early Oligocene fossil record; they display an evolutionary trend toward reduction in toes and fusion of limb bones related to specialized locomotion in open terrain. Yet not until the late Miocene did camels clearly manifest the specialized pattern of limb movement that characterizes living species. This specialized "pacing" is effective in open terrain where maneuverability is less important than in dense vegetation. Changes in camel limb structure in North America coincided with the spread of open grasslands (Webb 1972). The large, long-necked aepycameline camels of North America (late Miocene) were equivalent in their feeding adaptations to giraffes and other high browsers that evolved on other continents, particularly in wooded savannas (Janis and Damuth 1991).

The success of the ruminants (especially the bovids, cervids, antilocaprids, and giraffids) seems to relate to the development of complex guts that permit slow, more thorough extraction of nutrients from plant foods. Janis (1976, 1989) has discussed differences in diet and digestive physiology between hindgut fermenters (e.g., perissodactyls, proboscideans, hyracoids) and foregut fermenters (e.g., ruminant artiodactyls), which she believes allowed the latter to overshadow the former in diversity, and presumably abundance, during the late Cenozoic. Hindgut fermenters rely on rapid passage times and the processing of large quantities of high-fiber foods. Ruminants more thoroughly extract energy from fibrous food but must process it more slowly. Above 1000-kg body mass, there is no advantage to foregut fermentation, and there are no ruminants above this limit. Because of their ability to make maximal use of a limited quantity of vegetation, artiodactyls tend to function better than hindgut fermenters in seasonal habitats where vegetation periodically may be very reduced in abundance, whereas the digestive physiology of perissodactyls tends to function best with consistently large amounts of high-fiber vegetation in less-seasonal habitats (e.g., tropical browse for tapirs and smaller rhinos, fibrous parts of grasses for equids). Janis (1989) suggests that climatic changes during the late Cenozoic happened to favor the spread of vegetation more useful to foregut fermenters. The adaptive strategy of bulk feeding by horses would have restricted the number of grazing equid species that could coexist, while the selective feeding of artiodactyls permitted the diversification of many species that could coexist in the same general area. The maximum diversity (number of families and genera) of artiodactyls worldwide occurred during the Miocene; this coincided with the diversification of larger-bodied, horned forms, as vegetation shifted from forests to more open woodland, and with the development of savanna-mosaic habitats (Janis 1988, 1989).

In brief, the artiodactyls have exhibited (1) development of selenodont cheek teeth; (2) a tendency toward foregut breakdown of cellulose; (3) in parallel with the equids, a recurrent trend toward cursoriality, shown by the development of digitigrade features and controlled dorsoflexion and extension of the limbs (related to, for example, a "double pulley" modification of the astragalus and loss of the clavicle) (Eisenberg 1981; Janis 1976; Carroll 1988). Major ecological variants within the artiodactyls today range from nonruminant, omnivorous forms (certain suids); long-necked browsing specialists (giraffids, camelids); small, primitive ruminants (tragulids); numerous mixed browsing-grazing forms (some bovids, cervids, and antilocaprids); to the specialized grazing bovids.

Proboscideans.   The proboscideans (gomphotheres, elephantids, deinotheres, stegodonts, and mastodonts) are some of the most spectacular mammals of the late Cenozoic. Since the appearance of early forms in the Eocene of Africa and Asia, proboscideans have shown persistent tendencies (1) to increase in size by lengthening the proximal rather than distal limb bones; (2) to elongate the upper lip, which in many extinct and modern proboscideans combined with the nostrils to form a muscular trunk used in feeding; and (3) to increase the size of the second incisors to form tusks. As the gomphotheres diversified, first in Africa during the Oligocene and then in Eurasia and North and South America during the Miocene and Pliocene, a number of subfamilies ultimately expressed morphological trends similar to those of the elephantids, including loss of tusks, elongation of molars, and deepening of crania. The elephantids emerged by at least the late Miocene and are characterized by elongation of the molar teeth so that in later forms the anterior portion erupts as the posterior part is still forming. The mastodonts and stegodonts lost the lower tusks and showed elaborations of serial plate structure in their molars (transverse enamel ridges), which also is characteristic of the elephantids

(Carroll 1988; Eisenberg 1981). Deinotheres appeared in the early Miocene in Asia and continued in Africa and Southeast Asia into the Pleistocene (Savage and Russell 1983). They showed no trends toward hypsodonty related to grazing, as exhibited by other proboscidean groups (Harris 1983).

Creodonts and Carnivores.   Early in the Cenozoic two major groups of terrestrial carnivores diversified, the Creodonta and the early Carnivora. Although creodonts were common in the Eocene, they were greatly reduced in number at the end of the Eocene except in Africa, where the hyaenodontids were the dominant carnivores until the Miocene. They persisted into the late Miocene (8–9 Ma) only in southern Asia (J. C. Barry pers. comm. 1990). The hyaenodontids included genera with body proportions comparable to those of felids, canids, and hyaenids (Carroll 1988). Outside of Africa, the carnivore guild of the Oligocene included some of the earliest large fissiped carnivores. Despite the presence of the ursid-related amphicyonids, Van Valkenburgh (1985) concludes that there were no functionally equivalent bearlike predators in North America during the Orellan (early Oligocene), nor were there any wolflike predators. This implies that early carnivore guilds were ecomorphically less diverse than modern guilds. Oligocene carnivores apparently also were slower and had more robust limb bones than their modern counterparts (Van Valkenburgh 1985).

Most carnivores known from the Oligocene onwards belong to modern families that occupy distinctive adaptive zones on the basis of jaw mechanics and method of prey capture, which is thought to relate primarily to prey body size (Radinsky 1982). The two major extinct families of the Carnivora of the late Cenozoic are the amphicyonids and nimravids, represented by medium to very large genera from the early Oligocene to early Pliocene in North America, Africa, and Eurasia. Canids, recognized in the early Oligocene in North America, did not appear in Africa and South America until the Plio-Pleistocene, and hyaenids first occurred in the African Miocene. Procyonids and ursids emerged in the early and middle Oligocene, respectively. The giant panda (*Ailuropoda*), which became an important part of East Asian Pleistocene faunas, is first recognized during the late Miocene in Europe. Since the early Oligocene, the cats have exhibited two distinct dental morphologies: relatively short upper canines and highly developed saber-toothed upper canines (with an associated flange on the mandible). During the early Oligocene the catlike nimravids all possessed elongated upper canines, whereas the modern felids (known since the late Miocene) have exhibited both short and saber-toothed canine morphologies. The reiteration of the latter specialization in several lineages may be related to modes of attack and feeding that were directed at the soft tissues of relatively large prey. With the extinction of many of the large herbivores at the end of the Pleistocene, the last saber-toothed cats also disappeared.

Large carnivores such as the felids became specialized for short rush, leap, and quick-killing bite as their form of prey capture. The ability to kill large prey in this manner provided an opportunity for partial scavengers, including small forms (small canids) to moderately large forms (hyaenids). Several lineages of canids developed prey-capturing techniques that involved socially cooperative chasing of open-country, large ungulates (Eisenberg 1981).

As grasslands expanded during the late Miocene to Quaternary interval, the diversification of large cursorial grazing herbivores should have favored the evolution of carnivores that were larger and faster than their mid-Tertiary predecessors. Carnivores with limb proportions like those of modern forms that pursue prey, however, did not appear before the Plio-Pleistocene, which means that a time lag occurred between the appearance of cursorial ungulates and carnivores. Cursoriality in large ungulates and carnivores, rather than being tightly coupled, may have developed as these animals separately expanded their foraging areas in response to increasingly open habitats (Janis in press).

Small carnivores also have diversified greatly in diet and habitat since the late Miocene. Specialization for preying upon small herbivores may have been closely coupled with the radiation of cryptic rodents and birds, which was also favored as grasslands spread.

The absence of large carnivorous mammals in certain terrestrial ecosystems is associated with unusual developments in other organisms, especially birds. For instance, no cursorial carnivores evolved among the native marsupials of South America until late in the Cenozoic. Their role was presumably taken by large (almost 3 m tall in some forms) flightless birds (Phorusrhacidae) known from the early Oligocene to late Pliocene. During the Pleistocene, the flightless elephant birds of Madagascar and the moas of New Zealand appear to have been the dominant large herbivorous browsers, perhaps because of the absence of large herbivorous mammals. The absence of specializations for speedy escape in these birds may have been related to the original absence of large predators in their island habitats, in contrast with continental situations where the fast-running ostriches, rheas, and cassowaries evolved (Carroll 1988). Adaptation to closed, forest habitats could also have affected locomotor specializations in the *Moa* (Reif and Silyn-Roberts 1987) and other large avian herbivores.

2.3 Diversity Patterns

The number of taxa in the fossil record and the geographic pattern of alpha diversity have been related to several ecological factors. These include density-dependent origination and extinction, coevolution between taxa, and habitat structure and heterogeneity on local, regional, continental, and global scales (e.g., Gingerich 1984, 1987; Stucky 1990). Analysis of late Cenozoic fossil assemblages has helped to place long-term diversity patterns (for mammals only, so far) on an empirical basis.

Evaluation of taxonomic diversity during the Cenozoic has focused on North America because of the rich fossil record on this continent from the Paleocene to the Recent. Stucky (1990) documents North American diversity patterns that appear to be relevant to the general paleoecological history of the Cenozoic. After increasing greatly in the early Paleocene, the number of genera appears to have been stable on a continental scale from the mid-Paleocene to the Recent, when geographic and temporal sampling biases are taken into account. Despite this stability on a continental scale, any given late Cenozoic fossil assemblage tends to contain fewer taxa than does an average assemblage from the middle Paleocene through the middle Eocene. These same late Cenozoic assemblages, however, also tend to have greater variation between them in terms of taxa represented. Thus, there appears to be a distinct increase in faunal provinciality from the mid-Eocene onward. Overall climatic cooling and drying during the past 34 my probably had a complex influence on habitat diversity and niche space on local as opposed to continental scales. The overall effect based on mammalian diversity patterns may have been a reduction in niche space (number of niches) on a local scale but greater regional differences in ecology over the continent as a whole (Stucky 1990).

If such differences in diversity according to geographic scale are confirmed by further study, especially on other continents, it would be reasonable to relate these to (1) local decreases in productivity associated with climatic deterioration, (2) less-equable climates after the mid-Eocene (causing accentuated latitudinal vegetation zones), and (3) the enhanced effects of regional tectonic uplift during the late Cenozoic. The nonrandom temporal pattern of diversification of mammals during the Cenozoic appears to correspond to worldwide climatic change, notably cooling (Gingerich 1987). The potential for greater geographic coverage and resolution within the late Cenozoic and the effects of specimen sample sizes will need careful evaluation in testing the ecological significance of large-scale taxonomic diversity patterns.

3   ECOLOGICAL HISTORIES BY CONTINENT

3.1   North America

Introduction

North America abounds with Cenozoic fossil localities, and these have been subject to intensive study throughout the past century. Consequently, there is a rich faunal and floral record of regional and broader-scale ecological change that expands in quantity and improves in quality as we approach the present. The Cenozoic history of the floras and mammalian faunas of North America has been reviewed extensively in recent years (e.g., Wolfe 1981, 1987; Woodburne 1987; Savage and Russell 1983; Martin and Klein 1984; Graham et al. 1987; Leopold and Denton 1987), and our intention in the following pages is

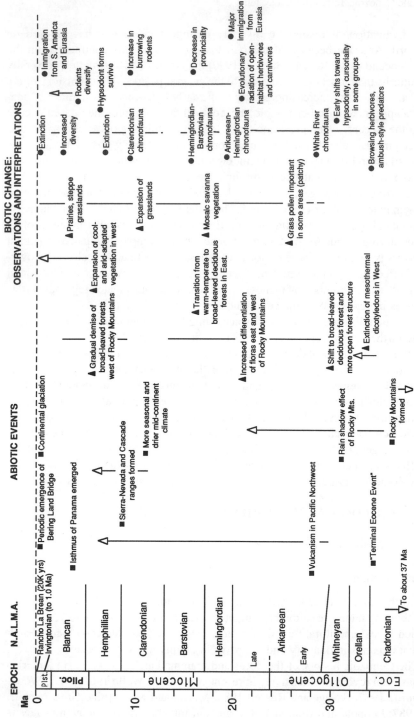

Fig. 7.3. Overview of abiotic and biotic changes in North America during the late Cenozoic. N.A.L.M.A. refers to North American Land Mammal Ages. Symbols same as in figure 7.2.

to highlight the ecological trends rather than to reiterate descriptive detail provided by other publications. An overview of changes in the physical environment and the floras and faunas that are discussed below is given in figure 7.3.

The North American continent has moved relatively little during the past 34 my (fig. 7.1). The major tectonic events involve the emergence of mountain ranges along the west coast, further evolution of the Basin and Range and Rocky Mountain provinces, continuing volcanism in the Pacific Northwest, and the progressive elevation of the midcontinent high plains. The modern flora consists primarily of conifer and deciduous forests and woodlands (east, west, north, and higher altitudes), midcontinent grasslands and steppes, Mediterranean-type sclerophylous woodlands (west coast), vast areas of tundra in the high latitudes, and regions of desert and subtropical forest in the lower latitudes. The faunas include a range of large herbivores (bovid, cervid, antilocaprid, caprid, tayassuid; see table A-7.2) and carnivores (canid, mustelid, ursid, felid), as well as a wide diversity of smaller vertebrates (especially cricetids). The modern mammalian megafauna is generally regarded as depauperate due to the late Quaternary loss of proboscideans, camelids, Xenarthra (ground sloths–megalonychids, megatheriids), and equids, although new species, including *Homo sapiens,* also immigrated across the Bering land bridge during the past 20–40 ka.

There are a number of recurring themes throughout the Oligocene-Quaternary interval. One is the important role played by periodically emergent bridges to Eurasia and to South America. North America has a long history of invasion and emigration of animals and plants, and the bridges also have acted as filters, at least in part because of their latitudinal positions. As a result of its contact with Beringia, North America has maintained faunas and floras with strong Holarctic affinities. Another feature of the late Cenozoic is the importance of the north-south-trending mountain belts, especially the Rocky Mountains and the Sierra Nevada, which have served as strong regional modifiers of climate as well as biogeographic barriers.

The ecomorphs that make up modern faunas were well established by the Miocene, and most of the North American mammal species that existed during the last part of the Cenozoic would appear familiar in general shape and form to a modern human observer. Our assessment of animals of any geologic age as "catlike" or "rhinolike" is based on this knowledge of Recent faunas, and taxonomic nomenclature at the family, tribe, or generic level often is used as shorthand for familiar ecomorphs in Cenozoic faunas. We will follow this practice in characterizing the Cenozoic mammal communities in terms of constituent families, and table A-7.2 provides a key to their ecomorphic features.

Pre-Quaternary Floras

Eocene floras record fluctuating temperatures following a global high in the late Early Eocene. The uplift of the Rocky Mountains helped to modify climates on a local to regional scale, and the midcontinent may have experienced drier conditions and considerable seasonal temperature fluctuations during the middle–late Eocene (Wing in press). In spite of this, broad-leaved evergreen forests persisted over much of the continent. During the late Eocene–early Oligocene (33–34 Ma), however, there was a marked cooling of mid-to-high latitudes, referred to as the "terminal Eocene event" (Pomerol and Premoli-Silva 1986a, 1986b; Swisher and Prothero 1990; Prothero and Berggren in press). The cooling event is correlated by Wolfe (1987, and references therein) with a phase of extinction for mesothermal evergreen and deciduous dicotyledons, particularly in western North America. The effects of the event are not obvious in midcontinent floras, however, perhaps because these taxa already were adapted to somewhat cooler and drier conditions. The shift from broad-leaved evergreen to broad-leaved deciduous forest and to a more open forest structure continued throughout the Oligocene.

From the Oligocene through the mid-Miocene, a major differentiation of vegetation types coincided with the Rocky Mountain axis. Areas immediately to the east of this axis bear evidence of limited, low-diversity forests, and xeric shrub–herb communities with little or no open grassland (e.g., the Kilgore flora of Nebraska: MacGinitie 1962; Leopold and Denton 1987; E. Leopold pers. comm. 1991; and the pollen record of Colorado: Leopold and MacGinitie 1972). Forests apparently were limited to sheltered valleys and riparian environments. The relatively open character of the region can be attributed to the rain-shadow effect of the Rockies. In contrast, west of the Rockies mesic forest mosaics were dominant, probably as a result of ample rainfall generated by moisture-laden winds from the Pacific. Open grass–forb "patches" were limited in scale except in areas of local volcanic activity, where they were extensive but of short duration. Such areas supported grazer-dominated faunas whose remains are more common than those of forest browsers, perhaps because active volcanism enhanced preservation (see chap. 2), as well as promoting early successional open vegetation. In eastern North America, the Oligocene record is very sparse, but apparently there were subtropical to warm-temperate forests with a mix of evergreen and deciduous species (Tiffney 1985). These gave way in the Miocene to broad-leaved deciduous forests similar to those of today, though the former were more diverse and included a number of warm-climate genera now found only in East Asia, Mexico, and Central America. Some of these taxa survived until they were eliminated by the cold phases of the Plio-Pleistocene climatic cycles (McCartan et al. 1990).

Grasses are documented as rare elements in Paleocene and Eocene pollen spectra and the megafossil record (Crepet and Feldman 1988) and were present much earlier than the spread of grasslands. By the Miocene, grass is an important component in some pollen records (Leopold and Denton 1987), although it is oddly rare in samples that supposedly document savanna vegetation in the midcontinent (Leopold and Denton 1987). In the Pacific Northwest it occurs only in volcanically disturbed areas from the Oligocene to the mid-Miocene. These areas preserve seral stages in plant communities that appear to have followed volcanically induced fires, with mixed mesophytic forms being the first to return after a major burn (Taggart et al. 1982).

In the Late Miocene, the Sierra Nevadan–Cascadian orogeny brought increasing aridity to the intermontane region west of the Rockies and led to the gradual demise of broad-leaved forests. Developing aridity in the southwest created an additional barrier, and extensive forests were restricted to a narrow zone west of the Cascades. Conifers that were able to exploit the moister, cooler uplands remained diverse, but the broad-leaved taxa suffered great losses. Cool-adapted sagebrush steppe vegetation developed in the northern part of the area. Many elements of the modern vegetation of northwestern North America were important components of the earlier, seral floras in the volcanically disturbed areas.

Thus it appears that modernization of the vegetation in western North America involved climatic exclusion of broad-leaved forest taxa, in situ evolution, and geographic expansion of xeric-adapted forms, including grasses, immigration of new taxa, including Eurasian microthermal forms (Wolfe 1987), and a shift in dominance of previous seral vegetation types, which became climax elements in the new xeric floral communities (Taggart et al. 1982).

The floral record of the mid-Miocene through the Pliocene is less well documented over much of the rest of the North American continent. Mid-Miocene Hemphillian floras in disturbed volcanic areas of Wyoming show a steppe-like seral sequence, indicating drier conditions than in the Pacific Northwest at the same time. Grasslands are indicated by macrofloral records in Nebraska, where a savanna-woodland environment has been inferred on the basis of the mid-Miocene Kilgore flora (MacGinitie 1962; Leopold and Denton 1987) and plant remains preserved in the mouths and abdominal cavities of a Pliocene *Teleoceros* herd that perished in a volcanic ash fall (Voorhies and Thomasson 1979; Thomasson et al. 1986).

Pre-Quaternary Faunas

The North American mammalian record is characterized by waves of immigration interspersed with periods of endemism throughout the past 34 my (Tedford et al. 1987). Only about 20% of the modern species can be traced to endemic Eocene ancestors; the familiar animals of today are largely an "expatriate" fauna with foreign origins. In general the periods of immigration were not marked by immediate replacement of endemic forms, and it appears that immigrants were able to find unused resources or otherwise become integrated with the existing fauna without disrupting it (Tedford et al. 1987). This contrasts with the pattern of rapid extinction and replacement in South America (at least for several groups) when immigrants arrived from North America (see § 3.2). Although immigration events are not obviously correlated in time with distinct pulses of extinction, the history of the savanna or grassland vertebrate fauna in North America in a broad sense is one of extinctions and replacements of similar ecomorphs from different lineages, with a directional trend toward increasing cursoriality and size. This pattern of change characterized the progressive adaptation of mammalian communities to open habitats.

The North American Land Mammal Ages (NALMA) are recognized primarily on the basis of first appearance datums and stratigraphic ranges of notable taxa. They reflect immigration, evolution, and community turnover, and they also have been influenced to some extent by the historical development of vertebrate paleontology in North America. Thus the major periods of ecological change in vertebrate faunas, as currently understood, often do not coincide with boundaries between Mammal Ages. These ages provide a temporal reference system, however, which is used in the following discussion of late Cenozoic faunal history (fig. 7.3).

OLIGOCENE   Fifteen mammalian orders recorded in the latest Eocene Chadronian NALMA include archaic groups such as multituberculates and pantolestids as well as diverse artiodactyls (10 families with as many as 32 genera) and perissodactyls (6 families with 18 genera) (Savage and Russell 1983). Merycoidodonts (oreodonts) were the most diverse artiodactyl group at this time, and rhinoceratids and brontotheres were the common perissodactyls. Early forms of amphicyonids, canids, camels, tayassuids, protoceratids, and anthracotheres were present. There were several suidlike lineages, including large entelodonts, anthracotheres, and tayassuids. Rodents were diverse, with 7–10 families, including early cricetids, heteromyids, and castorids. Although limb modifications for faster locomotion already were occurring in a number of groups by the early Oligocene, ungulates adapted to woodland browsing, such as oreodonts, and carnivores adapted to ambush-style predation, such as amphicyonids, dominated the large mammal fauna until the early Miocene (Van Valkenburgh 1988, in press). Much of the Oligocene record is derived from midcontinent deposits, especially the White River Group, which is typified by the Big Badlands exposures in South Dakota. In this region, artiodactyls, perissodactyls (especially horses), and fossorial rodents began to radiate in the early Oligocene, indicating a relative increase in the proportion of more open (e.g., woodland, bush) as opposed to forested habitats.

By the early Oligocene Orellan NALMA, some of the archaic orders had disappeared (pantolestids, multituberculates). Brontotheres went from approximately 6 genera to none, while oreodonts continued with high diversity (~13 genera compared with 11 in the Chadronian, probably inflated by oversplitting). The Orellan fauna has been recovered from the Orellan (or Scenic) Member of the Brule Formation, which has been extensively studied in terms of sedimentology, taphonomy, and paleosols (Clark et al. 1967; Retallack 1983, 1984). Although it spans a relatively short interval of time (approximately 2 my; Emry et al. 1987; Swisher and Prothero 1990), there was observable change within the rapidly evolving oreodonts. The late middle Oligocene Whitneyan NALMA was characterized by continued high diversity of oreodonts, morphological change and species turnover in some of the other groups (e.g., equids, camelids), and a reduction in creodont species. Rodents shifted from a dominance of eomyids (Chadronian) to cricetids (Orellan) to heteromyids (Whitneyan). The early Arikareean NALMA (late Oligocene) had diverse oreodonts (14 genera) as well as a large number of rodent groups.

Overall the Oligocene midcontinent sequence can be regarded as representing a stable chronofauna (the White River chronofauna) throughout this epoch (Emry et al. 1987). Even with the changes mentioned above, the fauna is a recognizable entity for approximately 10 my, until major reorganization in the middle of the Arikareean stage. This continuity is thought to indicate relatively stable ecological conditions (Emry et al. 1987).

There is continuing debate concerning the vegetation that characterized the midcontinental Oligocene ecosystems. Plant remains other than root casts and hackberry fruits are generally lacking in the White River Group, and reconstructions are based either on the functional morphology of the mammals or on interpretations of the paleosols. On the basis of the latter, Retallack (1983) infers widespread woodlands in Chadronian time, and more open patches of savanna with grass and small shrubs by the Whitneyan. Earlier work by Clark et al. (1967) supported similar inferences concerning increasingly open habitats based on the mammalian taxa and their distribution in the various floodplain facies. This evidence conflicts, however, with the palynological record, which indicates that grass was rare until the Miocene (E. Leopold, pers. comm., 1991).

Hypsodont oreodonts thought to be adapted to more open habitats (the leptauchenine oreodonts of 30–24 Ma; Schultz and Falkenbach 1968) are common in the late Oligocene of the midcontinent. They may have occurred even earlier in the volcanically disturbed areas in the Pacific northwest than in the midcontinent, suggesting that adaptation to more open habitats was taking place in drier, isolated areas west of the Rockies prior to grassland expansion in the midcontinent. In the midcontinent region, more burrowing mammals appeared in the early Miocene (Martin 1974, 1987) at about the same time that browsing oreodonts were declining in diversity. Oreodonts apparently survived longer in the Pacific Northwest (i.e., into the Barstovian NALMA) but were not dominant after the early Miocene.

EARLY TO MIDDLE MIOCENE   The Miocene fossil record in North America represents a variety of ecological settings, including coastal plains and mountains, streamsides and lake margins, hilly to submountain uplands, and open savannas or prairies. The western part of the continent experienced considerable fallout of volcanic ash, which forms a significant component of fossil-bearing deposits throughout the Miocene (Tedford et al. 1987; Hunt 1990). Resulting changes in soil texture and composition are thought to have contributed to the spread of grasslands, and increased grit may have promoted herbivore hypsodonty (Savage and Russell 1983). The Late Arikareean faunal substage represents the earliest Miocene, and the composite continental fauna included 7 orders and 31 families (Savage and Russell 1983). Oreodonts were still relatively numerous (18 genera), camelids and tayassuids had expanded in generic diversity, and cervids emerged as a new artiodactyl family. Perissodactyls included a number of new equid genera as well as tapirs, rhinos, and chalicotheres. Dominance in generic diversity in the Rodentia shifted to geomyids; sciurids and castorids increased as well (Savage and Russell 1983). Among the Carnivora, canids, amphicyonids, and ursids showed a marked increase in genera across the Oligocene-Miocene boundary, with more cursorial forms adapted to pursuit and capture of fast-running prey in open habitats (Van Valkenburgh 1988).

Evolutionary radiation in groups such as the camelids, equids, cervids, canids, and burrowing geomyid rodents indicate the expansion of more open vegetation at the beginning of the Miocene. The continued abundance of oreodonts and other less cursorial and browsing forms is evidence for the persistence of forest and woodland habitats as well. Miocene floras from the mid-continent show high variability in the proportion of grass in pollen, suggesting a mosaic of open and closed habitats that supported a diverse array of vertebrate ecomorphs. All evidence combined provides a picture of forest, woodland, and grassland biomes at an intermediate stage in the late Cenozoic trend toward more open habitats. This extended through the early Miocene (late Arikareean–early Hemingfordian), over a period of about 5 my. The higher silica content of grass, greater percentage of fiber during dry seasons, and also increased grit in ground-level forage required more durable herbivore teeth. Hypsodonty appears to have been the most typical early adaptation among mammalian herbivores to the expanding open-habitat plant communities. This was followed by increased cursoriality and by more complex social behavior, indicated by the evolution of various types of horns and antlers in some artiodactyl groups (Janis 1982), as well as herding, which is suggested by bone accumulations dominated by single species with age profiles similar to those of modern herding animals (e.g., Voorhies 1969; Voorhies and Thomasson 1979).

The early through middle Miocene is the first of two major periods of immigration into North America via a Bering land connection (the second being in the Pleistocene). Many of the successful invasions apparently occurred during periods of relatively low diversity in the endemic fauna (Webb 1989b). The interval between 23 and 15 Ma includes two major chronofaunas, separated by a turnover in the middle Hemingfordian at about 18 Ma (Tedford et al. 1987). The late Arikareean–early Hemingfordian chronofauna included a wave of immigrants that marks the lower Hemingfordian boundary. These taxa, consisting of various carnivores, a rhinoceros, shrews, and a lagomorph, apparently caused little disruption of the existing late Arikareean fauna (Tedford et al. 1987). The late Hemingfordian–early Barstovian chronofauna is characterized by the continued radiation of horses and cervids and a reduced diversity of oreodonts. Oreodonts and camelids occur in considerable abundance in individual assemblages, however, in spite of lower overall diversity (Savage and Russell 1983). This implies the continued success of these groups in localized portions of the overall habitat mosaic. Cat-like carnivores disappeared temporarily from North America in the early Miocene, their roles being partially filled by mustelids, amphicyonids, and canids. In the late Hemingfordian, antilocaprids and felids immigrated to North America, and new rhinoceratids and cricetid rodents joined the chronofauna as immigrants in the Barstovian. As in the preceding late Arikareean–early Hemingfordian fauna, the invasion or evolution of new forms was not accompanied by major turnover in the existing fauna. This chronofauna persisted for 4–5 my, until about 14.5 Ma (Tedford et al. 1987).

The invasion of proboscidea from Eurasia marks the beginning of the Clarendonian chronofauna, which persisted from the late Barstovian through the early Hemphillian, to about 6 Ma. This chronofauna spans the late Barstovian to Hemphillian NALMA and had the greatest generic diversity of mammals in the late Cenozoic of North America (Webb 1983a, 1989b; Stucky 1990). By the late Barstovian, 26 of the total of 35 families represented groups that still exist in North America (Savage and Russell 1983), but the ecological picture of diverse proboscidean, equid, camelid, rhinoceratid, and antilocaprid species occupying woodland savanna habitats contrasts strongly with modern North American mammal communities (Webb 1977, 1983a, 1989b). The Clarendonian fauna appears to have been remarkably similar throughout the continent as a whole, even though a wide range of local habitats is represented in the fossil record (Webb 1989b; Stucky 1990; Alroy in press).

During the 2.5 my spanned by the Clarendonian NALMA (11.5–9.0 Ma), 34 families of mammals are recorded in North America. Eight of them belonged to the Artiodactyla, including camelids, tayassuids, cervids, paleomerycids, antilocaprids, and 2 less diverse groups. Only one genus of oreodont survived into the Clarendonian. Within the large mammals, equids were most diverse, with 15 genera, followed by mustelids (11) and canids (9) (Savage and Russell 1983). Rodents comprised 9 families, with sciurids being generically dominant. The number of genera of savanna mammals for North America in the later Miocene is comparable to total diversity for the modern African savannas (Webb 1983b); the savannas include a mosaic of open to closed habitats rather than a continuous grassland. The Clarendonian fauna also was characterized by a number of striking morphological convergences on present-day members of the African savanna community. Among these were a long-necked giraffe-like camelid (*Aepycamelus*), a gazelle-like equid (*Nannipus*), and a hippo-like rhinoceratid (*Teleoceros*) (Webb 1983a; Savage and Russell 1983). Although this was the period of greatest ecomorphic similarity to African savanna faunas, there were important differences as well, such as fewer taxa with cranial appendages in the North American fauna (Janis 1982).

During the middle to late Miocene, members of most of the herbivore families exhibited parallel trends toward increased hypsodonty. The more hypsodont subgroups also tended to increase in diversity with time, indicating that they were favored by expanding grassland environments (Webb 1977). More burrowing rodents appeared around 10 Ma, concurrent with an increase in soil carbonate and caliches.

All of these trends indicate a long-term change toward drier climatic conditions (Webb 1977). Climatic changes that promoted the development of more open habitats probably included increases in the intensity of seasonal drought as well as changes in absolute amounts of rainfall. With drier seasonal conditions, fire probably also played a role in promoting grasslands (Axelrod 1985). It seems likely that during the transition to continuous expanses of open grassland, different kinds of vegetation (forest, woodland, grassland) expanded and contracted in area, probably with numerous local-to-regional reversals in the overall trend toward grassland. Inferences from the mammalian fauna suggest that phytogeographic diversity was highest in the mid-Miocene in North America (Webb 1977). In modern settings maximum diversity in vegetation occurs in heterogeneous settings. Although direct evidence for the small-scale spatial patterning of the Miocene paleofloras is lacking, the pollen record suggests that they were highly patchy (Leopold and Denton 1987), presenting many options for herbivore (and carnivore) evolution. The establishment of more homogeneous, extensive prairie communities appears to have been delayed until the latest Miocene to early Pliocene, although component taxa were present earlier in the record (Webb 1977, 1989b).

It is interesting to note, as Webb (1977) points out, that the overall trends toward hypsodonty and cursoriality continued through the succession of three faunas used by Tedford et al. (1987) to characterize the North American Miocene. Although turnovers and immigrations changed the taxonomic composition of the Miocene fauna as a whole, morphological responses to the overall transformation of climate and vegetation continued throughout the Miocene, during a period of approximately 17 my. This change toward dominance of savannas and grasslands was very slow by present-day ecological standards, and a diverse mosaic of habitats persisted throughout the early and middle Miocene. The most pronounced adaptations to open habitats occurred primarily within endemic North American lineages such as the equids, camels, canids, paleomerycids, and heteromyid and mylagaulid rodents. Of the immigrant lineages, only antilocaprids exhibit comparable cursoriality and hypsodonty (Webb 1977). Others, such as the cervids, ochotonid lagomorphs, cricetid and zapodid rodents, and proboscideans, successfully colonized the closed habitats of the savanna mosaic but did not radiate significantly into the more open environments during the Miocene. This suggests that the existing niches were already occupied by ecomorphs comparable to those of the immigrants and that ecological opportunities for the invaders were too limited to permit evolutionary radiation.

Much of the Miocene history of North America is based on the faunas of the Great Plains. In other parts of the continent, somewhat different floral and faunal associations attest to the provincial effects of climate and geography. The early Miocene coastal areas of the Gulf of Mexico included a number of endemic artiodactyls and lower vertebrates. Many of the endemic taxa in this region disappeared by the Barstovian, and the fauna became similar to that of the Great Plains (Webb 1977; Alroy in press). Some provincial faunal elements remained, including two dwarf, apparently forest-dwelling rhinoceroses from the middle Miocene Gulf Coast deposits of Texas (Prothero and Sereno 1982). The Miocene floras of the Pacific Northwest represent a warm-temperate, mesophytic forest with a fluctuating but progressive reduction in the proportion of entire-margined leaves, indicating cooler climatic conditions through the Miocene (Wolfe and Hopkins 1967; Wolfe 1987). Vertebrate faunas of this region imply a spectrum of habitats from forest to more open woodland, and the more open habitat faunal association is increasingly represented south of the Columbia Plateau (Webb 1977). The plant record from the interior Pacific Northwest indicates that there was no significant summer dryness (i.e., no loss of leaves for much of the flora) until the later Miocene, when many broad-leaved forms were lost and grasses became dominant (Leopold and Denton 1987). In Mexico and the southwest United States, middle Miocene faunas included a number of precociously hypsodont forms, and floral records suggest a savanna corridor linking the southwestern part of the continent with the Great Plains (Webb 1977). By Clarendonian times, however, increased aridity in the west may have begun to limit faunal interchange within this large biogeographic province. Central America had one apparently endemic gomphothere but no other record of a locally distinct fauna. Sloths and a toxodont emigrated from South America in the late Miocene, and sloths expanded into the Gulf Coast and southern Great Plains (Webb and Perrigo 1984; Tedford et al. 1987).

In the Greater Antilles, mammalian taxa derived from Late Cretaceous or Early Cenozoic forms contributed to unique island biotas. The eastward movement of the Caribbean lithospheric plate gradually cut off direct contact with North and Central America. Mainland representatives of the soricomorph insectivores became extinct by the end of the Oligocene, but island representatives (including *Nesophontes* and giant forms of *Solenodon*) survived and diversified throughout the Cenozoic (MacFadden 1980). They were part of a restricted island fauna consisting of primates, bats, rodents, edentates, and insectivores. Most of the fossil evidence for these faunas comes from Pleistocene cave deposits. At least 65 mammal species are recorded as extinct, based on the Quaternary record (Rosen 1975). The absence of mammalian carnivores in the faunas may be a result of sampling or taphonomic bias, but it is also possible that their absence on the islands was real, and that predatory birds occupied the carnivore niches (MacFadden 1980).

LATE MIOCENE-PLIOCENE   The vertebrate record for the period between about 12 and 4 Ma (late Miocene–Pliocene) shows a precipitous decrease in predator and herbivore diversity through a series of extinction events. The Clarendonian chronofauna lost a total of 44 genera of large mammals (Webb 1983a, 1983b). There were comparatively few immigrations during the period of extinction, and the endemic fauna did not generate replacements for most of the missing members of the savanna community (Webb 1983a, 1983b). The result was a large mammal fauna of considerably lower diversity by about 9 Ma (Tedford et al. 1987; Alroy in press). Extinction without replacement by ecologically equivalent taxa continued in this fauna, and by the end of the Hemphillian age, diversity had fallen to half of what it was 5 million years earlier (Webb 1983a). Stucky (1990) calculates an impressive extinction rate of 41 genera per million years (my) in the late Hemphillian, compared to 9/my in the early Hemphillian. In many of the herbivore groups, such as the equids and camelids, the lower-crowned, browsing forms disappeared, while the hypsodont forms survived. Equid genera dropped from over a dozen in the Clarendonian to less than 5 progressively hypsodont lineages in the late Hemphillian (Webb 1983a). The demise of the "Clarendonian chronofauna" is attributed by Webb (1983a) to the loss of the diverse mosaic of savanna vegetation and the progressive dominance of grassland and steppe, caused by increasingly cool and less equable climatic conditions over much of North America. Interestingly, the bird and herpetofaunas were not similarly affected, and most of the genera present in the Clarendonian are still extant today (Webb 1989b).

After the Late Miocene extinctions of large mammals, diversity increased again in the Plio-Pleistocene, largely through migrations across the Bering land bridge. The higher rate of immigration is attributed by Tedford et al. (1987) to the availability of dispersal routes from Asia and South America and also to increasing environmental instability at high latitudes. The early Blancan fauna (4.2–3.2 Ma) is striking for its limited number of camelids, tayassuids, cervids, antilocaprids, and equids, compared with the Hemphillian fauna. Tapirids, rhinoceratids, and four artiodactyl families had disappeared, including the last of the oreodonts (Savage and Russell 1983). Many of the early Blancan immigrants were carnivores and rodents. In the later Blancan, approximately 3.2−1.8 Ma, familiar Eurasian and African forms, such as hyaena, appeared, and a number of xenarthrans immigrated from South America (Savage and Russell 1983). Tapirids reappeared, presumably from Central America. Of a total of 76 new genera recognized in this interval, about 10% were immigrants from South or Central America and about 22% from Asia (Savage and Russell 1983). Rodents diversified further and became the dominant herbivores (in terms of numbers of species) in the grassland biome (Webb 1989b). Overall continental species diversity increased in the middle Blancan to a level comparable to that of the Clarendonian, before falling again at the end of the Blancan (Stucky 1990; Alroy in press).

Quaternary Floras

The early to middle Quaternary has only a few well-known floras and is poorly known in terms of continental-scale phytogeography. The late Quaternary spans the last 130 ka and is best known in terms of paleoclimate, glacial history, and paleobotany. It includes the last interglacial, the Wisconsin (Weichselian of Eurasia) glacial age, and the current interglacial. The Wisconsin glacial interval began about 70 ka, with glacial maxima at ca. 65 and 18 ka. The late Quaternary fossil record shows a massive flux of plant life, demonstrating the ephemeral nature of floral associations (communities) in an environment characterized by major climatic changes. In northern latitudes, boreal ecosystems were largely overridden by continental ice sheets; those that remained supported tundra vegetation rather than forest. At midlatitudes, temperate ecosystems gave way to boreal communities unlike those of today in terms of species composition. Farther south, the plant macrofossil record of the American West provides quantitative data on ecosystem changes far removed from the ice sheets (Spaulding et al. 1983; Thompson 1988). At any geographical point between 40° and 30° N latitude, the percentage similarity between full-glacial and modern floras is usually less than 30% and never exceeds 45% at the species level (e.g. Cole 1985; Spaulding 1990).

The dominant theme for the North American Quaternary plant record, which combines a fine level of time resolution with more complete spatial coverage than earlier records, is that species rather than communities moved in response to climatic change. If this ecological process was the same for pre-Quaternary vegetation, then what appears to be coordinated expansion and contraction of earlier plant communities may reflect our inability to resolve the separate, shorter-term histories of individual species that make up these communities. In this case, the Quaternary plant record magnifies what might be viewed as short-term noise in the older fossil record. It is also possible that there was less geographic movement of individual species prior to the Quaternary because climates fluctuated less radically, resulting in more unified responses of communities to changes that did occur. A second theme for the

Quaternary record is that very few plant taxa went extinct, even with the radical shifts in climate. It appears that the Pleistocene floras were composed of plants that already had been selected for their ability to respond to major environmental changes through extensive range displacement (e.g., Davis 1985; Van Devender et al. 1987; Webb 1987).

During the glacial ages there was intercontinental exchange of both floras and faunas across the Bering Land Bridge. These exchanges included cold-tolerant, steppe plants such as sagebrushes (*Artemisia* subgen. Tridentatae: McArthur 1979), which may have led to substantial enrichment of the steppe vegetation of North America. In some cases, endemic radiations in the arid-lands vegetation caused marked changes in the relative abundance of different species as represented in the palynological record (e.g., the apparent replacement of *Sarcobatus* (greasewood) by *Atriplex* (saltbush) species (Stutz 1978; Stutz and Sanderson 1983; Barnosky 1984).

Obviously habitat boundaries in the Pleistocene (and earlier) shifted geographically, expanding or contracting as floras changed. Some species enjoyed a wider range during glacial ages, while others were more abundant during interglacials (Spaulding et al. 1983). This can be attributed to many different factors. For instance, the development of nutrient-poor podzolic soils at the end of each interglacial is correlated with increased abundance of the American chestnut (*Castanea dentata*), which has its maximum competitive advantage on such soils (Paillet 1982). The developmental cycle of podzolization and acid-soil vegetation in response to Quaternary climatic cycling is so predictable that models of this process are easily constructed (Iversen 1958, 1964; Birks 1986). Thus there are strong links between particular environmental parameters (such as soil type) and plant species, and recurring associations of species often can be attributed to these parameters.

During the periods of climatic stress that characterized the Quaternary, organisms that did not or could not move geographically either evolved or went extinct. The fate of a plant species in a given region was a function of whether individual populations had access to growth sites within their range of environmental tolerance. Prior to the Quaternary, the North American floras were subjected to considerable change, and the fates of a number of conifer species provide evidence for the importance of migration and refugia in plant survival. For example, it is probable that *Taxodium* remained in eastern North America because it could move north and south, but it was excluded from similar latitudes in the west when there was no longer continuous access to suitable habitats. *Metasequoia* was common over western North America but was gradually excluded by environmental change; it exists today only in a limited refugium in China (Chaney 1948). The major period of plant extinctions in North America apparently occurred between 5 and 0.7 Ma, and the record of late Quaternary floral change involves the taxa that had already adapted to a new climatic regime, primarily through the ability to shift their geographic distributions.

Quaternary Faunas

The early to middle Pleistocene Irvingtonian fauna is characterized by the immigration of the mammoth, the bovid *Soergelia, Canis* (wolf), mustelids (ermine), *Alces* (elk), *Rangifer* (reindeer), *Panthera* (jaguar), and the cricetid genus *Clethrionomys* (Kurtén and Anderson 1980; Lundelius et al. 1987). Advanced *Equus, Smilodon* (saber-toothed cat), *Lepus* (hare) and *Microtus* (vole) typify this interval as well (Lundelius et al. 1987). There were five families of Artiodactyla, including the immigrant bovids, and camelids and antilocaprids remained relatively diverse. Xenarthrans were a conspicuous element in the Irvingtonian fauna, with glyptodonts, several species of armadillo, and three families of ground sloths. Proboscidea included single representatives of the gomphotheres and mastodonts in addition to several species of mammoth. Cricetids increased in diversity from 13 to 19 genera between the Blancan and the Irvingtonian (Savage and Russell 1983). The cricetids include at least 5 lineages that independently evolved microtine teeth as an adaptation for eating seeds and leaves in grassland habitats (Repenning 1987). Their explosive radiation in North America (and Europe as well) in the Plio-Pleistocene corresponds to the transition from savanna to grassland and steppe environments. The Irvingtonian fauna as a whole inhabited the extensive grasslands, deciduous woodlands, deserts, boreal forests, and tundra that characterized North America prior to and during the phases of continental glaciation. Provinciality of faunas was considerably greater than it had been in much of the Miocene (Webb 1989a).

The latter part of the Pleistocene, the Rancholabrean NALMA, was a time of further reduction in large mammal diversity, culminating in waves of extinction between 15 and 9 ka (Lundelius et al. 1987). Rodents continued to diversify, primarily at the species level, while proboscideans, perissodactyls, xenarthrans, and camelids experienced progressive losses. The immigration of *Bison* into North America marks the beginning of the Rancholabrean, and other notable immigrants included *Ovis* (mountain sheep), *Ovibos* (muskox), *Alces* (moose), *Panthera* (lion), and *Homo sapiens*. The pace of immigration and the blending of holarctic Pleistocene faunas continued throughout the latter third of the epoch, driven at least in part by the climatic fluctuations of the glacial episodes (Webb 1989a). The types of animals that were able to cross the Bering land bridge changed toward more cold-tolerant and steppe-adapted forms as the vegetation shifted to steppe and tundra (Webb 1989a). In addition to Eurasian immigrants, the Rancholabrean fauna was typified by *Smilodon* (saber-toothed cat), *Canis* (dire wolf), *Equus,* and *Mammuthus* (mammoth), and a diverse assemblage of sciurids and cricetids. Bovids were the dominant family of artiodactyls, with 10 genera. At the end of the Rancholabrean, about 10 ka, 43 genera of mammals disappeared in the second-largest pulse of extinction in the later Cenozoic (next to the Hemphillian pulse) (Webb 1989a). Mammoth, mastodont, gomphothere, saber-toothed cat, dire wolf, horse, sloths, camels, and a number of species of bear, deer, antilocaprid and other groups went extinct, apparently over a period of 5 ky or less (Webb 1989a, 1989b). This "megafaunal extinction" is attributed to a combination of factors, including human hunting and ecological changes accompanying the onset of the present interglacial period (Martin and Klein 1984; Fisher 1987; Webb 1989a). The complex features of megafaunal extinction in North America defy any single causal explanation and provide fertile ground for further study, particularly because this major turnover event is within the range of precise dating techniques and is accompanied by a wealth of paleoclimatic and paleobotanical data.

Glacial cycles were characterized by marked differences in the geographical distributions of mammals during warm and cold phases, and this is particularly well documented for small mammals. In the Holocene, many mammalian taxa that formerly occurred together now are separated, apparently because of increased patchiness and disjunction of habitats. These are referred to as "nonanalogue communities," or "disharmonious" faunas (Lundelius et al. 1983; Mead et al. 1983; Graham et al. 1987), since they include taxa that no longer coexist. The increased diversity that often accompanied these associations has been attributed to more equable climatic conditions and reduced ecological stress on the community as a whole (Graham and Semken 1976). The patterns of vertebrate distribution are similar to those of the floral assemblages in which species moved separately in response to environmental change, and communities of associated species changed radically during the course of the Quaternary.

Some General Themes of Ecological Change in North America

Over much of the past 34 my, endemic mammalian lineages in North America diversified and successfully adapted to the change from forest and woodland to savanna vegetation. The first major wave of Eurasian invaders in the early Miocene had relatively little impact on the endemic forms. The final transition from savanna mosaic to grasslands and steppes was, in contrast, accompanied by a wave of extinction of many endemic mammals and diversification of invading forms, such as the cricetid rodents. It appears that immigrants had an adaptive advantage during the last 6 my that previous invaders lacked, in comparison to the indigenous faunas. In the plant record, the later seral stages were eliminated as forests disappeared, and earlier seres became the new climax vegetation. This ecological retention of earlier seral stages results in increased annual productivity and turnover of nutrients in an ecosystem. With greater seasonal fluctuations in rainfall and temperature, as well as the general cooling and drying trend of the late Cenozoic, the overall biotic response maximized short-term productivity at the expense of stored biomass (e.g., grass as opposed to trees).

3.2 South America

Introduction

In contrast to its northern neighbor, South America was an island continent for much of the Cenozoic (fig. 7.1). Although a few waif-dispersal events introduced new types of plants and animals during the Miocene, it was isolated from wholesale invasion until the Pliocene, when a land bridge emerged at the Isthmus of Panama. This set the stage for the "Great American Interchange" (Webb 1976, 1978; Marshall 1981b, 1988), which was one of the most remarkable ecological events that occurred in the Cenozoic. The earlier history of South America provides evidence for morphological trends that developed in response to environmental change or other factors and that were phylogenetically independent of trends on other continents. The later history, over the past 3 my, records what happened when two distinct biotic provinces met and exchanged species while both were experiencing rapid climatic shifts culminating in glacial cycling.

The Cenozoic rise of the Andean mountains generated thick sequences of clastic deposits in intermontane basins and piedmont areas, many of which are very fossiliferous. The pre-Quaternary vertebrate fossil record in South America is distributed primarily along the eastern side of the Andes Mountains (Savage and Russell 1983). The plant record, as now known, is more restricted and occurs well outside of the modern forested belt, particularly in the northern coastal areas and in the arid south (Romero 1986). Paleomagnetic and radiometric dating now make it possible to correlate faunas and floras within South America and to compare the timing of ecological change with that on other continents. Major ecological trends during the late Cenozoic are summarized in figure 7.4.

Eocene to Miocene Floras

Botanists divide the modern vegetation of South America into the Neotropical and Antarctic botanical regions, with the former covering most of the continent. These in turn are divided into dominions that correspond to different types of forest and more open vegetation (Romero 1986). In the Upper Paleocene to Middle Eocene, a distinct vegetational assemblage referred to as the Mixed Paleoflora was present in the southern part of the continent. Romero (1986) hypothesizes that the Mixed Paleoflora was the source of taxa that

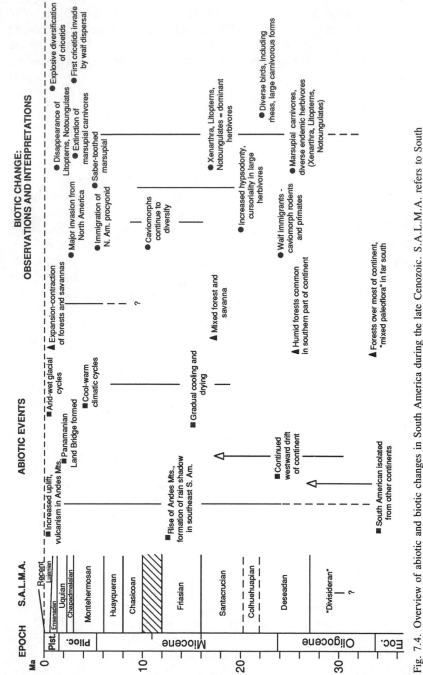

Fig. 7.4. Overview of abiotic and biotic changes in South America during the late Cenozoic. S.A.L.M.A. refers to South American biostratigraphic units (land mammal ages). Symbols same as in figure 7.2.

eventually formed the more open, dry-adapted vegetational dominions of today. Thus the precursor of later savanna floral communities in South America may have existed since early in the Cenozoic.

In the late Eocene, it appears that neotropical forests covered much of the continent except in the far south. Upper Eocene floras are recorded at both north and south ends of the continent. Taxa related to those of the Gulf Coast of North America occur in Venezuela; they indicate a warm and equable climate (Berry 1939; Romero 1986). In Patagonia, representatives of the Antarctic region are abundant, including *Nothofagus, Araucaria,* and *Lomanites*. With only 27% entire margin species, this flora represents a mixed mesophytic forest (Wolfe 1971, 1981) and a cool temperate moist climate (Dilcher 1973). This contrasts with the paratropical rain forest that occurred in the same area in the lower Eocene and indicates that change toward cooler climates in southern South America was underway by the middle to upper Eocene (Romero 1986).

The Oligocene plant record is restricted to the southern region and is similar in taxonomic composition to the upper Eocene, although the leaf sizes were smaller (Romero 1986). The Patagonian record thus demonstrates a change from rain forest to subtropical forest to mixed mesophytic forest through the Eocene and finally to a mixed hardwood forest in the lower Oligocene. The marked temperature decrease in the late Eocene (the Terminal Eocene Event) and a middle Oligocene change in high latitude floras, which are recorded in North America and elsewhere (Wolfe 1971, 1978, 1981), have not yet been detected in the South American paleofloras (Romero 1986).

Although the Oligocene–Miocene plant record is not sufficient to justify detailed climatic interpretations or correlations to global trends, some classic Miocene macrofloras record humid forests in areas now covered with dry-adapted vegetation types. These help to establish continuing trends toward desert and highland environments. Changes in climate generated by the rise of the Andes undoubtedly contributed to these trends (Axelrod 1979; Romero 1986).

Oligocene–Miocene Faunas

There is presently a 20 my gap in the Cenozoic faunal record between the middle Eocene Mustersan and the Deseadan South American Land Mammal Ages (SALMA), based initially on fossils recovered from Argentina by Simpson (1980) and currently dated as late Oligocene to early Miocene (MacFadden 1985; MacFadden et al. 1985) (fig. 7.4). Deseadan localities occur all along the Andean belt (Savage and Russell 1983) and document a mammalian fauna that consisted of 8 orders, 5 of which are now extinct. The Marsupialia, edentates (primarily Xenarthra), Notoungulata, Litopterna, Astrapotheria, Condylarthra, and Pyrotheria (see table A-7.3 for taxonomic key) continued from the Paleogene, either from Cretaceous origins or from early immigrants from other continents. Notoungulates were a very diverse group (46 species in 9 families in the Deseado fauna alone; Savage and Russell 1983) that produced a wide variety of body forms, including the large, hippo- and rhino-like toxodonts and rodent-like hegetotheres. Litopterns were horse- and camel-like, and the reduction of their lateral toes in the Pliocene was even more extreme than that of horses of the same age. Limb elongation and hypsodonty did not develop in this group to the same degree as in horses and camels, however (Carroll 1988). Astrapotheres resembled rhinos, and pyrotheres converged on elephants in size, incisors specialized as short tusks, bilophodont teeth, and retracted nasals indicating a proboscis. With the exception of the Marsupialia, all of these orders were gone by the late Quaternary, and diversity was already considerably reduced by the late Miocene, prior to the faunal interchange with North America (Marshall 1981b, 1988).

Marsupials included the borhyaenoids, which were somewhat similar in habit and body form to the medium and large carnivores of other continents, although they had shorter limbs and apparently were not strongly cursorial. The small insectivorous and omnivorous caenolestoids, another marsupial group, developed bladelike premolars that converged in form on those of the Mesozoic–early Tertiary multituberculates (Carroll 1988). The marsupial group Argyrolagidae converged in dental formula and other features on the heteromyids (e.g., kangaroo rats). Edentates were among the earliest placental mammals to appear in South America and were diverse through much of the Cenozoic, with numerous species of armadillos, giant armored glyptodonts, ground and tree sloths (Xenarthra), and anteaters. Members of both the Marsupialia and the Xenarthra participated in the faunal interchange with North America and have many living representatives.

The late Oligocene–early Miocene Deseadan record also includes two major groups that are thought to represent early waif dispersals from other continents. One of these, the caviomorph rodents (e.g., porcupines, capybaras, chinchillas, and a wide assortment of smaller forms), were the only group of rodents in South America until the Plio-Pleistocene. They diversified into 16 families, only 2 of which are now extinct (Carroll 1988). The second group of early immigrants were the primates. This group is first recognized on the basis of two genera, *Branisella* and *Szalatavus,* dated at about 26 Ma (MacFadden 1985; MacFadden et al. 1985; MacFadden 1990; Rosenberger et al. 1991). Their route of dispersal is uncertain, but affinities to monkeys in Africa (at the early Oligocene Fayum locality) and the relative ages may favor an African origin (Ciochon and Chiarelli 1980; Fleagle et al. 1986; Fleagle and Kay 1987; MacFadden 1990). The new continent presented a broad spectrum of opportunities for the invading primates, but extant and fossil forms indicate that monkeys diversified as strictly arboreal animals in South America and never produced the ground-adapted forms that evolved in Africa and Eurasia (MacFadden 1990; Pascual and Jaureguizar 1990; Ford 1990).

Birds are also an important component of the Oligocene through Miocene fossil record in South America. Fossil bird faunas include 61 families, with 8 (13%) extinct. Nine families are known from the Oligocene and Miocene, primarily from Argentina (Vuilleumier 1984). One of these families, the large, flightless, carnivorous phorusrhacoids, achieved considerable diversity during the Oligocene–Pliocene interval. The three-meter-high *Titanis,* recorded in the Pleistocene of Florida, was a member of this group that participated in the faunal interchange. The phorusrhacoids and the borhyaenid and didelphid marsupials apparently were the top predators in the larger vertebrate communities of South America through much of the Cenozoic, prior to the invasion of carnivores from North America in the Pliocene (Marshall 1988; Carroll 1988). Extinct genera of the ostrich-like Rheidae, as well as numerous other surviving families, are also known. The fossil record indicates that there was considerable exchange of flying birds between South and North America during the late Cenozoic (Rasmussen and Kay in press). Although many Recent families have a long Cenozoic history in South America, there has been considerable turnover in genera during the Plio-Pleistocene. The number of shared taxa between North and South America is large at this time, particularly among shore-dwelling forms (Vuilleumier 1984).

Reptiles must have contributed significantly to the faunal dynamics of Cenozoic South America, but reviews of their fossil record and its ecological implications are few. The largest known freshwater turtle, *Stupendemys* (carapace >2m long), occurs in the Pliocene of Venezuela (Wood 1976), and diverse crocodilians, including some giant species, are known from the later Cenozoic (Langston 1965). Many of these were piscivorous, but some of the larger, broad-snouted forms probably also preyed on terrestrial vertebrates. It is likely that snakes were important small vertebrate predators throughout this period, as they were on other continents. The fish of South America have unique adaptations to feeding on forest fruits, nuts, and litter during the part of the year when the major rivers flood their basins, which is often half of the annual cycle (Goulding 1980). Fossil evidence for the evolutionary history of one group of these fish (the characids) shows that diverse forms were present in the early Miocene (Lundberg et al. 1986), indicating that flooded forests have been a persistent habitat at least since that time.

Miocene–Pliocene Faunas

Early to middle Miocene faunas are well represented in South America, and localities occur along the length of the continent in Patagonia, Bolivia, Peru, Venezuela, and Colombia. The fauna from the Early Miocene Santa Cruz Formation in southern Argentina includes 76 genera representing (with a few notable exceptions) much of the same array of orders and morphologies as the Deseadan. The pyrotheres did not survive into the middle Miocene. A clawed, chalicothere-like form (*Homalodotherium*) developed with the Notoungulata. The carnivorous marsupials and rodents increased in diversity, while the notoungulates and astrapotheres declined. Edentates were the dominant group, with 8 families; the megalonychids and megatheriids (10 genera in all) are thought to have been generalized arboreal sloths that gave rise to later ground sloths as well as the living tree-dwelling forms (Carroll 1988). An anteater and several types of glyptodont also were extant at this time. The cursorial litopterns and notoungulates indicate open habitats, while the arboreal sloths and primates attest to the presence of forest or woodland. In combination, this evidence suggests a mixed forest to savanna vegetation. Middle Miocene Friasian faunas occur throughout South America, and some of these are similar to the preceeding Santacrucian, with differences between northern and southern localities. Didelphids, Chiroptera, a manatid (manatee), and several genera of primates are recorded in the La Venta fauna of Colombia, and litopterns and notoungulates appear to have been less diverse than in Patagonia (Savage and Russell 1983). The Caviidae (e.g., guinea pigs) occur in both northern and southern sites in the middle Miocene.

The Friasian spans the time interval between about 16 and 12 Ma, and the next documented interval, Chasicoan, begins at about 10 Ma. This early late Miocene fauna from Argentina continued to be dominated by edentates, notoungulates, and rodents, with a few litoptern and marsupial families also represented. The first hydrocherids (capybaras) appeared as the caviomorph rodents continued to diversify. The succeeding Huayquerian fauna (8–6 Ma) included a few immigrant taxon, a procyonid, the first true member of the Carnivora in South America. The large felid-like, saber-toothed marsupial *Thylacosmilus* evolved from the borhyaenids to join the array of late Miocene marsupial predators. Diversity in the edentates, litopterns, and notoungulates was maintained at a level similar to that in the Chasicoan stage.

The mammalian record continues in the Pliocene with the Montehermosan stage, based primarily on faunas from Argentina. The Montehermosan straddles the Mio-Pliocene boundary, and its upper limit is dated at 3.59 Ma (Marshall 1982). This fauna reflected the character of the mammalian community (*sensu lato*) immediately prior to the emergence of the land bridge to North America at about 3 Ma. Marsupials, edentates, and rodents were diverse, with a number of large marsupial carnivores, sloths, glyptodonts, armadillos, and 8 families of caviomorph rodents, including a giant capybara and a beaver-like form. Cricetids are first recorded in this faunal stage, representing waif-dispersal event(s) from North America, perhaps along the Greater Antilles island chain. Litopterns and notoungulates in the Montehermosan stage included 11 genera in 5 families (Marshall 1982), a reduction of 2 genera from the preceeding stage. The middle Pliocene Chapadmalalan stage is defined on the basis of abundant remains of North American taxa, reflecting the initial full-scale invasion across the Panamanian land bridge. Tayassuids, mustelids, and a number of cricetid genera began to mingle with the endemic fauna during this time. Litopterns and notoungulates declined to 7 genera in 5 families, but whether this was caused by the invasion from the north is debated. These two groups were already beginning a trend toward reduced diversity prior to the faunal interchange, and other factors such as climatic change may have contributed to their decline (Marshall 1988).

Through the Oligocene and Miocene, there was a general trend among the major ungulate groups toward increased hypsodonty, paralleling similar trends on other continents. When the Deseadan was regarded as lower Oligocene, it appeared that this trend might have started earlier in South America than elsewhere. With the presently available dates indicating a late Oligocene–early Miocene age for the Deseadan, however, the timing of the evolution of hypsodonty could be more in line with that in other faunas (MacFadden 1988), although the lack of a record prior to the Deseadan makes this hard to judge at present. In contrast to trends on other continents during the late Cenozoic, South American herbivores did not develop elaborate horns or specialized tusks, but a number of different lineages evolved proboscises. Evolution toward a more cursorial limb structure was characteristic of the litopterns and some notoungulates, but other ungulates did not emphasize elongation of their distal limb segments and were probably not strong runners. Possibly this was linked to vegetation structure and to the lack of cursorial adaptations among the marsupial predators.

Plio-Pleistocene Floras

The late Pliocene–Quaternary record of South American vegetation is based primarily on pollen from cores in lacustrine settings. This palynological evidence demonstrates marked expansions and contractions of forest and other types of vegetation during the past several million years (Flenley 1979; Van der Hammen 1982). The record from the 2550 m high Bogota basin in Colombia spans 3.5 my and includes 27 major climatic cycles, reflected in the fluctuations of forest along altitudinal gradients during warmer and cooler phases (Hooghiemstra 1984). Major fluctuations are interpreted as interglacial-glacial cycles, corresponding to variations in orbital eccentricity and obliquity. The taxonomic composition of the vegetation changed considerably during the 3.5 my interval, with North American invaders such as *Alnus* and *Quercus* replacing endemic forms and becoming important members of the forest communities. At lower altitudes, savanna vegetation expanded while forests contracted during the glacial periods, and as many as 20 cycles are postulated for the Quaternary (Van der Hammen 1982). In Argentina, *Nothofagus* retreated and dry steppe vegetation expanded during the last Northern Hemisphere glacial maximum, about 20 ka (Huesser and Wingenroth 1984). Although glacial maxima in the northern hemisphere appear to correlate with expansion of the South American savannas (Eden 1974), the relationship between glacial phases and vegetational patterns is complicated by local conditions and is not yet well understood. There is debate about the importance and placement of "forest refuges" during the periods of increased savanna and other more arid habitats (Van der Hammen 1982; Bradbury et al. 1981; Liu and Colinvaux 1985), but it seems highly probable that alternating contraction and expansion of different environmental conditions have had major effects on speciation, extinction, and overall diversity of organisms linked to specific climatic conditions and vegetation.

Plio-Pleistocene Faunas

The Plio-Pleistocene Uquian, earlier Pleistocene Ensenadan, and later Pleistocene Lujanian stages document radical change in the South American mammalian community compared to the Chapadmalalan. Litopterns and notoungulates were reduced to one genus each (*Macrauchenia* and *Toxodon*), and the large marsupial carnivores apparently were extinct prior to the main period of invasion of carnivorous forms from North America. Ursids, canids, felids, and various mustelids from North America occupied the predator niches, while tayassuids, camelids (llamas), cervids, equids, tapirids and gomphotherids were the dominant large herbivores (Pascual et al. 1966, Savage and Russell 1983). In contrast to the pattern of extinction of indigenous large mammals, native rodent groups did not decline significantly either prior to or during the cricetid invasion and subsequent radiation in the later Pleistocene (Marshall 1988).

The northward invasion of South American mammals had, in contrast, relatively little impact on the fauna of North America. Opossums, sloths, armadillos, glyptodonts, hydrochoerids, the South American porcupine *Erethizon*, and possibly some southern Chiroptera are recorded in North American Pleistocene faunas. One notoungulate, the rhino-like *Toxodon*, was present in southern central America (Webb and Perrigo 1984). Ground sloths included three genera, one of which (*Megalonyx*) evolved in North America from an early invader. The glyptodont *Glyptotherium* was a heavily armored giant approximately 2.5 m in length that successfully occupied southern North America for much of the Pleistocene, producing several species (Gillette and Ray 1981; Carroll 1988). None of these larger South American derivatives survived the wave of late Pleistocene extinctions in North America.

The invasion in both directions probably was facilitated by Quaternary climatic cycling, which had two major effects on the habitats along the route across the Panamanian land bridge. During glacial maxima, cooler and drier climates in the southern latitudes led to expansion of grasslands and restriction of forest, opening corridors for the savanna-adapted species in both Central and South America (Marshall 1988). At the same time, lowered sea levels created broad expanses of exposed continental shelf, which also may have been covered with fairly open vegetation. The northern species that crossed the land bridge included many cursorial forms, and it is difficult to imagine that animals with this adaptation would have been successful invaders if Central and northern South America were as densely forested throughout the Pleistocene as they are today (Marshall 1988).

The Great American Interchange (Webb 1976, 1978; Marshall 1981b, 1988) is remarkable for its rapidity and the asymmetrical effects on the two Americas. During the Huayaquerian stage, at about 7.5 Ma, the mammalian fauna of South America consisted of 60% purely endemic families and 40% that were derived from waif-dispersal events. Today half of the families and genera in South America belong to Holarctic groups that immigrated during the past 3 Ma (Marshall 1988). The success of this invasion, from the standpoint of the emigrants from the north, has been attributed to (1) competitive superiority (Webb 1985); (2) ability to occupy unexploited or newly vacant niches (Patterson and Pascual 1972; Hershkovitz 1966); (3) emergence of new habitat types with geological and/or climatic changes; (4) a combination of the first three, but stressing the relatively greater size of the holarctic "staging area" and the diversity of its faunas (Webb 1985; Marshall 1988). North America alone had 60% greater generic diversity during the time of the interchange, and the apparent imbalance in the numbers of invaders from the north and from the south is in fact proportional to the size of the source faunas (Marshall 1982; Marshall 1988).

Both caviomorph rodents and primates seem to have been little affected by the Plio-Pleistocene faunal exchange with North America. Primates emigrated to Central America, but there is little fossil record of this expansion. The southward invasion and spectacular diversification of cricetid rodents in the Plio-Pleistocene was not accompanied by extinction of native South American forms. This implies that numerous ecological opportunities were available to small mammals in the southern ecosystems, in spite of the presence of native forms that apparently had failed to fully exploit these opportunities themselves.

The great radiation of northern-derived forms in South America stands in sharp contrast to the lack of diversification in southern-derived forms in North and Central America. Ecological opportunities in North America may have been too limited to permit diversification among the sloths, glyptodonts, aquatic rodents, and other southern forms. This implies that there was a different overall balance in the two continents between available resources, existing organisms that utilized these resources, and numbers and adaptability of invaders. These differing balances set the stage for very different ecological and evolutionary consequences of faunal exchange.

3.3   Africa

Introduction

In a seminal paper, Moreau (1966) proposed that the present-day distribution of plants and animals in Africa is the result of past climatic (and vegetational) change, especially during the Quaternary, when sequences of contraction, expansion, and fragmentation of forest habitats occurred in response to alternating wet and arid periods. Considering the entire period from the Oligocene through the Quaternary, the biotic history of Africa exemplifies the basic ecological trends summarized in §2. Because of the number of fossil localities and their distribution through time, much of the floral, faunal, and climatic history of this continent is disproportionately derived from eastern Africa, and to a lesser extent, southern and northern Africa. Major abiotic environmental changes and ecological trends are summarized in figure 7.5. A guide to taxonomic names is given in table A-7.4. The faunas are summarized by Maglio and Cooke (1978) and Savage and Russell (1983).

Early Cenozoic and Oligocene

Africa, including Arabia, was isolated from contact with other continents for much of the Eocene and Oligocene (Coryndon and Savage 1973). Older faunas from Morocco demonstrate that interchange did occur between Africa and Europe during the early Cenozoic (Gheerbrant 1990). But the highly endemic character of Eocene and Oligocene faunas strongly suggests that a barrier to dispersal, possibly a shallow seaway covering the northern part of the continent, was in effect until northward drift and orogeny in the Tethyan region established land bridges to Eurasia in the early Miocene (fig. 7.1). Africa itself was relatively stable tectonically prior to the Miocene. The absence of eastern highlands, the presence of a northern seaway, and the overall greater warmth of the Atlantic than during the late Cenozoic meant that significant amounts of monsoonal moisture were able to penetrate far into the African continent throughout the first half of the Cenozoic.

Axelrod and Raven (1978) speculate that the continent was covered primarily by lowland tropical to temperate rain forest at this time. Areas of sclerophyll woodland and dry scrub also are inferred for south-central and southwestern Africa, primarily on the basis of records of such floras in non-African localities. A small flora from Banke, Namaqualand (Rennie 1931; Axelrod and Raven 1978), considered to be transitional Eocene-Oligocene, possesses what are believed to be coarse, sclerophyllous leaves suggesting the existence of a dry season by the late Eocene in this area (when it was 35°S; it is now approximately 29°S). This flora appears to represent an ecotonal area between temperate rain forest and sclerophyllous vegetation.

By the Paleocene/Eocene boundary, prosimian primates, creodont carni-

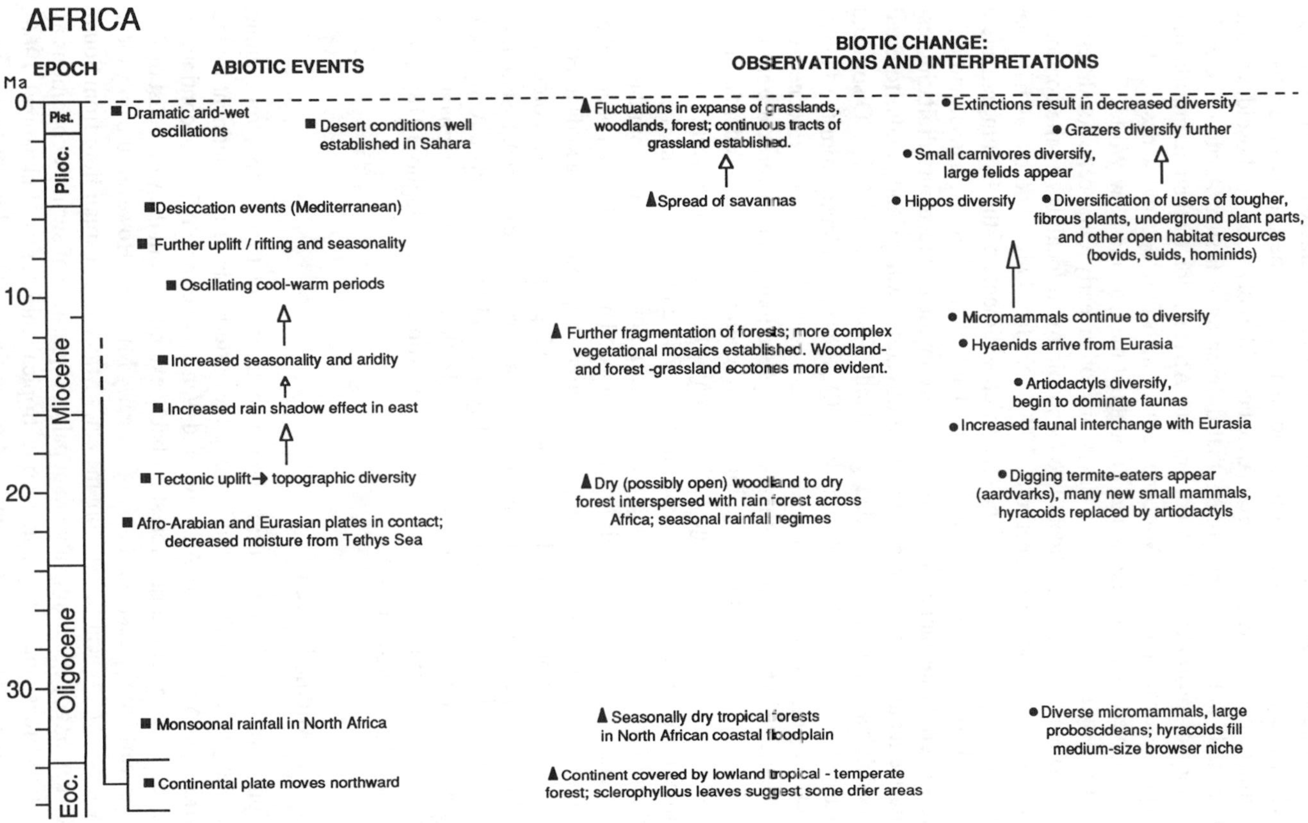

Fig. 7.5. Overview of abiotic and biotic changes in Africa during the late Cenozoic. Symbols same as in figure 7.2.

vores, and several lineages of early condylarths had entered Africa. Artiodactyls, perissodactyls, insectivores, and members of the Carnivora evolved on northern continents and entered Africa considerably later. Proboscideans and hyracoids are among the groups that differentiated first in Africa. The early history of proboscideans is confined to Africa prior to their migration into Eurasia during the early Miocene, after which they became widespread and common elements in the faunas throughout the Old and New World.

The Fayum Depression in Egypt and other early Oligocene localities in Arabia are the only significant late Cenozoic fossil localities on the continent prior to the Miocene (Gheerbrant 1990). No late Oligocene biotas have yet been reported from Africa. The fossil-rich sediments of the Fayum indicate a tropical coastal floodplain that supported abundant tall trees, lianas, and a diverse vertebrate fauna. The overall environment included interdigitating estuarine mangroves, fresh and brackish fluviatile swamps and ponds, forested floodplains, and moist wooded grasslands (Bown et al. 1982; Olson and Rasmussen 1986; contra Kortlandt 1980). Present evidence from the Fayum indicates that only two mammalian families were derived from Eocene stock endemic to Africa (gomphotheres and hyaenodonts), and over a dozen new families related to non-African lineages were added, including rodents, insectivores, and bats. This suggests that a major faunal turnover occurred sometime in the Eocene-Oligocene transition. Primates appeared in the African record for the first time during the Oligocene, creodonts were abundant, and two endemic orders had their first appearance. The Embrithopoda occurred as a single genus that filled the large herbivore niche, and the hyracoids filled the medium-sized browser niche and showed the highest diversity this order ever achieved (7 genera known from northeastern Africa alone). A rich assemblage of proboscideans is known from the Fayum, and these animals already were large, with most of the postcranial specializations of modern elephants. Two genera of anthracotheres were the main artiodactyls (Maglio 1978; Bown et al. 1982; Carroll 1988).

Early to Middle Miocene

Uplift, volcanism, and rifting began to transform topography, orographic rainfall, and plant distributions along the eastern side of the continent by the Miocene. According to Andrews and Van Couvering (1975), a continuous band of broad-leaved rain forest probably extended from the west coast of Africa nearly to the east coast during the early Miocene. However, the only clear evidence of tropical forest from the early Miocene is a small flora from South Kivu, Congo, containing abundant palms suggestive of swampy conditions as well as trees and a liana indicating tropical rain forest (Lakhampal 1966; Bonnefille 1985). All other early Miocene fossil leaf and wood floras (e.g., Bugishu, Mount Elgon; Karugamania beds near Lake Albert; Rusinga Island, Lake Victoria) together indicate either savanna-woodland, dry-adapted woodland, or dry forest floras related to rain forest or gallery forest taxa in a seasonal rainfall regime (Axelrod and Raven 1978; Bonnefille 1984, 1985; Chaney 1933; Lakhampal and Prakash 1970). Hence, by 20–18 Ma, rain forest apparently was discontinuous across Africa, interrupted by areas of dry (and perhaps open) woodland or other nonforest vegetation.

The numerous and geographically widespread early Miocene vertebrate localities of East Africa suggest that a major faunal shift occurred sometime between the early Oligocene and early Miocene. Maglio (1978) reports that, only 14 families of mammals persisted from the early Oligocene, whereas 29 new families and 79 new genera are first recorded in Africa during the early Miocene. The timing of this faunal transition is unclear, however, because of the absence of late Oligocene localities. About 40% of the new taxa were micromammals. Aardvarks (Tubulidentata), exhibiting clear adaptations for digging and feeding on termites, appeared in the early Miocene and probably were endemically derived. Perissodactyls entered Africa from Eurasia in the form of chalicotheriids and rhinocerotids, with the latter becoming widespread. Hyracoids were greatly reduced in diversity and continued to decrease during the late Cenozoic; they were replaced by similarly-sized artiodactyls (first anthracotheriids and then bovids). Proboscideans continued to radiate, with mammutids probably derived from Oligocene gomphotheres. The first appearance of proboscideans in Eurasia during the early Miocene indicates faunal emigration was occurring along with immigration to Africa. Creodont carnivores remained diverse, even as the first fissiped carnivores (viverrids and felids) arrived from Eurasia. Among the artiodactyls, archaic (probably omnivorous) suids migrated into Africa from Eurasia, small browsers with primitive selenodont teeth radiated, but only a few ruminants are known (two endemic giraffids—the palaeotragines and sivatheriincs—and a few bovids) (Maglio 1978; Carroll 1988; Butzer and Cooke 1982; Dawson and Krishtalka 1984).

Researchers have presented varying ideas about Africa's movements and the timing of contact with Eurasia. Some paleontologists accept the view that Africa lay 15° south of its present position during the early Miocene, and about 6° south during the middle Miocene (e.g., Van Zinderen Bakker and Mercer 1986; Smith et al. 1981). Others state that Africa was in its present position throughout the Miocene, and land bridges with Eurasia were established by the late Oligocene, though little faunal interchange occurred before the early Miocene (e.g., Bonnefille 1984). Laporte and Zihlman (1983) report that Africa impinged upon Eurasia in the latter half of the early Miocene (17 Ma) and that about 7–8° latitude of movement northward took place during the Miocene. Overall the various lines of evidence, especially faunal interchange, suggest that definite contact between Africa and Eurasia occurred by at least 20 Ma. Northward drift of the continent continued into the middle

Miocene, enabling more vigorous biotic exchanges (Bernor 1983; Maglio 1975, 1978).

During the Miocene several factors contributed to increasing aridity. Interruption and contraction of the Tethys Sea by the union of the African-Arabian plate with Eurasia (in the vicinity of Iran) depleted a major source of atmospheric moisture for rainfall and also reduced the ameliorating effects of sea temperatures on land climates. Tectonic uplift and tilting of major portions of the continent, especially in eastern and southern Africa, began by the early Miocene and created significant topographic diversity by the middle Miocene. Major uplift in East Africa related to rifting continued during the late Neogene and into the Quaternary, creating a rain shadow between wet Central-West Africa and the drier East African plateau. Moist air masses evidently penetrated to the Congo Basin over the past 16 million years, but not much further east than the western rift valley (in contrast with the early Miocene). In addition, Antarctic glaciation had a controlling influence on atmospheric and oceanic circulation patterns that have influenced the climate of southern Africa. The broad latitudinal circulation system that existed prior to the early Miocene and that had distributed moisture over large tracts changed during the Miocene to a more restricted north-south cellular circulation pattern; the latter anticyclonic circulation system became intensified as colder climates developed in the Northern Hemisphere. All of these factors resulted in increases in seasonality and aridity, enhancing the effects of global cooling (relative to the early Cenozoic).

With global-scale change toward decreased temperature and moisture, topography appears to have increased its role in moderating continental climates. The effect on the African biota was further fragmentation of closed forests and the subdivision of habitats into complex mosaics of moist and dry zones, including forests, woodlands, grasslands, wetlands, and various montane zones within a geomorphological mosaic of fluvial systems spread out over floodplains and feeding into perennial lakes. Seasonality and climatic drying during the Miocene would have selected for drought-resistant taxa, whether in rain forests, savanna, scrub, or semidesert areas (Axelrod and Raven 1978; Butzer and Cooke 1982; Laporte and Zihlman 1983; Van Zinderen Bakker and Mercer 1986).

A pollen flora from the eastern rift locality of Fort Ternan (14 Ma) contains 54% grass pollen, some montane forest taxa, but no lowland rain forest taxa, all indicating relatively open vegetation (Bonnefille 1984, 1985). This evidence agrees with analyses of the Fort Ternan fauna and suggests a drier, more seasonal climate in the middle Miocene of East Africa than in the early Miocene (e.g., Andrews and Nesbit Evans 1979; Shipman 1982; Shipman et al. 1981). Functional analysis of the bovid postcrania (Kappelman 1991) and isotopic analysis of paleosol carbonates from Fort Ternan (Ambrose et al. 1991), however, indicate a woodland or woodland-forest habitat at this locality during the middle Miocene. Possibly both open woodland with grasses and forested habitats occurred at different times, as has been suggested for the Miocene of the Pacific Northwest of North America (Taggart et al. 1982; see also §3). On the other hand, a 12.2 Ma leaf flora from the Ngorora Formation, in the eastern rift, indicates the presence of a lowland-type rain forest of West African affinities (on the basis of identified taxa and a high percentage of entire-margined leaves) (Jacobs and Kabuye 1987). Although the fauna from other parts of the Ngorora Formation (14–8.5 Ma) points toward more open woodland to savanna mosaic (Van Couvering 1980; Gentry 1978; Pickford 1978; Thomas 1981), this botanical evidence for lowland-type rain forest in East Africa around 12 Ma suggests that there was not a simple, straightforward replacement of rain forest by savanna-woodland during the Neogene in this region of Africa. Other evidence of rain forest conditions in East Africa derives from the late Miocene and Pliocene, though continuous tracts with Central and West African forests are not necessarily implied (Bonnefille and Letouzey 1976; Yemane et al. 1987; Williamson 1985; Bonnefille 1985; Jacobs and Kabuye 1987).

Although the controlling effects of late Neogene and Quaternary climatic oscillations are emphasized in most accounts of African (and global) environmental history, fluctuations in temperature and rainfall probably occurred throughout the late Cenozoic, leading to changing relationships between rain forest, woodland, and savanna vegetation. In southern Africa, for instance, savanna-woodland covered the interior plateau perhaps as early as 16–10 Ma, while lowland subtropical forests occurred at the same time in coastal areas (Van Zinderen Bakker and Mercer 1986). Paleoecological analyses of middle Miocene faunas tend to conclude that East Africa had more open vegetation than during earlier times, but all such studies also emphasize that woodland and forest-grassland ecotones were far more important than they are in modern environments. Ecological diversity spectra (see §1 and chaps. 1 and 4), for instance, illustrate that large ground-dwelling browser species indicative of woodlands are far more numerous at certain middle Miocene localities (e.g., Fort Ternan) than during the early Pleistocene (e.g., Olduvai) or in modern savannas (e.g., Serengeti) (Andrews et al. 1979; Van Couvering 1980; also see Shipman 1982). These woodlands may represent vegetational structures that do not have any modern analogues. The impression that open habitats were widespread in the middle Miocene may also be affected by a preservational bias toward a more prolific vertebrate fossil record from the drier habitats (Kidwell and Behrensmeyer 1988).

Mid-Miocene to Pliocene

From the middle to late Miocene, artiodactyls began to dominate the African mammalian fauna; nine distinct tribes of bovids occur, mostly with Asiatic ties, though several genera from the late Miocene are unknown outside of Africa. Hippopotamids evolved endemically by this time, and more advanced forms of suids occurred than in the early Miocene. The first African equid, *Hipparion,* arrived during the middle Miocene from Eurasia. Rhinoceroses remained diverse, and primitive elephants evolved from earlier gomphotheres, though short-jawed versions of the latter persisted. Mustelid and hyaenid carnivores arrived from Eurasia, and micromammals continued to diversify (Maglio 1978; Butzer and Cooke 1982). Faunas from the Tugen Hills succession, central Kenya, provide a record of these changes 13–6 Ma. During this interval, an archaic fauna containing creodonts, *Climacoceras* (an early artiodactyl), caprines, and boselaphines was replaced by a more modern one containing equids, advanced tribes of bovids, elephantids, and leporids. The first phase of this change involved the appearance of *Hipparion,* murid rodents, and reduncine bovids, apparently between 11 and 9.5 Ma. The faunas before this time were similar to those of southern Asia (Chinji Fm.) (Barry et al. 1985). Subsequent changes were different in character from those in South Asia (§ 3.4), involving the disappearance of archaic lineages. Consequently the faunas by 6 Ma were quite different not only from older ones in East Africa but also from contemporaneous faunas in southern Asia (Hill et al. 1985; Hill et al. 1986; Hill 1987; see also Kalb et al. 1982).

Although the late Miocene had pronounced cooling periods (e.g., 10–9 Ma) and warming periods (e.g., 9–8 Ma) (Shackleton and Kennett 1975b; Van Zinderen Bakker and Mercer 1986), the late Miocene biota of Africa was most greatly affected by aridification that took place at the end of the Miocene. Between 6.4 and 4.6 Ma (the Messinian), the Mediterranean dried out as sea level dropped approximately 70 m. Multiple desiccation events probably resulted from a series of southern hemisphere glaciations that went along with continued cooling of Eurasian landmasses and greater aridity on all continents. A decline in global temperature at 5.2 Ma affected the ocean around Antarctica; glacial conditions were circumpolar in the Southern Hemisphere, and the West Antarctic ice sheet developed (Mercer 1983; Stein and Sarnthein 1984; Van Zinderen Bakker and Mercer 1986). Further uplift and rifting in East Africa caused additional restriction of forest and woodland and favored plants that made up the open savanna-mosaic vegetation. Spread of savannas, interrupted by pockets of montane forest, characterized eastern and northern Africa (and southern Europe, which was also affected by the Messinian drying of the Mediterranean). In the southwestern interior and southern coastal areas of Africa, savanna was distributed with sclerophyllous, thorn scrub and laurel forest vegetation. Seasonal rainfall made the availability of plants more variable, which meant an adaptive advantage for large mammals that could exploit choice patches of food through nomadic foraging or seasonal migration. In adapting to dry seasons, plants evidently acquired tougher outer coverings or developed underground storage organs. In eastern to southern

Africa, radiations of bovids (especially grazers) and suids and the development of hominids coincided roughly with the expansion of savanna-mosaic habitats from late Miocene to early Quaternary (Laporte and Zihlman 1983; Axelrod and Raven 1978; Vrba 1985a).

Faunas of North Africa during the late Miocene and early Pliocene indicate extensive interchanges between Eurasian and African biotas. For instance, the fauna from Sahabi, Libya, exhibits varied faunal relationships with the rest of Africa (e.g., the suids), Eurasia (e.g., giraffids), and combined Eurasian and East African ties (e.g., bovids) (Bernor and Pavlakis 1987).

From the late Miocene through the Pliocene, African faunas experienced their greatest change since the Oligocene-Miocene transition. At the generic level, 76% of the land mammals evident in the Pliocene were new, and about 53% of these taxa were endemic (Maglio 1978). Ruminants diversified actively, two endemic genera of aardvarks and a large deinothere appeared, and three new genera of elephants emerged from a Mio-Pliocene radiation in this group. Common in Africa since the early Oligocene, anthracotheres became extinct during the Pliocene, perhaps in part because of the diversification of hippopotamids between 6 and 4 Ma (Carroll 1988). New arrivals from Eurasia included a camel, a chalicothere, the modern *Giraffa* and *Equus* (*Hipparion* also persisted), and a saber-toothed carnivore. During the Pliocene the felids *Felis* and *Panthera* joined the saber-toothed forms as significant meat eaters; canids and viverrids also radiated and became diverse components of the Carnivora (Maglio 1978; Barry 1987). Dramatic turnover and radiations in the carnivores, ungulates, and ground-dwelling primates in East Africa after 5 Ma paralleled a similar but earlier radiation during the late Orleanian–late Turolian faunas of Eurasia. In both regions, faunal changes corresponded with the development of more open, savanna-like communities (Bernor 1983). Sharp changes in fauna also occurred in southern Africa during the Pliocene, suggesting further disappearance of closed forests and woodlands with the expansion of savannas and grasslands (Van Zinderen Bakker and Mercer 1986).

Late Pliocene to Quaternary.

Around 2.5 Ma, desert conditions in the Sahara were apparently well established, coinciding with climatic cooling on a global scale. Although Axelrod and Raven (1978) propose that the Sahara (and other desert regions) are no older than the Pliocene, coarse-grain eolian sands thought to be transported from continental shelves during periods of aridity and low sea level are concentrated in marine sequences off North Africa ca. 38–34 Ma, 23–20 Ma, 13–12 Ma, and 3–2 Ma (and others of middle and late Quaternary age). These data strongly suggest that a desert area existed periodically in North Africa since at least the early Miocene. This arid zone was displaced southward as the continent moved northward, and subsequent expansions coincided with dry phases throughout the Neogene and Quaternary (Sarnthein 1978; Sarnthein and Diester-Haass 1977; Diester-Haass and Schrader 1979; Street 1981).

Pollen samples from various Plio-Pleistocene localities in East Africa document dry-adapted *Acacia-Commiphora* deciduous bushland, semidesert scrub, grassland, and mesic Afromontane forest. Bonnefille (1984, 1985) documents four main periods during which grass pollen increased and tree pollen decreased in abundance, again indicating increasing aridity in East Africa. Pollen samples from Laetoli, Tanzania (3.7 Ma), reflect evergreen bushland and *Acacia-Commiphora* grassland, a habitat only slightly more arid than that represented by pollen samples from Hadar, Ethiopia (3.3–2.9 Ma) (Bonnefille et al. 1987). Grass pollen increased in the record approximately 2.5–2.3 Ma (recorded in the Omo Shungura Fm., southern Ethiopia), and again around 1.8 Ma (recorded in the Turkana basin). These periods coincided with considerable species turnover in bovids and rodents, also suggesting spread of grasslands. A fourth change, at about 1.7 Ma (recorded at the top of Bed I, Olduvai, Tanzania), involved a decrease in forest tree pollen and an increase in subdesertic pollen taxa (Bonnefille et al. 1982)

The presence of localized rain forest taxa indicating warm, wet conditions prior to 2.5 Ma (see above) and abundant Quaternary evidence for climatic and vegetational conditions that alternated between being more humid and drier than today demonstrate that grasslands, woodlands, and forests have had a complicated history on the African continent. Although there is no one point when savanna replaced forest, the overall trend has been toward restriction of forests and the development of continuous tracts of grass. The exact nature of such savanna-grasslands prior to the mid-Pleistocene, however, remains to be determined. On the basis of a study of fossil rodent faunas, for example, Denys (1985) infers the development of communities between 3 and 1.8 Ma in East Africa that were essentially the same as those of the Serengeti savanna today. In contrast, the isotopic composition of paleosol carbonates at East African localities indicates a significant expansion of plants (essentially grasses) that possess a C_4 photosynthetic pathway around 1.7 Ma, but not until 0.6 Ma (Masek Beds at Olduvai Gorge) is there evidence of a virtually 100% C_4 biomass that typifies the present Serengeti grasslands (Cerling et al. 1977; Cerling and Hay 1986; Cerling et al. 1988; Cerling et al. 1989). This pace of shift to C_4 grasslands thus apparently is later in East Africa than in southern Asia (see § 3.4).

Furthermore, Andrews's (1989) analysis of paleocommunity structure based on the Laetoli fauna (3.8–3.5 Ma) appears to contradict the assumption of several investigators (e.g., Leakey and Harris 1987) that modern savannas were well established in East Africa by the Pliocene. Although similarities with the Serengeti are certainly evident, the Laetoli fauna exhibits a spectrum of browsing herbivores, carnivores, and other dietary types that is similar to the woodland component of the present Serengeti ecosystem. Thus, Andrews suggests that the Laetoli habitat overall may have been far more wooded than is suggested by the term savanna. Ecological diversity diagrams of the faunas from the younger Olduvai Gorge locality (1.8 Ma) differ only in minor ways from those representing the grass plains of the Serengeti today (Andrews et al. 1979). Differences in the composition of large-mammal faunas do occur between Olduvai and the Serengeti, but these appear to result from taxonomically biased processes of bone accumulation at Olduvai, not necessarily from a significant dissimilarity between Plio-Pleistocene and modern savanna-mosaic environments (Potts 1988).

During the first half of the Quaternary, the fauna of Africa showed considerable continuity with the Pliocene, though new taxa appeared and disappeared. Bovids were the most abundant of all large mammals; gazelles (Antilopini) were important in North African faunas, whereas the grazing alcelaphines were especially diverse and numerous in sub-Saharan areas. Archaic genera of bovids declined during the early Quaternary, replaced by bovids of living genera. Cervids appeared in North Africa but not south of the Sahara. Suids were a fast-evolving group during this interval. They diversified during the Plio-Pleistocene and then were characterized through the middle Pleistocene by two associated trends: increase in body size and hypsodonty (Harris and White 1979; Harris 1983). In contrast, hippopotamids were reduced in diversity during the first half of the Quaternary. According to Coryndon (1978), the restricted terrestrial habitus of the pygmy form of hippo and the aquatic specialization of *Hippopotamus* were partly a response to competition with the rapidly diversifying bovids during the late Cenozoic. The Plio-Pleistocene diversity of giraffids decreased by the mid-Pleistocene, when the short-necked *Sivatherium* and several species of *Giraffa* became extinct. *Hipparion* and *Equus* both persisted, though the former met its demise during the mid-Pleistocene. It is generally assumed that *Equus* continued to thrive because its extreme hypsodonty and monodactyly made it better suited to open grassland (Harris 1983). The survival of the browsing rhinoceros *Diceros* from the Pliocene is perhaps related to the persistence of bush-browse patches, while the emergence of the grazing, open habitat *Ceratotherium* in the Plio-Pleistocene can be linked to the development of grasslands (Harris 1983; Carroll 1988). The specialized browsing chalicotheres became extinct during the early Quaternary. Deinotheres and gomphotheres also disappeared from Africa during this interval, leaving only modern genera of elephants (e.g., *Elephas recki*, known from the Pliocene to mid-Pleistocene especially in East Africa; *Loxodonta atlantica* in North Africa). The most abundant primates in the fossil record are large monkeys (e.g., giant *Theropithecus*, which became extinct and was replaced by *Papio* baboons). Carnivores changed considerably during the early Quaternary, with several genera entering the continent from Eurasia (Maglio 1978).

During the latter half of the Quaternary, relatively little faunal change occurred in Africa at the generic level, and minor changes involved modern species replacing archaic ones. Maglio's data (1975) suggest that an essentially modern fauna (at the generic level) was present in Africa by the early Pleistocene, whereas the same degree of modernization did not occur in Europe and northern Asia until the mid-Pleistocene.

As on other continents, dramatic oscillations in climate and vegetation were experienced in Africa during the Quaternary. Over the past 1 million years in particular, vegetational belts contracted and expanded in response to changes in temperature and moisture that were possibly tied to Milankovitch cycles. An especially good, precisely dated record of fossil pollen and lake levels during the past 30,000 years indicates that relatively moist conditions and high lake levels prevailed during interglacial periods, whereas glacial conditions in the Northern Hemisphere tended to coincide with a reduction in moisture content of air masses that moved over Africa, thus leading to aridity (Street and Grove 1979; van Zinderen Bakker 1976; Hamilton 1982; Street-Perrot and Harrison 1984). This generalization is the reverse of the original "pluvial hypothesis" championed by Leakey (Leakey 1931; Leakey and Cole 1952), and invalidated by Cooke (1958), Flint (1959), and others. There is some evidence, however, to indicate that the moist and dry intervals in the late Quaternary of Africa persisted only for 1000–5000 years, not as long as glacial, interglacial, or even stadial periods at higher latitudes.

The extent, phase, and duration of arid-moist changes in Africa evidently are not perfectly correlated with glacial-interglacial environments of Eurasia, and they varied extensively from region to region. Synchroneity of climatic change through the African continent was complicated by variations in atmospheric and oceanic circulation patterns. Thus, no general climatic-chronological framework similar to the glacial-interglacial framework of Europe has been worked out for Africa. Although glacial-arid and interglacial-moist correspondences do indeed occur there in the late Quaternary, important exceptions have been found (e.g., the Kalahari and Vaal-Orange basins in southern Africa: Butzer and Cooke 1982). Jansen et al. (1986) point out an intriguing situation during the mid-Brunhes (0.3–0.4 Ma), when warmer mean sea surface temperatures and wetter conditions in the Southern Hemisphere appear correlated with a trend toward glacial conditions in the Northern Hemisphere. Considerable change in vegetation accompanied the arid-moist alternations and increases in seasonality during the Quaternary. In the temperate zones of southern Africa, glacial periods would have involved an influx of polar air, which decreased temperatures. Alpine vegetation probably replaced temperate grass veld of the inland plateau, forests were replaced by grassland on the Cape coastal plain, and the semidesertic Karoo vegetation moved northward during these cooler, more arid periods (Van Zinderen Bakker 1982). Species that make up the unusual succulent semidesert vegetation of the Karoo (ecomorphic counterparts of the succulent vegetation of Baja California or northern

Chile) appear to represent evolved survivors of savanna and sclerophyllous vegetation that had formerly occupied southern Africa; that is, taxa that were adapted structurally and physiologically to drought conditions diversified and became dominant during the pulses of dry and wet climate during the Quaternary (Axelrod and Raven 1978).

During the latter half of the Quaternary, the effects of glacial aridity on tropical Africa appear to have been quite severe. For example, the Senegal River ceased to flow into the Atlantic (Diester-Haass 1976), and parts of the Niger River were blocked by dunes during the last glacial interval (Talbot 1980). Equatorial lowland forests probably became restricted at the last glacial maximum to enclaves in Liberia, the Ivory Coast, Nigeria, southern Cameroon, Gabon, eastern Zaire, and the East African coast. Refugia of montane forests occurred in the Cameroon highlands, the Tanganyika-Malawi highlands, and the highlands of eastern Zaire. Impoverished islands of montane forests also have persisted in northern Ethiopia and northern Kenya (Street 1981; Hamilton 1976). It is believed that during arid phases, savanna occupied much of what had been rain forest. During wetter periods, forests again occupied a more or less continuous band across western and central Africa, while savanna moved northward, occupying what was desert (Roberts 1984; Hamilton 1982). Such vast changes in the distribution of rain forest are suggested by the taxonomic impoverishment of African rain forests relative to those on other continents. For example, the flora of tropical West Africa is very poor in Annonaceae, palms are rare, bamboos are insignificant, orchids are not abundant, and epiphytes and lianas are not as plentiful as in the rain forests of South America, Malaysia, and Madagascar. In addition, vegetational change during the Quaternary may have eliminated many rain forest species, since the total number of plant taxa in tropical Africa today is low compared with that in Southeast Asia (see Axelrod and Raven 1978). Finally, only the areas named above as probable enclaves for rain forest during the last glacial maximum are rich in endemic plants and animals, suggesting that these were stable areas of rain forest during the late Cenozoic. Other areas that are now forested (e.g., the Congo basin) are poor in endemic species, indicating that forest fully retreated from such areas during the late Quaternary (Hamilton 1982).

Despite this dramatic vegetational flux, the scale of change was apparently not sufficient to completely eliminate the complex mosaic of ecological zones in Africa; thus animals could survive by migrating with shifting vegetational types. The disjunct distributions of many animal species today (e.g., isolated patches of certain bird and insect species, distributions of *Oryx*, *Ceratotherium* [white rhino], *Giraffa*, Madoquinae [dik-dik], and many small mammals) corroborate the idea that there was significant disruption and reformation of formerly continuous habitats during the late Cenozoic (Butzer and Cooke 1982).

High rates of large mammal extinction seem to have occurred during various intervals prior to the Quaternary. Generic extinctions for large mammals declined to a low level during the Pliocene. Although extinctions during the Quaternary rose somewhat, the majority of these events took place during the early Pleistocene: 59% of 56 extinct genera had their last appearance during the early Pleistocene, as opposed to 21% during the middle Pleistocene, and 20% during the late Pleistocene (Maglio 1978). Increased evidence for late Pleistocene to Holocene faunas from archeological assemblages indicates that dominant herbivores in East African savanna communities differed from those of the present day, consisting of now extinct or geographically displaced taxa (Marean and Gifford-Gonzalez 1991; Potts and Deino MS). In spite of this, the absence of a pronounced megafaunal extinction event during the late Pleistocene may partly explain the present high diversity of large ungulates in the African savannas (in contrast to those of the Americas and Australia).

Uniquely important features of the present African mammalian fauna are (1) the widest spectrum of large ungulates known in the modern world; (2) more species of bovids than found on any other continent; (3) numerous endemic insectivores, including kangaroolike elephant shrews, fossorial golden "moles," and aquatic otter shrews; (4) an extreme ecological and morphological diversity of rodents, although not nearly as many species occur as in South America; and (5) a great diversity of large hunting and scavenging carnivores and an even wider diversity of the viverrid carnivores, which fill most of the small carnivore and omnivore niches on the continent (Bigalke 1978). Each of these features are particular present-day expressions in African faunas of the primary global ecological trends of the late Cenozoic.

3.4 Eurasia

Introduction

Europe and Asia form a huge continental mass that experienced a regionally diverse ecological history during the late Cenozoic. While similar-age biotas of Africa have been studied from an ecological viewpoint, this perspective is absent in most accounts of the fossil faunas and floras of Eurasia. The faunas have been used primarily in systematic and biostratigraphic analyses, to study paleoclimates and vegetation, or to correlate changes in the terrestrial biota with marine isotopic indicators of climate. Rather than offering comprehensive coverage of each region, this review aims only to exemplify the changes in the biota and climates through time in certain Eurasian locales. Paleontological emphasis on biochronology and correlation means that ecological shifts must be inferred largely from the appearance and disappearance of certain taxa rather than by the analysis of entire faunas and floras. A summary of major ecological events in Eurasia is given in figure 7.6, and taxonomic names are explained in appendix tables A-7.1, A-7.2, and A-7.4.

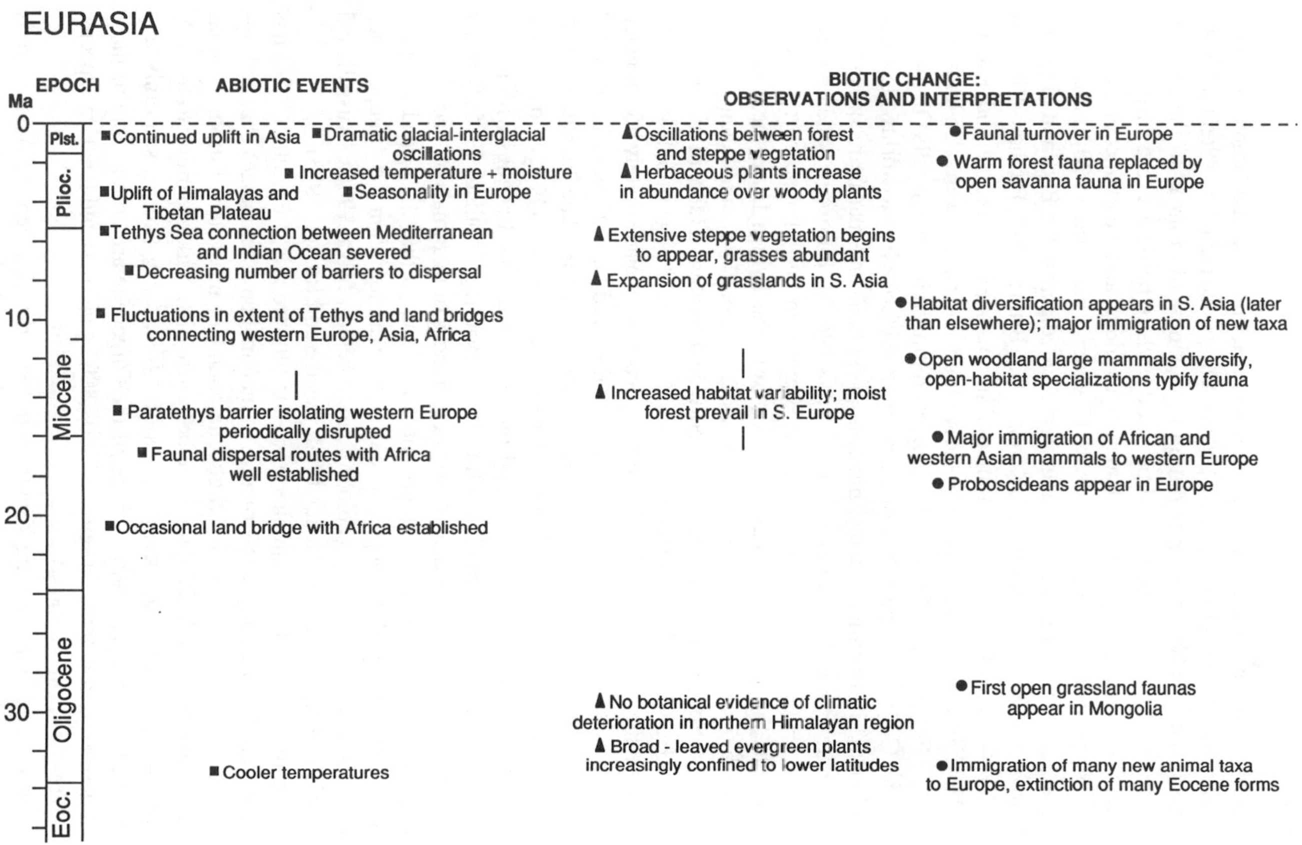

Fig. 7.6. Overview of abiotic and biotic changes in Eurasia during the late Cenozoic. Symbols same as in figure 7.2.

Oligocene

An extraordinary, sudden change in the fauna, known as the Grand Coupure, occurred in western Europe at the Eocene-Oligocene boundary (Stehlin 1909). This event involved the immigration (apparently from areas to the east) of many new taxa, artiodactyls and perissodactyls in particular (e.g., rhinocerotoids, chalicotheriids, anthracotheres, tayassuids), and the extinction of many Eocene genera and species. At least 17 generic extinctions, 20 first appearances, and 25 unaffected genera are represented across the Eocene-Oligocene boundary in western Europe (McKenna 1983; Brunet 1977; Prothero 1985). A wider sample of European mammals, summarized by Savage and Russell (1983), indicates that 34% of 114 early Oligocene genera were also known in the late Eocene, and 57% of these genera continued into the middle Oligocene.

The Grande Coupure has often been characterized as a worldwide event evidenced in mammal faunas, macroflora, pollen, and marine records (e.g., Legendre 1986; Pomerol and Premoli-Silva 1986b). European pollen records indeed suggest that dramatic shifts in vegetation and climate corresponded with the change in mammals. Samples from the British Isles, for instance, indicate that the Oligocene of northwestern Europe was significantly cooler than the Eocene and exhibited markedly cold winters (Hubbard and Boulter 1983). However, the occurrence of a sudden biotic change of unusually large scope at the Eocene/Oligocene boundary has not been adequately documented outside of Europe, and it is now clear that the word "event" must be qualified (Russell and Tobien 1986). Moreover faunas from the late Eocene to mid-Oligocene in Europe and North America appear to have changed more by a progressive turnover in species and morphological modifications within lineages associated with climatic cooling (e.g., dwarfing: Prothero 1985; Swisher and Prothero 1990). In Europe, first occurrences of mammals outpace last occurrences near the Eocene/Oligocene boundary, so that the Grand Coupure apparently involved the appearance of new taxa more than the disappearance of taxa already in place (Pomerol and Premoli-Silva 1986a; Russell and Tobien 1986; Lopez and Thaler 1974; Savage and Russell 1983).

On the basis of fossil samples from karst fissure fills, Legendre (1986) has proposed that the body size structure of mammalian assemblages changed significantly across the Eocene-Oligocene boundary in western Europe. Faunas preceding the Grande Coupure have body size distributions similar to those of well-vegetated tropical environments of Africa today (i.e., a smoothly distributed size array), whereas those of the Oligocene are comparable to those of desert environments (absence of 500–8000 g animals and few large species relative to forms <500 g).

Although few faunas are known from the early Oligocene of Asia, there is little to suggest that eastern Asia was involved in the large migrational events known in Europe. Instead the faunas of eastern Asia appear to have maintained a strong endemic character (Savage and Russell 1983). Paleobotanical evidence from China indicates that the climate also became markedly cooler there relative to the Eocene. This is true particularly in northern China, where microphyllous, broad-leaved deciduous forest mixed with coniferous forest was present and warm-adapted plants of the Eocene were absent. A warm and wet tropical climate with broad-leaved evergreen forests of Lauraceae and Fagaceae remained in southern China (Wang 1984). On the basis of data from eastern and Southeast Asia, Wolfe (1985) concludes that various types of broad-leaved evergreen vegetation were more confined to lower latitudes during the Oligocene than during the Eocene or even in recent times. Spruce, hemlock, and alder occurred in equatorial Borneo during the Oligocene, though these taxa apparently disappeared from there in the early Miocene (Muller 1966).

The middle Oligocene Hsanda Gol fauna from Mongolia is believed to represent an open-country habitat because of the presence of a hypsodont ruminant, ctenodactylid rodents, and ochotonid lagomorphs that are considered to be grass-adapted (Van Couvering 1980). If these interpretations are correct, then this fauna represents the earliest open-grass-dominated community recorded in the Old World. Bernor (1983) notes that the very early occurrence of the advanced bovid *Oioceros* in northern China supports the idea that open habitats occurred earlier in eastern Asia than in other areas of the Old World.

On the basis of a good pollen record in the northern Himalayas, there is little to suggest any floral change in this region related to environmental deterioration at the Eocene/Oligocene boundary and into the Oligocene. Instead vegetation remained tropical (upland forests, swamps and marshes, and abundant ferns), and rainfall may even have increased (Mathur 1984).

Miocene

It is estimated that at the beginning of the Miocene Asia was about 3–5° south of its present position. At that time the Sea of Japan, the Yellow Sea, the East China Sea, and the South China Sea had not yet appeared (they began to open by the early Pliocene) (Xu 1984a). During the earliest Miocene in Europe and western Asia, the Paratethys Sea extended from the southwestern limit of the Alps, eastward north of the Alps through central Europe, to the area of the present Aral Sea. The Paratethys had narrow connections with the Mediterranean basin; the latter formed part of the Tethys Sea, which connected with the Indian Ocean (Bernor 1983; Whybrow 1984) (figs. 7.1, 7.6.). The Indian subcontinent reached the southern edge of Eurasia by the middle Eocene (fig. 6.4). Although crustal compaction began 40–50 Ma, closure of the Tethys Sea and significant elevation of the Himalayas and the Tibetan (Qinghai-Xizang) plateau were a phenomenon of the late Neogene. Estimates of the rate and timing of uplift of the Himalayas and the Tibetan plateau during the Neogene and Quaternary vary (e.g., compare articles in Whyte 1984 and Laporte and Zihlman 1983), but it is generally agreed that global and regional climate tended to be cooler and drier as a result of the uplift. In eastern Asia this trend is indicated by the occurrence of temperate plants during the early and middle Miocene of Japan (compared with subtropical and warm temperate plants of Paleogene forests) (Xu 1984a) and by the eventual development of steppes and semidesert during the late Miocene in northwest China (though the rest of northern China exhibited broad-leaved deciduous forests, and southeast China retained warm and moist subtropical forests) (Wang 1984; Xu 1984b; Wu and Olsen 1985).

The earliest Miocene (25–20 Ma) faunas of Europe are considered to be largely an impoverished Oligocene fauna, showing little change except for some first appearances due to in situ evolution (Mein 1979). The Tethyan complex of seas probably acted as a barrier between most of Europe, on the one hand, and eastern Europe, Asia, and Africa, on the other. It is likely that this barrier buffered Europe from the extensive faunal changes seen in Africa and perhaps the rest of Eurasia (Bernor 1983). At least an occasional land bridge between Arabia and Iran, which connected Africa and Eurasia, existed during the period 26–23 Ma, though this connection was low-lying and sometimes submerged. The faunas of eastern Arabia during this time show similarities to faunas from Pakistan, China, and Mongolia, while a poor floral record from Arabia suggests grassy scrubland and palms growing close to freshwater lakes (Whybrow 1984). African proboscideans have been reported in Europe from deposits estimated at 20 Ma and are well-documented at 18.3 Ma in the Siwalik sequence of Pakistan (Barry et al. 1985). Certainly by 17.5–15 Ma, dispersal routes out of Africa and into most of Eurasia (via Arabia) were well established, reaching at least into central Europe. By this time European faunas were also connected with eastern Asia and include North American immigrants (Bernor 1983). One early Miocene immigrant to Eurasia was the equid *Anchitherium*, which diversified into species that persisted as late as the Pliocene in China; all species of this equid retain primitive features of the limbs and dentition (Carroll 1988). *Chalicotherium* also appears in Europe during the early Miocene; this perissodactyl apparently possessed limb proportions analogous to those of a gorilla and was able to assume a bipedal stance that may have enabled it to browse on tall foliage (Coombs 1983). About 17.5–15 Ma, the Paratethys linked up with the Tethys Sea further west than in the earliest Miocene, but periodically it still formed a barrier that isolated western Europe (Bernor 1983).

Around 15 Ma, western Europe experienced a major immigration of mammals (involving proboscideans, bovids, suids, and hominoid primates) from Africa and western Asia (Mein 1979). Bernor (1983) notes that in southern France, two distinct kinds of mammal species associations have been identified from middle Miocene faunas, one aligned with closed forest and the other with subtropical woodland (see also Nagatoshi 1987). Floral and faunal evidence from the middle and late Miocene indicates that western Europe was characterized by tropical to subtropical vegetation (Gregor 1982) and associated large mammals (forest dwellers, grassland dwellers, and ecotonal forms). Moist forest prevailed in this region up to about 5 Ma (Meon et al. 1979).

European localities from the latter part of the middle Miocene possess faunas and floras that appear to indicate greater habitat variation than existed earlier in the Miocene, including ecotones between forest, open vegetation, and marshy areas (as shown, e.g., by the association of such forms as *Anchitherium* [forest], *Hipparion* [open vegetation], and *Tapirus* [marsh]) (Nagatoshi 1987). During this period the Paratethys, Mediterranean basin, and Tethys became disconnected, forming a number of land dispersal routes utilized by the North American equid *Hipparion* and other mammalian species. Western Europe experienced a major immigration that included *Hipparion*, a murid rodent from Southeast Asia, several carnivores, the rhino *Diceros*, and several giraffids. This immigration is thought to have enriched the fauna rather than caused a major extinction, since archaic insectivores, rodents, primates, carnivores, proboscideans, cervids, and suids survived. According to Bernor (1983): "Ecologically, this pattern of immigration of carnivores and browsing ungulates superimposed upon the retention of several Astaracian [15–12 Ma] forest denizens suggests the possibility that provincial environments were enriched by the shift toward more open woodlands." On the other hand, the faunas of central and north central Europe appear to have been less affected by such open woodland immigrants, despite being accessible to mammals that probably went from Asia to western Europe. The faunas of central Europe in fact retained a more closed vegetation aspect.

In contrast with middle Miocene faunas of Europe that are believed to be indicative of relatively closed vegetation, those of western Asia and Arabia (Bernor's sub-Paratethyan region) record the evolution of open-country large mammals. The sub-Paratethyan region was evidently the center of what Bernor (1983) calls the open-country woodland chronofauna, characterized by the presence of species of large body size with open habitat specializations (e.g., hypsodonty, cursoriality). Known throughout Africa and Eurasia, this mammalian chronofauna began to take shape in Eurasia by the early Miocene. It developed gradually, at different paces in different regions, and had its maximum geographic extent during the late Miocene (though each region maintained its own distinctive taxonomic character). Consisting of a rich array of bovids, equids, giraffids, rhinos, felids, hyaenids, and proboscideans, this fauna by the late Miocene (10–5 Ma) in Eurasia apparently had a diversity far exceeding that of extant savanna communities in the Old World, especially in the number of very large taxa (giraffe–elephant size) (Bernor 1983, 1984). The peak in generic diversity exhibited by the late Miocene chronofauna of Eurasia parallels and exceeds that of the Clarendonian chronofauna of North America (broadly defined, between 14–7 Ma) (Savage and Russell 1983).

The interplay between forest and more open habitats during the middle–late

Miocene is exemplified by the ecological-evolutionary contrast between hominoid primates and hipparionine horses (Bernor 1984). During the middle Miocene thin-enameled hominoids (e.g., *Dryopithecus*) in western Europe appear to be closely associated with moist and closed woodland to forest environments. These hominoids disappeared around 10 Ma as environments in western Europe became more temperate and in some cases drier; the radiation of cercopithecid monkeys in Eurasia immediately followed these changes. Middle and late Miocene hominoids with thick occlusal enamel (e.g., *Sivapithecus, Kenyapithecus, Lufengpithecus*) survived and radiated in areas of eastern Europe, eastern Africa, and western to eastern Asia in open and/or dry habitats (e.g., Paşalar, Çandir localities) and in more closed woodland areas (e.g., Lufeng, Siwaliks). The diversification of these hominoids followed by the initial radiation of cercopithecid monkeys may be related to the ability of both groups to thrive in either open woodland or more closed vegetation. It has been proposed by Bernor that in addition to the Paratethys Sea, the progressive shift to more open woodlands provided an effective environmental barrier to the continuity of later Miocene hominoid populations. While Miocene hominoids thrived in closed to open woodlands, hipparions diversified in open-country habitats, especially savanna-mosaics. Although low-crowned hipparions may have coincided with closed-woodland habitats (and with *Dryopithecus* in western Europe), the appearance of higher-crowned species across Eurasia and Africa appears to be related to their increased ability to disperse as open country expanded. The diachroneity of their first appearance across Europe, southwestern Asia, and North Africa (12 Ma) as opposed to Africa (11–9.5 Ma) and the Siwaliks (10–9 Ma) may reflect different rates of change in habitats from relatively closed to more open vegetation, or more subtle shifts in other ecological barriers.

Much of what is known about Neogene vertebrates in southern Asia is based on abundant fossils from the Siwalik formations of Pakistan and Indian. The record of early Miocene faunas in this region is still relatively poor but shows clear evidence of large-bodied mammals, including a considerable diversity of anthracotheres, rhinoceratids, and a deinothere; the diversity of suids was low and the carnivores represented a mixture of Eurasian (amphicyonids) and African (creodonts) elements (Bernor 1984; Barry et al. 1985). By 16.1 Ma, the oldest-known bovids occur in this sequence, along with other ruminants and rodents. By 13.5 Ma, the diversity of bovids rose dramatically to include at least 5 genera. During the period 16–13.5 Ma, bovids, muroid rodents, and to a lesser extent tragulids and gomphotheres radiated independently in southern Asia and Africa. Overall, however, after 16 Ma, in situ evolution is believed to have had a minor influence over faunal change in the Siwaliks relative to the immigration of species from eastern and southeastern Asia, Europe, and Africa (Barry et al. 1985).

The early and middle Miocene faunas from southern Asia have a lower diversity of browsing ruminants and a greater diversity of higher primates, suids, tragulids, and anthracotheres than middle Miocene faunas of western Asia (Bernor 1984). Archaic groups such as the adapids and creodonts survived longer there than anywhere else, with the last-dated creodonts occurring at about 9 Ma in the Siwalik sequence of Pakistan (J. C. Barry pers. comm. 1990). Indeed, throughout the Miocene the fauna of southern India retained its own distinctive, provincial character reminiscent of archaic faunas from other parts of the Old World. Bernor (1983) associates this archaic retention with Axelrod's (1975) inference that sclerophyllous vegetation appeared relatively late (late Miocene) in southwestern India and Pakistan. An analysis employing ecological diversity spectra of 13–11 Ma faunas from the Siwaliks of India indicates that browsers dominate the fauna and that woodlands were probably more closed than those of the present day (Gaur and Chopra 1983).

The middle and late Miocene faunas of the Siwaliks are quite similar to one another, in contrast with faunas of Europe and Africa, which underwent considerable change at the end of the middle Miocene. The immigration of equids, suids, and large giraffids from Eurasia marks a late Miocene faunal transition on the Indian subcontinent, commencing around 9.5 Ma. The appearance of hipparionine equids at this time occurred somewhat later than in Europe and Africa. Other new rodent and tragulid taxa were derived from Africa, Southeast Asia, or locally. The faunal shift initiated by these immigrant taxa involved an almost complete replacement of the most common elements in the older Siwalik fauna (Barry et al. 1982, 1985). By 8 Ma, a mosaic of vegetation types (forest, woodland-savanna, open grassy areas) was inhabited by an ungulate fauna that was richer than any recorded today on the Indian subcontinent. Browsing and grazing ungulates of small to medium size predominated, implying the occurrence of a heterogenous and well-partitioned mosaic of habitats. This paleocommunity consisted of a minimum of 36 genera and 39 species >5 kg, and at least 20 smaller mammalian taxa (Badgley and Behrensmeyer 1980). Another turnover event appears to have occurred 7.4 Ma; it involved a slow change (over at least 400 ky) in the ecological, not just taxonomic, makeup of the fauna. This coincides with a shift in carbon isotopes from paleosol carbonates, indicating that a predominantly C_3 vegetation (trees, shrubs, some types of cool growing season grasses) gave way to open habitats with C_4 grasses (warm growing season, associated with more arid climates) between about 7.4 and 5 Ma (Quade et al. 1989; Cerling et al. 1989). The vegetation appears to have been virtually pure C_4 grasses on the river floodplains as early as about 5 Ma. The proposed timing for the expansion of grasslands between 8 and 5 Ma in the Siwaliks implies that this important habitat change occurred earlier in southern Asia than in Africa.

Corresponding with this inferred vegetational change, the fauna prior to 7.3 Ma is dominated by browsers, including tragulids, early *Hipparion*, *Okapia*-like giraffids, and low-crowned bovids, whereas grazers, including a large, hypsodont bovid, are better represented after this period of change (Barry et al. 1985; Quade et al. 1989; Quade et al. in press). Fossil rodents from the Siwalik sequence also show evidence of increasing aridity through time (Jacobs and Flynn 1981). Despite these ecological changes, Siwalik faunas from the late Miocene bear a greater resemblance to earlier Miocene faunas than is the case in Africa or western Asia. A faunal association of gomphotheres, rhinocerotids, tragulids, boselaphines, chalicotheres, anthracotheres, sivatheres, and amphicyonids persists in the Siwaliks from about 12 Ma to at least 5 Ma, even in the presence of new animal taxa (e.g., hipparionines), shifts in relative abundance within these groups, and changes in the vegetational mosaic (Hill et al. 1985; Barry et al. 1982).

Palynological evidence supports inferences from faunas and geochemistry concerning the timing of habitat changes in southern Asia. On the basis of the occurrence of fern spores and pollen of Palmae, Rutaceae, and other warm and moist habitat plants, early Miocene environments are believed to have consisted mainly of wet upland forests, swamps and marshes. Grasses are also known in the early Miocene and increase in the record during the early part of the middle Miocene. At that time the vegetation around lakes, swamps, and streams in the Himalaya region consisted of herbaceous and tree ferns and tall mesophyllous grasses. The presence of tropical lowland forests is also evident in the pollen record, which consists of taxa of Malayan and Southeast Asian distribution. Open grasslands with complexes of Gramineae and Compositae developed in Nepal during the latter part of the middle Miocene, though grasses did not become very widespread through the Himalayas and northern India until the end of the Miocene. During the late Miocene, conifer pollen also enters the fossil record as a result of the uplift of the Tibetan plateau and Himalayas, though such vegetation was apparently sparse in lowland areas (Mathur 1984; Song et al. 1984; Vishnu-Mittre 1984).

Pollen samples indicate that extensive steppe vegetation had developed in central Asia and Siberia by the late Miocene. By 5 Ma, western Turkey experienced an influx of *Artemisia* and the first massive occurrence of Gramineae, two of the main indicators of steppe. This change coincided with the disappearance of many of the Miocene woody plants from the pollen record. A continuous core record of pollen from the Black Sea indicates that forest declined in favor of steppe vegetation (*Artemisia*, Chenopodiaceae, and Gramineae) from 10 Ma to the Olduvai magnetic event (1.8 Ma) (Traverse 1988).

Evidence concerning species distributions during the late Miocene, when the open-woodland large mammal chronofauna of Eurasia reached its maximum diversity and extent, suggests that there were minimal physiographic or environmental barriers across Eurasia. However, some ecological provinciality continued, and faunal changes due to immigration were not synchronous across regions. Western Asia possessed the richest savanna-mosaic fauna. Although western Europe received some of the advanced species from this region during the late Miocene, it also retained some archaic carnivores and ungulates and showed local evolutionary changes in small mammal families. Immigration of advanced open-woodland ungulates and carnivores into central Europe was more limited than into western and southern Europe, which suggests that different European regions experienced the overall shift toward open woodland or savanna somewhat independently (Bernor 1983, 1984). It appears that faunas of the Indian subcontinent maintained their endemic character through the late Miocene, prior to the time when several African taxa, as well as *Equus* and cervids from Eurasia, entered this area between 5.3 and 2.9 Ma (Barry et al. 1982, Azzaroli 1983).

Tectonic activity had an important effect on the climate and vegetation of the Mediterranean region during the Miocene and Pliocene. The marine connection between the Indian Ocean and the Mediterranean was severed during the mid-Miocene, and this probably contributed to increasing aridity in southern Europe. The desiccation of the Mediterranean and Black Sea basins during the Messinian (late Miocene) corresponds with southward expansion of relatively xerophytic vegetation in the Balkan area (Pantic and Mihajlovic 1977). Toward the end of the Messinian (5 Ma), western Europe developed a Mediterranean climate and flora with evidence of a distinctive dry season (Meon et al. 1979; Gregor 1982). However, the habitats of Syria and Iran during the Messinian appear to have been wet and probably cool, indicating that aridification was not an effect in all areas surrounding the Mediterranean (Whybrow 1984).

Pliocene to Quaternary

At the outset of the Pliocene, older Tertiary plant associations, especially palms and other warmth-loving plants, began to drop out of the Black Sea pollen record, and the Gramineae made their first massive appearance. During the Pliocene, conifers of many genera dominated in the Black Sea basin, marking a dramatic cooling (Traverse 1988). The tendency in western Europe toward a distinctive dry season was amplified up to about 3.3–2.5 Ma, when floral and faunal evidence both indicate warm-temperate moist forest, with summer aridity, though xerophytic Mediterranean vegetation occurred at high latitudes (Meon et al. 1979). Azzaroli (1983) notes that a faunal dispersal event took place 2.6–2.5 Ma in Europe that involved the disappearance of a warm forest assemblage of mammals and its replacement with an assemblage indicative of more open savanna. In central Europe, the pollen families that became more important in the Pliocene all point toward an emphasis on herbaceous over woody plants during a period of cooling (Gregor 1982; Traverse 1988). In Mediterranean Greece, conifers such as *Pinus* and certain Taxodiaceae (which were well established in the early Pliocene) dominated up through the Pliocene-Pleistocene boundary, and at that point the family Compositae was added to the pollen floras (Sauvage 1979). Between 1.9 and 1.4 Ma, western European faunas and floras examined by Meon et al. (1979) suggest the presence of both forests and herbaceous vegetation corresponding to a considerably cooler, but still humid, climate. Pollen records from France studied by Suc (1980) indicate the first predominance of xerophytic steppe plants toward the end of the Pliocene, and then alternation between steppe and forest pollen floras.

Beginning especially in the late Neogene, China developed a highly diversified topography because of differential uplift of the Himalayas and the Tibetan plateau, creating a wealth of new habitats for high altitude plants. The approximately equal elevation that occurred across Tibet and the Himalayas during the Miocene became a high altitude barrier to dispersal during the Quaternary. The climatic effects of this orogeny, along with worldwide temperature decrease, led to increased drying on the plateau northeast of the Himalayas (by the early Pliocene) and blocking of the southwesterly monsoon during the summer months. Late Pliocene floras of China reflect an overall cooler climate than occurred during the Miocene. Paleobotanical records for the coastal areas of east China suggest the presence of evergreen, cold-resistant conifers with some broad-leaved deciduous forests, indicating a more humid and warm temperate climate than today. In addition, uplift of the Qin Ling mountains during the early Quaternary created a barrier preventing the passing of monsoon winds from the south into northern China. Since the mid-Pleistocene, this barrier accentuated the contrast between the cool-dry north and the warm-wet south (Xu 1984a, 1984b; Guo 1984). (Evidence of a north-south vegetational and climatic contrast, however, extends into the Miocene. Tobien et al. [1984] point out that the presence of bunodont mastodonts [e.g., *Synconolophus*] in south central China suggests a more tropical, forested habitat in comparison with more northerly representatives of Miocene mastodonts with cement-covered, relatively high-crowned molars [e.g., *Anancus, Tetralophodon*].)

The main vegetational changes during the Quaternary of China include the following: warm temperate deciduous broad-leaved forest moved southward; Siberian cold temperate conifers moved southward to form taiga in northern China; xeromorphic vegetation of southwest and north China expanded; herbaceous communities much simpler in structure and taxonomic composition became dominant in northern China during the late Pleistocene; alpine meadows and desert formed on the Tibetan plateau; and alpine forests expanded downward during the cold glacial phases in eastern China (Wang 1984).

The loess sequence of central China provides a record of environmental change over the past 2.5 Ma. Deposition of the earliest loess strata (2.5–2.3 Ma) marks a major shift from a warm-humid environment to harsh continental steppe, like that which typified the middle and late Pleistocene of northern temperate latitudes in Eurasia. Although pollen records pose problems of contamination by long-distance wind transport, such records imply that landscapes of loess had a highly restricted vegetational cover of grass, shrubs, weedy herbs, and very few trees. However, the soil layers that interfinger with loess layers exhibit pollen profiles indicative of relatively dense vegetation (woodlands present at least near the loess plateau) (Kukla 1987).

The earliest Quaternary faunas of China (Nihewan) exhibit continuity with Neogene faunas, though some new genera appear for the first time (e.g., *Equus, Camelus, Elephas, Bison,* and *Ovis*). Quaternary faunas of China have been categorized for years by such terms as the "*Gigantopithecus* Fauna" and "*Ailuropoda-Stegodon* Fauna," suggesting that long-term associations occur among certain taxa. However, it is now believed that the mammalian faunas associated with the large primate *Gigantopithecus* underwent considerable change during the early and middle Pleistocene. Although the proboscidean *Stegodon* and the giant panda *Ailuropoda* are especially recurrent in faunal assemblages of southern China, these taxa are associated with a changing and increasingly modern fauna from early to late Pleistocene (Han and Xu 1985). The middle Pleistocene faunas of Zhoukoudian cave in northern China (0.5–0.25 Ma), which have been studied extensively from a climatic perspective, show that assemblages of forest-dominated and grassland-dominated mammals alternate stratigraphically (e.g., Xu and Lian 1982).

Neogene macrofossil floras from Sumatra studied by Krausel (1929) contain dicotyledonous leaves with entire margins, suggesting that the floras were similar to modern rain forests (Flenley 1979). The absence of open-environment animals such as camelids, equids, and giraffoids during the Quaternary of Southeast Asia suggests that tropical forest rather than savanna or other open vegetation predominated during most of the past 2 million years (Pope 1984, 1988). The few published pollen data from Sumatra and Java suggest that lowland and highland forest did indeed persist during the Quaternary, even during periods when desiccation and demise of forests occurred in New Guinea and Australia (Flenley 1979).

In the Himalaya-Siwalik region, climatic oscillations during the Pliocene and Quaternary are reflected by shifts in the pollen record between forests in warm times and *Artemisia*/chenopod/grass steppe during cool times, though it is not known whether the latter periods were also associated with regional glaciations prior to the middle Pleistocene (Vishnu-Mittre 1984). Uplift of the Himalayas apparently provided refuges for plants that may have otherwise been exterminated by changing climatic conditions. Although xerophytic plants increased and rain forest plants were lost during the late Neogene, some moisture-loving groups of Indian plants survived in the Himalayas (Traverse 1982). Significant uplift of the Himalayas during the early Quaternary helped to create the arctic-alpine belt to which central Asian floras contributed (Vishnu-Mittre 1984). The Quaternary faunas of India (mainly late Pleistocene) are summarized by Badam (1984).

On the basis of correlations of terrestrial sequences, including loess deposits and terraces, with variations in oxygen isotopic composition in deep-sea sequences, at least seventeen glacial-interglacial cycles are known in Europe over the past 1.7 my (Kukla 1977). Eight of these occurred just in the past 0.8 my. Climatic oscillations during the latter period were apparently associated with temperate forests during the interglacials, interrupted by invasions of grass and herbs and intervals of predominantly steppelike vegetation during glacial phases (Birks and Birks 1980; Kukla 1977; Roberts 1984). By 0.9 Ma, floras and faunas of western Europe reflect truly cold intervals accompanied by temperate hot and cold seasons; both forest and prairie species are preserved in the fossil biotas (Meon et al. 1979). This date appears to mark a revolutionary event in European faunas that involved the modernization of the fauna and the emergence of larger-bodied mammals than previously known in certain artiodactyl families (e.g., bovids such as *Bos, Bison, Ovibos,* and the giant cervid *Megaceros*). Over 50% of late Villafranchian species apparently disappeared without descendants, and most other lineages that survived this event may have done so with considerable evolutionary change, some involving increase in body size (Azzaroli 1983).

In response to glacial-interglacial oscillations, mammals either migrated or expanded their ecological tolerance to withstand extreme climatic and vegetational changes. Associations of taxa with certain types of environments and vegetation types during the Pleistocene of Europe suggest that some have changed their ecological preferences or tolerances. For example, the hamster *Cricetus cricetus* is now found on the Asian steppe but has been recovered from sediments deposited in forested environments during the Cromerian interglacial. While the genus *Equus* typically is associated with open environments, modern wild *Equus cabalus* also occurs in forested habitats, and fossil *Equus* is found in forested environments during the interglacials and in glacial steppe habitats (Birks and Birks 1980; Sutcliffe 1985).

During the past 10 ky, most European pollen records have experienced a decline in Gramineae and *Artemisia* pollen (abundant during the glacials) and notable increases in tree pollen. The decline of steppe pollen began earlier in the Mediterranean region than farther north (Traverse 1988; see Huntley and Birks 1983).

Extinctions in Eurasia during the late Quaternary did not reach the dramatic peak seen in the Americas or Australia. In Asia, as in Africa, no late Pleistocene families were lost. Out of 37 genera of large mammals that became extinct in Europe over the past 3 my, 13 became extinct near the end of the Pleistocene (Martin 1984). The relatively gradual disappearance of large mammal genera from Eurasia during the Quaternary therefore does not seem to require special explanation, though both climatic change and the appearance of human hunters have been proposed as factors responsible for the demise of certain taxa (Martin and Klein 1984).

3.5 *Australia*

Introduction

Australia separated from Antarctica at about 45 Ma and has stayed completely isolated from other major land masses (fig. 7.1). The development of Australian ecosystems since the Paleogene thus affords intriguing comparisons to the ecological histories of other continents. Australia is well known for its unusual fauna and flora from a taxonomic standpoint, and this distinctiveness also appears in certain ecological characteristics of the biota (e.g., the radiation of large saltatory marsupials and the development of plant communities not recognized elsewhere). Despite these dissimilarities, however, Australia has undergone ecological changes over the past 34 my that parallel those seen on other continents (fig. 7.7; see appendix table A-7.4 for a guide to animal and plant names).

The main theme in the evolution of the Australian flora over the past 34 my has been a change from closed broad-leaved forests to more open, dry sclerophyll forests, grasslands, and deserts. Scleromorphy, the most distinctive quality of the flora, refers to leathery, hard, and reduced leaves (Beadle 1967). Over time, the dry-adapted sclerophyll plants became dominant in the Australian flora, and Pleistocene aridity along with human impact favored the fire-resistant forms that make up much of the present-day vegetation. The main exceptions are in broad-leaved rain forests limited to the eastern edge of the continent and the wet sclerophyll forests of the southeastern and southwestern high country, as well as scattered semideciduous tropical woodlands (Kershaw 1984).

The diverse adaptations of the modern sclerophyll flora include: hard, woody fruits and seeds that require fire to germinate; trees with thick, insulating bark that protects living tissues from overheating; flammable oils in tree leaves (eucalypts), which encourage intense but brief fires that minimally damage woody parts of a tree; buds produced below the bark so that when the leafy crown of a tree is destroyed by fire, new growth commences over the entire trunk; shrubs that possess woody underground tubers capable of withstanding hot fires and that send up new branches after ashes have cooled; and seeds that lie dormant for years in soil and germinate only after a fire has produced nutrients from the ashes (White 1986; Kershaw 1986). As an important factor in the rejuvenation of sclerophyll woodland-forest communities, fire evidently has helped to maintain plant diversity and, secondarily, the diversity of insects and birds that are the most conspicuous inhabitants of recently burned areas. Many of these adaptations to fire appear to be a relatively recent response to arid Pleistocene climates and aboriginal use of fire. The preexisting non-fire-adapted sclerophyll and conifer dry-adapted flora was replaced over much of the continent during this time (Kershaw 1984, 1986).

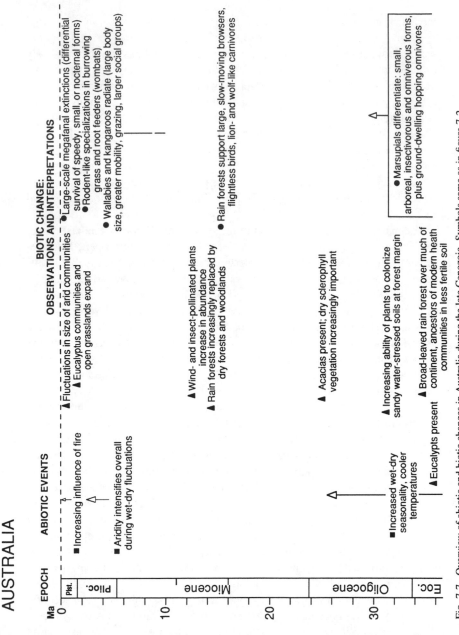

AUSTRALIA

BIOTIC CHANGE:
OBSERVATIONS AND INTERPRETATIONS

● Large-scale megafaunal extinctions (differential survival of speedy, small, or nocternal forms)
● Rodent-like specializations in burrowing grass and root feeders (wombats)
● Wallabies and kangaroos radiate (large body size, greater mobility, grazing, larger social groups)

● Rain forests support large, slow-moving browsers, flightless birds, lion- and wolf-like carnivores

● Marsupials differentiate: small, arboreal, insectivorous and omniverous forms, plus ground-dwelling hopping omnivores

▲ Fluctuations in size of arid communities
▲ Eucalyptus communities and open grasslands expand

▲ Wind- and insect-pollinated plants increase in abundance
▲ Rain forests increasingly replaced by dry forests and woodlands

▲ Acacias present; dry sclerophyll vegetation increasingly important

▲ Increasing ability of plants to colonize sandy water-stressed soils at forest margin
▲ Broad-leaved rain forest over much of continent, ancestors of modern heath communities in less fertile soil
▲ Eucalypts present

ABIOTIC EVENTS

■ Increasing influence of fire

■ Aridity intensifies overall during wet-dry fluctuations

■ Increased wet-dry seasonality, cooler temperatures
▲ Eucalypts present

EPOCH

Plst.
Plioc.
Miocene
Oligocene
Eoc.

Ma
0
10
20
30

Fig. 7.7. Overview of abiotic and biotic changes in Australia during the late Cenozoic. Symbols same as in figure 7.2.

In contrast to floras of Africa and the Americas, stem succulents are generally absent in Australia; this is something of an enigma because other structural adaptations common in the world's deserts do occur, including leaf succulence. As in the warm American and African deserts, spiny plants are common in arid Australia, although spinescence is developed independently in different lineages (e.g., spinifex grasses). Drought-tolerant members of the Leguminosae (*Cassia, Acacia*) are also widespread here, as they are in many parts of arid to semiarid America and Africa. Besides scleromorphy, these dry-adapted plants of Australia are typified by rapid growth and efficient seed dispersal under adverse environmental conditions.

Eucalyptus and *Acacia* are the most dry-adapted arborescent genera. Represented today by more than 500 species, eucalypts occur in broad stands of woodland and forests covering approximately 25% of the continent, and they range from alpine to desertic areas. The Myrtaceae family (to which *Eucalyptus* belongs) today includes 14 genera that possess fleshy fruits, which are believed to represent a pre-Neogene element of the flora. Dominant in the family, however, are 55 genera characterized by dry capsular fruits, which are thought to have developed in response to increased aridity over time. Because of their physiological adaptations to fire, eucalypt woodlands and forests rejuvenate quickly after burning.

Most of the acacias and cassias of Australia possess phyllodinous leaves (stems with a flattened leaflike morphology). Retention and flattening of the central rachis (the phyllode) and loss of the multiple leaflets that were attached to that rachis in the presumed ancestors of these leguminous genera have resulted in decreased water loss resulting from decreased leaf surface area. These two genera are the dominant shrubs over vast regions of central Australia where sands or gravels cover richer soils. Ant-*Acacia* interactions dominate the animal-plant interactions in mulga-dominated regions, and insect predation is also intense upon eucalypts.

Spinifex grasslands and heath include present-day communities dominated by sclerophylls. "Spinifex" refers to a uniquely Australian group of desert grasses, composed of about 50 species and representing the dominant plants of much of the arid lands of the central and northern parts of the interior. Insects, especially ants and termites, are the dominant animals (most abundant and highest biomass) in spinifex habitats. This results from the fact that spinifex is digestible to termites but is too tough for most mammalian and avian herbivores. Heath is another modern plant community type that initially developed in response to low soil fertility and later was able to expand into arid environments. Although dominant heath plants include banksias and other taxa of the Proteaceae which had their origin in Mesozoic rain forests, these plants today exhibit arid habitat characteristics (e.g., short internodal growth; leaves that are densely crowded, small, sharp, and stiff) (Vandenbeld 1988).

A wide belt of *Nothofagus* (southern beech) forest is traditionally thought to have covered the landmasses of Gondwana during the early Paleogene. Patches of beech forest persist in Australia, Tasmania, and the highlands of Southeast Asia. Such forest has been used as an analogue for the dominant late Paleogene–early Neogene environments of Australia (Specht 1981; Lange 1982). On the basis of paleobotanical evidence, however, this broad-leaved rain forest dominated by beech is now seen to have been discontinuous and spatially limited during the Eocene. During the Paleogene, large-leaved species of *Nothofagus* were replaced by modern-type, small-leaved forms in southern portions of the continent (Hill 1990). Analysis of fossil leaf morphology indicates warm and cool temperate to humid subtropical (and possible tropical) conditions; Eocene rain forests over the continent mirrored this climatic variation and generally were diverse taxonomically (Christophel and Greenwood 1989; Christophel 1989).

Oligocene to Pliocene

During the Oligocene, lower-diversity, smaller-leaved sclerophyllous taxa replaced the high-diversity leaf floras of the Eocene (Christophel 1989; Hill 1990). This change appears to reflect climatic deterioration related to the separation of Australia from Antarctica, formation of Antarctic ice caps, and development of cold circumpolar currents (Kemp 1978). Leaf physiognomy indicates that heightened seasonality in addition to a decrease in mean annual temperature was an important influence on vegetational change (Christophel and Greenwood 1989; Hill and MacPhail 1983). Fossil pollen indicates that *Eucalyptus* was abundant by the Oligocene, though diversification of this genus began in the Eocene (Lange 1978, 1982; Ambrose et al. 1979; Greenwood and Christophel pers. comm. 1991 on revised date). *Casuarina* (she-oaks) and *Callitris* are believed to have typified the earliest dry sclerophyll communities. *Callitris,* a gymnosperm native to Australia, is an ecomorphic parallel to the drought-adapted, western North American *Juniperus*. Like the latter genus in western North American deserts, *Callitris* was apparently widespread across the now-desert lowlands of Australia during the last glacial age. It now occurs in both mesic and arid habitats but is restricted because of the influence of fire (Singh and Luly in press).

According to one popular view, eucalypts and other sclerophyll taxa may have originally specialized in colonizing sandy, less nutritious soils and possibly water-stressed areas at the rain forest margins (Beadle 1967; Christophel and Greenwood 1989; Hill 1990). The Proteaceae and Casuarinaceae contributed significantly to the early dry sclerophyll vegetation. Since these families are characterized by proteoid roots related to colonizing sand, the early evolution of scleromorphic traits was likely related to impoverished soils, whereas adaptation to arid conditions was a later result. As a result of their ability to survive increasingly arid conditions during the Cenozoic, casuarinas, eucalypts, and other sclerophyllous taxa contributed importantly to new plant communities that replaced rain forests. Although Christophel (1989) notes that the dramatic increase in sclerophylly during the Oligocene and Miocene was related to climatic deterioration, this does not conflict with the view that the initial adaptations of sclerophylls were to impoverished soils.

The Oligocene fauna of Australia is poorly known. The first occurrence of marsupial fossils on the continent is late Oligocene, though marsupials must have a considerably longer fossil record. Ancestral diprotodontoids and dasyuroids were undoubtedly differentiated by the Oligocene, and most modern families had diversified by the Miocene (Carroll 1988; Tedford et al. 1975). Possible analogues for the early Dasyuroidea are the modern phascogales: small (e.g., 200 g), arboreal, with a diet composed of spiders and centipedes (Vandenbeld 1988) (table A-7.4). Represented today by small- to medium-sized insectivores, carnivores, and omnivores, the dasyuroids have occupied a functional role similar to that of the didelphoids in South America (Carroll 1988). Pygmy possums (*Ceratetus*) are believed to provide a reasonable analogue for the early Diprotodontoidea, as very small-bodied, arboreal nest builders, forest floor and tree foragers, with an omnivorous diet (insects, nectar, buds, and fruits). The wombats, koalas, kangaroos, and wallabies were derived from such forms. Inhabitants today of isolated patches of Queensland lowland rain forest, such macropods as the musky rat kangaroo probably represent analogues for the early macropods. They possess a possumlike hallux and an omnivorous diet of invertebrates and fruits. Protokangaroo groups similar to the bettongs and potoroos lived among shrubs and ground cover of early Australian forests. Their hind feet were used for hopping and their forelimbs for manipulating food. This trend toward terrestriality, as well as the trend of possums and dasyuroids toward colonizing higher levels of the forest, meant that by the Oligocene, the stage was set for the radiation of marsupials that occurred during the Cenozoic isolation of Australia (Vandenbeld 1988).

During the early to mid-Miocene, Australia met with increased seasonality of rainfall and periods of dryness. The appearance of *Acacia* pollen in the fossil record attests to the growing importance of dry sclerophyll vegetation at the expense of broad-leaved rain forest. Sluiter and Kershaw (1982) emphasize that early Neogene rain forests were not replaced directly by modern, eucalypt-dominated communities. Instead xerophytic forests of conifers and dry-adapted sclerophylls became dominant over much of the continent (especially northern and eastern portions), while eucalypt communities were relatively restricted until as recently as the late Quaternary.

During the late Cenozoic, wind- and bee-pollinated plants increased dramatically over plants pollinated by birds and nonflying mammals, and this also appears to be related to increasingly open environments. The Compositae (daisy family) first appears in the pollen record in the mid-Miocene and increases rapidly by the Pleistocene. Many members of this family have wind-dispersed seeds and provide many of the ephemerals that bloom after rainfalls in the deserts of Australia. Grass pollen first appeared in the Eocene but is very rare until the Pliocene and becomes abundant only in the Pleistocene (Martin 1978).

Climatic drying during the late Cenozoic was not uniform in time or space. The 15 Ma fossil locality of Riversleigh (northern Queensland) provides ample evidence of a warm, lush rain forest at a time when most of the continental interior was under water. A huge inland sea supported large crocodiles, freshwater dolphins, turtles, and lungfish. There were wading flamingoes in shallow waters. The forests supported large flightless birds (forerunners of cassowaries and emus) and a diverse marsupial fauna. The latter included a wide variety of possum- and cuscus-like taxa; hunting carnivores such as the lion-like *Wakaleo,* wolf-like thylacines (similar to the borhyaenids of South America), and quolls; proto-potoroos (a minor component, but one that would eventually give rise to the kangaroos); and diprotodonts. The Miocene diprotodonts were a varied group, consisting mostly of large, slow-moving browsers, some of which had specialized lips or snouts used to grasp plants on which they fed. Some were tapirlike, possessing well-developed proboscises (e.g., *Palorchestes*) (Carroll 1988; Vandenbeld 1988). The thylacine carnivores, known up until this century, developed specializations for eating flesh and also evolved into bone-crushing carrion feeders, such as *Sarcophilus* (Tasmanian devil).

After 15 Ma, rain forests of varying types apparently gave way to drier forests and woodland at an increasing rate. Aridity intensified after 6 Ma, though as on other continents, the Pleistocene was characterized by shifts between wetter and drier climates. By the late Miocene, the remaining forests were becoming more temperate in character, and the diverse rain forest flora had been replaced by a less diverse set of sclerophyllous species; many of these are thought to be derived from earlier forest margin habitats (Hill 1983, 1990). The evolution of floras and faunas associated with the increasing aridity of the late Neogene and Quaternary is summarized in Christophel and Greenwood (1989) and examined in detail in Barker and Greenslade (1982).

As sclerophyll communities began to replace rain forests, the teeth of marsupials such as koalas (first known in the Miocene) and of possums that inhabit present-day sclerophyll communities became increasingly adapted to grinding coarse leaves. The guts of these animals also must have adjusted to processing the oils and toxins that characterize such leaves. Although the family Macropodidae is recognized in the mid-Miocene, the wallabies and true kangaroos began to radiate only in the Pliocene. The 28 fossil and living species of *Macropus* have probably differentiated within the last 5 million years (Rich 1982). The shift from browse plants to grasses apparently set the stage for their rapid expansion. Like placental ruminants, the kangaroos possess gut bacteria that aid in digesting cellulose (Carroll 1988). With the radiation of the kangaroos, we see the development of a correlated set of traits evident in placental mammals in other areas of the world: larger body size, greater mobility, greater reliance on grasslands, and larger aggregations of individuals. The lengthening of the hind foot and increased power in the hind limb in kangaroos are related to sustained and rapid bipedal hopping (Dawson 1977). The kangaroos today comprise 19 genera and about 60 species.

The wombats are burrowers the size of small bears that feed on grass and roots. Their fossil record also extends back to the Miocene, but only from the Pliocene onwards have they been characterized by rootless, continuously growing, bilobate cheek teeth and a single pair of upper and lower incisors. These rodentlike specializations parallel that of the South American Argyrolagidae (Carroll 1988).

Quaternary

The latest Tertiary and Quaternary radiation and dominance of *Eucalyptus*, a hallmark of Australian ecological history, is usually attributed to the superior ability of *Eucalyptus* to withstand an increased incidence of widespread fires (started chiefly from lightening strikes) associated with drier, more open conditions during the Neogene (Kershaw 1986). (See Kershaw 1984 for a review of *Eucalyptus* radiation.) By the Pleistocene, the flora of Australia was essentially modern, though the spatial ranges of the major vegetation types have fluctuated widely, probably coinciding with global glacial conditions.

Megafaunal extinctions during the late Pleistocene appear to have been of the same magnitude as those in North and South America. Thirteen genera of large marsupials (>44 kg) met their demise, including 3 of 6 families represented in the Pleistocene (Marshall 1981a; Martin 1984). The giant birds (emu, cassawary, and the ostrich-like *Genyornis*), a giant varanid lizard, a horned tortoise, and a python-sized snake may also be included in the late Pleistocene extinction of Australian megafauna. Within their lineages, the surviving mammals appear to be selected for either rapid bounding locomotion, gracile body form, smaller body size, or nocturnal habits (Martin 1984). Human occupation, which to varying degrees has been implicated in these extinctions, occurred by at least 35 ka and probably more than 40 ka (Bowler and Thorne 1976).

Considerable climatic change has occurred over the past 100 ky. In Queensland the major wet phases appear to coincide with $\delta^{18}O$ minima, including evidence that a complex rain forest dominated by angiosperms (and indicative of high rainfall) occurred during isotope stage 5 (Kershaw 1985). After 80 ka, a drier rain forest containing high percentages of the conifer *Podocarpus* and other gymnosperms appeared, which then gave way to an open woodland dominated by sclerophyllous taxa during the last glacial. Kershaw (1978) suggests that a rainfall of 600 mm/yr would have accompanied this glacial vegetation, compared with 2500 mm today and 3500 mm during the last interglacial (isotope stage 5e). Sclerophyll vegetation in Queensland ultimately was replaced by rain forest again 9500–7000 yrs ago. Although the present rain forest of Queensland comprises many species of primitive angiosperms, its age can be measured in only thousands of years (Christophel 1989).

In apparent contrast with the evidence from Queensland, Jones and Bowler (1980) report that about 30 ka, the continent overall was much wetter than today. Lakes that are now dry or nearly dry were approximately ten times larger. Expansion of Australia's arid center commenced by 26 ka, and by 18 ka the large lakes had almost entirely disappeared. The arid period, with an annual rainfall of perhaps 250 mm around 12 ka (Horton 1984), lasted until about 10 ka. Horton (1984) makes a case that the selective death of megafauna was based upon a need for a consistent water supply. A long distance between water sources during the dry phase meant that species were geographically and ecologically restricted, and this restriction provided the background for local extinctions during periods of food scarcity. Megafaunal extinctions mostly occurred during this period of climatically induced ecological stress. Presumably water requirements also would have induced human populations to confine their activities to the same geographic zones as those where animals were concentrated (i.e., the continental margins). However, archeological evidence suggests that the aboriginal economic system seen in recent times, which focuses on small mammals and rarely exploits large animals, operated in the Pleistocene as well. No mass kill sites have been documented, and late Pleistocene archeological sites preserve few specimens of megafaunal species, while small forms are abundant (Horton 1984; Martin 1984).

Aboriginal use of fire may well have helped change the late Pleistocene landscape, and this influence also has been invoked as contributing to megafaunal extinctions (e.g., Merrilees 1968). However, judging from the present, the main effect of fire is the destruction of shelter for small mammals, whereas fires often aid the rejuvenation of food supplies in both the grasslands and the eucalypt woodland-forests of Australia (Horton 1984).

Although Australian floras and faunas possess many distinctive traits, we have noted that certain taxa are also similar in morphology and in their ecological roles to taxa from other continents. In other cases the morphology and ecology of certain plants and animals are indeed unique. The absence of succulent-stem desert plants means that certain ecomorphic (functional) traits typical of deserts on other continents are unknown in Australian desert floras. Similarly, the specific type of rapid, sustained locomotion that has evolved among large mammals (the macropods) in Australia differs considerably from the cursorial locomotion that developed in large herbivores on other continents. Yet the Australian biota experienced selective regimes and adaptive trends very similar to those on other continents. The development of certain forms of scleromorphy and bipedal hopping connote a similar adaptive trajectory expressed through morphological traits somewhat different from those that appeared on other continents during the late Cenozoic. As a result, the histories of the ecological communities that occur on these diverse continental masses show significant parallelisms.

This observation leads to a potentially important implication concerning the comparative study of ancient ecosystems. In studying biotic communities from disparate times and places, it is important to take into account the historical contexts and trends of ecomorphic change when judging how similar or different these communities are on the basis of the specific morphological attributes of their component taxa. Although the taxa comprising two communities may possess dissimilar traits, documentation of parallel ecological trends through time (acting on unrelated lineages at the outset) can illuminate fundamental ecological similarities between those communities.

PATTERNS AND PROCESSES OF ECOLOGICAL CHANGE

As illustrated in the previous sections, open woodlands, deciduous forests and eventually savannas, prairies, and deserts expanded over continental surfaces during the past 34 my, while rain forest and closed woodlands decreased in area and became more patchy in distribution. Occurring gradually throughout the late Cenozoic, these changes were neither catastrophic nor synchronous on a worldwide scale (figs. 7.2–7.7). The glacial cycles of the Quaternary generated marked short-term changes in species distributions but little taxonomic modification of floras. In contrast to plants and small vertebrates (in most areas), large mammals suffered greatly increased extinction rates during (in Australia) or near the end of (in the Americas) the last glaciation. A dense fossil record pertaining to the late Cenozoic biota and climates provides an opportunity to examine the ecological implications of these major changes more closely. In this section we summarize the main processes of biotic and biome change that are discussed by late Cenozoic paleoecologists. We then consider a recurrent theme in paleoecological studies: the relationship between climatic and biotic change, including the implications for a paleoecological concept of community. Finally the causes of late Quaternary extinctions and the impact of humans on recent ecosystems are discussed.

4.1   Overall Processes of Biotic Change

The comparatively dense fossil record of the late Cenozoic enables close examination of an example of biome replacement, i.e., the overall (though complex and mosaic) replacement through time of closed vegetation by relatively open biomes (with associated ecomorphic changes in faunas). Paleoecologists working on different time periods of the late Cenozoic have adopted varied perspectives concerning change in biotic communities. Quaternary studies have emphasized that plant and animal associations break down relatively rapidly, and communities that include new associations of organisms are successively formed. In contrast, earlier time periods appear to be characterized by more stable associations of organisms (e.g., Oligocene and Miocene chronofaunas). Across all continents, each with its own distinctive geographic, tectonic, and climatic settings, we have noted evidence of strikingly similar changes in habitat structure (from closed to open habitats) and in the ecomorphic characteristics of faunas and floras. Although different processes and circumstances enter into accounts of biome change on different continents, we may glean from these accounts a general list of factors that are believed to have influenced the shift from closed to open vegetation biomes; these factors may have more general applicability to biome changes in the terrestrial ecological record.

Factors Contributing to Biotic Change

Table 7.1, which is based on the late Cenozoic literature, summarizes the major processes and events that generally are considered to have contributed to biotic change and to have shaped ecosystem structure through time. The effects of some of these factors (e.g., niche partitioning, fire) are inferred largely from present-day ecological information deemed appropriate to extrapolate to ecological systems in the past. Other factors (e.g., immigration

Table 7.1   Major Processes Affecting Biotic Change and Ecosystem Structure through Time

Biotic Events and Processes	Abiotic Environmental Events and Processes	
Varied Time Scales	Long-Term	Short-Term
Habitat diversification	Climatic change (including glaciation)	Meteor impacts
Niche partitioning	Sea level rise and fall	Meteorological phenomena (e.g., hurricanes)
Guild or niche expansion	Eustatic movements	Fire
Biogeographical rearrangement (immigration/emigration)	Continental movements and connections	Volcanism
Speciation	Tectonic movements	
Extinction	Soil development	
Coevolution		
Disease		

and emigration, continental drift) may be inferred from the fossil and geologic record itself. Biotic processes appear to have been complexly intertwined with one another and with both short- and long-term abiotic processes throughout the history of terrestrial ecosystems. The division between short- and long-term abiotic factors is artificial, since the intervals over which they have affected ecosystems comprise a continuum.

In cases where mass extinctions are not involved, short-term or local factors (such as nutrient depletion or volcanic eruption) may affect the original set of biomes, causing turnover in species that may alter the structure of some of the constituent communities. Subsequently these altered communities may either expand or contract in response to longer-term abiotic environmental changes, eventually creating a new type of biome or set of biomes (e.g., the forest to savanna transition). The nature of the short-term processes (and the original ecological contexts in which they occur) may determine which lineages of organisms have the opportunity to adapt to a new environment and contribute to the structure of a new biome (e.g., trophic shape, ecological diversity, vegetational architecture, resource complexity). It may be the case that the organisms that "get there first" (occupy or radiate into a new environmental regime) have a disproportionate influence on subsequent community structure. As patches of new habitats expand and interconnect, opportunities for migration of the newly adapted organisms are likely to increase, and fewer opportunities may occur for those organisms in habitats that become patchier and decrease in size.

One illustration of how this general model of biome change operated in practice comes from volcanically generated patches of open habitat in the Oligocene of the Pacific Northwest of North America (Taggart et al. 1982). Local and regional volcanic activity caused widespread and recurrent disruption of the landscape. Forest communities were destroyed repeatedly; each time, grass-forb communities became established on the barren landscape; these then were replaced by pine-parkland communities, and eventually by forests again. These inferences suggest that during the middle Miocene, grasses, forbs, and shrubs played important roles as successional plants, which coincidentally adapted them to the regional aridity of the late Tertiary. That is, xeric plants that evolved as successional species in mid-Cenozoic times became increasingly important and widespread as rainfall declined and became more seasonal in distribution during the late Neogene. As a result, "elements of the Neogene successional communities have become major components of the modern climax vegetation of the region" (Taggart et al. 1982). In other regions and continents (e.g., Australia), it also appears that grassland, heath, and other xeric communities were spatially limited, or at least were not involved in replacing forests in a progressive manner, until such replacement was favored by the climatic deterioration of the late Tertiary (caused by both global climatic and regional tectonic events).

Local versus Exogenous Factors

Biotic change occurs through a combination of local evolution of indigenous species, extinction, and introduction of species from other regions. Endemism and immigration provide other illustrations of the interrelationship between short- and long-term biotic and abiotic factors in producing general ecological change during the late Cenozoic. During the middle–late Miocene, the assimilation of immigrants into local communities was a common phenomenon in the Americas and Africa-Eurasia. We may also note that faunas of this time appear to be richer than their modern or earlier counterparts, which suggests that the richness of the biotic inventory, the diversity of the ecological mosaic, and the assimilation of immigrants are closely related. By analogy with island ecosystems, where the biotic inventory is depauperate, an ecosystem may be ripe for invasion by adventive species when the system is no longer isolated. Either large- or small-scale immigration of species may be accompanied by extinction of indigenous organisms. As the middle–late Miocene record from the Siwaliks indicates, turnover of taxa in response to immigration does not necessarily entail loss of previous distinctive elements of the biota. The mosaic forest-woodland-grassland settings of parts of Eurasia, North America, and Africa during the Miocene apparently also were able to assimilate immigrants without immediate extinction of prior inhabitants. It seems likely that factors involving local climate, habitat patchiness, and overlap in the use of resources by indigenous and immigrant organisms have strongly influenced (i.e., buffered or slowed) biotic change. Of course the distant climatic, eustatic, and tectonic events that create dispersal opportunities also have exerted significant controls on the history of ecological communities at local and continental scales. Circum-Mediterranean land connections determined the passage of species between Europe, Africa, and Asia, thereby affecting the degree of endemism and the opportunities for ecological assimilation over time.

During the late Cenozoic, the repeated appearance of dispersal routes may have tended to homogenize the biotic events and ecological history of certain areas (e.g., Africa and Eurasia during the Neogene). On the other hand, terrestrial dispersal routes to some areas (e.g., Australia, South America) were completely absent during much of the late Cenozoic. Nonetheless, as we have already noted, there were intriguing parallels in the ecological histories of all continental masses. The fact that Australia and other continents all experienced the development of large-bodied grazers adapted to long-distance and speedy locomotion, as well as the expansion of grass-dominated vegetation during the late Cenozoic, indicates that global factors, especially climate, held an overarching influence upon the history of terrestrial ecosystems over at least the past 34 million years.

4.2   Relationships between Climatic and Ecological Changes

Response of Plants to Quaternary Climatic Change

Since the mid-1970s, much of the literature on Quaternary vegetation has emphasized that rapid shifts in the distribution of vegetation types and in species assemblages probably characterize the last $10^5 - 10^6$ years of ecological history. Climate and pollen maps for the Northern Hemisphere since the last glacial maximum, for example, illustrate that plant populations have continually migrated in response to complex changes in climate (Delcourt et al. 1983; Huntley and Webb 1989). Populations of trees have responded largely by migration, at rates of 100–1000 meters/year over periods of thousands of years. Orbital forcing of periodic, global climatic change is believed to have caused climate to force vegetational change; these periodic changes have occurred predominantly on a wavelength of 100,000 yrs over the last 750,000 yrs, corresponding to glacial-interglacial cycles. Pollen distribution maps suggest that different trees responded independently to the environmental changes of the late- and postglacial periods. Thus, late Quaternary associations of plants have been in continuous flux, and plant communities over the past several 10^5 yrs have been ephemeral aggregates of species that disassociated and reassociated in new combinations as a result of climatic change (e.g., West 1964; Huntley 1988; Huntley and Webb 1989).

New combinations of species may have resulted repeatedly from delays in the migration of certain plants during climatic change (e.g., Cole 1985; Davis 1976; Spaulding 1990). Possible causes of this lag effect include differential response of species to soil maturity or successional state; persistence of reproductive advantage in resident plants over invading species (Cole 1985); and inherent differences in plant species reproduction, dispersal, and population increase. On the other hand, supposed lag effects may actually result from the varied (but immediate) responses of plants to shifts in temperature and moisture, possibly linked ultimately to orbital variations (see Huntley and Webb 1989).

Rapid fluctuations in Quaternary plant communities are thought to have occurred in the tropics as well as in the temperate regions. Independent behavior of plant species in response to climatic change also has been suggested for Southeast Asian forest species (Flenley 1979). The present Amazon rain forest has been referred to as a "pioneer community" (Tricart 1977), suggesting that this equatorial forest is much younger than traditionally believed and not necessarily in stable equilibrium (Street 1981; Flenley 1979). If this is true, however, some zones of rain forest must represent long-term optimum areas for supporting such vegetation. According to recent assessments, evidence for such refugia and last-glacial aridity in the Amazon is less secure than it once appeared (Eden 1974; Colinvaux 1989; Nelson et al. 1990;). As we discussed earlier, the geomorphologic, pollen, and biogeographic evidence for aridity and refugia in Africa is more compelling. In contrast to zones where rain forests disappeared during glacial phases (e.g., Queensland, large areas of equatorial Africa), forests persisted apparently without significant change in Sumatra and in African refugia during periods of desiccation.

The concept of communities as ephemeral entities appears to be closely linked to the high-amplitude, short-wavelength climatic shifts that have characterized the past one million years. Terrestrial floras and faunas (e.g., chronofaunas) of other times prior to the Quaternary do, however, exhibit long-lived associations of particular taxa or of forms that are consistent in their ecomorphic characteristics (see also chaps. 4 and 5). This contrast does not imply that late Quaternary palynology gives a false impression of the nature of paleocommunities. Rather we surmise that the temporal nature of paleocommunities may shift according to climatic, tectonic, and other influential parameters that vary considerably in frequency and amplitude over geologic time. Because Quaternary studies offer the finest resolution of ecological events during a period of rapid climatic shifts, paleoecological data from this time interval may ultimately provide useful hypotheses concerning the processes of ecosystem change or stasis during earlier but similar periods of climatic fluctuation (e.g., Paleozoic glaciations).

The Turnover-Pulse Hypothesis

Although the direct effect of both long- and short-term climatic change on the structure and composition of plant communities is recognized (e.g., Watt 1947; Grime 1979), considerable attention has been paid recently to causal relationships between climatic and faunal change during the late Cenozoic. Brain (1981), Vrba (1985a, b, c, 1988), and Denton (1985; Prentice and Denton 1988) have stimulated discussion of the idea that nearly synchronous turnover across diverse groups of organisms tends to be associated with change in the physical environment. Building upon ideas fundamental to the neo-Darwinian synthesis regarding speciation and environmental effects on animal populations, the formal name "turnover-pulse" now has been ascribed to this correlation by Vrba (1985c). According to this idea, significant evolutionary change within lineages occurs only when changes in the physical environment create habitat fragmentation and subdivision leading to vicariance of species populations. Climatic change in particular is thus a special initiating cause of new species originations. Since speciation, extinction, and immigration may all be involved in faunal turnover, Vrba does include the removal of dispersal barriers in this concept of turnover-pulse (e.g., habitat variables related to physiological tolerance or physical movement of an organism). However, the main point of the hypothesis is that climatic change forces evolutionary events to occur. Climatic and other shifts in the physical environment cause speciation, while extinction may result either directly from climatic change or from ecological interactions among newly associated species.

According to the hypothesis, communities of organisms will tend to reorganize taxonomically and presumably ecologically during major shifts in the abiotic environment. New species originations, climatic change, and major peaks in faunal turnover should coincide. Temporally restricted shifts in global climate during the Neogene are deemed to have caused synchronous first appearances of taxa throughout the world's land ecosystems. Support for such a turnover-pulse episode is derived from pulses in the origination of bovid taxa that have been identified during the late Neogene and Quaternary (Vrba 1985a). Three such pulses appear to correlate with global climatic shifts (5 Ma, 2.5 Ma, and 0.9—0.7 Ma); species origins and extinctions in the human fossil record also appear to coincide with these events (Vrba et al 1989; Prentice and Denton 1988). Vrba (1985a) points out that ecological features such as resource breadth, mobility, and social behavior may determine the response of a mammalian lineage to major climatic (and accompanying vegetational) change.

The turnover-pulse idea makes very clear the pivotal role temporal correlations play in virtually all discussions of paleoecological history. A number of factors influence how effectively correlations between events (turnover in different lineages, climatic shifts) can be documented. Chronological precision (whether in absolute age determinations or in correlations among fossil localities) has the most obvious effect. One difficulty pertinent to the late Cenozoic relates to the Milankovitch cycles, which influenced global climatic change at a much higher amplitude starting 0.9 Ma. If these cycles were responsible for lower-amplitude change through much of the Cenozoic, the correlation between climatic and turnover events may require a chronological precision somewhat finer than 100,000 years. Although partial confirmation of climatic causality could be derived from longer time intervals, such as from progressive deterioration of the environment over several hundred thousand years, the lumping of several distinct climatic and turnover events into one thwarts adequate testing of any hypothesis concerning a one-to-one relationship between large-scale biotic and climatic change. In the marine realm, Cronin (1985) suggests that long-term, directional climatic change is far more effective in causing turnover, especially in allowing time for speciation to occur, than are rapid, cyclical fluctuations. Latest Cenozoic (Quaternary) sequences in which finer chronological precision is possible may be the only adequate testing ground for relating climatic events to biotic turnover pulses.

Furthermore, sampling density (e.g., number of specimens or localities over a given interval) affects the recognition of first and last appearances of taxa in the fossil record, and local or regional tectonic or volcanological history may have strong effects on migrations or in situ evolution of organisms (Hill 1987; Barry et al. 1990). In an analysis of Neogene faunal change in the Siwaliks, Barry et al. (1985) emphasize that pulses of immigration following climatic, tectonic, and eustatic events can play a large role in the turnover of mammalian communities within a region. Extinctions may correlate with immigrations because old species must interact in new ways with newly arrived species. In situ evolution appears to play only a minor role in overall faunal change in the Siwalik mammalian community after 16 Ma. In contrast with the expectations of the climatic forcing model of speciation, faunal change due to immigration depends solely on the opening of migration routes or the removal of barriers, which are not necessarily related to climatic change. Barry et al. (1990) also note that different components of a community can change in different ways; certain groups (e.g., bovids and muroid rodents in the Siwaliks 16–13.5 Ma) exhibit some in situ evolution, while others increase in diversity because of immigration, or undergo extinction. Climatic forcing is one process that can be included in the abiotic events that control faunal change by immigration, but isolating its role in turnover events is problematic. Moreover, because the mechanism of faunal turnover proposed by Barry et al. (1985, 1990) includes taxa that may have existed for some time in other biogeographic provinces, it contrasts with Vrba's turnover pulse hypothesis, which emphasizes the causal relationship and synchroneity between climatic forcing and speciation and extinction events. The prediction that speciation and extinction pulses should be contemporaneous, or nearly so, is not borne out by the Siwalik data (Barry et al. 1990). Distinguishing between new appearances by speciation or by immigration remains a major problem in testing the turnover-pulse hypothesis.

A final comment on turnover-pulse concerns the role of climatically induced fragmentation of habitats. If speciation tends to be caused by division of a species' preferred habitat, the likelihood of speciation should correspond to the degree to which a species exhibits specialized use of its habitat. Building on her analysis of African bovids, Vrba (1987) posits that habitat specialists (e.g., specialized grazers) are expected to contribute more than habitat generalists (e.g., mixed browsers and grazers) to an episode of taxonomic turnover. Although this has been shown to apply to African bovids during the late Cenozoic, Janis and Damuth (1991) offer a potential counterexample in which presumed dietary specialists, browsing equids of the early Cenozoic of North America, exhibited little speciation in comparison with Miocene equids that were presumably more generalized in their use of habitat (mixed feeders). Although the ecological attributes of species may well strongly influence taxonomic turnover and accordingly ecological change, the generality of any relationship between species ecology, habitat fragmentation, and turnover during the late Cenozoic remains to be determined.

Tectonic Forcing

Discussions about climatic forcing of biotic change during the late Cenozoic usually focus on variations in the earth's orbit as an initiating cause of significant change in climate. General circulation models, however, illustrate that the complex climatic deterioration of the late Cenozoic may have resulted from many interacting factors. This point parallels our emphasis upon multiple factors and processes implicated in the late Cenozoic history of biotic change (table 7.1).

In addition to seasonal insolation heating and glacial conditions—both of which are coupled to orbital variations—long-term tectonic uplift probably had significant effects on late Cenozoic climates. Simulation modeling indicates that uplift of the Tibetan plateau and a broad region of the American West had plausible cooling and drying effects on Africa, Eurasia, and North America (Ruddiman et al. 1989; Ruddiman and Kutzbach 1989). Consideration of global-scale linkage between tectonic and climatic change is prompted by long-term records of dust influx and other evidence that suggests that global and regional aridity and possibly cooling during the late Cenozoic preceded significant, orbitally forced climatic change. Massive tectonic uplift in Asia and North America is modeled as a gradual process since the late Neogene. Its effect therefore appears to mirror an overall climatic deterioration and is consistent with evidence of increasing aridity and cooling from 10 Ma to the present (Ruddiman and Kutzbach 1989). The lack of precise chronology for these uplift episodes, however, precludes dissecting the effects of tectonics, glaciation, insolation, and other factors that modified global climate and led to biotic change. In any case, the potential importance of massive tectonic uplift implies that climatic forcing entailed more than a single ultimate causal variable that will express precise correlations with major episodes of biotic change.

4.3 Megafaunal Extinctions

Although the late Cenozoic experienced numerous periods of biotic turnover of varying scale, the extinction of large vertebrate taxa at the end of the Pleistocene has received special attention. One reason for this is that human hunters have been implicated as a cause for these megafaunal extinctions. Alternative causes, especially climatic change, also have been proposed. The chronological precision enabled by carbon 14 analysis permits refined correlations between ecological variables and evidence of a marked extinction event. The late Pleistocene experienced a loss in overall animal diversity that was much less significant than that exhibited by certain earlier extinction episodes, which affected diverse components of the fauna. What is intriguing about the late Pleistocene extinction event is that it preferentially affected animals possessing certain ecomorphic traits, in particular, species with body size greater than 40–50 kg. The differential extinction of megafauna occurred on several continents. Among birds, generally the largest were lost. Proboscideans, edentates, and equids were especially affected, and the diversity of cervids, bovids, camelids, and antilocaprids also was reduced. The large diurnal primates of Madagascar were greatly diminished as well. In all, some 200 genera of mammals, mostly over 50 kg, disappeared within a period of several thousand years, especially during 12–10 ka (Martin and Klein 1984; Padian and Clemens 1985).

Martin (1967, 1984) has been a primary proponent of the "overkill hypothesis," which invokes human predation as the main cause of late Pleistocene extinction of large mammals and birds. Originally proposed as an alternative to climatic change, the overkill hypothesis argues that continental variation in the timing and degree of megafaunal extinction during the late Pleistocene was the direct result of the dispersal of human hunters to different continents at different times. Late Pleistocene extinctions in Africa, where hominids have the longest history, were minor, possibly because of earlier predator-prey coevolution. On that continent very large mammals, large carnivores, and other species (e.g., cercopithecoid monkeys) that were potential competitors of hominids became extinct during the early Pleistocene. According to the overkill hypothesis, heightened predatory activities of (and competition from) Plio-Pleistocene humans explains why this period of megafaunal extinction occurred earlier in Africa than elsewhere. Megafaunal extinctions in North and South America, on the other hand, were concentrated between 15 and 8 ka, coinciding with radiocarbon dates for early definite evidence of human activity on those continents. Europe experienced fewer extinctions during the late Pleistocene, though the disappearance of equids and of cold-adapted animals such as mammoth, woolly rhino, and musk ox at the end of the last glacial is notable. In comparison to Europe, massive extinctions of the largest vertebrates on Australia and many large oceanic islands (e.g., Madagascar, New Zealand) appear to correspond more closely to the earliest occupation by humans in these areas.

Thus, those regions where late Pleistocene megafaunal extinctions were most marked were also those previously unexposed to modern human populations. Furthermore, as Martin (1984) emphasizes, the overkill hypothesis implies that taxa of large body size and lacking in escape mechanisms (e.g., speed, nocturnal habits) should have experienced greater extinction during the spread of technologically advanced humans into a particular area. Although these are testable implications, the data are still ambiguous in many areas of the world. Australia, for example, provides excellent evidence of differential loss of large, slow-moving marsupials, but archeological kill sites with remains of these animals are lacking, and the correlation between extinction and the arrival of humans on the continent is still poorly understood. The overkill hypothesis, however, also encompasses human alteration of native habitats, an indirect cause of animal extinction. The burning of portions of the landscape, generally believed to have been an important aspect of land use by Australian aborigines, is a prime example of human destruction of habitats that could have led to megafaunal extinctions. Human involvement at the death sites of mammoths (Fisher 1987) and other large mammals during the late Pleistocene of North America suggests that humans did play a direct predatory role in the demise of certain megafauna. Nonetheless, evidence from some regions of North America (e.g., the High Plains) argues against mass killing of large mammals at a level necessary to identify humans as the sole cause of extinction (e.g., Frison 1987).

Environmental change, by itself, is considered to be either an alternative explanation or a contributing factor to account for late Pleistocene extinctions. The late Pleistocene was a period of extensive climatic change, often coinciding with the spread, or increase in density, of modern humans on new land masses (e.g., North America). Authors who have favored climatic causes of megafaunal extinctions typically invoke climatically induced reduction in habitat diversity or increase in seasonality to account for the extirpation of megafauna at the end of the Pleistocene (see Martin and Klein 1984). Martin (1984), however, points out that while decreased equability is invoked as a mechanism for animal extinctions in North America, extinctions in South America during the early Holocene appear to coincide with climatic amelioration, expansion of tropical forests, and increased equability. In addition, shifts between glacial and interglacial conditions occurred throughout the latter half of the Quaternary without triggering extinctions of large mammals. According to a review of these issues (Martin and Klein 1984), the majority of researchers now attempt to explain late Pleistocene extinctions of large vertebrates with a combination of climatic and human causes; human hunting had the drastic effects it did only because it occurred at a time of ecological stress induced by climatic change. Moreover, the exact combination of causes may differ from one region to another (see Marshall 1984). For example, an increase in seasonality may have contributed significantly to megafaunal extinctions in North America, whereas the replacement of open-vegetation habitats (which tend to support large mammals) by tropical forest may have been the important factor in South America. The idea that climatically induced water stress posed an extreme ecological difficulty for Australian mammals (Horton 1984) is a further example of the context-specific explanations that are now favored for late Pleistocene extinctions on different continents.

Two other hypotheses put forward to account for late Pleistocene extinctions, coevolutionary disequilibrium and taxonomic equilibrium, have challenged the direct role of humans maintained by the overkill hypothesis. Graham and Lundelius (1984) propose that environmental change during the late Pleistocene disrupted particular dependencies that had evolved between animals and plants during earlier times. This disruption of coevolved relationships created an ecological disequilibrium; both plant and animal species responded individualistically to environmental change, and "disharmonious" associations of plants and animals (relative to the present) occurred. Organisms did not migrate as well-integrated, stable communities. As suggested by palynologists studying late Pleistocene plant communities, change in the composition of animal communities from the late Pleistocene to today entailed accidental association or reorganization of species assemblages. According to this reasoning, environmental change reduced the predictability of community structure and composition and consequently decreased the fitness of both specialist and generalist taxa, contributing to their extinction. Exactly how ecological disequilibrium could cause animal extinction is not completely specified; however, these causes may have included novel competitive interactions and problems of plant detoxification by animals as a result of new biotic associations (Graham and Lundelius 1984).

The idea of taxonomic equilibrium, as developed by Gingerich (1977, 1984), is based on the observation that taxonomic (e.g., generic) diversity remained stable within particular taxonomic groups through the Cenozoic, even though these groups experienced periodic turnover. A high correlation appears to exist between origination and extinction curves (rates) for Tertiary rodents, carnivores, primates, and artiodactyls. To maintain equilibrium, periods of relatively high faunal extinction would require high rates of taxonomic origination during previous or contemporaneous times (Gingerich 1977). The peak extinctions during the Pliocene and Pleistocene of North and South America imply preceding periods of greatly increased species richness due to immigrations and autochthonous diversification (Webb 1976, 1977). In developing this idea as a worldwide model, Gingerich (1977, 1984) generalizes that extraordinary extinctions necessitate favorable prior conditions for supporting a heightened taxonomic diversity compared with earlier and eventually later periods. Since extinctions and originations proceed hand in hand over the long term, taxonomic equilibrium is maintained. As the idea is applied to the late Pleistocene, the high rate of extinction is considered to be an inevitable consequence of a very high rate of early Pleistocene originations and of the turnover that happened to result during the major environmental oscillations of the latter half of the Pleistocene (Gingerich 1984).

Cifelli (1981) has noted that generic extinctions and originations among artiodactyls and perissodactyls track one another during the Cenozoic. However, Padian and Clemens (1985) comment with regard to both Cifelli's and Gingerich's studies that extinction and origination curves tend to be similar to one another (and similar across different groups of organisms) as a result of peaks in the sample sizes of fossils during certain time periods. Sampling, rather than faunal equilibrium, is perhaps a more important factor in the apparent peaks and drops in the extinction and origination curves. Sampling effects also have played a significant role in apparent faunal turnovers in the Eocene of North America (Badgley and Gingerich 1988) and the Miocene of Pakistan (Barry et al. 1990). Nonetheless Padian and Clemens adopt Gingerich and Cifelli's idea that the key to understanding extinctions is in the pattern of originations, which generally has been poorly investigated. Thus, besides traditional hypotheses concerning overkill and climate, an additional hypothesis to test regarding late Pleistocene extinctions is that they are truly anomalous, as opposed to being part of predicted long-term stability trends in diversity and the expected lifetimes of taxa.

4.4   Post-Pleistocene Impact of Humans

Human activities represent another biotic process that has influenced terrestrial ecosystems increasingly over the past 10,000 years. In particular, the control of plant and herbivore productivity by people engaged in agriculture and pastoralism has had significant impact on terrestrial biotas and environments. The amount of land devoted to agriculture is estimated at about 10% of the total land surface on earth, though estimates range up to 15.4% (Atjay et al. 1979). While this may seem to be a relatively low proportion of the earth's terrestrial area, it includes a much larger percentage of the land that has soil and climate suitable for high biological productivity. Agricultural ecosystems occur widely in tropical, subtropical, and temperate zones and on all continents except Antarctica. Humans have inhabited virtually all terrestrial biomes, and all expanding human populations are supported ultimately by controlled productivity of plant and animal populations.

The way human food production has affected terrestrial ecosystems varies from region to region. In many areas a diversity of nondomesticated plant and animal species continues to exist where people cultivate plants and maintain livestock. However, much of the surface area devoted today to human food production involves growth of one or a very few species of plants (e.g., cereals) and/or the maintenance of one or a few species of herbivores. The latter consume low-quality, high-fiber plants (i.e., grasses) that cannot be digested by humans, though people, in turn, eat the herbivores as well as manage the herds.

The development and intensification of agricultural ecosystems has involved dramatic changes in the dynamics and overall structure of local food webs. According to faunal and paleobotanical evidence uncovered mainly by archeologists, food webs as portrayed in figure 7.8a were associated with the initial phases of food production (which involved a broadening of the dietary base of people) (Flannery 1969, 1986; Hole et al. 1969; Stark 1986; Redding 1988; Smith 1989). A broad spectrum of plants, herbivores, and carnivores were part of these ecosystems much as they are today where humans have little impact. Later, intensification of food production involved selection by

a. Post-Permian Terrestrial Ecosystems   **b. Human-dominated Ecosystems**

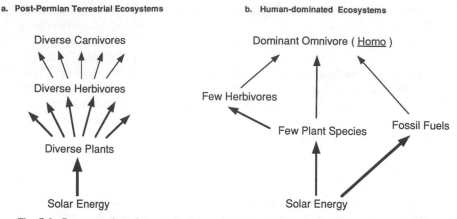

Fig. 7.8. Representation of energy flow through trophic levels, *a* in most terrestrial ecosystems after the Permian diversification of herbivores, and *b* in recent ecosystems in which human populations control biological productivity through intensive food production.

humans of a very small number of plant and herbivore species at the expense of other species. Undomesticated species have not necessarily disappeared from such ecosystems, but their biomass and distribution have been diminished greatly compared with cultivated species. Since the latter are protected by people, carnivores also have decreased greatly in numbers and diversity, and humans have become the top carnivore in many areas. Particularly over the brief period of the past 100 years, energy derived by humans from fossil fuels also has contributed significantly to the development of agricultural systems. The type of ecosystem dominated by human food production is diagrammed in figure 7.8b.

Our point in drawing attention to agricultural ecosystems (or more generally to food production ecosystems) is that from the broad perspective of the Paleozoic to the present, the intensification of food production represents a dramatic shift in ecosystem structure and dynamics. The ecological change embodied in this system is comparable in scale to major changes noted in the Paleozoic and Mesozoic (see chaps. 5 and 6) and is extraordinary relative to the ecosystem developments that took place from the Oligocene through the Quaternary. By the end of the Permian, the pathways for energy transfer within terrestrial biotic systems had been established, and these were characterized by a diverse assemblage of herbivore species feeding off a rich array of plant species. Since the Permian, plant-herbivore dynamics has played a paramount role in terrestrial ecosystem evolution (see chap. 5). This type of ecosystem, familiar to all ecologists from food web and energy pyramid diagrams in textbooks, is portrayed in figure 7.8a. During the Oligocene through Quaternary interval, the structure of food webs remained largely stable, and although cursorial grazing animals came to dominate the newly established savannas and grasslands, there were no major changes in animal or plant ecomorphs comparable in scale to those discussed by Paleozoic and Mesozoic paleoecologists.

The structural changes between post-Permian ecosystems and human-dominated ecosystems involve the emergence of a dominant omnivore (humans), reduction of plant diversity with increased frequency of monospecific subhabitats, dominance of grasses over other plants, reduction in herbivore diversity, and reduction (removal) of predators, among other modifications. This new type of ecosystem now occupies 10–15% of the Earth's land surface and is controlled by a single species that successfully manipulates most components of the food chain.

It might be argued that this appraisal of human-dominated ecosystems is anthropocentric, that the changes represent extremely recent phenomena, and that these ecosystems would revert back to the type seen in figure 7.8a if humans ceased to exist. Yet the perspective we seek here is one offered by the entire history of terrestrial ecosystems rather than a narrow slice of anthropological time. Through food production and its consequences, humans have created a fundamentally different type of ecosystem structure. Ecosystems other than those dominated by human food production still occupy most of the land surface, and it is difficult to project the long-term impact of such a recent, unique event. Nonetheless the development of food production ecosystems is comparable in scale to the restructuring of local and regional food webs that is inferred for the Permian. In the latter, the change involved the radiation of herbivores at the expense of detritivores; over the past 10,000 years, restructuring of ecosystems has involved domination by a single omnivore and the domesticates that support it. The new type of ecosystem has also proved to be extremely effective in invading new terrain and replacing other ecosystems with unprecedented speed (e.g., Spurr and Vaux 1976).

The fact that ecosystem structure can be influenced in such a dramatic way by a single species suggests something about ecosystem dynamics that is relevant to the more distant past. In this instance, we see that a dominant species can constrain diversity, modify structural properties, and alter energetic pathways of a community. These influences may represent a highly exaggerated version of the effects that other species (past and present) can have upon the communities in which they live. Evidence of more subtle examples of these effects might be sought in the fossil record.

5 CONCLUSION

In contrast with the Paleozoic and Mesozoic–early Tertiary intervals, during which organisms appeared that represented entirely new adaptive themes in the terrestrial realm (chaps. 5 and 6), the paleoecological history of the late Cenozoic mainly involved new emphases upon ecomorphic themes established by the end of the Eocene. This does not mean that the radiation and demise of particular groups of late Cenozoic plants and animals were not ecologically important, especially on a regional scale. Worldwide, however, the late Cenozoic is characterized by shifts in the frequency and spatial distribution of structural and functional characteristics that were already in existence by the beginning of the Oligocene. In a sense, the ecomorphs and community structures that became typical of the latter half of the Cenozoic were "waiting in the wings," in spatially restricted habitats, until environmental changes allowed them to expand their ranges. The Miocene-Pleistocene radiation of grasses and grazing adaptations, which were present in a limited number of taxa prior to the Oligocene, exemplifies this pattern.

It can be argued, however, that the habitat structures and ecological associations that developed were new and not simply equivalent to habitats and communities that had existed previously on even a small spatial or temporal scale. Indeed major shifts in the distribution of biomes (resulting especially from the radiation and geographic spread of open vegetation plants and animals) did occur during this time interval, as did numerous shifts in animal-plant associations. The long-term stability of ecological associations may itself have changed during the late Cenozoic. The fine degree of chronological precision that can be obtained in the late Quaternary enables the relative instability and loose organization of plant and animal communities to be detected. The evidence for community reorganization, chance associations of species, and disequilibrium implies that biotic communities only 10^4–10^5 years apart may differ greatly in taxonomic composition and interspecific interactions. Whether this is the result of accentuated environmental instability throughout the late Cenozoic or precise chronological analysis of a continued Cenozoic trend is yet to be determined. The very brief time perspective on which this view of communities is based and the emphasis upon present-day taxonomic associations as comparisons for earlier biotas provide an insufficient temporal and ecological perspective for judging stability and change in Cenozoic communities overall. Functional and ecological (rather than taxonomic) characterization of early and late Cenozoic biotas will be necessary to evaluate patterns of change and stasis in Oligocene-Quaternary faunas and floras.

In the light of research on biodiversity and conservation efforts today, it should be noted that the very recent impact of humans on ecosystems may be a dramatic exception to the idea that adaptive innovation with sweeping ecological consequences is absent from the late Cenozoic. The technological capacities developed by *Homo sapiens* may ultimately have an influence, however brief, on ecological diversity and habitat structure comparable to major ecomorphic adaptations evolved by entire clades of organisms prior to the late Cenozoic. As highlighted in the debates about Pleistocene extinctions, human activities and climatic change during the late Quaternary appear to have had mutually reinforcing effects upon biotic communities, leading to reduction in diversity of both habitats and species compared with peaks in biotic diversity during the late Miocene and early Pleistocene.

Late Cenozoic studies bring into sharp focus the fact that much of the significant work in paleoecology depends upon temporal correlations, from which processes or causes of ecological change may be inferred. Through the analysis of correlations among such variables as taxonomic diversity, turnover, climatic fluctuation, and the emergence of land bridges, detailed interpretations of biotic stability and change are possible for the late Cenozoic fossil record. The precision of correlations between events, however, varies depending on accessible levels of time resolution (see chap. 2, §8), which in turn is largely dependent on the precision of geochronological methods. For example, at the outset of the late Cenozoic interval, events that occurred within an 8-million-year time span (one or more major turnover events and global cooling from 40–32 Ma) have been considered to be correlated, as though they represent one long-term climatic trend and a related episode of biotic turnover. Yet in the late Quaternary, extinctions, originations, migrations, and climatic shifts that are only a few hundreds or thousands of years apart may be considered nonsynchronous or uncorrelated. Corresponding with this difference in resolving correlations, paleoecological interpretations prior to the Quaternary tend to emphasize coincidence (and causal relationships) between climatic trends and ecosystem stability or change. In late Quaternary studies, lack of precisely synchronous correlations indicate the complexity of short-term cause and effect and help to foster concepts of the ephemeral nature of communities and time lags between climatic change and organic response. Although the high-amplitude climatic oscillations of the Quaternary may be partly responsible for these latter concepts, the level of temporal correlation also permits much closer looks at the synchrony or asynchrony of ecological events because of relatively precise dating techniques. A critical question involves the relationship of the short- and long-term patterns of environmental change and biotic response. Do the late Quaternary patterns simply indicate "noise" that would also occur in the earlier records if we could resolve them? (In this case, the documented long-term associations of taxa imply that species consistently returned to a particular ecological community in spite of short-term fluctuations, and that these associations are preferentially represented in the fossil record.) Or were pre-Quaternary communities actually more stable and subject to longer-term climatic forcing with fewer short-term fluctuations?

Despite generally greater chronological precision relative to earlier time periods, many questions remain concerning the causes of biotic and ecological change during the late Cenozoic. Although the continents explored in this chapter all had similar trends in vegetational and faunal history, the time intervals that possess the richest fossil records show that significant variations may occur between regions in the pace and pattern of biotic change and associated climatic and geologic events. This level of spatial and temporal resolution provides essential clues about the processes involved in major faunal and floral turnovers, as well as their relative importance. If viewed from 100 my in the future, these variations might not be resolved, however, and the shift from forest to grassland biomes could appear as a synchronous global event. The existence of such variations in the late Cenozoic warns against making generalizations about the processes and causes of ecological change on the basis of a single region or continent, or a small sample of fossil localities spread over the globe. The late Cenozoic record illustrates that worldwide comparisons of numerous, carefully reconstructed local sequences of biotic, climatic, and geologic history are essential in investigating the behavior of ecological systems over macroevolutionary time.

APPENDIX

Table A-7.1   Guide to Botanical Terms Used in Text

Ecological Terms	
Climax	Final stage of successional vegetation
Sere or seral	Early to middle stage vegetation in a plant succession
Macrophyllous	Size categories for leaves; macrophyllous implies high humidity
Mesophyllous	and temperature; microphyllous implies water stress and cool
Microphyllous	temperature
Mesothermal	Adapted to warm climate with Mean Annual Temp. $>13°$
Microthermal	Adapted to temperatures $<13°C$ MAT
Habitat Types	
Bushland	Shrub-dominated
Forest	Tree-dominated; trees form continuous canopy, often multistoried
Grassland	General term for short and/or long grass-dominated open habitat
Savanna	General term for grass-dominated open habitats with varying amounts of tree and bush cover
Steppe	Open grassland, generally short-grass species
Woodland	Tree-dominated; trees form single story, noncontinuous canopy

Table A-7.2   Guide to North American Animal Names Used in Text

Taxa	Examples, Description, Ecological Analogue
Families: Extant	
Antilocapridae	Pronghorn antelope
Bovidae	Bison, cow, muskox, African antelope
Camelidae	Camel, llama
Canidae	Wolf, coyote, jackal
Capridae	Goat
Castoridae	Beaver
Cervidae	Deer, elk, reindeer
Cricetidae	Vole
Equidae	Horse, donkey
Felidae	Cat, lion
Geomyidae	Pocket gopher
Heteromyidae	Kangaroo rat, pocket mouse
Hominidae	Human
Leporidae	Hare, rabbit
Muridae	Rat, mouse
Mustelidae	Badger, otter, weasel
Ochotonidae	Pica
Ovidae	Sheep
Rhinoceratidae	Rhinoceros
Sciuridae	Squirrel (ground and tree)
Suidae	Pig, warthog
Tapiridae	Tapir
Tayassuidae	Peccary
Ursidae	Bear
Zapodidae	Jumping mouse
Families: Extinct	
Amphicyonidae	Large carnivore; bear-like
Anthracotheriidae	Medium to large herbivore; pig- and hippo-like
Brontotheriidae	Large, horned herbivores; rhino-like
Entelodontidae	Medium to large herbivore; pig-like
Eomyidae	Early gopher- and mouse-like rodents
Gomphotheriidae	Mastodont; large, elephant-like herbivore
Megalonychidae	Ground sloth
Megatheriidae	Ground sloth
Merycoidodontidae (Oreodontidae)	"Oreodonts"; small to medium herbivores; capybara or pig-like
Mylagaulidae	Medium to large rodents, some horned
Paleomerycidae	Small to medium herbivore; antelope-like
Pantolestidae	Carnivores, otter-like
Protoceratidae	Camel- and antelope-like artiodactyls

Table A-7.2   (*continued*)

Taxa	Examples, Description, Ecological Analogue
Orders: Extant	
Artiodactyla	Bovid, suid, giraffid
Perissodactyla	Equid, rhinoceros, tapir
Proboscidea	Elephants
Soricomorpha (suborder)	Insectivores: moles and shrews
Xenarthra	Sloth, armadillo, glyptodont
Orders: Extinct	
Multituberculata	Small omnivore or herbivore; rodent-like
Creodonta	Medium carnivores; cat, bear, and hyena-like

Note: These names are used in §3.1 and later sections.

Table A-7.3   Guide to South American Animal Names Used in Text

Taxa	Examples, Description, Ecological Analogue
Families: Extant	
Caviidae	Guinea pig
Cricetidae	Vole
Didelphidae	Opossum
Hydrocheridae	Capybara
Manatidae	Manatee
Procyonidae	Raccoon, coatamundi
Rheidae	Rhea (ostrich)
Tapiridae	Tapir
Teratornithidae	Vulture, condor
Families: Extinct	
Argyrolagidae	Marsupial rodent-like herbivore
Borhyaenidae	Marsupial carnivore, ambush adapted
Caenolestidae	Rodent or multituberculate-like
Hegetotheriidae	Rodent-like notoungulate
Glyptodontidae	Giant armadillo-like herbivore
Megalonychidae	Arboreal sloth (early forms), ground sloth (later forms)
Megatheriidae	Arboreal sloth (early), ground sloth (later)
Phorusrhacidae	Large, carnivorous ground bird
Toxodontidae	Rhino-like large herbivore
Orders: Extant	
Caviomorpha (infraorder)	Porcupine, capybara, chinchilla, guinea pig
Chiroptera	Bat
Marsupialia	Opossum
Xenarthra	Sloth, armadillo
Orders: Extinct	
Astrapotheria	Large, rhino-like herbivore
Condylarthra	Generalized small to medium-sized carnivores, herbivores, omnivores
Litopterna	Horse- and camel-like herbivore
Notoungulata	Hippo-, rhino-, and rodent-like herbivores; also chalicothere-like
Pyrotheria	Elephant-like herbivore, large, with proboscis

Note: Includes only terms not in table A-7.2; these names are used in §3.2 and later sections.

Table A-7.4   Guide to Additional Plant and Animal Names for Africa, Eurasia, and Australia

	Examples, Description, Ecological Analogue
Animal Taxa	
Families and Tribes: Extant	
Alcelaphini	Bovids: grazers such as wildebeest
Antilopini	Bovids: gazelle, impala
Boselaphine	Bovid tribe: large, e.g., bluebuck
Cercopithecidae	African monkeys
Ctenodactylidae	Rodents: guinea pig– or gopher-like
Hippopotamidae	Hippopotamus
Madoquinae	Bovid subfamily: small, e.g., dik-dik
Reduncini	Bovid tribe: waterbuck, reedbuck
Tragulidae	Primitive artiodactyl, small
Families and Tribes: Extinct	
Adapidae	Lemur-like arborial omnivore-herbivore
Chalicotheriidae	Large-bodied clawed perissodactyl: sloth-like
Deinotheriidae	Proboscidean with downturned lower tusks, bilophodont molars
Diprotodontidae	Large, bear-like marsupial herbivores
Hyaeonodontidae	Creodonts: hyaena-, cat-, bear-like
Paleotraginae	Early giraffids
Sivatheriinae	Short-limbed giraffe with palmate horns
Thylacinidae	Tasmanian wolf
Orders: Extant and Extinct	
Creodonta	Cat-, dog-, bear-like carnivores, extinct
Embrithopoda	Extinct Arsinoitheres, rhino-like
Hyracoidea	Hyraxes: extant small, hoofed herbivores; extinct, small-large
Tubulidentata	Aardvark
Australian Marsupials	
Cuscus	Rain forest adapted, arboreal, bush-baby-like frugivore
Dasyuroidea (Dasyuridae)	Small–medium insectivores, omnivores, carnivores: e.g., shrew- and cat-like.
Macropodidae	Kangaroos, wallabies
Phascogales	Small, shrew-like dasyurid
Potoroo	Small forest-dwelling macropod
Quoll	Dasyurid—"native cat"—arboreal cat-like insectivore/carnivore

Table A-7.4 (*continued*)

	Examples, Description, Ecological Analogue
Plant Taxa	
Families	
Annonaceae	Pawpaw
Cactaceae	Cactuses
Casuarinaceae	She-oak
Chenopodiaceae	Herbs, including goosefoot
Compositae	Sunflowers, asters, daisies, ragweed
Cyperaceae	Sedges
Dipterocarpaceae	Rain forest canopy trees
Fagaceae	Oaks, beeches
Graminae	Grasses
Lauraceae	Avocados
Leguminosae	Bean family; includes acacias
Moraceae	Figs, mulberries
Myrtaceae	Eucalyptus
Palmae	Palms
Proteaceae	*Banksia* (heath community plants, fire resistant)
Rutaceae	Citrus family
Taxodiaceae	Bald cyprus

Note: These names are used in §3.3–3.5

REFERENCES

Alroy, J. In press. Conjunction among species distributions and the Miocene mammalian biochronology of the Great Plains. *Paleobiology* 18.

Ambrose, G. J., R. A. Callen, R. B. Flint, and R. T. Lange. 1979. *Eucalyptus* fruits in stratigraphic context in Australia. *Nature* 280:387–89.

Ambrose, S. H., N. E. Sikes, T. E. Cerling, and J. Quade. 1991. Paleosol carbon isotope evidence for Middle Miocene woodland and forest at Fort Ternan, Kenya. *American Journal of Physical Anthropology*, Supplement 12:43.

Andrews, P. J. 1989. Palaeoecology of Laetoli. *Journal of Human Evolution* 18: 173–81.

Andrews, P., J. M. Lord, and E. M. Nesbit Evans. 1979. Patterns of ecological diversity in fossil and modern mammalian faunas. *Biological Journal of the Linnean Society* 11:177–205.

Andrews, P., and E. Nesbit Evans. 1979. The environment of *Ramapithecus* in Africa. *Paleobiology* 1:22–30.

Andrews, P., and J. A. H. Van Couvering. 1975. Palaeoenvironments in the East African Miocene. 62–103 in F. S. Szalay, ed., *Approaches to Primate Paleobiology*. Basel: Karger.

Artemiou, C. 1983. Mammalian community palaeoecology: A review of recent methods with special reference to Miocene mammalian faunas of Europe. *Paléobiologie Continentale* 14:91–109.

Atjay, G. L., P. Ketner, and P. Duvigneaud. 1979. Terrestrial primary production and phytomass. 129–82 in B. Bolin, E. T. Degens, S. Kempe, and P. Ketner, eds., *The Global Carbon Cycle*. New York: Wiley.

Axelrod, D. I. 1958. Evolution of the Madro-Tertiary geoflora. *Botanical Review* 24:433–509.

———. 1975. Evolution and biogeography of Madrean-Tethyan sclerophyll vegetation. *Annals of the Missouri Botanical Garden* 62:280–334.

———. 1979. Desert vegetation, its age and origin. 1–72 in J. R. Goodin and D. K. Northington, eds., *Arid Plant Resources*. Lubbock, Tex.: International Center for Arid and Semi Arid Land Studies.

———. 1985. Rise of the grassland biome, central North America. *Botanical Review* 51:164–201.

Axelrod, D. I., and P. H. Raven. 1978. Late Cretaceous and Tertiary vegetation history of Africa. 77–130 in M. J. A. Werger, ed., *Biogeography and Ecology of Southern Africa*. The Hague: D. W. Junk.

———. 1985. Origins of the Cordilleran flora. *Journal of Biogeography* 12:21–47.

Azzaroli, A. 1983. Quaternary mammals and the "End-Villafranchian" dispersal event: A turning point in the history of Eurasia. *Palaeogeography, Palaeoclimatology, Palaeoecology* 44:117–39.

Badam, G. L. 1984. Pleistocene faunal succession of India. 746–75 in R. O. Whyte, ed., *The Evolution of the East Asian Environment*, vol. 2. Hong Kong: Centre of Asian Studies, University of Hong Kong.

Badgley, C. E., and A. K. Behrensmeyer. 1980. Paleoecology of middle Siwalik sediments and faunas, Northern Pakistan. *Palaeogeography, Palaeoclimatology, Palaeoecology* 30:133–55.

Badgley, C. E., and P. D. Gingerich. 1988. Sampling and faunal turnover in early Eocene mammals. *Palaeogeography, Palaeoclimatology, Palaeoecology* 63: 141–58.

Barker, W. R., and P. J. M. Greenslade, eds., 1982. *Evolution of the Flora and Fauna of Arid Australia*. Frewville, South Australia: Peacock Publications.

Barnosky, C. W. 1984. Late Miocene vegetational and climatic variations inferred from a pollen record in northwest Wyoming. *Science* 223:49–51.

Barry, J. C. 1987. Large carnivores (Canidae, Hyaenidae, Felidae) from Laetoli. 235–58 in M. D. Leakey and J. M. Harris, eds., *Laetoli, a Pliocene Site in Northern Tanzania*. Oxford: Clarendon Press.

Barry, J. C., L. J. Flynn, and D. R. Pilbeam. 1990. Faunal diversity and turnover in a Miocene terrestrial sequence. 381–421 in R. M. Ross and W. D. Allmon, eds., *Causes of Evolution: A Paleontological Perspective*. Chicago: University of Chicago Press.

Barry, J. C., N. M. Johnson, S. M. Raza, and L. L. Jacobs. 1985. Neogene mammalian faunal changes in southern Asia: Correlations with climatic, tectonic, and eustatic events. *Geology* 13:637–40.

Barry, J. C., E. H. Lindsay, and L. L. Jacobs. 1982. A biostratigraphic zonation of the middle and upper Siwaliks of the Potwar Plateau of northern Pakistan. *Palaeogeography, Palaeoclimatology, Palaeoecology* 37:95–130.

Beadle, N. C. W. 1967. Soil phosphate and its role in molding segments of the Australian flora and vegetation, with special reference to xeromorphy and sclerophylly. *Ecology* 47:992–1007.

Benson, L. V. 1978. Fluctuation in the level of pluvial lake Lahontan during the last 40,000 years. *Quaternary Research* 9:300–18.

Berggren, W. A., D. V. Kent, J. J. Flynn, and J. A. Van Couvering. 1985. Cenozoic geochronology. *Geological Society of America Bulletin* 96:1407–18.

Berggren, W. A., and J. A. Van Couvering. 1974. The Late Neogene: Biostratigraphy, geochronology, and paleoclimatology of the last 15 million years in marine and continental sequences. *Palaeogeography, Palaeoclimatology, Palaeoecology* 16:1–216.

Bernor, R. L. 1983. Geochronology and zoogeographic relationships of Miocene Hominoidea. 21–64 in R. L. Ciochon and R. S. Corruccini, eds., *New Interpretations of Ape and Human Ancestry*. New York: Plenum.

———. 1984. A zoogeographic theater and biochronologic play: The time/biofacies phenomena of Eurasian and African Miocene mammal provinces. *Paléobiologie Continentale* 14:121–42.

Bernor, R. L., and P. P. Pavlakis. 1987. Zoogeographic relationships of the Sahabi large mammal fauna (early Pliocene, Libya). 349–83 in N. T. Boaz, A. El-Arnauti, A. W. Gaziry, J. De Heinzelin, and D. D. Boaz, eds., *Neogene Paleontology and Geology of Sahabi*. New York: Alan R. Liss.

Bernor, R. L., H. Tobien, and M. O. Woodburne. 1990. Patterns of Old World Hipparionine evolutionary diversification and biogeographic extension. 263–319 in E. H. Lindsay, et al., eds., *European Neogene Mammal Chronology*. New York: Plenum Press.

Berry, E. W. 1939. Eocene plants from a well core in Venezuela. *Johns Hopkins University Studies in Geology* 13:157–62.

Bigalke, R. C. 1978. Present-day mammals of Africa. 1–16 in V. J. Maglio and

H. B. S. Cooke, eds., *Evolution of African Mammals*. Cambridge: Harvard University Press.

Birks, H. J. B. 1986. Late Quaternary biotic changes in terrestrial and lacustrine environments, with particular reference to north-west Europe. 3–65 in B. E. Berglund, ed., *Handbook of Holocene Paleoecology and Paleohydrology*. New York: John Wiley and Sons.

Birks, H. J. B., and H. H. Birks. 1980. *Quaternary Palaeoecology*. London: Edward Arnold.

Bishop, W. W. 1968. The evolution of fossil environments in East Africa. *Transactions of the Leicester Literary and Philosophical Society* 62:22–44.

Bonnefille, R. 1984. Cenozoic vegetation and environments of early hominids in East Africa. 579–612 in R. O. Whyte, ed., *The Evolution of the East Asian Environment*, vol. 2. Hong Kong: Centre of Asian Studies, University of Hong Kong.

————. 1985. Evolution of the continental vegetation: The palaeobotanical record from East Africa. *South African Journal of Science* 81:267–70.

Bonnefille, R., and P. Letouzey. 1976. Fruits fossiles d'*Antrocaryon* dans la Vallée de l'Omo (Ethiopie). *Adansonia* 2:65–82.

Bonnefille, R., D. Lobreau, and G. Riollet. 1982. Fossil pollen of *Ximenia* (Olacaceae) in the Lower Pleistocene of Olduvai, Tanzania: Palaeoecological implications. *Journal of Biogeography* 9:469–86.

Bonnefille, R., A. Vincens, and G. Buchet. 1987. Palynology, stratigraphy, and palaeoenvironment of a Pliocene hominid site (2.9–3.3 my) at Hadar, Ethiopia. *Palaeogeography, Palaeoclimatology, Palaeoecology* 60:249–81.

Bourliere, F., ed. 1983. *Ecosystems of the World*. Vol. 13, *Tropical Savannas*. Amsterdam: Elsevier.

Bowler, J. M., and A. G. Thorne. 1976. Human remains from Lake Mungo: Discovery and excavation of Lake Mungo III. In D. L. Kirk and A. G. Thorne, eds., *The Origin of the Australians*. Canberra: Australian Institute of Aboriginal Studies.

Bown, T. M., M. J. Kraus, S. L. Wing, J. G. Fleagle, B. H. Tiffney, E. L. Simons, and C. F. Vondra. 1982. The Fayum primate forest revisited. *Journal of Human Evolution* 11:603–32.

Bradbury, J. P., B. Leyden, M. Salgado-Labouriau, W. M. Lewis, Jr., C. Schubert, M. W. Binford, D. G. Frey, D. R. Whitehead, and F. H. Weibezahn. 1981. Late Quaternary environmental history of Lake Valencia, Venezuela. *Science* 214: 1299–1305.

Brain, C. K. 1981. The evolution of Man in Africa: Was it a consequence of Cainozoic cooling? *Annals of the Geological Society of South Africa* 84:1–19.

Brunet, M. 1977. Les mammifères et le probleme de la limite Eocene-Oligocene en Europe. *Géobios Mémoire Spéciale* 1:11–27.

Butzer, K. W., and H. B. S. Cooke. 1982. The palaeo-ecology of the African continent: The physical environment of Africa from the earliest geological to Later Stone Age times. 1–69 in J. D. Clark, ed., *The Cambridge History of Africa*, vol. 1. Cambridge: Cambridge University Press.

Carpenter, F. M., and L. Burnham. 1985. The geological record of insects. *Annual Review of Earth and Planetary Science* 13:297–314.

Carroll, R. L. 1977. Patterns of amphibian evolution: An extended example of the incompleteness of the fossil record. 405–37 in A. Hallam, ed., *Patterns of Evolution*. Amsterdam: Elsevier Scientific Publications.

———. 1988. *Vertebrate Paleontology and Evolution*. New York: W. H. Freeman.

Cerling, T. E., J. R. Bowman, and J. R. O'Neil. 1988. An isotopic study of a fluvial-lacustrine sequence: The Plio-Pleistocene Koobi Fora sequence, East Africa. *Palaeogeography, Palaeoclimatology, Palaeoecology* 63:335–56.

Cerling, T. E., and R. L. Hay. 1986. An isotopic study of paleosol carbonates from Olduvai Gorge. *Quaternary Research* 25:63–78.

Cerling, T. E., R. L. Hay, and J. R. O'Neil. 1977. Isotopic evidence for dramatic climatic changes in East Africa during the Pleistocene. *Nature* 267:137–38.

Cerling, T. E., J. Quade, Y. Wang, and J. R. Bowman. 1989. Carbon isotopes in paleosol carbonates as paleoecology indicators. *Nature* 341:138–39.

Chaney, R. W. 1933. A Tertiary flora from Uganda. *Journal of Geology* 41:702–9.

———. 1948. The bearing of the living *Metasequoia* on problems of Tertiary paleobotany. *Proceedings of the National Academy of Sciences* 34:503–15.

Christophel, D. C. 1989. Evolution of the Australian flora through the Tertiary. *Plant Systematics and Evolution* 162:63–78.

Christophel, D. C., and D. R. Greenwood. 1989. Changes in climate and vegetation in Australia during the Tertiary. *Review of Palaeobotany and Palynology* 58:95–109.

Cifelli, R. L. 1981. Patterns of evolution among the Artiodactyla and Perissodactyla (Mammalia). *Evolution* 35:433–40.

Ciochon, R. L., and A. B. Chiarelli, eds. 1980. *Evolutionary Biology of the New World Monkeys and Continental Drift*. New York: Plenum.

Clark, J., J. R. Beerbower, and K. K. Kietzke. 1967. Oligocene sedimentation, stratigraphy, paleoecology, and paleoclimatology in the Big Badlands of South Dakota. *Fieldiana, Geology Memoirs* 5:1–158.

Cole, K. 1985. Past rates of change, species richness, and a model of vegetational inertia in the Grand Canyon, Arizona. *American Naturalist* 125:289–303.

Colinvaux, P. A. 1989. Ice Age Amazon revisited. *Nature* 340:188–89.

Cooke, H. B. S. 1958. Observations relating to Quaternary environments in East and Southern Africa. *Transactions of the Geological Society of South Africa* 61 (annex.):1–73.

Coombs, M. C. 1983. Large mammalian clawed herbivores: A comparative study. *Transactions of the American Philosophical Society* 73:1–96.

Coope, G. R. 1967. The value of Quaternary insect faunas in the interpretation of ancient ecology and climate. 359–80 in E. J. Cushing and H. E. Wright, eds., *Quaternary Paleoecology*. New Haven: Yale University Press.

Corliss, B. H., M. P. Aubry, W. A. Berggren, J. M. Fenner, L. D. Keigwin, and G. Keller. 1984. The Eocene/Oligocene boundary event in the deep sea. *Science* 226:806–10.

Coryndon, S. C. 1978. Hippopotamidae. 483–95 in V. J. Maglio and H. B. S. Cooke, eds., *Evolution of African Mammals*. Cambridge: Harvard University Press.

Coryndon, S. C., and R. J. G. Savage. 1973. The origin and affinities of African mammal faunas. *Special Papers in Palaeontology* 12:121–35.

Crepet, W. L., and G. D. Feldman. 1988. Paleocene/Eocene Grasses from the southwestern USA. *American Journal of Botany* 76:161.

Cronin, T. M. 1985. Speciation and stasis in marine ostracoda: Climatic modulation of evolution. *Science* 227:60–63.

Davis, M. B. 1976. Pleistocene biogeography of temperate deciduous forests. *Geoscience and Man* 13:13–26.

———. 1985. Climatic instability, time lags, and community disequilibrium. 269–84 in J. Diamond and T. J. Case, eds., *Community Ecology.* New York: Harper and Row.

Dawson, M. R., and L. Krishtalka. 1984. Fossil history of the families of recent mammals. 11–57 in S. Anderson and J. K. Jones, eds., *Orders and Families of Recent Mammals of the World.* New York: Wiley.

Dawson, T. J. 1977. Kangaroos. *Scientific American* 237 (2):78–89.

Delcourt, H. R., P. A. Delcourt, and T. Webb III. 1983. Dynamic plant ecology: The spectrum of vegetational change in space and time. *Quaternary Science Reviews* 1:153–75.

Denton, G. H. 1985. Did the Antarctic ice sheet influence Late Cenozoic climate and evolution in the southern Hemisphere? *South African Journal of Science* 81:224–29.

Denton, G. H., R. I. Armstrong, and M. Stuivier. 1971. The late Cenozoic glacial history of Antarctica. 267–306 in K. K. Turekian, ed., *The Late Cenozoic Glacial Ages.* New Haven: Yale University Press.

Denys, C. 1985. Palaeoenvironmental and palaeogeographical significance of the fossil rodent assemblages of Laetoli (Pliocene, Tanzania). *Palaeogeography, Palaeoclimatology, Palaeoecology* 52:77–97.

Diester-Haass, L. 1976. Late Quaternary climatic variations in northwest Africa deduced from East Atlantic sediment cores. *Quaternary Research* 6:299–314.

Diester-Haass, L., and H. J. Schrader. 1979. Neogene coastal upwelling history off northwest and southwest Africa. *Marine Geology* 29:39–53.

Dilcher, D. 1973. A paleoclimatic interpretation of the Eocene floras of southeastern North America. 39–59 in A. Graham, ed., *Vegetation and Vegetational History of Northern Latin America.* Amsterdam: Elsevier.

Dodson, J. 1977. Late Quaternary palaeoecology of Wyrie Swamp, southeastern South Australia. *Quaternary Research* 8:97–114.

Eden, M. J. 1974. Palaeoclimatic influences and the development of savanna in southern Venezuela. *Journal of Biogeography* 1:95–109.

Eisenberg, J. F. 1981. *The Mammalian Radiations.* Chicago: University of Chicago Press.

Emry, R. J., L. S. Russell, and P. R. Bjork. 1987. The Chadronian, Orellan, and Whitneyan North American Land Mammal Ages. 118–52 in M. O. Woodburne, ed., *Cenozoic Mammals of North America.* Berkeley: University of California Press.

Fisher, D. C. 1987. Mastodont procurement by Paleoindians of the Great Lakes region: Hunting or scavenging? 309–421 in M. H. Nitecki and D. V. Nitecki, eds., *The Evolution of Human Hunting.* New York: Plenum.

Flannery, K. V. 1969. Origins and ecological effects of early domestication in Iran and the Near East. 73–100 in P. J. Ucko and G. W. Dimbleby, eds., *The Domestication and Exploitation of Plants and Animals.* London: Gerald Duckworth.

Flannery, K. V., ed. 1986. *Guila Naquitz: Archaic Foraging and Early Agriculture in Oaxaca, Mexico.* Orlando: Academic Press.

Fleagle, J. G. 1988. *Primate Adaptation and Evolution.* San Diego: Academic Press.

Fleagle, J., T. M. Bown, J. D. Obradovich, and E. L. Simons. 1986. Age of earliest African anthropoids. *Science* 234:1247–49.

Fleagle, J. G., and R. F. Kay. 1987. The phyletic position of the Parapithecidae. *Journal of Human Evolution* 16:483–532.

Flenley, J. R. 1979. *The Equatorial Rain Forest: A Geological History.* London: Butterworth.

Flint, R. F. 1959. Pleistocene climates in eastern and southern Africa. *Bulletin of the Geological Society of America* 70:343–74.

Ford, S. M. 1990. Locomotor adaptations of fossil platyrrhines. *Journal of Human Evolution* 1/2:141–74.

Frakes, L. A. 1979. *Climates throughout Geologic Time.* Amsterdam: Elsevier.

Frison, G. C. 1987. Prehistoric, plains-mountain, large-mammal, communal hunting strategies. 177–223 in M. H. Nitecki and D. V. Nitecki, eds., *The Evolution of Human Hunting.* New York: Plenum.

Gaur, R., and S. R. K. Chopra. 1983. Palaeoecology of the Middle Miocene Sivalik sediments of a part of Jammu and Kashmir State (India). *Palaeogeography, Palaeoclimatology, Palaeoecology* 43:313–27.

Gentry, A. W. 1978. The fossil Bovidae of the Baringo area, Kenya. 293–308 in W. W. Bishop, ed., *Geological Background to Fossil Man.* Edinburgh: Scottish Academic Press.

Gheerbrant, E. 1990. On the early biogeographical history of the African placentals. *Historical Biology* 4:107–16.

Gillette, D. G., and C. E. Ray. 1981. Glyptodonts of North America. *Smithsonian Contributions to Paleobiology* 40:1–255.

Gingerich, P. D. 1977. Patterns of evolution in the mammalian fossil record. 469–500 in A. Hallam, ed., *Patterns of Evolution As Illustrated by the Fossil Record.* Amsterdam: Elsevier.

———. 1984. Pleistocene extinctions in the context of origination-extinction equilibria in Cenozoic mammals. 211–22 in P. S. Martin and R. G. Klein, eds., *Quaternary Extinctions.* Tucson: University of Arizona Press.

———. 1987. Evolution and the fossil record: Patterns, rates, and processes. *Canadian Journal of Zoology* 65:1053–60.

Goulding, M. 1980. *The Fishes and the Forest.* Berkeley: University of California Press.

Graham, R. W., and E. L. Lundelius, Jr. 1984. Coevolutionary disequilibrium and Pleistocene extinctions. 223–49 in P. S. Martin and R. G. Klein, eds., *Quaternary Extinctions.* Tucson: University of Arizona Press.

Graham, R. W., and H. A. Semken. 1976. Paleoecological significance of the short-tailed shrew (*Blarina*) with a systematic discussion of *Blarina ozarkensis. Journal of Mammalogy* 57:433–49.

Graham, R. W., H. A. Semken, and M. A. Graham, eds. 1987. *Late Quaternary Mammalian Biogeography and Environments of the Great Plains and Prairies.* Springfield: Illinois State Museum.

Gregor, H.-J. 1982. *Die jungtertiären Floren Süddeutschlands: Paläokarpologie, Phytostratigraphie, Paläookologie, Paläoklimatologie.* Stuttgart: Ferdinand Enke Verlag.

Grime, J. P. 1979. *Plant Strategies and Vegetation Processes.* London: John Wiley.

Guo Shuangxing. 1984. The Tertiary climates of China, as indicated by plant mega-fossils. 467–69 in R. O. Whyte, ed., *The Evolution of the East Asian Environment,* vol. 2. Hong Kong: Centre of Asian Studies, University of Hong Kong.

Hamilton, A. 1976. The significance of patterns of distribution shown by forest plants and animals in tropical Africa for the reconstruction of Upper Pleistocene palaeoenvironment. *Palaeoecology of Africa* 9:63–97.

Hamilton, A. C. 1982. *Environmental History of East Africa.* London: Academic Press.

Han, D., and C. Xu. 1985. Pleistocene mammalian faunas of China. 267–89 in Rukang Wu and J. W. Olsen, eds., *Palaeoanthropology and Palaeolithic Archaeology in the People's Republic of China.* Orlando: Academic Press.

Haq, B. U., J. Hardenbol, and P. R. Vail. 1987. Chronology of fluctuating sea levels since the Triassic. *Science* 235:1156–67.

Harris, J. M. ed. 1983. *Koobi Fora Research Project,* vol. 2. Oxford: Clarendon Press.

Harris, J. M., and T. D. White. 1979. Evolution of the Plio-Pleistocene African Suidae. *Transactions of the American Philosophical Society* 69:1–128.

Hays, J. D., J. Imbrie, and N. J. Shackleton. 1976. Variations in the earth's orbit: Pacemaker of the ice ages. *Science* 194:1121–32.

Hershkovitz, P. 1966. Mice, land bridges, and Latin America faunal interchange. 725–47 in R. L. Wenzel and V. J. Tipton, eds., *Ectoparasites of Panama.* Chicago: Field Museum of Natural History.

Heusser, L. E., and N. J. Shackleton. 1979. Direct marine-continental correlation: 150Ka oxygen isotope pollen record from North Pacific. *Science* 204:837–39.

Heusser, L. E., and M. Wingenroth. 1984. Late Quaternary continental environments of Argentina: Evidence from pollen analyses of the upper 2 meters of deep-sea core RC 12-241 in the Argentine Basin. 79–92 in J. Rabassa, ed., *Quaternary of South America and Antarctic Peninsula.* Rotterdam: A. A. Balkema.

Hill, A. 1987. Causes of perceived faunal change in the later Neogene of East Africa. *Journal of Human Evolution* 16:583–96.

Hill, A., G. Curtis, and R. Drake. 1986. Sedimentary stratigraphy of the Tugen Hills, Baringo District, Kenya. 285–95 in L. E. Frostick, R. Renaut, I. Reid, and J. J. Tiercelin, eds., *Sedimentation in the African Rift System.* Oxford: Blackwell.

Hill, A., R. Drake, L. Tauxe, M. Monaghan, J. C. Barry, A. K. Behrensmeyer, G. Curtis, B. Fine Jacobs, L. Jacobs, N. Johnson, and D. Pilbeam. 1985. Neogene palaeontology and geochronology of the Baringo Basin, Kenya. *Journal of Human Evolution* 14:759–73.

Hill, R. S. 1983. Evolution of *Nothofagus cunninghamii* and its relationship to *N. moorei* inferred from Tasmanian macrofossils. *Australian Journal of Botany* 31:453–66.

———. 1990. Evolution of the modern high latitude southern hemisphere flora: Evidence from the Australian macrofossil record. 31–42 in J. G. Douglas and D. C.

Christophel, eds., *Proceedings of the Third International Organization for Paleobotany Conference, Melbourne.* Publication no. 2.

Hill, R. S., and M. K. MacPhail. 1983. Reconstruction of the Oligocene vegetation at Pioneer, northeast Tasmania. *Alcheringa* 7:281–99.

Hole, F., K. V. Flannery, and J. A. Neely. 1969. Prehistory and human ecology of the Deh Luran Plain: An early village sequence from Khuzistan, Iran. *Museum of Anthropology University of Michigan Memoir* 1.

Hooghiemstra, H. 1984. Vegetational and climatic history of the High Plain of Bogota, Colombia: A continuous record of the last 3.5 million years. *Dissertationes Botanicae,* Band 79. Vaduz: J. Cramer.

Horton, D. R. 1984. Red kangaroos: Last of the Australian megafauna. 639–80 in P. S. Martin and R. G. Klein, eds., *Quaternary Extinctions.* Tucson: University of Arizona Press.

Hubbard, R. N. L. B., and M. C. Boulter. 1983. Reconstruction of Palaeogene climate from palynological evidence. *Nature* 301:147–50.

Hunt, R. M. 1990. Taphonomy and sedimentology of Arikaree (lower Miocene) fluvial, eolian, and lacustrine paleoenvironments, Nebraska and Wyoming: A paleobiota entombed in fine-grained volcaniclastic rocks. 69–112 in M. Lockley and A. Rice, eds., *Volcanism and Fossil Biotas: Implications for Preservation.* Geological Society of America Special Paper 244.

Huntley, B. 1988. Glacial and Holocene vegetation history: Europe. 341–83 in B. Huntley and T. Webb III, eds., *Vegetation History.* Dordrecht: Kluwer.

Huntley, B., and H. J. B. Birks. 1983. *An Atlas of Past and Present Pollen Maps for Europe: 0–13,000 Years Ago.* Cambridge: Cambridge University Press.

Huntley, B., and T. Webb III. 1989. Migration: Species' response to climatic variations caused by changes in the earth's orbit. *Journal of Biogeography* 16:5–19.

Imbrie, J., J. D. Hays, D. G. Martinson, A. McIntyre, A. C. Mix, J. J. Morley, N. G. Pisias, W. L. Prell, and N. J. Shackleton. 1984. The orbital theory of Pleistocene climate: Support from revised chronology of the marine $\delta^{18}O$ record. 269–305 in A. L. Berger, J. Imbrie, J. Hays, G. Kukla, and B. Saltzman, eds., *Milankovitch and Climate,* part 1. Higham, Mass.: Reidel.

Iversen, J. 1958. The bearing of glacial and interglacial epochs on the formation and extinction of plant taxa. *Uppsala Universiteit Arssk* 6:210–15.

———. 1964. Retrogressive vegetational succession in the postglacial. *Journal of Ecology* 52 (Suppl.):421–28.

Jacobs, B. F., and C. H. S. Kabuye. 1987. A middle Miocene (12.2 m.y. old) forest in the East African Rift Valley, Kenya. *Journal of Human Evolution* 16:147–55.

Jacobs, L. L., and L. J. Flynn. 1981. Development of the modern rodent fauna of the Potwar Plateau, northern Pakistan. *Proceedings of the Neogene-Quaternary Boundary Field Conference, India:* 79–81.

Janis, C. M. 1976. The evolutionary strategy of the Equidae and the origins of rumen and cecae digestion. *Evolution* 30:757–74.

———. 1982. The evolution of horns in ungulates: Ecology and paleoecology. *Biological Reviews* 57:261–318.

———. 1988. New ideas in ungulate phylogeny and evolution. *Trends in Ecology and Evolution* 3:291–97.

————. 1989. A climatic explanation for patterns of evolutionary diversity in ungulate mammals. *Palaeontology* 32:463–81.

————. In press. Do legs support the arms race in mammalian predator/prey relationships? In J. R. Horner and L. Neill, eds., *Vertebrate Behavior As Derived from the Fossil Record.* New York: Columbia University Press.

Janis, C. M., and J. Damuth. 1991. Mammals. 301–45 in K. J. McNamara, ed., *Evolutionary Trends.* London: Belhaven Press.

Jansen, J. H. F., A. Kuipers, and S. R. Troelstra. 1986. A mid-Brunhes climatic event: Long-term changes in global atmosphere and ocean circulation. *Science* 232:619–22.

Jones, R., and J. M. Bowler. 1980. Struggle for the savanna: Northern Australia in ecological and prehistoric perspective. 3–31 in R. Jones, ed., *Northern Australia.* Canberra: Research School of Pacific Studies.

Jouzel, J., C. Lorius, J. R. Petit, C. Genthon, N. I. Barkov, V. M. Kotlykov, and V. N. Petrov. 1987. Vostok ice core: A continuous isotope temperature record over the last climatic cycle (160,000 years). *Nature* 329:403–8.

Kalb, J. E., C. J. Jolly, S. Tebedge, A. Mebrate, C. Smart, E. B. Oswald, P. F. Whitehead, C. B. Wood, T. Adefris, and V. Rawn-Schatzinger. 1982. Vertebrate faunas from the Awash Group, Middle Awash Valley, Afar, Ethiopia. *Journal of Vertebrate Paleontology* 2:237–58.

Kappelman, J. 1991. The paleoenvironment of *Kenyapithecus* at Fort Ternan. *Journal of Human Evolution* 20:95–130.

Kemp, E. M. 1978. Tertiary climatic evolution and vegetation history in the southeastern Indian Ocean region. *Palaeogeography, Palaeoclimatology, Palaeoecology* 24:169–208.

Kennett, J. P. 1978. The development of planktonic biogeography in the Southern Ocean during the Cenozoic. *Marine Micropaleontology* 3:301–45.

————. 1985. *The Miocene Ocean: Paleoceanography and Biogeography.* Geological Society of America Memoir 163.

Kershaw, A. P. 1978. Record of last interglacial-glacial cycle from northeastern Queensland. *Nature* 272:159–61.

————. 1984. Late Cenozoic plant extinctions in Australia. 691–707 in P. S. Martin and R. G. Klein, eds., *Quaternary Extinctions.* Tucson: University of Arizona Press.

————. 1985. An extended late Quaternary vegetation record from northeastern Queensland and its implications for the seasonal tropics of Australia. *Proceedings of the Ecological Society of Australia* 13:179–89.

————. 1986. Climatic change and aboriginal burning in north-east Australia during the last two glacial/interglacial cycles. *Nature* 322:47–49.

Kidwell, S. M., and A. K. Behrensmeyer. 1988. Overview: Ecological and evolutionary implications of taphonomic processes. *Palaeogeography, Palaeoclimatology, Palaeoecology* 63: 1–14.

Kortlandt, A. 1980. The Fayum primate forest: Did it exist? *Journal of Human Evolution* 9:277–97.

Kräusel, R. 1929. Fossile Pflanzen aus dem Tertiär von Süd-Sumatra. *Geol. Mijnb. Genootschap Nederlanden Kolonien, Geol. Ser.* 9:335–78.

Kukla, G. J. 1977. Pleistocene land-sea correlations 1. Europe. *Earth-Science Reviews* 13:307–74.

———. 1987. Loess stratigraphy in central China. *Quaternary Science Reviews* 6:191–219.

———. 1989. Long continental records of climate. *Palaeogeography, Palaeoclimatology, Palaeoecology* 72:1–225.

Kurtén, B., and E. Anderson. 1980. *Pleistocene Mammals of North America*. New York: Columbia University Press.

Lakhampal, R. N. 1966. Some middle Tertiary plant remains from south Kivu, Congo. *Musée Royal de l'Afrique central, Tervuren, Belgique, Annales Sci. Géol.* 52:21–30.

Lakhampal, R. N., and U. Prakash. 1970. Cenozoic plants from Congo. 1. Fossil woods from the Miocene of Lake Albert. *Musée Royal de l'Afrique Centrale, Tervuren, Belgique, Annales Sci. Géol.* 64:1–20.

Lange, R. T. 1978. Carpological evidence for fossil *Eucalyptus* and other Leptospermeae (subfamily Leptospermoideae of Myrtaceae) from a Tertiary deposit in the South Australian arid zone. *Australian Journal of Botany* 26:221–33.

———. 1982. Australian Tertiary vegetation, evidence and interpretation. 44–89 in J. M. G. Smith, ed., *A History of Australasian Vegetation*. Sydney: McGraw Hill.

Langston, W., Jr. 1965. Fossil crocodiles from Colombia and the Cenozoic history of crocodilians in South America. *University of California Publications in Geological Sciences* 52:1–157.

Laporte, L. F., and A. L. Zihlman. 1983. Plates, climate, and hominoid evolution. *South African Journal of Science* 79:96–110.

Leakey, L. S. B. 1931. *Stone Age Cultures of Kenya Colony*. Cambridge: Cambridge University Press.

Leakey, L. S. B., and S. Cole, eds. 1952. *Proceedings of the Pan-African Congress on Prehistory 1947*. Oxford: Oxford University Press.

Leakey, M. D., and J. M. Harris, eds. 1987. *Laetoli: A Pliocene Site in Northern Tanzania*. Oxford: Oxford University Press.

Legendre, S. 1986. Analysis of mammalian communities from the Late Eocene and oligocene of southern France. *Palaeovertebrata* 16:191–212.

Lemasurier, W. E. 1972a. Volcanic record of Antarctic glacial history: Implications with regard to Cenozoic sea levels. 59–74 in R. J. Price and D. E. Dugden, eds., *Polar Geomorphology*. Special Publications of the Institute of British Geography, no. 4.

———. 1972b. Volcanic record of Cenozoic glacial history of Marie Byrd Land. 251–60 in R. J. Adie, ed., *Antarctic Geology and Geophysics*. Oslo: Universitetsforlaget.

Leopold, E. B. 1966. Late Cenozoic palynology. 377–438 in R. H. Tschudy and R. A. Scott, eds., *Aspects of Palynology*. New York: John Wiley & Sons.

Leopold, E. B., and M. F. Denton. 1987. Comparative age of grassland and steppe east and west of the northern Rocky Mountains. *Annals of the Missouri Botanical Garden* 74:841–67.

Leopold, E. B., and MacGinitie, H. D. 1972. Development and affinities of Tertiary floras in the Rocky Mountains. 147–200 in A. Graham, ed., *Floristics and Paleofloristics of Asia and Eastern North America*. Amsterdam: Elsevier.

Leopold, E. B., and V. C. Wright. 1985. Pollen profiles of the Plio-Pleistocene transition in the Snake River plain, Idaho. 323–48 in D. J. Smiley, ed., *Late Cenozoic History of the Pacific Northwest*. Pacific Division, American Association for the Advancement of Science.

Liu, K., and P. A. Colinvaux. 1985. Forest changes in the Amazon Basin during the last glacial maximum. *Nature* 318:556–57.

Livingstone, D. A., and T. Van der Hammen. 1978. Palaeogeography and palaeoclimatology. *Natural Resources Research* 14:61–90.

Lopez, N., and L. Thaler. 1974. Sur le plus ancien lagomorphe européen et al "Grande Coupure" Oligocene de Stehlin. *Palaeovertebrata* 6:243–51.

Lundberg, J. G., A. Machado-Allison, and R. F. Kay. 1986. Miocene characid fishes from Colombia: Evolutionary stasis and extirpation. *Science* 234:208–9.

Lundelius, E. L., Jr., T. Downs, E. H. Lindsay, H. A. Semken, R. J. Zakrzewski, C. S. Churcher, C. R. Harington, G. E. Schultz, and S. D. Webb. 1987. The North American Quaternary sequence. 211–35 in M. O. Woodburne, ed., *Cenozoic Mammals of North America*. Berkeley: University of California Press.

Lundelius, E. L., Jr., R. W. Graham, E. Anderson, J. Guilday, J. A. Holman, D. W. Steadman, and S. D. Webb. 1983. Terrestrial vertebrate faunas. 311–53 in S. C. Porter, ed., *Late Quaternary environments of the United States*. Vol. 1, *The Late Pleistocene*. Minneapolis: University of Minnesota Press.

MacFadden, B. J. 1980. Rafting mammals or drifting islands? Biogeography of the Greater Antillean insectivores *Nesophontes* and *Solenodon*. *Journal of Biogeography* 7:11–22.

————. 1985. Drifting continents, mammals, and time scales: Current developments in South America. *Journal of Vertebrate Paleontology* 5:169–74.

————. 1988. Horses, the fossil record, and evolution: A current perspective. 131–58 in M. K. Hecht, B. Wallace, and G. T. Prance, eds., *Evolutionary Biology*, vol. 22. New York: Plenum.

————. 1990. Chronology of Cenozoic primate localities in South America. *Journal of Human Evolution* 1/2:7–22.

MacFadden, B. J., K. E. Campbell, Jr., R. L. Cifelli, O. Siles, N. M. Johnson, C. W. Naeser, and P. K. Zeitler. 1985. Magnetic polarity stratigraphy and mammalian fauna of the Deseadan (Late Oligocene–Early Miocene) Salla Beds of northern Bolivia. *Journal of Geology* 93:223–50.

MacGinitie, H. D. 1962. The Kilgore flora: A late Miocene flora of northern Nebraska. *University of California Publications in Geological Sciences* 35:67–158.

Maglio, V. J. 1975. Pleistocene and faunal evolution in Africa and Eurasia. 419–76 in K. W. Butzer and G. L. Isaac, eds., *After the Australopithecines*. The Hague: Mouton.

————. 1978. Patterns of faunal evolution. 603–19 in V. J. Maglio and H. B. S. Cooke, eds., *Evolution of African Mammals*. Cambridge: Harvard University Press.

Maglio, V. J., and H. B. S. Cooke, eds. 1978. *Evolution of African Mammals*. Cambridge: Harvard University Press.

Maloney, B. K. 1980. Pollen analytical evidence for early forest clearance in North Sumatra. *Nature* 287:324–26.

Marean, C. W., and D. Gifford-Gonzalez. 1991. Later Quaternary extinct ungulates of East Africa and palaeoenvironmental implications. *Nature* 350:418–20.

Marshall, L. G. 1981a. The families and genera of Marsupialia. *Fieldiana, Geology* n.s. 8:1–65.

———. 1981b. The Great American Interchange: An invasion-induced crisis for South American mammals. 133–229 in M. H. Nitecki, ed., *Biotic Crises in Ecological and Evolutionary Time.* New York: Academic Press.

———. 1982. Calibration of the Age of Mammals in South America. 427–37 in E. Buffetaut, P. Janvier, J.-C. Rage, and P. Tassy, eds., Livre Jubilaire en l'Honneur de Robert Hofstetter. *Géobios, Mémoire Spéciale* 6.

———. 1984. Who killed Cock Robin? An investigation of the extinction controversy. 785–806 in P. S. Martin and R. G. Klein, eds., *Quaternary Extinctions.* Tucson: University of Arizona Press.

———. 1988. Land mammals and the Great American Interchange. *American Scientist* 96:380–88.

Martin, H. A. 1978. Evolution of the Australian flora and vegetation through the Tertiary: Evidence from pollen. *Alcheringa* 2:181–202.

Martin, L. D. 1974. New rodents from the lower Miocene Gering Formation of western Nebraska. *Kansas University Museum of Natural History Occasional Paper* no. 32:1–12.

———. 1987. Beavers from the Harrison Formation (Early Miocene) with a revision of *Euhapsis. Dakoterra* (South Dakota School of Mines and Technology, Museum of Geology) 3:73–91.

Martin, P. S. 1967. Prehistoric overkill. 75–120 in P. S. Martin and H. E. Wright, Jr., eds., *Pleistocene Extinctions.* New Haven: Yale University Press.

———. 1984. Prehistoric overkill: The global model. 354–403 in P. S. Martin and R. G. Klein, eds., *Quaternary Extinctions.* Tucson: University of Arizona Press.

Martin, P. S., and R. G. Klein, eds. 1984. *Quaternary Extinctions.* Tucson: University of Arizona Press.

Mathur, Y. K. 1984. Cenozoic Palynofossils, vegetation, ecology, and climate of the north and northwestern Sub-Himalayan region, India. 504–52 in R. O. Whyte, ed., *The Evolution of the East Asian Environment.* Hong Kong: Centre of Asian Studies, University of Hong Kong.

McArthur, E. D. 1979. Sagebrush systematics and evolution. 14–22 in Natural Resources Alumni Association, eds., *The Sagebrush Ecosystem: A Symposium.* Logan: Utah State University, College of Natural Resources.

McBrearty, S. 1990. Consider the humble termite: Termites as agents of postdepositional disturbance at African Archaeological sites. *Journal of Archaeological Science* 17:111–43.

McCarten, L., B. H. Tiffney, J. A. Wolfe, T. A. Ager, S. L. Wing, L. A. Sirkin, L. W. Ward, and J. Brooks. 1990. Late Tertiary floral assemblage from upland gravel deposits of the southern Maryland Coastal Plain. *Geology* 18:311–14.

McKenna, M. C. 1983. Holarctic landmass rearrangement, cosmic events, and Cenozoic terrestrial organisms. *Annals of the Missouri Botanical Garden* 70:459–89.

Mead, J. I., T. R. Van Devender, and K. L. Cole. 1983. Late Quaternary small mammals from Sonoran Desert packrat middens, Arizona and California. *Journal of Mammalogy* 64:173–80.

Mein, P. 1979. Rapport d'activité du group de travail vertébrés mise a jour de la biostratigraphie de Neogene basée sur les mammifères. *Annales Géologique des Pays Helleniques* 3:1367–72.

Meon, H., R. Ballesio, C. Guerin, and P. Mein. 1979. Approche climatologique du Neogene supérieur (Tortonien à Pléistocéne moyen ancien) d'après les faunes et les flores d'Europe occidentale. *Mémoires, Muséum National d'Histoire Naturelle* 27:182–95.

Mercer, J. H. 1983. Cenozoic glaciation in the Southern Hemisphere. *Annual Reviews of Earth and Planetary Sciences* 11:99–132.

Merrilees, D. 1968. Man the destroyer: Late Quaternary changes in the Australian marsupial fauna. *Journal of the Royal Society of Western Australia* 51:1–24.

Miller, K. G., R. G. Fairbanks, and G. S. Mountain. 1987. Tertiary oxygen isotope synthesis, sea level history, and continental margin erosion. *Paleooceanography* 2: 1–19.

Moreau, R. E. 1966. *The Bird Faunas of Africa and Its Islands.* New York: Academic Press.

Muller, J. 1966. Montane pollen from the Tertiary of North West Borneo. *Blumea* 14:231–35.

———. 1981. Fossil pollen records of extant angiosperms. *Botanical Review* 47(1): 1–142.

Nagotoshi, K. 1987. Miocene hominoid environments of Europe and Turkey. *Palaeogeography, Palaeoclimatology, Palaeoecology* 61: 145–54.

Nelson, B. W., C. A. C. Ferreira, M. F. da Silva, and M. L. Kawasaki. 1990. Endemism centres, refugia, and botanical collection density in Brazilian Amazonia. *Nature* 345:714–16.

Nocchi, M., G. Parisi, P. Monaco, S. Monechi, and M. Madile. 1988. Eocene and early Oligocene micropaleontology and paleoenvironments in southeast Umbria, Italy. *Palaeogeography, Palaeoclimatology, Palaeoecology* 67:181–244.

Olson, S. L. 1985. The fossil record of birds. 79–238 in D. Farner, J. King, and K. Parks, eds., *Avian Biology,* vol. 8. Orlando: Academic Press.

Olson, S. L., and D. T. Rasmussen. 1986. Paleoenvironment of the earliest hominoids: New evidence from the Oligocene avifauna of Egypt. *Science* 233:1202–4.

Padian, K., and W. A. Clemens. 1985. Terrestrial vertebrate diversity: Episodes and insights. 41–96 in J. W. Valentine, ed., *Phanerozoic Diversity Patterns.* Princeton: Princeton University Press.

Paillet, F. L. 1982. The ecological significance of American chestnut (*Castanea dentata* (Marsh) Borkh.) in the Holocene forests of Connecticut. *Bulletin of the Torrey Botanical Club* 109:457–73.

Pantic, N. K., and D. S. Mihajlovic. 1977. Neogene floras of the Balkan land areas and their bearing on the study of paleoclimatology, paleobiogeography, and biostratigraphy (part 2). *Annales Géologiques de la Péninsule Balkanique* (Belgrade) 41:159–73.

Pascual, R., and E. O. Jaureguizar. 1990. Evolving climates and mammal faunas in Cenozoic South America. *Journal of Human Evolution* 1/2:23–60.

Pascual, R., E. J. Ortega-H., D. Gondar, and E. Tonni. 1966. *Paleontografia Bo-naerense.* Fasc. 4. Vertebrata. Commision de Investigaciones Cientificas de la Provincia de Buenos Aires.

Patterson, B., and R. Pascual. 1972. The fossil mammal fauna of South America. 247–309 in A. Keast, F. C. Erk, and B. Glass, eds., *Evolution, Mammals, and Southern Continents.* Albany: State University of New York Press.

Pickard, J., D. A. Adamson, D. M. Harwood, G. H. Miller, P. G. Quilty, and R. K. Dell. 1988. Early Pliocene marine sediments, coastline, and climate of East Antarctica. *Geology* 16:158–61.

Pickford, M. H. L. 1978. Geology, paleoenvironments, and vertebrate faunas of the mid-Miocene Ngorora Formation, Kenya. 237–62 in W. W. Bishop, ed., *Geological Background to Fossil Man.* Edinburgh: Scottish Academic Press.

———. 1983. Sequence and environments of the Lower and Middle Miocene hominoids of western Kenya. 421–39 in R. L. Ciochon and R. S. Corruccini, eds., *Interpretations of Ape and Human Ancestry.* New York: Plenum.

Pomerol, C., and I. Premoli-Silva. 1986a. The Eocene-Oligocene transition: Events and boundary. 1–24 in C. Pomerol and I. Premoli-Silva, eds., *Terminal Eocene Events.* Amsterdam: Elsevier.

Pomerol, C., and I. Premoli-Silva, eds. 1986b. *Terminal Eocene Events.* Amsterdam: Elsevier.

Pope, G. G. 1984. The antiquity and paleoenvironment of the Asian Hominidae. 822–47 in R. O. Whyte, ed., *The Evolution of the East Asian Environment,* vol. 2. Hong Kong: Centre of Asian Studies, University of Hong Kong.

———. 1988. The paleoenvironment of East Asia from the mid-Tertiary. 1097–1123 in R. O. Whyte, ed., *The Palaeoenvironment of East Asia from the Mid-Tertiary,* vol. 2. Hong Kong: Centre of Asian Studies, University of Hong Kong.

Potts, R. 1988. *Early Hominid Activities at Olduvai.* New York: Aldine de Gruyter.

Potts, R., and A. Deino. MS. Mid-Pleistocene change in large mammal faunas of eastern Africa.

Pratt, D. J., P. J. Greenway, and M. D. Gwynne. 1966. A classification of East African rangeland, with an appendix on terminology. *Journal of Applied Ecology* 3:369–82.

Prentice, M. L., and G. H. Denton. 1988. The deep-sea oxygen isotope record, the global ice sheet system, and hominid evolution. 383–403 in F. E. Grine, ed., *Evolutionary History of the "Robust" Australopithecines.* New York: Aldine de Gruyter.

Prothero, D. R. 1985. North American mammalian diversity and Eocene-Oligocene extinctions. *Paleobiology* 11:389–405.

Prothero, D. R., and W. A. Berggren. In press. *Eocene-Oligocene Climatic and Biotic Evolution.* Princeton: Princeton University Press.

Prothero, D. R., and P. C. Sereno. 1982. Allometry and paleoecology of medial Miocene dwarf rhinoceroses from the Texas Gulf Coastal Plain. *Paleobiology* 8:16–30.

Quade, J., T. E. Cerling, and J. R. Bowman. 1989. Development of Asian monsoon revealed by marked ecological shift during the latest Miocene in northern Pakistan. *Nature* 342:163–66.

Quade, J., T. Cerling, M. M. Morgan, J. C. Barry, J. Lee-Thorp, N. J. van der

Merwe. In press. The carbon and oxygen isotopic composition of carbonate in fossil tooth enamel: Is it altered? *Isotope Geoscience.*

Radinsky, L. B. 1982. Evolution of skull shape in carnivores. 3. The origin and early radiation of the modern carnivore families. *Paleobiology* 8:177–95.

———. 1984. Ontogeny and phylogeny in horse skull evolution. *Evolution* 38:1–15.

Rasmussen, D. T., and R. F. Kay. In press. A Miocene Anhinga from Colombia, and comments on the zoogeographical relationships of the South American Tertiary avifauna. 137–40 in K. E. Campbell, ed., *Avian Paleontology*. Special Publication of the Natural History Museum of Los Angeles County.

Redding, R. W. 1988. A general explanation of subsistence change: From hunting and gathering to food production. *Journal of Anthropological Archaeology* 7:56–97.

Reif, W.-E., and H. Silyn-Roberts. 1987. On the robustness of moa's leg bones. An exercise in functional morphology of extinct organisms. *Neues Jahrbuch für Geologie und Paläontologie, Monatshefte:* 155–60.

Rennie, J. V. L. 1931. Note of fossil leaves from the Banke Clays. *Transactions of the Royal Society of South Africa* 19:251–53.

Repenning, C. A. 1987. Biochronology of the microtine rodents of the United States. 236–68 in M. O. Woodburne, ed., *Cenozoic Mammals of North America*. Berkeley: University of California Press.

Retallack, G. J. 1983. A paleopedological approach to the interpretation of terrestrial sedimentary rocks: The mid-Tertiary fossil soils of Badlands National Park, South Dakota. *Geological Society of America Bulletin* 94:823–40.

———. 1984. Completeness of the rock and fossil record: Some estimates using fossil soils. *Paleobiology* 10:59–78.

Rich, T. H. 1982. Monotremes, placentals, and marsupials: Their record in Australia and its biases. 385–488 in R. V. Rich and E. M. Thompson, eds., *The Fossil Vertebrate Record of Australasia*. Clayton: Monash University Press.

Roberts, N. 1984. Pleistocene environments in time and space. 25–53 in R. Foley, ed., *Hominid Evolution and Community Ecology*. London: Academic Press.

Robin, G. de Q. 1988. The Antarctic ice sheet, its history and response to sea level and climatic changes over the past 100 million years. *Palaeogeography, Palaeoclimatology, Palaeoecology* 67:31–50.

Romero, E. J. 1986. Paleogene phytogeography and climatology of South America. *Annals of the Missouri Botanical Garden* 73:449–61.

Rose, K. D., and J. G. Fleagle. 1981. The fossil history of nonhuman primates in the Americas. 111–67 in A. F. Coimbra-Filho and R. A. Mittermeier, eds., *Ecology and Behavior of Neotropical Primates*. Rio de Janeiro: Academia Brasileira de Ciencias.

Rosen, D. E. 1975. A vicariance model of Caribbean biogeography. *Systematic Zoology* 24:431–64.

Rosenberger, A. L., W. C. Hartwig, and R. G. Wolff. 1991. *Szalatarus attricuspis,* an early platyrrhine primate. *Folia Primatologia* 56:225–33.

Ruddiman, W. F., and J. E. Kutzbach. 1989. Forcing of Late Cenozoic Northern Hemisphere climate by plateau uplift in southern Asia and the American west. *Journal of Geophysical Research* 94 D 15, 409–18, 427.

Ruddiman, W. F., M. Sarnthein, J. Backman, J. G. Baldauf, W. Curry, L. M. Dupont, T. Janecek, E. M. Pokras, M. E. Raymo, B. Stabell, R. Stein, and T. Tiedemann.

1989. Late Miocene to Pleistocene evolution of climate in Africa and the low-latitude Atlantic: Overview of leg 108 results. *Proceedings of the Ocean Drilling Program, Scientific Results* 108: 463–84.

Russell, D. E., and H. Tobien. 1986. Mammalian evidence concerning the Eocene-Oligocene transition in Europe, North America, and Asia. 299–308 in C. Pomerol and I. Premoli-Silva, eds., *Terminal Eocene Events.* Amsterdam: Elsevier.

Sarnthein, M. 1978. Neogene sand layers off northwest Africa: Composition and source environment. 939–59 in Y. Lancelot, E. Siebold, et al., eds., *Initial reports of the Deep Sea Drilling Project 41.* Washington, D.C.: U.S. Government Printing Office.

Sarnthein, M., and L. Diester-Haass. 1977. Eolian sand turbidites. *Journal of Sedimentary Petrology* 47: 868–90.

Sauvage, J. 1979. The palynological subdivisions in Hellenic Cainozoic and their stratigraphical correlations. *Annales Géologique des Pays Helleniques* 3: 1091–95.

Savage, D. E., and D. E. Russell. 1983. *Mammalian Paleofaunas of the World.* Reading, Mass.: Addison-Wesley.

Schultz, C. B., and C. H. Falkenbach. 1968. The phylogeny of the oreodonts, parts 1 and 2. *Bulletin of the American Museum of Natural History* 109: 377–482.

Shackleton, N. J. 1987. Oxygen isotopes, ice volume, and sea level. *Quaternary Science Review* 6: 183–90.

Shackleton, N. J., J. Backman, H. Zimmerman, D. V. Kent, M. A. Hall, D. G. Roberts, D. Schnitker, J. G. Baldauf, A. Desprairies, R. Homrighausen, P. Huddlestun, J. B. Keene, A. J. Kaltenback, K. A. O. Krumsiek, A. C. Morton, J. W. Murray, and J. Westberg-Smith. 1984. Oxygen isotope calibration of the onset of ice-rafting and history of glaciation in the North Atlantic region. *Nature* 307: 620–23.

Shackleton, N. J., and J. P. Kennett. 1975a. Late Cenozoic oxygen and carbon isotope changes at DSDP Site 284: Implications for glacial history of the Northern Hemisphere and Antarctica. 801–6 in J. P. Kennett, R. E. Houtz, P. B. Andrews, A. R. Edwards, V. A. Gostin, M. Hajos, M. A. Hampton, D. G. Jenkins, S. V. Margolis, A. T. Ovenshine, and K. Perch-Nielsen, eds., *Initial Reports of the Deep Sea Drilling Project,* 29. Washington, D.C.: U.S. Government Printing Office.

———. 1975b. Paleotemperature history of the Cenozoic and the initiation of Antarctic glaciation: Oxygen and carbon isotope analyses in DSDP sites 277, 279, and 281. 743–55 in J. P. Kennett, R. E. Houtz, P. B. Andrews, A. R. Edwards, V. A. Gostin, M. Hajos, M. A. Hampton, D. G. Jenkins, S. V. Margolis, A. T. Ovenshine, and K. Perch-Nielsen, eds., *Initial Reports of the Deep Sea Drilling Project,* 29. Washington, D.C.: U.S. Government Printing Office.

Shackleton, N. J., and N. D. Opdyke. 1977. Oxygen isotope and palaeomagnetic evidence for early Northern Hemisphere glaciation. *Nature* 270: 216–19.

Shipman, P. 1982. *Taphonomy of* Ramapithecus wickeri *at Fort Ternan, Kenya.* Museum Brief no. 26. Columbia: University of Missouri.

———. 1986. Paleoecology of Fort Ternan reconsidered. *Journal of Human Evolution* 15: 193–204.

Shipman, P., A. Walker, J. A. Van Couvering, P. J. Hooker, and J. A. Miller. 1981. The Fort Ternan hominoid site, Kenya: Geology, age, taphonomy, and paleoecology. *Journal of Human Evolution* 10: 49–72.

Simpson, G. G. 1980. *Splendid Isolation: The Curious History of South American Mammals*. New Haven: Yale University Press.

Singh, G., and J. Luly. In press. Changes in vegetation and seasonal climate since the last full glacial at Lake Frome, South Australia. *Palaeogeography, Palaeoclimatology, Palaeoecology*.

Sluiter, I. R., and A. P. Kershaw. 1982. The nature of late Tertiary vegetation in Australia. *Alcheringa* 6:211–22.

Smith, A. G., A. M. Hurley, and J. C. Briden. 1981. *Phanerozoic Paleocontinental World Maps*. Cambridge: Cambridge University Press.

Smith, B. D. 1989. Origins of agriculture in eastern North America. *Science* 246: 1566–71.

Smith, G. I. 1984. Paleohydrologic regimes in the southwestern Great Basin, 0–3.2 my ago, compared with other long records of "global" climate. *Quaternary Research* 22:1–17.

Solunias, N., and B. Dawson-Saunders. 1988. Dietary adaptations and paleoecology of the late Miocene ruminants from Pikermi and Samos in Greece. *Palaeogeography, Palaeoclimatology, Palaeoecology* 65: 149–72.

Solunias, N., M. Teaford, and A. Walker. 1988. Interpreting the diet of extinct ruminants: The case of the non-browsing giraffid. *Paleobiology* 14: 287–300.

Song, Z., H. Li, Y. Zheng, and G. Liu. 1984. The Miocene floristic regions of east Asia. 448–60 in R. O. Whyte, ed., *The Evolution of the East Asian Environment*, vol. 2. Hong Kong: Centre of Asian Studies, University of Hong Kong.

Spaulding, W. G. 1990. Vegetation dynamics during the last deglaciation, southeastern Great Basin, USA. *Quaternary Research* 33:188–203.

Spaulding, W. G., and L. J. Graumlich. 1986. The last pluvial climatic episodes in the deserts of southwestern North America. *Nature* 320:441–44.

Spaulding, W. G., E. B. Leopold, and T. R. Van Devender. 1983. Late Wisconsin paleoecology of the American Southwest. 259–92 in S. C. Porter, ed., *The Late Pleistocene*. Minneapolis: University of Minnesota Press.

Specht, R. L. 1981. Evolution of the Australian flora: Some generalizations. 783–805 in A. Keast, ed., *Ecological Biogeography of Australia*, vol. 1. The Hague: Dr. W. Junk.

Spurr, S. H., and H. J. Vaux. 1976. Timber: Biological and economic potential. *Science* 191:752–56.

Stark, B. L. 1986. Origins of food production in the New World. 277–321 in D. Meltzer, D. Fowler, and J. Sabloff, eds., *American Archaeology Past and Future*. Washington, D.C.: Smithsonian Institution Press.

Stehlin, H. G. 1909. Remarques sur les faunules de mammifères des couches éocenes et oligocenes du Bassin de Paris. Bulletin de la Société Géologique de France 9: 488–520.

Stein, R., and M. Sarnthein. 1984. Late Neogene events of atmospheric and oceanic circulation off shore of northwest Africa: High resolution record from deep-sea sediments. 9–36 in J. A. Coetzee and E. M. Van Zinderen Bakker, eds., *Palaeoecology of Africa*, vol. 16. Rotterdam: Balkema.

Street, F. A. 1981. Tropical palaeoenvironments. *Progress in Physical Geography* 5:157–85.

Street, F. A., and A. T. Grove. 1979. Global maps of lake-level fluctuations since 30,000 BP. *Quaternary Research* 12:83–118.

Street-Perrott, F. A., and S. P. Harrison. 1984. Global maps of lake level fluctuations since 30,000 yr B.P.: An index of the global hydrological cycle. 118–29 in J. E. Hansen and T. Takahashi, eds., *Climate Processes and Climate Sensitivity.* American Geophysical Union Monograph 29.

Stucky, R. K. 1990. Evolution of land mammal diversity in North America during the Cenozoic. 375–432 in H. H. Genoways, ed., *Current Mammalogy.* New York: Plenum.

Stutz, H. C. 1978. Explosive evolution of perennial *Atriplex* in North America. 161–68 in K. T. Harper and J. L. Reveal, eds., *Intermountain Biogeography: A Symposium.* Great Basin Naturalist Memoirs 2.

Stutz, H. C., and S. C. Sanderson. 1983. Evolutionary studies of *Atriplex:* Chromosome races of *A. confertifolia* (shadscale). *American Journal of Botany* 70: 1536–47.

Suc, J. P. 1980. Contribution à la connaissance du Pliocene et du Pleistocene inférieur des régions méditerranéennes d'Europe occidentale par l'analyse palynologique de dépôts du Languedoc-Rousillon (Sud de la France) et de la Catalogne (Nord-est de l'Espagne). Languedoc: Université des Sciences et Techniques du Languedoc.

Sutcliffe, A. J. 1985. On the Track of Ice Age Mammals. Cambridge: Harvard University Press.

Swisher, C. C., III, and D. R. Prothero. 1990. Single crystal $^{40}Ar/^{39}Ar$ dating of the Eocene-Oligocene transition in North America. *Science* 249: 760–62.

Szalay, F. S., and E. Delson. 1979. *Evolutionary History of the Primates.* New York: Academic Press.

Taggart, R. E., A. T. Cross, and L. Satchell. 1982. Effects of periodic volcanism on Miocene vegetation distribution in eastern Oregon and western Idaho. 535–40 in B. Mamet and M. J. Copeland, eds., *Third North American Paleontological Convention, Montreal.*

Talbot, M. R. 1980. Environmental responses to climatic change in the West African Sahel over the past 20,000 years. 37–62 in M. A. J. Williams and H. Faure, eds., *The Sahara and the Nile: Quaternary Environments and Prehistoric Occupation in Northern Africa.* Rotterdam: Balkema.

Tedford, R. H., M. R. Banks, N. Kemp, I. McDougal, and F. L. Sutherland. 1975. Recognition of the oldest known fossil marsupials from Australia. Nature 255: 141–42.

Tedford, R. H., M. F. Skinner, R. W. Fields, J. M. Rensberger, D. P. Whistler, T. Galusha, B. E. Taylor, J. R. MacDonald, and S. D. Webb. 1987. Faunal succession and biochronology of the Arikareean through Hemphillian interval (Late Oligocene through earliest Pliocene Epochs) in North America. 153–210 in M. O. Woodburne, ed., *Cenozoic Mammals of North America.* Berkeley: University of California Press.

Thomas, B. A., and R. A. Spicer. 1987. *The Evolution and Palaeobiology of Land Plants.* London: Croom Helm.

Thomas, H. 1981. Les bovides Miocenes de la formation de Ngorora du Bassin de Baringo (Rift Valley, Kenya). *Palaeontographica, B* 84: 335–410.

Thomasson, J. R., M. E. Nelson, and R. J. Zakrzewski. 1986. A fossil grass (Gramineae: Chloridoideae) from the Miocene with Kranz Anatomy. *Science* 233: 876–78.

Thompson, R. S. 1988. Western North America: Vegetation dynamics in the western

United States; modes of response to climatic fluctuations. 415–58 in B. Huntley and T. Webb II, eds., *Vegetation History*. Boston: Kluwer Academic Publishers.

Tiffney, B. H. 1984. Seed size, dispersal syndromes, and the rise of the angiosperms: Evidence and hypothesis. *Annals of the Missouri Botanical Garden* 71:551–76.

———. 1985. Perspectives on the origin of the floristic similarity between eastern Asia and eastern North America. *Journal of the Arnold Arboretum* 66: 73–94.

Tobien, H., G. Cheng, and Y. Li. 1984. The Mastodonts (Proboscidea, Mammalia) of China: Evolution, palaeobiogeography, palaeoecology. 689–96 in R. O. Whyte, ed., *The Evolution of the East Asian Environment,* vol. 2. Hong Kong: Centre of Asian Studies, University of Hong Kong.

Traverse, A. T. 1982. Response of world vegetation to Neogene tectonic and climatic events. *Alcheringa* 6:197–209.

———. 1988. *Paleopalynology.* Boston: Unwin Hyman.

Tricart, J. 1977. Aperçus sur le Quaternaire amazonien: Recherches françaises sur le Quaternaire hors de France. *Bulletin de l'Association Française pour l'Educe du Quaternaire* supplement 1977-1, 50:265–71.

Unwin, D. M. 1988. Extinction and survival in birds. 295–318 in G. P. Larwood, ed., *Extinction and Survival in the Fossil Record.* Oxford: Clarendon Press.

Vail, P. R., and B. U. Haq. 1988. Sea level history. *Science* 241:599.

Van Campo, E., J. C. Duplessy, W. L. Prell, N. Barratt, and R. Sabatier. 1990. Comparison of terrestrial and marine temperature estimates for the past 135 kyr off southeast Africa: A test for GCM simulations of palaeoclimate. *Nature* 348: 209–12.

Van Couvering, J. A. H. 1980. Community evolution in East Africa during the Late Cenozoic. 272–98 in A. K. Behrensmeyer and A. P. Hill, eds., *Fossils in the Making.* Chicago: University of Chicago Press.

Vandenbeld, J. 1988. *Nature of Australia.* New York: Facts on File.

Van der Hammen, T. 1982. Paleoecology of tropical South America. 60–66 in G. T. Prance, ed., *Biological Diversification in the Tropics.* New York: Columbia University Press.

Van der Hammen, T., T. A. Wijmstra, and W. H. Zagwijn. 1971. The floral record of the late Cenozoic of Europe. 391–424 in K. K. Turekian, ed., *The Late Cenozoic Glacial Ages.* New Haven: Yale University Press.

Van Devender, T. R., R. S. Thompson, and J. L. Betancourt. 1987. Vegetation history of the deserts of southwestern North America: The nature and timing of the Late Wisconsin–Holocene transition. 323–52 in W. F. Ruddiman and H. E. Wright, Jr., eds., *The Geology of North America.* Vol. K-3, *North America and Adjacent Oceans during the Last Glaciation.* Boulder: Geological Society of America.

Van Valkenburgh, B. 1985. Locomotor diversity within past and present guilds of large predatory mammals. *Paleobiology* 11:406–28.

———. 1988. Trophic diversity in past and present guilds of large predatory mammals. *Paleobiology* 14:155–73.

———. In press. Ecomorphological analysis of fossil vertebrates and paleocommunities. In P. C. Wainwright and S. J. Reilly, eds., *Ecological Morphology: Integrative Organismal Biology.* Chicago: University of Chicago Press.

Van Zinderen Bakker, E. M. 1976. The evolution of late Quaternary palaeoclimates of southern Africa. *Palaeoecology of Africa* 9: 160–202.

———. 1978. Quaternary vegetation changes in southern Africa. 131–43 in M. J. A.

Werger, ed., *Biogeography and Ecology of Southern Africa*. The Hague: Dr W. Junk.

———. 1982. African palaeoclimates 18,000 years B.P. *Palaeoecology of Africa* 15:77–99.

Van Zinderen Bakker, E. M., and J. H. Mercer. 1986. Major late Cainozoic climatic events and palaeoenvironmental changes in Africa viewed in a world wide context. *Palaeogeography, Palaeoclimatology, Palaeoecology* 56:217–35.

Vishnu-Mittre, 1984. Floristic change in the Himalaya (southern slopes) and Siwaliks from the mid-Tertiary to recent times. 483–503 in R. O. Whyte, ed., *The Evolution of the East Asian Environment*, vol. 2. Hong Kong: Centre of Asian Studies, University of Hong Kong.

Voorhies, M. R. 1969. Taphonomy and population dynamics of an early Pliocene vertebrate fauna, Knox County, Nebraska. *University of Wyoming Contributions to Geology Special Paper* 1:1–69.

Voorhies, M. R., and J. R. Thomasson. 1979. Fossil grass anthoecia within Miocene rhinoceros skeletons: Diet in an extinct species. *Science* 206:331–33.

Vrba, E. S. 1985a. African Bovidae: Evolutionary events since the Miocene. *South African Journal of Science* 81:263–66.

———. 1985b. Ecological and adaptive changes associated with early hominid evolution. 63–71 in E. Delson, ed., *Ancestors: The Hard Evidence*. New York: Alan R. Liss.

———. 1985c. Environment and evolution: Alternative causes of the temporal distribution of evolutionary events. *South African Journal of Science* 81:229–36.

———. 1987. Ecology in relation to speciation rates: Some case histories of Miocene-Recent mammal clades. *Evolutionary Ecology* 1:283–300.

———. 1988. Late Pliocene climatic events and hominid evolution. 405–26 in F. E. Grine, ed., *Evolutionary History of the "Robust" Australopithecines*. New York: Aldine de Gruyter.

Vrba, E. S., G. H. Denton, and M. L. Prentice. 1989. Climatic influences on early hominid behavior. *Ossa* 14:127–56.

Vuilleumier, F. 1984. Faunal turnover and development of fossil avifaunas in South America. *Evolution* 38:1384–96.

Wang Xianzeng, 1984. The palaeoenvironment of China from the Tertiary. 472–82 in R. O. Whyte, ed., *The Evolution of the East Asian Environment*, vol. 2. Hong Kong: Centre of Asian Studies, University of Hong Kong.

Watt, A. S. 1947. Pattern and process in the plant community. *Journal of Ecology* 35:1–22.

Webb, P. N. 1990. The Cenozoic history of Antarctica and its global impact. *Antarctic Science* 2:3–21.

Webb, P. N., D. M. Harwood, B. C. McKelvey, J. H. Mercer, and L. D. Stott. 1984. Cenozoic marine sedimentation and ice-volume variation on the East Antarctic craton. *Geology* 12:287–91.

Webb, S. D. 1972. Locomotor evolution in camels. *Forma et Functio* 5:99–112.

———. 1976. Mammalian faunal dynamics of the great American interchange. *Paleobiology* 2:220–34.

———. 1977. A history of savanna vertebrates in the New World. Part 1: North America. *Annual Review of Ecology and Systematics* 8:355–80.

———. 1978. A history of savanna vertebrates in the New World. Part 2: South

America and the great interchange. *Annual Review of Ecology and Systematics* 9:393–426.

———. 1983a. The rise and fall of the late Miocene ungulate fauna in North America. 267–306 in M. H. Nitecki, ed., *Coevolution*. Chicago: University of Chicago Press.

———. 1983b. On two kinds of rapid faunal turnover, 417–36 in W. Berggren and J. Van Couvering, eds., *Uniformitarianism, Catastrophes, and Earth History*. Princeton: Princeton University Press.

———. 1985. Late Cenozoic mammal dispersals between the Americas. 357–86 in F. G. Stehli and S. D. Webb, eds., *The Great American Biotic Interchange*. New York: Plenum.

———. 1989a. Faunal dynamics of Pleistocene mammals. *Annual Reviews of Earth and Planetary Sciences* 17:413–38.

———. 1989b. The fourth dimension in North American terrestrial mammal communities. 181–203 in D. W. Morris, Z. Abramsky, B. J. Fox, and M. R. Willig, eds., *Patterns in the Structure of Mammalian Communities*. Special Publication Museum Texas Tech University no. 28.

Webb, S. D., and S. C. Perrigo. 1984. Late Cenozoic vertebrates from Honduras and El Salvador. *Journal of Vertebrate Paleontology* 4:237–54.

Webb, T., III. 1987. The appearance and disappearance of major vegetational assemblages: Long-term vegetational dynamics in eastern North America. *Vegetatio* 69:177–87.

Wells, P. V. 1979. An equable glaciopluvial in the west: Pleniglacial evidence of increased precipitation on a gradient from the Great Basin to the Sonoran and Chihuahuan deserts. *Quaternary Research* 12:311–25.

West, R. G. 1964. Inter-relations of ecology and Quaternary palaeobotany. *Journal of Ecology* 52 (supplement):47–57.

White, M. E. 1986. *The Greening of Gondwana*. Sydney: Reed Books.

Whybrow, P. J. 1984. Geological and faunal evidence from Arabia for mammal "migrations" between Asia and Africa during the Miocene. *Courier Forschungsinstitut Senckenberg* 69:189–98.

Whyte, R. O., ed. 1984. *The Evolution of the East Asian Environment*. Hong Kong: Centre of Asian Studies, University of Hong Kong.

Williamson, P. G. 1985. Evidence of an early Plio-Pleistocene rainforest expansion in East Africa. *Nature* 315:487–89.

Wing, S. L. 1987. Eocene and Oligocene floras and vegetation of the Rocky Mountains. *Annals of the Missouri Botanical Garden* 74:748–84.

———. In press. Tertiary vegetational history of North America as a context for mammalian evolution. In C. Janis, L. L. Jacobs, and K. Scott, eds., *Tertiary Mammals of North America*. Cambridge: Cambridge University Press.

Wolfe, J. A. 1971. Tertiary climatic fluctuations and methods of analysis of Tertiary floras. *Palaeogeography, Palaeoclimatology, Palaeoecology* 9:27–57.

———. 1978. A paleobotanical interpretation of Tertiary climates in the Northern Hemisphere. *American Scientist* 66:694–703.

———. 1980. Tertiary climates and floristic relationships at high latitudes in the Northern Hemisphere. *Palaeogeography, Palaeoclimatology, Palaeoecology* 30:313–23.

————. 1981. Paleoclimatic significance of the Oligocene and Neogene floras of the northwestern United States. 79–101 in K. Niklas, ed., *Paleobotany, Paleoecology, and Evolution.* New York: Praeger Press.

————. 1985. Distribution of major vegetational types during the Tertiary. 357–75 in E. T. Sundquist and W. S. Broecker, eds., *The Carbon Cycle and Atmospheric CO_2.* American Geophysical Union Monograph 32.

————. 1987. An overview of the origins of the modern vegetation and flora of the northern Rocky Mountains. *Annals of the Missouri Botanical Garden* 74:785–803.

Wolfe, J. A. and D. M. Hopkins. 1967. Climatic changes recorded by Tertiary land floras in northwestern North America. 67–76 in K. Hatai, ed., *Tertiary Correlations and Climatic Changes in the Pacific.* Tokyo: Symposium of the Eleventh Pacific Scientific Congress.

Wood, R. C. 1976. *Stupendemys geographicus,* the world's largest turtle. *Breviora* 436:1–31.

Woodburne, M. O., ed. 1987. *Cenozoic Mammals of North America.* Berkeley: University of California Press.

Woodruff, F., S. M. Savin, and R. G. Douglas. 1981. Miocene stable isotope record: A detailed deep Pacific Ocean study and its paleoclimatic implications. *Science* 212:665–68.

Wu, R., and J. W. Olsen, eds. 1985. *Palaeoanthropology and Palaeolithic Archaeology in the People's Republic of China.* Orlando: Academic Press.

Xu, Q., and O. Lian. 1982. Climatic changes during Peking Man's time. *Acta Anthropologica Sinica* 1:80–90.

Xu, R. 1984a. Changes of the palaeoenvironment of southern East Asia since the late Tertiary. 419–25 in R. O. Whyte, ed., *The Evolution of the East Asian Environment,* vol. 2. Hong Kong: Centre of Asian Studies, University of Hong Kong.

————. 1984b. Changes of the vegetation in China since the late Tertiary. 426–32 in R. O. Whyte, ed., *The Evolution of the East Asian Environment.* Hong Kong: Centre of Asian Studies, University of Hong Kong.

Yemane, K., C. Robert, and R. Bonnefille. 1987. Pollen and clay mineral assemblages of a late Miocene lacustrine sequence from the northwestern Ethiopian highlands. *Palaeogeography, Palaeoclimatology, Palaeoecology* 60:123–41.

Appendix:
Phanerozoic Geologic Time Scale

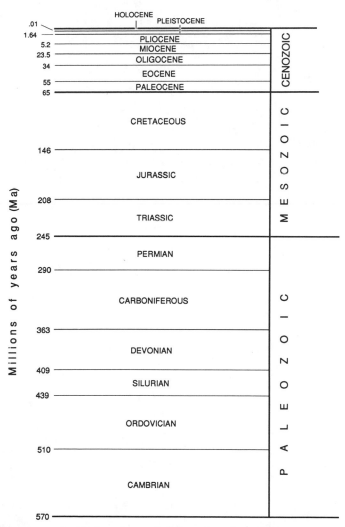

Note: Boundary ages for Phanerozoic eras, periods, and epochs are from:
1. Harland, W., R. L. Armstrong, A. V. Cox, L. E. Craig, A. G. Smith, and D. G. Smith. 1990. *A Geologic Time Scale, 1989*. Cambridge: Cambridge University Press.
2. Paleocene-Eocene: Wing, S. L., T. M. Bown, and J. D. Obradovich. 1991. Early Eocene biotic and climatic change in interior western North America. *Geology* 19:1189–92.
3. Eocene-Oligocene: Prothero, D. R. and W. A. Berggren. In press. *Eocene-Oligocene Climatic and Biotic Evolution*. Princeton: Princeton University Press.

Index of Systematic Names

Index of Subjects